Methods in Enzymology

Volume 422
TWO-COMPONENT SIGNALING SYSTEMS,
PART A

METHODS IN ENZYMOLOGY

EDITORS-IN-CHIEF

John N. Abelson   Melvin I. Simon

DIVISION OF BIOLOGY
CALIFORNIA INSTITUTE OF TECHNOLOGY
PASADENA, CALIFORNIA

FOUNDING EDITORS

Sidney P. Colowick and Nathan O. Kaplan

Methods in Enzymology

Volume 422

Two-Component Signaling Systems, Part A

EDITED BY

Melvin I. Simon

CALIFORNIA INSTITUTE OF TECHNOLOGY
DIVISION OF BIOLOGY
PASADENA, CALIFORNIA

Brian R. Crane

DEPARTMENT OF CHEMISTRY AND CHEMICAL BIOLOGY
CORNELL UNIVERSITY
ITHACA, NEW YORK

Alexandrine Crane

DEPARTMENT OF CHEMISTRY AND CHEMICAL BIOLOGY
CORNELL UNIVERSITY
ITHACA, NEW YORK

AMSTERDAM • BOSTON • HEIDELBERG • LONDON
NEW YORK • OXFORD • PARIS • SAN DIEGO
SAN FRANCISCO • SINGAPORE • SYDNEY • TOKYO
Academic Press is an imprint of Elsevier

ELSEVIER

Academic Press is an imprint of Elsevier
525 B Street, Suite 1900, San Diego, California 92101-4495, USA
84 Theobald's Road, London WC1X 8RR, UK

This book is printed on acid-free paper. ∞

Permissions may be sought directly from Elsevier's Science & Technology Rights
Department in Oxford, UK: phone: (+44) 1865 843830, fax: (+44) 1865 853333,
E-mail: permissions@elsevier.com. You may also complete your request on-line
via the Elsevier homepage (http://elsevier.com), by selecting "Support & Contact"
then "Copyright and Permission" and then "Obtaining Permissions."

For information on all Elsevier Academic Press publications
visit our Web site at www.books.elsevier.com

ISBN-13: 978-0-12-373851-6
ISBN-10: 0-12-373851-2

PRINTED IN THE UNITED STATES OF AMERICA
07 08 09 10   9 8 7 6 5 4 3 2 1

Table of Contents

Section I. Computational Analyses of Sequences and Sequence Alignments

Section II. Biochemical and Genetic Assays of Individual Components of Signaling Systems

Section III. Physiological Assays and Readouts

Contributors to Volume 422

Article numbers are in parentheses following the names of contributors.
Affiliations listed are current.

ROGER P. ALEXANDER (1), *School of Biology, Georgia Institute of Technology, Atlanta, Georgia*

BURTON W. ANDREWS (6), *Department of Electrical and Computer Engineering, Johns Hopkins University, Baltimore, Maryland*

DAVID ARPS (25), *Department of Biological Chemistry, University of Michigan Medical School, Ann Arbor, Michigan*

STEPHEN ATKINS (25), *Department of Biological Chemistry, University of Michigan Medical School, Ann Arbor, Michigan*

MARIETTE R. ATKINSON (25), *Department of Biological Chemistry, University of Michigan Medical School, Ann Arbor, Michigan*

RICHARD G. BAKKER (21), *Departments of Microbiology, Immunology, and Cell Biology, R. C. Byrd Health Sciences Center, West Virginia University, Morgantown, West Virginia*

CARL E. BAUER (8, 9), *Department of Biology, Indiana University, Bloomington, Indiana*

BRAD M. BINDER (13), *Department of Horticulture, University of Wisconsin, Madison, Wisconsin*

CORY J. BOTTONE (16), *Department of Chemistry and Biochemistry, University of North Carolina, Wilmington, North Carolina*

ERIC S. CASPER (16), *Department of Chemistry and Biochemistry, University of North Carolina, Wilmington, North Carolina*

JOHN CAVANAGH (7), *Department of Molecular and Structural Biochemistry, North Carolina State University, Raleigh, North Carolina*

NYLES W. CHARON (21, 22), *Departments of Microbiology, Immunology, and Cell Biology, R. C. Byrd Health Sciences Center, West Virginia University, Morgantown, West Virgina*

YOUNG-HO CHUNG (9), *Department of Biology, Indiana University, Bloomington, Indiana; Systems Biology Team, Korea Basic Science Institute, Daejeon, South Korea*

PIERRE-DAMIEN COUREUX (15), *Department of Biochemistry, Brandeis University, Waltham, Massachusetts*

TATSUYA FUKUSHIMA (20), *Division of Cellular Biology, Department of Molecular and Experimental Medicine, The Scripps Research Institute, La Jolla, California*

EIJI FURUTA (19), *Department of Bioscience, Graduate School of Agriculture, Kinki University, Nara, Japan*

MICHAEL Y. GALPERIN (3), *National Center for Biotechnology Information, National Library of Medicine, National Institutes of Health, Bethesda, Maryland*

HAO GENG (23), *Departments of Environmental and Biomolecular Systems, OGI School of Science and Engineering, Oregon Health & Science University, Beaverton, Oregon*

ULRICH K. GENICK (15), *Nestlé Research Center, Vers-chez-les-Blanc, Lausanne, Switzerland*

YASUHIRO GOTOH (19), *Department of Bioscience, Graduate School of Agriculture, Kinki University, Nara, Japan*

EDUARDO A. GROISMAN (18), *Department of Molecular Microbiology, Howard Hughes Medical Institute, Washington University School of Medicine, St. Louis, Missouri*

R. MATTHEW HAAS (16), *Department of Chemistry and Biochemistry, University of North Carolina, Wilmington, North Carolina*

TOSHIO HAKOSHIMA (14), *Structural Biology Laboratory, Nara Institute of Science and Technology, and CREST, Japan Science and Technology Agency, Ikoma, Nara, Japan*

CHRISTOPHER J. HALKIDES (16), *Department of Chemistry and Biochemistry, University of North Carolina, Wilmington, North Carolina*

OSCAR HARARI (18), *Department of Computer Science and Artificial Intelligence, University of Granada, Granada, Spain*

JAMES A. HOCH (4, 20), *Division of Cellular Biology, Department of Molecular and Experimental Medicine, The Scripps Research Institute, La Jolla, California*

TERENCE HWA (4), *Center for Theoretical Biological Physics and Department of Physics, La Jolla, California*

HISAKO ICHIHARA (14), *Institute for Chemical Research, Kyoto University, Kyoto, Japan*

PABLO A. IGLESIAS (6), *Department of Electrical and Computer Engineering, Johns Hopkins University, Baltimore, Maryland*

MARK S. JOHNSON (10), *Division of Microbiology and Molecular Genetics, Loma Linda University, Loma Linda, California*

CHIKARA KAITO (11), *Laboratory of Microbiology, Graduate School of Pharmaceutical Sciences, The University of Tokyo, Bunkyo-ku, Tokyo, Japan*

LINDA J. KENNEY (17), *Department of Microbiology and Immunology, University of Illinois-Chicago, Chicago, Illinois*

S. THOMAS KING (17), *Department of Microbiology and Immunology, University of Illinois-Chicago, Chicago, Illinois*

BRUCE S. KLEIN (24), *Departments of Medical Microbiology, Immunology, Pediatrics, and Internal Medicine, University of Wisconsin-Madison School of Medicine and Public Health, Madison, Wisconsin*

DOUGLAS J. KOJETIN (7), *Department of Molecular Genetics, Biochemistry, and Microbiology, University of Cincinnati College of Medicine, Cincinnati, Ohio*

KENJI KUROKAWA (11), *Laboratory of Microbiology, Graduate School of Pharmaceutical Sciences, The University of Tokyo, Bunkyo-ku, Tokyo, Japan*

CHUNHAO LI (21, 22), *Departments of Microbiology, Immunology, and Cell Biology, R. C. Byrd Health Sciences Center, West Virginia University, Morgantown, West Virginia; Department of Oral Biology, State University of New York, Buffalo, New York*

SHINJI MASUDA (9), *Graduate School of Bioscience and Biotechnology, Tokyo Institute of Technology, Yokohama, Japan*

Avi Mayo (25), *Department of Biological Chemistry, University of Michigan Medical School, Ann Arbor, Michigan*

Kenneth McAdams (16), *Department of Chemistry and Biochemistry, University of North Carolina, Wilmington, North Carolina*

Michael R. Miller (21, 22), *Department of Biochemistry and Molecular Pharmacology, R. C. Byrd Health Sciences Center, West Virginia University, Morgantown, West Virginia*

Md. A. Motaleb (21, 22), *Departments of Microbiology, Immunology, and Cell Biology, R. C. Byrd Health Sciences Center, West Virginia University, Morgantown, West Virginia*

Michiko M. Nakano (23), *Department of Environmental and Biomolecular Systems, OGI School of Science and Engineering, Oregon Health and Science University, Beaverton, Oregon*

Julie C. Nemecek (24), *Departments of Pediatrics, Medical Microbiology and Immunology, University of Wisconsin-Madison School of Medicine and Public Health, Madison, Wisconsin*

Anastasia N. Nikolskaya (3), *Protein Information Resource, Georgetown University Medical Center, Washington, DC*

Alexander J. Ninfa (25), *Department of Biological Chemistry, University of Michigan Medical School, Ann Arbor, Michigan*

Ario Okada (19), *Department of Bioscience, Graduate School of Agriculture, Kinki University, Nara, Japan*

Nicolas Perry (25), *Department of Biological Chemistry, University of Michigan Medical School, Ann Arbor, Michigan*

Mircea Podar (2), *Department of Biology, Portland State University, Portland, Oregon*

Kathleen Postle (12), *Department of Biochemistry and Molecular Biology, The Pennsylvania State University, University Park, Pennsylvania*

G. Eric Schaller (13), *Department of Biological Sciences, Dartmouth College, Hanover, New Hampshire*

Kazuhisa Sekimizu (11), *Laboratory of Microbiology, Graduate School of Pharmaceutical Sciences, The University of Tokyo, Bunkyo-ku, Tokyo, Japan*

Stephen Selinsky (25), *Department of Biological Chemistry, University of Michigan Medical School, Ann Arbor, Michigan*

Qi Xiu Song (25), *Department of Biological Chemistry, University of Michigan Medical School, Ann Arbor, Michigan*

Daniel M. Sullivan (7), *Department of Molecular and Structural Biochemistry, North Carolina State University, Raleigh, North Carolina*

Lee R. Swem (8), *Department of Molecular Biology, Princeton University, Princeton, New Jersey*

Danielle L. Swem (8), *Department of Molecular Biology, Princeton University, Princeton, New Jersey*

Hendrik Szurmant (4, 20), *Division of Cellular Biology, Department of Molecular and Experimental Medicine, The Scripps Research Institute, La Jolla, California*

Barry L. Taylor (10), *Division of Microbiology and Molecular Genetics, Loma Linda University, Loma Linda, California*

Richele J. Thompson (7), *Department of Molecular and Structural Biochemistry, North Carolina State University, North Carolina*

Ryutaro Utsumi (19), *Department of Bioscience, Graduate School of Agriculture, Kinki University, Nara, Japan*

KOTTAYIL I. VARUGHESE (5), *Department of Physiology and Biophysics, University of Arkansas for Medical Sciences, Little Rock, Arkansas*

VIDYA HARINI VELDORE (5), *Department of Physiology and Biophysics, University of Arkansas for Medical Sciences, Little Rock, Arkansas*

TAKAFUMI WATANABE (19), *Department of Bioscience, Graduate School of Agriculture, Kinki University, Nara, Japan*

KYLIE J. WATTS (10), *Division of Microbiology and Molecular Genetics, Loma Linda University, Loma Linda, California*

ROBERT A. WHITE (4), *Division of Cellular Biology, Department of Molecular and Experimental Medicine, The Scripps Research Institute, La Jolla, California*

PETER WOOLF (25), *Department of Biological Chemistry, University of Michigan Medical School, Ann Arbor, Michigan*

JIANG WU (8), *Department of Biology, Indiana University, Bloomington, Indiana*

KRISTIN WUICHET (1), *School of Biology, Georgia Institute of Technology, Atlanta, Georgia*

MARCEL WÜTHRICH (24), *Department of Pediatrics, University of Wisconsin-Madison School of Medicine and Public Health, Madison, Wisconsin*

KANEYOSHI YAMAMOTO (19), *Department of Bioscience, Graduate School of Agriculture, Kinki University, Nara, Japan*

TAU-MU YI (6), *Department of Developmental and Cell Biology, Center for Complex Biological Systems, Irvine, California*

JAMES ZAPF (5), *Department of Physiology and Biophysics, University of Arkansas for Medical Sciences, Little Rock, Arkansas*

HAIYAN ZHAO (5), *Department of Physiology and Biophysics, University of Arkansas for Medical Sciences, Little Rock, Arkansas*

IGOR B. ZHULIN (1), *Joint Institute for Computational Sciences, University of Tennessee, Oak Ridge National Laboratory, Oak Ridge, Tennessee*

PETER ZUBER (23), *Departments of Environmental and Biomolecular Systems, OGI School of Science & Engineering, Oregon Health & Science University, Beaverton, Oregon*

IGOR ZWIR (18), *Department of Molecular Microbiology, Howard Hughes Medical Institute, Washington University School of Medicine, St. Louis, Missouri; Department of Computer Science and Artificial Intelligence, University of Granada, Granada, Spain*

METHODS IN ENZYMOLOGY

VOLUME 226. Metallobiochemistry (Part C: Spectroscopic and
Physical Methods for Probing Metal Ion Environments in Metalloenzymes
and Metalloproteins)
Edited by JAMES F. RIORDAN AND BERT L. VALLEE

VOLUME 227. Metallobiochemistry (Part D: Physical and Spectroscopic
Methods for Probing Metal Ion Environments in Metalloproteins)
Edited by JAMES F. RIORDAN AND BERT L. VALLEE

VOLUME 228. Aqueous Two-Phase Systems
Edited by HARRY WALTER AND GÖTE JOHANSSON

VOLUME 229. Cumulative Subject Index Volumes 195–198, 200–227

VOLUME 230. Guide to Techniques in Glycobiology
Edited by WILLIAM J. LENNARZ AND GERALD W. HART

VOLUME 231. Hemoglobins (Part B: Biochemical and Analytical Methods)
Edited by JOHANNES EVERSE, KIM D. VANDEGRIFF, AND ROBERT M. WINSLOW

VOLUME 232. Hemoglobins (Part C: Biophysical Methods)
Edited by JOHANNES EVERSE, KIM D. VANDEGRIFF, AND ROBERT M. WINSLOW

VOLUME 233. Oxygen Radicals in Biological Systems (Part C)
Edited by LESTER PACKER

VOLUME 234. Oxygen Radicals in Biological Systems (Part D)
Edited by LESTER PACKER

VOLUME 235. Bacterial Pathogenesis (Part A: Identification and Regulation of
Virulence Factors)
Edited by VIRGINIA L. CLARK AND PATRIK M. BAVOIL

VOLUME 236. Bacterial Pathogenesis (Part B: Integration of Pathogenic
Bacteria with Host Cells)
Edited by VIRGINIA L. CLARK AND PATRIK M. BAVOIL

VOLUME 237. Heterotrimeric G Proteins
Edited by RAVI IYENGAR

VOLUME 238. Heterotrimeric G-Protein Effectors
Edited by RAVI IYENGAR

VOLUME 239. Nuclear Magnetic Resonance (Part C)
Edited by THOMAS L. JAMES AND NORMAN J. OPPENHEIMER

VOLUME 240. Numerical Computer Methods (Part B)
Edited by MICHAEL L. JOHNSON AND LUDWIG BRAND

VOLUME 241. Retroviral Proteases
Edited by LAWRENCE C. KUO AND JULES A. SHAFER

VOLUME 242. Neoglycoconjugates (Part A)
Edited by Y. C. LEE AND REIKO T. LEE

VOLUME 243. Inorganic Microbial Sulfur Metabolism
Edited by HARRY D. PECK, JR., AND JEAN LEGALL

Section I

Computational Analyses of Sequences and Sequence Alignments

[1] Comparative Genomic and Protein Sequence Analyses of a Complex System Controlling Bacterial Chemotaxis

By KRISTIN WUICHET, ROGER P. ALEXANDER, and IGOR B. ZHULIN

Abstract

Molecular machinery governing bacterial chemotaxis consists of the CheA–CheY two-component system, an array of specialized chemoreceptors, and several auxiliary proteins. It has been studied extensively in *Escherichia coli* and, to a significantly lesser extent, in several other microbial species. Emerging evidence suggests that homologous signal transduction pathways regulate not only chemotaxis, but several other cellular functions in various bacterial species. The availability of genome sequence data for hundreds of organisms enables productive study of this system using comparative genomics and protein sequence analysis. This chapter describes advances in genomics of the chemotaxis signal transduction system, provides information on relevant bioinformatics tools and resources, and outlines approaches toward developing a computational framework for predicting important biological functions from raw genomic data based on available experimental evidence.

Introduction

Signal transduction systems link internal and external cues to appropriate cellular responses in all organisms. Prokaryotic signal transduction can be classified into three main families based on the domain organization and complexity: one-component systems, classical two-component systems anchored by class I histidine kinases, and multicomponent systems anchored by class II histidine kinases often referred to as chemotaxis systems (Bilwes *et al.*, 1999; Dutta *et al.*, 1999; Stock *et al.*, 2000; Ulrich *et al.*, 2005). As their name suggests, one-component systems consist of a single protein that is capable of both sensing a signal and directly affecting a cellular response, either through a single domain (such as a DNA-binding domain that senses a signal through its metal cofactor) or multiple domains (separate input and output domains) (Ulrich *et al.*, 2005). As a consequence of their single protein nature and typical lack of transmembrane regions, one-component systems are predicted to primarily sense the internal cellular environment, while the division of input and output between two or more proteins and association of the sensor with the membrane in two-component systems

METHODS IN ENZYMOLOGY, VOL. 422   0076-6879/07 $35.00
DOI: 10.1016/S0076-6879(06)22001-9

allows them to detect both internal and external signals (Ulrich *et al.*, 2005). The chemotaxis system centered around the class II histidine kinase CheA contains multiple proteins separating input and output, along with additional regulatory components that are not present in class I histidine kinase-containing two-component systems. There are many common input (sensing) modules among all three families of prokaryotic signal transduction; one-component systems and two-component systems also share common outputs (Ulrich *et al.*, 2005), whereas two-component systems and chemotaxis systems share several common signaling modules (Dutta *et al.*, 1999; Stock *et al.*, 2000).

The chemotaxis system is classically portrayed as a network of interacting proteins, which senses environmental stimuli to regulate motility. The system consists of two distinct pathways: an excitation pathway that has the downstream result of interacting with the motility organelle and an adaptation pathway that provides a mechanism for molecular memory (Baker *et al.*, 2006; Wadhams and Armitage, 2004). The excitation pathway involves methyl-accepting chemotaxis proteins (MCPs) for sensing environmental signals that are transmitted to a scaffolding protein, CheW, and a histidine kinase, CheA, via a highly conserved cytoplasmic signaling module of the MCPs. The signals regulate the kinase activity of CheA and the phosphorylation state of its cognate response regulator CheY controls its affinity for the motor. Many chemotaxis systems have one or more phosphatases (CheC, CheX, and/or CheZ) involved in the excitation pathway that aid in dephosphorylating CheY (Szurmant and Ordal, 2004). Signal propagation through the MCPs is further controlled in most systems by an adaptation pathway that regulates their methylation state via the CheB methylesterase, a response regulator that is phosphorylated by CheA to stimulate the removal of methyl groups from the receptors, and the CheR methyltransferase that constitutively methylates specific glutamate residues of the receptors. Many chemotaxis systems have an additional adaptation protein, CheD, for the deamidation of particular amino acid side chains of many MCPs prior to their methylation, and in some of these systems CheD also interacts with CheC to increase its dephosphorylation activity (Chao *et al.*, 2006; Kristich and Ordal, 2002). The final characterized chemotaxis protein is CheV, a fusion of CheW and a CheY-like receiver domain, which affects the signaling state of the MCP based on its phosphorylation state as controlled by the CheA kinase (Karatan *et al.*, 2001; Pittman *et al.*, 2001).

In addition to component diversity between chemotaxis systems, there are also functional differences between their outputs. Historically, the focus of detailed molecular investigation is on the chemotaxis system that controls flagellar motility, but studies have demonstrated that chemotaxis systems are also involved in regulating type IV pili-based motility (Bhaya *et al.*, 2001; Sun *et al.*, 2000; Whitchurch *et al.*, 2004). Even more recently, chemotaxis systems were implicated in controlling diverse cellular functions, such as intracellular levels of cyclic di-GMP, transcription, and other functions (Berleman and Bauer, 2005; D'Argenio *et al.*, 2002; Hickman *et al.*, 2005; Kirby and Zusman, 2003). Many organisms have multiple chemotaxis systems that can have both overlapping and/or unrelated functional outputs (Berleman and Bauer, 2005; Guvener *et al.*, 2006; Kirby and Zusman, 2003; Martin *et al.*, 2001; Wuichet and Zhulin, 2003). Beyond the functional diversity of the system outputs, there can be significant mechanistic diversity within these functional classes. For example, the signaling and adaptation mechanisms in *Escherichia coli* and *Bacillus subtilis* differ markedly. In *E. coli*, positive stimuli inhibit CheA activity, whereas in *B. subtilis* the opposite is true. In *E. coli*, MCP demethylation increases in response to negative stimuli only, whereas in *B. subtilis*, it occurs in response to both positive and negative stimuli (Szurmant and Ordal, 2004).

The diversity found among chemotaxis systems cannot be efficiently addressed by experimental means alone, nor can the questions about the function and origin of this system. Initial genomic studies have already identified the core set of chemotaxis proteins as CheA, CheW, CheY, and MCP, which are present in all chemotaxis systems (Zhulin, 2001), unlike the sporadic distributions of CheC, CheD, and CheZ (Kirby *et al.*, 2001; Szurmant and Ordal, 2004; Terry *et al.*, 2006) and the occasional absence of CheB and CheR (Terry *et al.*, 2006; Zhulin, 2001). Diversity within the CheA domain organization was also reported (Acuna *et al.*, 1995; Bhaya *et al.*, 2001; Whitchurch *et al.*, 2004), as well as the broad repertoire of MCP sensor domains (Aravind and Ponting, 1997; Shu *et al.*, 2003; Taylor and Zhulin, 1999; Ulrich and Zhulin, 2005; Zhulin, 2001; Zhulin *et al.*, 2003) and their evolutionary trends (Wuichet and Zhulin, 2003), and the length variability of the MCP signaling module (Alexander and Zhulin, 2007; LeMoual and Koshland, 1996). Motivating factors to further study the chemotaxis system using comparative genomic methods are the wealth of genomic data available for prokaryotes, the large evolutionary distances between prokaryotes that have this system, and the propensity for its components to be encoded in gene clusters. The extensive molecular and biochemical characterizations of the system and its components and the availability of three-dimensional structures for most of the components provide most valuable information for comparison and validation of findings obtained through computational analysis. Although this chapter focuses on the chemotaxis system, the methodology of this research is applicable to all signal transduction systems, prokaryotic and eukaryotic, with the caveats that certain thresholds (e.g., sequence conservation) must be altered to suit the evolutionary rate of a given protein or domain and that some techniques (e.g., gene neighborhood analysis) are best applied to prokaryotic systems.

Bioinformatics Tools and Resources for Identifying and Analyzing Chemotaxis Components

Many tools and databases are available to aid comparative genomic analyses. The SMART (Letunic *et al.*, 2006) and Pfam (Finn *et al.*, 2006) databases are primary sources for Hidden Markov Models (HMMs) that can identify conserved domains and domain combinations within protein sequences. Each model captures the key sequence features of a specific domain, based on the multiple alignments from which it is built. When a model for a given domain is not available or is inadequate (e.g., poor quality, artificial relationship between sequences), the sequence of a representative protein family member can be used to search against common sequence databases such as those at the National Center for Biotechnology Information (NCBI) (Wheeler *et al.*, 2006) using various versions of the Basic Local Alignment Search Tool (BLAST) algorithm (McGinnis and Madden, 2004). For comparative analysis we often focus on completely sequenced genomes, which make the RefSeq and microbial databases of NCBI (Wheeler *et al.*, 2006) ideal to search against, but even within these searches there are many ways to further narrow down search results, for example, by retrieving only sequences of a certain length range (McGinnis and Madden, 2004). While a single search iteration is standard in BLAST, the Position-Specific Iterative BLAST (PSI-BLAST) program enables iterative searches by updating a position-specific score matrix (PSSM) with each iteration. PSI-BLAST enables identifying many divergent members of a particular protein family (Altschul *et al.*, 1997). Typical sequence similarity searches compare DNA to DNA, DNA to protein, protein to DNA, or protein to protein. Ideally protein-to-protein searches should be performed because the greater number and diversity of sequence characters in proteins (20 amino acids versus 4 nucleotide bases) make them more sensitive. Searching a database using a PSSM output by PSI-BLAST produces more sensitive results than a search using a single protein sequence. This approach can be useful for finding protein family members in newly released genomes.

Although BLAST and PSI-BLAST searches are invaluable in comparative genomic analyses, they can be time-consuming and the results need to be analyzed carefully (which is particularly true for PSI-BLAST analyses). HMM domain models of well-characterized protein families can quickly identify new family members and help understand the functions of all proteins in the family. Regularly updated databases of protein domain architectures, such as SMART and Pfam, are important tools to begin to understand protein function at the individual or family level. The recently developed Microbial Signal Transduction (MiST) database expands this concept by extracting signal transduction profiles for all complete microbial proteomes, taking advantage of both SMART and Pfam models and the wealth of knowledge generated in the area of microbial signal transduction (Ulrich and Zhulin, 2007; Ulrich *et al.*, 2005). Figure 1 shows representative members of each chemotaxis protein as visualized in MiST, and the following sections discuss the best way to identify each of these proteins in public databases and other sources of genomic data.

Once protein family members are identified, other bioinformatics tools need to be employed in order to derive meaningful information about their function and relationships. Because multiple sequence alignments are the essential backbone of most comparative analyses, building high-quality multiple alignments is critically important. There are many programs currently available to build initial multiple alignments, including most popular Clustal (Chenna *et al.*, 2003), MUSCLE (Edgar, 2004), PCMA (Pei *et al.*, 2003), and T-COFFEE (Notredame *et al.*, 2000). ClustalW, MUSCLE, and PCMA are very fast programs suitable for a large number of sequences, particularly for a set of sequences that is highly conserved. T-COFFEE is slower, but has a higher accuracy for the alignment of sets of sequences that are not highly conserved. For a given set of sequences one program may produce better initial results than the others, but manual analysis and editing are then needed in most cases. Manual editing with a program such as SeaView (Galtier *et al.*, 1996) can help resolve the gap regions and poorly conserved alignment regions that are not handled well by the alignment software. The VISSA program aids manual editing by visualizing the secondary structure of each protein sequence in a given multiple alignment (Ulrich and Zhulin, 2005). Although some regions of a protein may display poor sequence conservation, there is often still pressure on these regions to maintain their secondary structure. Identifying unstructured regions that link secondary structure elements can also aid in the placement of gaps during the editing process.

A multiple sequence alignment provides immediate information about potentially important functional regions and individual amino acids by revealing highly conserved positions. The CONSENSUS script (http://coot. embl.de/Alignment/consensus.html) and the WebLogo program (Crooks *et al.*, 2004) can further aid in the identification of highly conserved positions in multiple alignments. If the structure of a protein is available, conserved sites can be easily visualized on the structure with structure viewing packages, such as DeepView (Schwede *et al.*, 2003) and PyMol (http://www.pymol.org). In order to cluster related protein sequences,

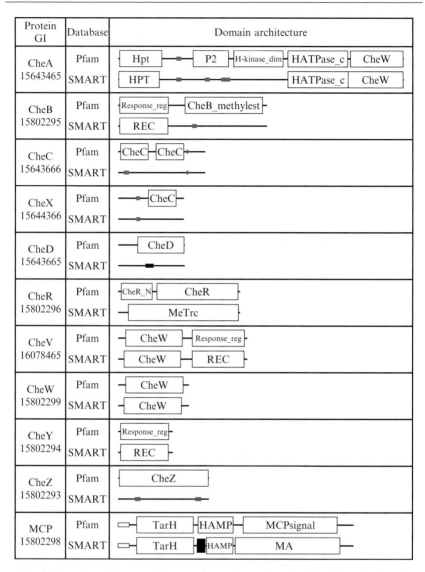

Fig. 1. Domain architecture of chemotaxis proteins as visualized in MiST. The MiST database (Ulrich and Zhulin, 2007) uses the domain models from both Pfam and SMART databases. Domains are shown as white boxes with their names inside. Small black, gray, and white boxes indicate predicted transmembrane, low complexity, and signal peptide regions, respectively. The NCBI database GI (GenBank identifier) numbers corresponding to each protein sequence are given under their respective protein names.

phylogenetic trees can be built from a multiple alignment with many different methods and programs. The MEGA program (Kumar *et al.*, 2004) is a user-friendly tool used to build neighbor-joining trees for the quick identification of protein subfamilies based on sequence similarity. MEGA can also be used to easily view and edit trees produced by different multiple alignment programs. Sequence similarity is not always a reflection of the evolutionary history of a protein, as there can be multiple mutation events that obscure origins. For more precise evolutionary analysis, maximum likelihood trees are more appropriate and can be built using the ProML program of the PHYLIP package (Felsenstein, 1989). Because trees can often vary depending on the methods used to build them, it is best to validate them by independent means. Unless the gene encoding a protein is a subject of frequent horizontal transfer, it is expected that the most closely related proteins in a tree will be from closely related organisms. We also expect proteins with similar domain architectures to cluster together in a tree. Most importantly, because chemotaxis genes are encoded in conserved gene clusters, closely related proteins are predicted to be encoded in similar gene neighborhoods. Gene neighborhood (or genome context) analysis (Overbeek *et al.*, 1999) can become a very useful approach in elucidating specific interactions when multiple chemotaxis systems are encoded within a genome. Occasionally, distinct protein subfamilies can be correlated to specific motifs within the alignment as well as insertions or deletions that might be specific to their structure and function.

Defining MCP Membrane Topology

Methyl-accepting chemotaxis proteins are the receptors at the beginning of the chemotaxis signal transduction cascade that process environmental and intracellular sensory (input) signals and alter the activity of the CheA histidine kinase. MCP sequences typically consist of a sensory domain, a HAMP linker domain, and a signaling domain that interacts with CheA (Fig. 1). The HAMP and signaling domains are always cytoplasmic, but the membrane topology of the sensory domain varies. Figure 2 shows the four main classes of MCP membrane topology (Zhulin, 2001). Sensory class I MCPs have a periplasmic sensory domain anchored by an N-terminal transmembrane (TM) helix and connected by an internal TM helix to the HAMP linker and signaling domains. Most MCPs, including the Tar, Tsr, Trg, and Tap receptors of *E. coli*, have this sensory topology (Ulrich and Zhulin, 2005; Zhulin, 2001). Sensory class II MCPs have an N-terminal cytoplasmic sensory domain connected by an internal TM helix to the HAMP linker and signaling domains. The Aer aerotaxis receptor of *E. coli* is an example

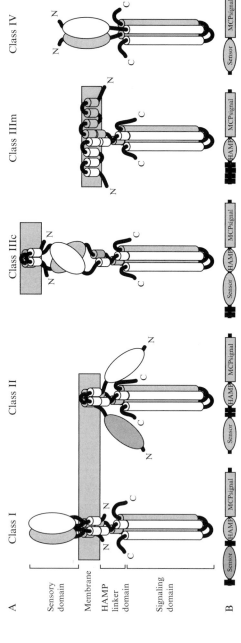

Fig. 2. MCP membrane topology classes. Differing membrane topology divides MCPs into four main classes. (A) Schematic representation of the three-dimensional structure of MCP dimers of different sensor classes. Oval domains are sensory domains of varied secondary structure. Cylinders represent α-helical and coiled coil regions. MCP monomers are differentiated by gray and white coloring. Class I, transmembrane MCPs with extracellular sensory domains; class II, membrane-bound MCPs with N-terminal cytoplasmic sensory domains; class III, membrane-bound MCPs with cytoplasmic sensory domains located C-terminally to the last transmembrane regions (IIIc) or with membrane-bound sensory domains (IIIm); class IV, cytoplasmic MCPs. (B) MCP sensor class can be determined from domain architecture where transmembrane regions and domains are well predicted. Transmembrane regions are indicated by black boxes.

of a class II sensor (Bibikov *et al.*, 1997; Rebbapragada *et al.*, 1997). Since the previous classification of MCP sensor classes (Zhulin, 2001), many more MCP sequences have become available, and we now split sensor class III into two subgroups. Sensory class IIIc MCPs are anchored at their N terminus by a TM helix, downstream of which are a cytoplasmic sensory domain and the HAMP linker domain and cytoplasmic signaling domain. Sensory class IIIm MCPs are like class IIIc MCPs except that the sensory domain is membrane bound rather than cytoplasmic. The Htr8 aerotaxis receptor of *Halobacterium salinarum* (Brooun *et al.*, 1998) is an example of a sensory class IIIm receptor. Some MCPs are hybrids of class II and class III, containing a periplasmic sensory domain separated by a TM helix from an additional cytoplasmic sensory domain (Wuichet and Zhulin, 2003). Sensory class IV MCPs are entirely cytoplasmic; they lack TM helices and usually also HAMP domains. The oxygen sensor HemAT from *B. subtilis* is an example of a class IV sensor (Hou *et al.*, 2000; Zhang and Phillips, 2003).

Methyl-accepting chemotaxis protein sensor class and membrane topology can be easily determined by visual inspection of a two-dimensional domain model that includes TM regions (Fig. 2B). Transmembrane regions can be identified in MCPs and other proteins using various TM prediction programs, such as Phobius (Kail *et al.*, 2004) and DAS-TMfilter (Cserzo *et al.*, 2002). These two programs give similar results and are amenable to high-throughput scripting. It should be noted that because DAS-TMfilter is a modification of the Dense Alignment Surface (DAS) algorithm (Cserzo *et al.*, 1997) to screen out false positives, if one suspects underprediction of TM regions in an MCP of interest, the original DAS algorithm can be used on a case-by-case basis. Both DAS and Phobius can generate graphical TM prediction plots for visual inspection.

Diversity of Input (Sensory) Domains in MCPs

The MCP signaling domain is highly conserved because it maintains multiple protein–protein interactions within the chemoreceptor–kinase complex. MCP sensory domains, however, evolve rapidly, being subject to frequent domain birth and death events, and are quite variable in sequence (Wuichet and Zhulin, 2003). In fact, the lack of good sensory domain models is still an unsolved problem not only in chemotaxis, but in microbial signal transduction in general (Ulrich and Zhulin, 2005). Figure 3 shows an array of well-defined sensory domains found in MCPs. PAS (Taylor and Zhulin, 1999) and GAF (Aravind and Ponting, 1997) are ubiquitous sensory domains with the similar protein fold (currently known simply as the PAS/GAF fold) that can be found throughout prokaryotic and eukaryotic signal transduction. Most members of these domain families are

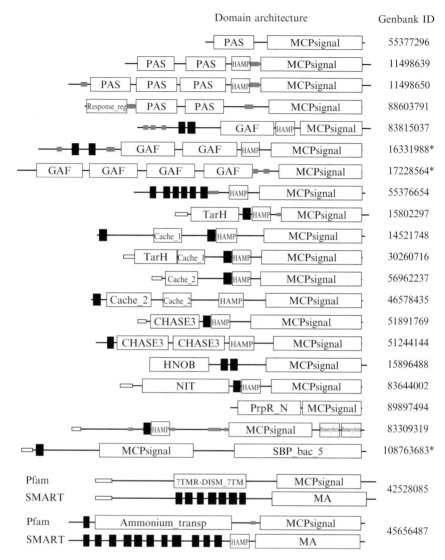

FIG. 3. Diversity of sensory domains in MCPs. All sensory domains are Pfam domain models, except the GAF domain, which is the SMART model (it is slightly longer than the Pfam domain model). HAMP domains are the SMART domain model. MCPs containing hemerythrin and SBP_bac_5 sensory domains represent the atypical topology where the MCP signaling domain is N-terminal of the sensory domain. The Pfam TarH model has shown to be erroneous and will soon be replaced by a correct model termed 4HB_MCP (Ulrich and Zhulin, 2005). Both Pfam and SMART domain architectures are shown for two MCPs with class IIIm membrane topology. Small gray and white boxes indicate predicted low complexity and signal peptide regions, respectively. Black boxes represent transmembrane regions. Long sequences marked by an asterisk (*) were shortened for display and are not to scale.

cytoplasmic, although a divergent PAS subfamily is exclusively extracellular (Reinelt *et al.*, 2003). In addition to MCPs where they are located exclusively extracellularly, Cache family domains are also found in extracellular subunits of eukaryotic Ca^{2+} channels that are implicated in signal transduction (Anantharaman and Aravind, 2000). For some sensory domains, their signal specificity can be predicted to be in a narrow range, for example, the nitrate- or nitrite-responsive NIT domain (Shu *et al.*, 2003); however, in most instances, the MCP signal spectrum cannot be readily predicted by the sequence conservation of their sensory domains. Furthermore, for a significant number of MCPs, while the sensory topology can be determined from the pattern of TM helices, no known domains are identified by current domain models. These regions contain either known domains that are not recognized by low-sensitivity models or novel, uncharacterized domains. Further computational and experimental work is necessary to identify and understand the function of novel sensory domains in MCPs.

HAMP Domain Identification

The HAMP linker domain is an important module, which is present in many membrane-bound signal transduction proteins, including MCPs and the sensor histidine kinases of two-component systems (Aravind and Ponting, 1999). The HAMP domain is about 60 amino acids long and consists of two amphipathic α helices (AS1 and AS2) separated by a loop. Because of its structural flexibility, the mechanism of signal transmission by the HAMP domain has been difficult to characterize (Williams and Stewart, 1999); however, the nuclear magnetic resonance structure of a stable archaeal HAMP domain has been determined (Hulko *et al.*, 2006) and should lead to new developments in the field. Because some MCPs contain multiple HAMP domains, understanding of its mechanism should involve modularity and the possibility of self-interaction.

The domain models of the HAMP linker domain in Pfam and SMART (Pfam, HAMP; SMART, HAMP) are slightly different from each other and are of relatively poor quality. The Pfam domain model includes a fully lipophilic helix at its N terminus upstream of AS1, which overlaps the TM regions that determine MCP sensory topology (see example shown in Fig. 1). The SMART HAMP domain model extends three residues past the Pfam model, which is important to keep in mind when trying to establish the boundaries of the signaling domain. Most importantly, both the Pfam and the SMART domain models fail to identify HAMP domains in many sequences, where they are obviously present (Fig. 4). If the domain organization of an MCP of interest contains a short region free of identified

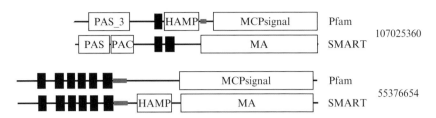

Fig. 4. HAMP domain models are imperfect. Both the Pfam and SMART HAMP domain models have low sensitivity; however, implementation of both models in MiST enables the identification of HAMP domains in many cases when one of the domain database models misses the target. Note that the Pfam HAMP domain models often (but not always) overlap with one of the transmembrane regions.

domains downstream of the membrane and upstream of the signaling domain, a PSI-BLAST search should be performed, which in many instances will lead to the detection of the HAMP domain.

MCP Signaling Domain

The cytoplasmic signaling domain of MCPs is a coiled coil with a hairpin at its base that is highly conserved in sequence. The presence of this highly conserved domain (HCD) in MCPs makes it possible to extract all MCP sequences from a genome with high confidence using the Pfam or SMART domain models of the cytoplasmic signaling domain (Pfam, MCPsignal; SMART, MA). However, the Pfam and SMART domain models do a poor job of delineating the exact boundaries of the signaling domain because of the significant variability of its length. LeMoual and Koshland (1996) identified three classes of the MCP signaling domain that were different in length by multiples of seven residues, or exactly two turns of an α helix in a coiled coil protein. The signaling domains of MCPs from *E. coli* have four 14-residue gaps relative to those from *B. subtilis*, a total of 56 residues difference in length. Recent in-depth computational analysis resulted in the identification of seven major and several minor length classes of the MCP signaling domain, revealing the subdomain organization and unusual evolutionary history of this important signaling module (Alexander and Zhulin, 2007).

MCP Pentapeptide Tether

Alexander and Zhulin (2007) collected 2125 MCP sequences from 152 bacterial and archaeal genomes and analyzed their C-terminal five residues. In *E. coli*, this C-terminal pentapeptide has been shown to bind to the adaptation enzymes CheB (Barnakov *et al.*, 2001) and CheR (Djordjevic *et al.*, 1998). The pentapeptide motif in *E. coli* MCPs is NWETF, but it was found that the motif could be generalized with an emphasis on two aromatic residues (-x-[HFWY]-x(2)-[HFWY]-). Only 217 MCPs from 67 of 152 genomes contained sequences that matched this motif. All of these MCPs belong to the same major class of the signaling domain as the five MCPs of *E. coli*. All but two of the organisms where pentapeptide-containing MCPs are found are proteobacterial, implying that the pentapeptide tether is a recently evolved mode of interaction between MCPs and adaptation enzymes. The pentapeptide can be easily identified in newly available MCP sequences by visual inspection or simple scripting in Perl.

The CheA Histidine Kinase: Domain Organization, Conservation, and Diversity

The CheA histidine kinase is an essential component of the chemotaxis system and has a complex multidomain architecture (Bilwes *et al.*, 1999; Stock *et al.*, 2000). Five domains were identified in CheA from model organisms *E. coli* and *B. subtilis*, but analysis of CheA sequences from more recent experimental studies have revealed that its domain architecture can be highly variable (Fig. 5) (Acuna *et al.*, 1995; Bhaya *et al.*, 2001; Porter and Armitage, 2004; Whitchurch *et al.*, 2004). CheA has a conserved core of four domains, a histidine phosphotransfer domain (Pfam, Hpt; SMART, HPT) that is autophosphorylated by ATP (Kato *et al.*, 1997), a dimerization domain (Pfam, H-kinase_dim) (Bilwes *et al.*, 1999), an ATPase domain (Pfam and SMART, HATPase_c) (Bilwes *et al.*, 1999), and a CheW scaffolding domain (Pfam and SMART, CheW) that is homologous to the CheW protein (Bilwes *et al.*, 1999). Although the dimerization domain (Pfam, H-kinase_dim) was not initially identified in some of the CheA proteins, this is the result of a poor domain model, as revealed by a multiple alignment of CheA sequences. The crystal structure of the dimerization, ATPase, and CheW domains of the CheA from *Thermotoga maritima* (Bilwes *et al.*, 1999) shows that the domain model does not cover the full domain length. Despite the dimerization domain discrepancy, the two-dimensional domain model for CheA does capture the overall information about the three-dimensional state of the protein (Fig. 6). Domain searches for proteins with both the HATPase_c and the CheW domains should easily identify CheA homologs without the need for additional search tools.

While the majority of CheA proteins contain all of the core domains, a few chemotaxis systems have the HPT domain detached from the other three core domains as a separate protein. An unusual split CheA in *Rhodobacter sphaeroides* has both parts found in the same gene neighborhood

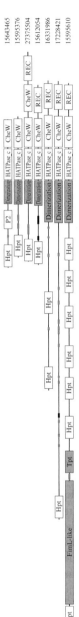

FIG. 5. A common core and diversity of CheA homologs. The domain architectures of selected CheA proteins are shown with their corresponding NCBI GI numbers to the right. All shown domains are from Pfam except for the REC domain (SMART domain model). The dimerization domains shown in gray were delineated by PSI-BLAST analysis; current dimerization domain models have very low sensitivity and fail to predict the domain in many instances. Our analysis shows that the dimerization domain is present in all CheA homologs identified to date (K. Wuichet, unpublished data). Small black, gray, and white boxes indicate predicted transmembrane, low complexity, and signal peptide regions, respectively. The FimL-like domain shows similarity to the FimL pili motility protein, and the Tpt domain shows similarity to Hpt domains, but it has a threonine in place of the conserved histidine (the phosphorylation site). Despite diverse domain architectures, all CheA proteins contain Hpt, dimerization, HATPase_c, and CheW domains, with the latter three forming in a tight protein core. CheA–CheC fusion proteins were also identified; see Fig. 8.

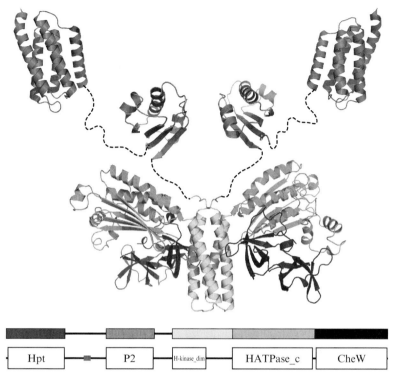

FIG. 6. The relationship between the domain architecture and the structure of CheA. The domain architecture of the CheA protein directly relates to its structure. The Pfam domain model of CheA (GI 15643465) and its two-dimensional color scheme are shown below the three-dimensional model that has a matching color code. The three-dimensional model consists of three different crystal structures: the Hpt (or P1) domain (PDB identifier 1I5N), the P2 domain (1UOS), and the three core domains (PDB, 1BDJ)—dimerization (or P3) (Pfam, H-kinase_dim), HATPase_c (or P4), and CheW (or P5), respectively, with the linker regions hand drawn. The first two linker regions found in the domain architecture are predicted to be loops between the globular Hpt and P2 domains. The third predicted linker region of CheA suggests that the H-kinase_dim domain model does not capture the entire dimerization domain.

(Porter and Armitage, 2004). One gene encodes the dimerization domain, ATPase, and CheW domains while the other has the HPT and CheW domains. Neither protein is able to undergo autophosphorylation, but they are able to interact together for transphosphorylation. In *Synechocystis* sp. PCC 6803, the HPT domain protein is found in a region of the genome distant from its partner protein, which contains dimerization, ATPase, CheW, and REC (response regulator receiver) domains (Yoshihara *et al.*, 2002).

Both proteins are necessary for chemotaxis similarly to the split CheA in *R. spharoides*. Although the core domains of CheA are conserved in all chemotaxis systems, the extensive domain architecture diversity implies that there is a high level of functional and mechanistic differences that must be addressed by comparative analysis and experiment.

The P2 domain (Pfam, P2) is absent from many CheA proteins, while many CheA have an additional carboxyl-terminal REC domain (Acuna *et al.*, 1995; Bhaya *et al.*, 2001; Whitchurch *et al.*, 2004). Histidine kinases that contain the REC domain are termed "hybrid." For any CheA sequences that have extended undefined regions, PSI-BLAST searches should be performed in order to find out whether they contain known domains missed by current models or novel domains. Such searches identified multiple divergent HPT domains, and new domains not previously associated with CheA proteins, in undefined regions of a CheA homolog in *Pseudomonas aeruginosa* that regulates the type IV pili-based motility (Whitchurch *et al.*, 2004). Because the P2 domain has particularly low sequence conservation in comparison to the rest of the CheA domains, undefined regions between HPT and dimerization domains that are not characterized by low complexity are potential P2 domains. Crystal structures of the P2 domains of *T. maritima* and *E. coli* CheA proteins show that the *E. coli* P2 domain is reduced in size and missing some structural elements present in the P2 domain of *T. maritima* (McEvoy *et al.*, 1998; Park *et al.*, 2004a). Our sequence analysis of P2 domains and unidentified regions between HPT and dimerization domains revealed three classes of the P2 domain. Class I is represented by the *T. maritima* P2 domain. Class II shows structural similarities to class I, but it has an extra insertion. Class II also shows sequence similarities to the reduced P2 domain of class III, which is represented by the *E. coli* P2 domain. The three classes can be aligned accurately with the help of VISSA (Ulrich and Zhulin, 2005), and the gap regions of the alignment clearly show the three classes (Fig. 7). Despite the poor domain model for P2, all CheA proteins that contain a P2 domain have the typical domain architecture shown in Fig. 1 with the exception of the unusual HPT–CheW protein of *R. sphaeroides* (Porter *et al.*, 2002), where we find a previously unidentified P2 domain in the long undefined region between the two domains, and some archaeal CheA proteins that have tandem P2 domains.

Interestingly, many CheA proteins that lack P2 domains contain the C-terminal REC domain (Fig. 5). Conversely, none of the CheA proteins that contain the C-terminal REC domain has the P2 domain (the P2 domain was proposed in the hybrid CheA from *Synechocystis* [Bhaya *et al.*, 2001]; however, our computational analysis failed to support it). This observation can lead to several experimentally testable hypotheses.

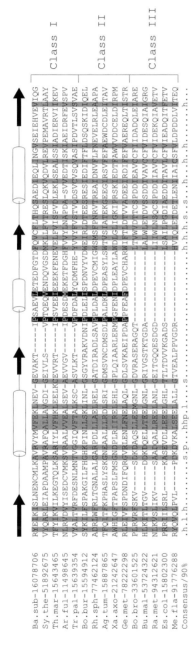

Fig. 7. Multiple alignment of the P2 domain and its classification. Three subclasses of the P2 domain were identified. A multiple alignment with representative members of each class of P2 sequences. Positions conserved at 90% or more in an alignment of 116 P2 sequences are shown in gray. Black columns show conserved proline and hydrophobic positions in classes I and II. The secondary structure elements are shown above the alignment based on crystal structures from *E. coli* and *T. maritima* (McEvoy, 1998; Park, 2004a,b). Black arrows represent β strands. White cylinders represent α helices. Species abbreviations and NCBI GI numbers for each sequence are given at the left (full species name can be found by searching the NCBI nonredundant database with the corresponding GI number).

The CheY Response Regulator: Big Problems of the Small Protein

Although essentially all CheY proteins can be identified by domain searches, such searches cannot identify CheY proteins exclusively because there is no specific domain model for CheY. CheY is a single domain protein, which is a variant of the ubiquitous receiver domain (Pfam, response_reg; SMART, REC) that is found in response regulators of classic two-component signal transduction systems as well as chemotaxis systems (Galperin, 2006; West and Stock, 2001). In order to find stand-alone REC domains (CheY candidates), BLAST searches can be restricted to retrieve sequences of only a certain length or they can be identified by extensive domain architecture queries. Unfortunately, identifying CheY proteins in a set of stand-alone REC domains is still a serious challenge. In some two-component systems, stand-alone REC domains serve as middlemen in extended phosphotransfer relays (Hoch, 2000; Stock *et al.*, 2000).

Stand-alone REC domains that are more similar to experimentally characterized CheY proteins rather than components of phosphorelay systems are predicted to be CheY proteins. Gene neighborhood analysis is a powerful technique used to confidently identify CheY proteins because they are often encoded in a cluster of other chemotaxis genes. We can begin to identify CheY proteins computationally by building phylogenetic trees of stand-alone REC domains and searching for subfamilies that can be linked to chemotaxis by experimental evidence and gene neighborhood analysis. Once CheY proteins are predicted there is the added confusion of identifying what type of motility they regulate. Given that CheY proteins have been shown to regulate both flagellar and type IV pili-based motility, the whole-genome search for the presence or absence of genes encoding these motility organelles can aid in delineating functional subfamilies of CheY proteins.

CheB and CheR

The CheB methylesterase and CheR methyltransferase work together to regulate the methylation state of MCPs (Li and Hazelbauer, 2005). Although there are examples of flagellar and pili-based chemotaxis systems that lack CheB and CheR (Terry *et al.*, 2006; Whitchurch *et al.*, 2004; Zhulin, 2001), they are present in the vast majority of chemotaxis systems that have been studied experimentally or deduced from genome sequence. Unexpectedly, some chemotaxis systems that contain all core components may lack CheR but not CheB (Whitchurch *et al.*, 2004) or lack CheB but not CheR (e.g., the genomes of *Listeria innocua, Listeria monocytogenes, Bacillus cereus, Bacillus anthracis,* and *Bacillus thuringiensis*; K. Wuichet, unpublished observation). The genome of *Hyphomonas neptunium* that lacks all core chemotaxis components still contains the *cheR* gene, although no chemotaxis was detected in this organism (Badger *et al.*, 2006).

CheB is typically defined by the presence of the catalytic domain (Pfam, CheB_methylest) fused to a regulatory amino-terminal REC domain. In Pfam, CheR is defined by two domain models, CheR_N and CheR, whereas both of the regions corresponding to these domains are encompassed in the SMART MeTrc domain model (Fig. 1). The Pfam domains are a better reflection of the three-dimensional structure of the protein, which consists of two globular subdomains corresponding to the domain models; however, the N-terminal subdomain (CheR_N) is not highly conserved, which has resulted in a poor domain model. The CheR and MeTrc domains, which include the highly conserved catalytic region, are the best models to search for CheR proteins in genome databases.

Comparative genomics has already identified a variety of CheR domain architectures, including CheR fusions with C-terminal tetratricopeptide repeats (SMART, TPR) and N-terminal CheW domains (Shiomi *et al.*, 2002), as well as CheB association with class I histidine kinase domains (SMART: HisKA and Pfam: HWE_HK) (Karniol and Vierstra, 2004). TPR domains are known to promote protein–protein interactions (Lamb *et al.*, 1995). The CheR–TPR fusion proteins have been shown to be involved in chemotaxis, and their TPR domains are found to interact with their N-terminal CheR regions and with CheY (Bustamante *et al.*, 2004). A simple domain search for CheB reveals the histidine kinase fusion proteins along with stand-alone CheB catalytic domains that lack a regulatory REC domain. The roles of the adaptation domains of the kinase fusion proteins and the stand-alone CheB catalytic domains have not been determined experimentally. Comparative genomic analyses have the potential to reveal new insight into this adaptation system and identify specific targets for further experimental analysis.

CheC and CheX

The crystal structures of the closely related CheC and CheX proteins reveal distinct differences in their structures and interactions (Park *et al.*, 2004b). These two CheY phosphatases share sequence similarity, but have different structures and domain architectures. The CheC/CheX homolog FliY, a component of the flagellar motor, can be clearly discriminated from CheC and CheX by the presence of a C-terminal SpoA domain (Pfam, SpoA) that is involved in structural assembly. An exception from this rule is the split FliY protein of *T. maritima* (Park *et al.*, 2004b; Szurmant *et al.*, 2004). The CheC domain model (Pfam, CheC) is built from the active site of the enzyme. CheC and FliY have two homologous active sites (Fig. 1).

The CheX phosphatase is closely related to CheC, but has only the second active site of CheC (Fig. 1). CheX was found to act as a dimer in the crystal structure, unlike CheC. CheX also differs from CheC in its length and secondary structure (Park *et al.*, 2004b). CheC has been cocrystallized with CheD, an interaction that has been shown to increase the phosphatase activity of CheC (Chao *et al.*, 2006). CheC and CheX are poorly conserved even at the very small active site region, which makes their identification by domain models rather difficult. Similarity searches, such as BLAST, are better suited for finding CheC and CheX homologs, although these searches are unable to discriminate between CheC and CheX (Fig. 8). Multiple sequence alignments are needed to confirm the validity of CheC/CheX/FliY protein family members identified by similarity searches. The VISSA program (Ulrich and Zhulin, 2005) can aid in distinguishing CheC and CheX given the distinct differences in their secondary structures. We find new CheC/CheX protein domain architectures including a fusion with CheA, fusions via duplication, and fusions with REC domains in a variety of Proteobacteria. Given that CheC and CheX act to dephosphorylate the CheY REC domain, it is possible that the REC–CheC/CheX fusion proteins promote dephosphorylation of the REC domain in the fusion. Experimental analysis is needed to clarify the function of these fusion proteins.

CheC is often encoded in the genome near CheD and/or a CheY-like protein, whereas CheX is typically not encoded near chemotaxis components with the exception of Spirochetes. CheC is also found in some genomes that lack CheD, for example, *Vibrio parahaemolyticus* and *Myxococcus xanthus*. The CheX protein has been shown to dephosphorylate CheY-P (Motaleb *et al.*, 2005; Park *et al.*, 2004b) and interact with CheA (Sim *et al.*, 2005); however, it remains to be seen whether CheX acts as a phosphatase and plays a role in chemotaxis in more distantly related organisms.

CheD

In addition to playing a role in the excitation pathway by aiding CheY-P dephosphorylation by CheC, CheD also plays a role in the adaptation pathway by deamidating key glutamine residues of MCPs into glutamate residues so they can be methylated by CheR (Kristich and Ordal, 2002). Similarity searches reveal that CheD is highly conserved and can be easily identified solely by queries for its domain model (Pfam, CheD). The phyletic distribution of CheD and CheC showed that many organisms that have CheD lack CheC (Kirby *et al.*, 2001). CheD is a single domain protein, but its fusion with CheB can be seen in *Bdellovibrio bacteriovorans*.

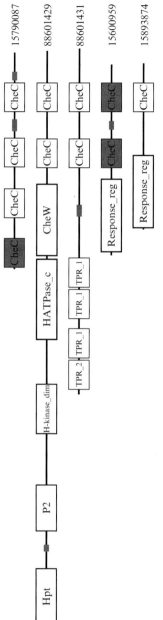

FIG. 8. Diversity of CheC homologs. CheC and CheX proteins can be fused to different domains and proteins. Domains shown in gray were missed by the current domain models and were found by PSI-BLAST searches. Their approximate position in corresponding protein sequences is shown. Domain models are from Pfam. Small gray boxes indicate predicted low complexity regions. The NCBI GI number associated with each sequence is shown at the right.

Our gene neighborhood analysis showed that the overwhelming majority of CheD proteins are encoded in the genomes near other chemotaxis proteins, implicating their involvement in chemotaxis regardless of the presence of CheC.

CheZ

Although the CheZ phosphatase of CheY was previously found only in some representatives of β/γ-Proteobacteria (Szurmant and Ordal, 2004), experiments have identified a divergent CheZ protein, which was not detected by the current Pfam domain model (Pfam, CheZ) in the member of ε-Proteobacteria, *Helicobacter pylori* (Terry *et al.*, 2006). We performed PSI-BLAST searches against completely sequenced genomes to identify many other previously undetected members of the CheZ family from different species, including representatives of α- and δ-Proteobacteria (Fig. 9).

A multiple alignment reveals that all of the sequences identified in α- and δ-Proteobacteria form a specific CheZ subfamily, which can be distinguished by the conserved catalytic glutamine residue and high conservation of positions surrounding the catalytic residue. The phylogenetic tree built from the multiple alignment suggests three subfamilies of CheZ proteins based on sequence features and taxonomy (Fig. 9). Although CheZ has been shown to interact with both CheY and CheA, the subfamily of the α- and δ-Proteobacterial sequences lacks the CheA-binding region entirely, and thus these CheZ proteins are predicted not to interact with CheA. None of these proteins have been experimentally characterized, but the CheZ of *Caulobacter crescentus* is located near a chemotaxis locus containing *cheA*, *cheB*, *cheR*, *cheW*, and *cheY*, which supports the hypothesis that representatives of this subfamily play a role in chemotaxis. The experimentally characterized CheZ of *E. coli* is found in the β/γ-Proteobacteria subfamily. The CheZ of *E. coli* interacts with CheY-P at two distinct regions (Zhao *et al.*, 2002), and a third region interacts with CheA (Cantwell *et al.*, 2003). Both the CheY and the CheA interaction regions are found in all members of the β/γ subfamily except *Xanthomonas axonopodis* and *Xanthomonas campestris*, which lack the CheA-binding region. The ε-Proteobacteria subfamily has an elongated CheA-binding region that shares no sequence similarity with the CheA-binding region of the β/γ subfamily, but the presence of this subdomain suggests that it may still be involved in binding CheA. The CheZ of *H. pylori* has been implicated in chemotaxis, but direct interaction studies have not been carried out (Terry *et al.*, 2006). Although the ε- and α/δ-proteobacterial subfamilies are quite divergent, the conservation of catalytic residues suggests that they

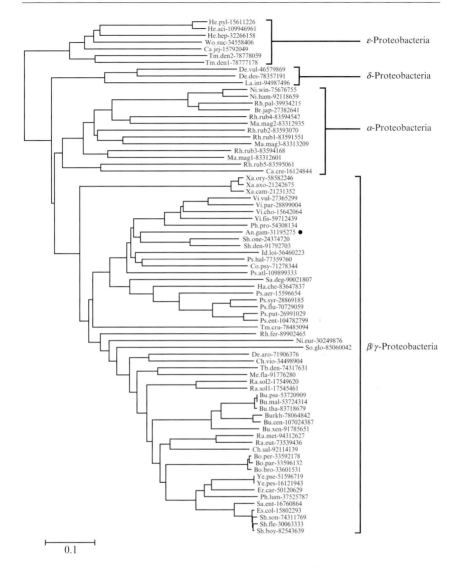

FIG. 9. Neighbor-joining tree of the extended CheZ protein family. The CheZ protein family has members present in all classes of Proteobacteria, and the phylogenetic tree suggests its vertical evolution. The sequence identified by a black circle comes from a likely contamination with prokaryotic DNA in the genome of the mosquito *Anopheles gambiae*.

are involved in dephosphorylation of CheY-P proteins, and the phyletic distribution suggests that CheZ originated in a common ancestor of Proteobacteria.

CheW and CheV

CheW and CheV have both been shown to be involved in sensory lattice scaffolding by interacting with CheA and MCPs (Gegner *et al.*, 1992; Rosario *et al.*, 1994). As seen in Fig. 1, the CheW protein is a single domain (Pfam, CheW; SMART, CheW), but domain queries with CheW will identify multiple components of the chemotaxis system, as it is homologous to domains found in all CheV and CheA proteins in addition to an unusual CheW–CheR fusion protein found exclusively in Spirochetes. Searches that include the CheW domain while excluding REC, HATPase_c and MeTrc domains should identify true CheW proteins. Phyletic distribution shows that the CheW protein is essential to all chemotaxis systems with the exceptions of *L. innocua, L. monocytogenes, B. cereus, B. anthracis,* and *B. thuringiensis,* which appear to exclusively use CheV in place of CheW. CheW proteins are typically found in major chemotaxis loci together with CheA. CheW is a subject of frequent domain duplication events. On a phylogenetic tree the duplicate CheWs often fall into different subgroups, which raise questions as to the function of these multiple CheW proteins. There are also a few proteins that contain multiple CheW domains that are significantly diverged in sequence. The lack of distinct subfamilies that can be grouped by consistent gene neighborhoods shows that more detailed sequence and structure analysis is needed to derive meaningful conclusions about the functional implications of CheW diversity. One attractive hypothesis is that various CheW paralogs recognize specific classes of MCP signaling domain and thus link particular MCPs to individual signal transduction pathways.

The CheV protein is composed of a CheW domain and a C-terminal REC domain. The CheA kinase regulates the phosphorylation state of the REC domain in order to modulate CheV function (Karatan *et al.*, 2001). Domain searches for proteins that have CheW and REC domains, but not HATPase_c domains, should clearly identify CheV proteins. The phyletic distribution shows that CheV are only present in a few representatives of Firmicutes and ε, δ, and β/γ classes of Proteobacteria. The disparate distribution of CheV does not allow us to clearly delineate its evolutionary origins. Specific subfamilies and duplication events can be identified from a phylogenetic tree, but there are no sequence features that can be related to functional differences among the subfamilies. CheV proteins are sometimes found near flagellar proteins and CheR, and rarely near CheA, but they are not typically encoded near other chemotaxis proteins.

Because both CheW and CheV have primary roles in scaffolding with fewer dynamic interactions than the other chemotaxis proteins, it is possible that the evolutionary pressures on these components have resulted in more divergence at the individual sequence level rather than easily identifiable insertion and deletion events. This is supported by the observation that such divergence in sequence between distant CheW homologs does not prevent functional complementation (Alexandre and Zhulin, 2003).

References

Acuna, G., Shi, W., Trudeau, K., and Zusman, D. R. (1995). The 'CheA' and 'CheY' domains of *Myxococcus xanthus* FrzE function independently *in vitro* as an autokinase and a phosphate acceptor, respectively. *FEBS Lett.* **358**, 31–33.

Alexandre, G., and Zhulin, I. B. (2003). Different evolutionary constraints on chemotaxis proteins CheW and CheY revealed by heterologous expression and protein sequence analysis. *J. Bacteriol.* **185**, 544–552.

Alexander, R. P., and Zhulin, I. B. (2007). Evolutionary genomics reveals conserved structural determinants of signaling and adaptation in microbial chemo-receptors. *Proc. Natl. Acad. Sci. USA* **104**, 2885–2890.

Altschul, S. F., Madden, T. L., Schaffer, A. A., Zhang, J., Zhang, Z., Miller, W., and Lipman, D. J. (1997). Gapped BLAST and PSI-BLAST: A new generation of protein database search programs. *Nucleic Acids Res.* **25**, 3389–3402.

Anantharaman, V., and Aravind, L. (2000). Cache: A signaling domain common to animal Ca(2+)-channel subunits and a class of prokaryotic chemotaxis receptors. *Trends Biochem. Sci.* **25**, 535–537.

Aravind, L., and Ponting, C. P. (1997). The GAF domain: An evolutionary link between diverse phototransducing proteins. *Trends Biochem. Sci.* **22**, 458–459.

Aravind, L., and Ponting, C. P. (1999). The cytoplasmic helical linker domain of receptor histidine kinase and methyl-accepting proteins is common to many prokaryotic signaling proteins. *FEMS Microbiol Lett.* **176**, 111–116.

Badger, J. H., Hoover, T. R., Brun, Y. V., Weiner, R. M., Laub, M. T., Alexandre, G., Mrazek, J., Ren, Q., Paulsen, I. T., Nelson, K. E., Khouri, H. M., Radune, D., et al. (2006). Comparative genomic evidence for a close relationship between the dimorphic prosthecate bacteria *Hyphomonas neptunium* and *Caulobacter crescentus*. *J. Bacteriol.* **188**, 6841–6850.

Baker, M. D., Wolanin, P. M., and Stock, J. B. (2006). Signal transduction in bacterial chemotaxis. *Bioessays* **28**, 9–22.

Barnakov, A. N., Barnakova, L. A., and Hazelbauer, G. L. (2001). Location of the receptor-interaction site on CheB, the methylesterase response regulator of bacterial chemotaxis. *J. Biol. Chem.* **276**, 32984–32989.

Berleman, J. E., and Bauer, C. E. (2005). Involvement of a Che-like signal transduction cascade in regulating cyst cell development in *Rhodospirillum centenum*. *Mol. Microbiol.* **56**, 1457–1466.

Bhaya, D., Takahashi, A., and Grossman, A. R. (2001). Light regulation of type IV pilus-dependent motility by chemosensor-like elements in *Synechocystis* PCC6803. *Proc. Natl. Acad. Sci. USA* **98**, 7540–7545.

Bibikov, S. I., Biran, R., Rudd, K. E., and Parkinson, J. S. (1997). A signal transducer for aerotaxis in *Escherichia coli*. *J. Bacteriol.* **179**, 4075–4079.

Bilwes, A. M., Alex, L. A., Crane, B. R., and Simon, M. I. (1999). Structure of CheA, a signal-transducing histidine kinase. *Cell* **96**, 131–141.

Brooun, A., Bell, J., Freitas, T., Larsen, R. W., and Alam, M. (1998). An archaeal aerotaxis transducer combines subunit I core structures of eukaryotic cytochrome c oxidase and eubacterial methyl-accepting chemotaxis proteins. *J. Bacteriol.* **180**, 1642–1646.

Bustamante, V. H., Martinez-Flores, J., Vlamakis, H. C., and Zusman, D. R. (2004). Analysis of the Frz signal transduction system of *Myxococcus xanthus* shows the importance of the conserved C-terminal region of the cytoplasmic chemoreceptor FrzCD in sensing signals. *Mol. Microbiol.* **53**, 1501–1513.

Cantwell, B. J., Draheim, R. R., Weart, R. B., Nguyen, C., Stewart, R. C., and Manson, M. D. (2003). CheZ phosphatase localizes to chemoreceptor patches via CheA-short. *J. Bacteriol.* **185**, 2354–2361.

Chao, X., Muff, T. J., Park, S. Y., Zhang, S., Pollard, A. M., Ordal, G. W., Bilwes, A. M., and Crane, B. R. (2006). A receptor-modifying deamidase in complex with a signaling phosphatase reveals reciprocal regulation. *Cell* **124**, 561–571.

Chenna, R., Sugawara, H., Koke, T., Lopez, R., Gibson, T. J., Higgins, D. G., and Thompson, J. D. (2003). Multiple sequence alignment with the Clustal series of programs. *Nucleic Acids Res.* **31**, 3497–3500.

Crooks, E. G., Hon, G., Chandonia, J. M., and Brenner, S. E. (2004). WebLogo: A sequence logo generator. *Genome Res.* **14**, 1188–1190.

Cserzo, M., Eisenhaber, F., Eisenhaber, B., and Simon, I. (2002). On filtering false positive transmembrane protein predictions. *Protein Eng.* **15**, 745–752.

Cserzo, M., Wallin, E., Simon, I., von Heijne, G., and Elofsson, A. (1997). Prediction of transmembrane alpha-helices in prokaryotic membrane proteins: The dense alignment surface method. *Protein Eng.* **10**, 673–676.

D'Argenio, D. A., Calfee, M. W., Rainey, P. B., and Pesci, E. C. (2002). Autolysis and autoaggregation in *Pseudomonas aeruginosa* colony morphology mutants. *J. Bacteriol.* **184**, 6481–6489.

Dutta, R., Qin, L., and Inouye, M. (1999). Histidine kinases: Diversity of domain organization. *Mol. Microbiol.* **34**, 633–640.

Djordjevic, S., and Stock, A. M. (1998). Chemotaxis receptor recognition by protein methyl-transferase CheR. *Nat. Struct. Biol.* **5**, 446–450.

Edgar, R. C. (2004). MUSCLE: Multiple sequence alignment with high accuracy and high throughput. *Nucleic Acids Res.* **32**, 1792–1797.

Felsenstein, J. (1989). PHYLIP—Phylogeny Inference Package (Version 3.2). *Cladistics* **5**, 164–166.

Finn, R. D., Mistry, J., Schuster-Bockler, B., Griffiths-Jones, S., Hollich, V., Lassmann, T., Moxon, S., Marshall, M., Khanna, A., Durbin, R., Eddy, S. R., Sonnhammer, E. L., *et al.* (2006). Pfam: Clans, web tools and services. *Nucleic Acids Res.* **34**, D247–D251.

Galperin, M. Y. (2006). Structural classification of bacterial response regulators: Diversity of output domains and domain combinations. *J. Bacteriol.* **188**, 4169–4182.

Galtier, N., Gouy, M., and Gautier, G. (1996). SEAVIEW and PHYLO_WIN: Two graphic tools for sequence alignment and molecular phylogeny. *Comput. Appl. Biosci.* **12**, 543–548.

Gegner, J. A., Graham, D. R., Roth, A. F., and Dahlquist, F. W. (1992). Assembly of an MCP receptor, CheW, and kinase CheA complex in the bacterial chemotaxis signal transduction pathway. *Cell* **70**, 975–982.

Guvener, Z. T., Tifrea, D. F., and Harwood, C. S. (2006). Two different *Pseudomonas aeruginosa* chemosensory signal transduction complexes localize to cell poles and form and remold in stationary phase. *Mol. Microbiol.* **61**, 106–118.

Hickman, J. W., Tifrea, D. F., and Harwood, C. S. (2005). A chemosensory system that regulates biofilm formation through modulation of cyclic diguanylate levels. *Proc. Natl. Acad. Sci. USA* **102**, 14422–14427.

Hoch, J. A. (2000). Two-component and phosphorelay signal transduction. *Curr. Opin. Microbiol.* **3**, 165–170.

Hou, S., Larsen, R. W., Boudko, D., Riley, C. W., Karatan, E., Zimmer, M., Ordal, G. W., and Alam, M. (2000). Myoglobin-like aerotaxis transducers in Archaea and Bacteria. *Nature* **403**, 540–544.

Hulko, M., Berndt, F., Gruber, M., Linder, J. U., Truffault, V., Schultz, A., Martin, J., Schultz, J. E., Lupas, A. N., and Coles, M. (2006). The HAMP domain structure implies helix rotation in transmembrane signaling. *Cell* **126**, 929–940.

Kail, L., Krogh, A., and Sonnhammer, E. L. (2004). A combined transmembrane topology and signal peptide prediction method. *J. Mol. Biol.* **338**, 1027–1036.

Karatan, E., Saulmon, M. M., Bunn, M. W., and Ordal, G. W. (2001). Phosphorylation of the response regulator CheV is required for adaptation to attractants during *Bacillus subtilis* chemotaxis. *J. Biol. Chem.* **276**, 43618–43626.

Karniol, B., and Vierstra, R. D. (2004). The HWE histidine kinases, a new family of bacterial two-component sensor kinases with potentially diverse roles in environmental signaling. *J. Bacteriol.* **186**, 445–453.

Kato, M., Mizuno, T., Shimizu, T., and Hakoshima, T. (1997). Insights into multistep phosphorelay from the crystal structure of the C-terminal HPt domain of ArcB. *Cell* **88**, 717–723.

Kirby, J. R., Kristich, C. J., Saulmon, M. M., Zimmer, M. A., Garrity, L. F., Zhulin, I. B., and Ordal, G. W. (2001). CheC is related to the family of flagellar switch proteins and acts independently from CheD to control chemotaxis in *Bacillus subtilis*. *Mol. Microbiol.* **42**, 573–585.

Kirby, J. R., and Zusman, D. R. (2003). Chemosensory regulation of developmental gene expression in *Myxococcus xanthus*. *Proc. Natl. Acad. Sci. USA* **100**, 2008–2013.

Kristich, C. J., and Ordal, G. W. (2002). *Bacillus subtilis* CheD is a chemoreceptor modification enzyme required for chemotaxis. *J. Biol. Chem.* **277**, 25356–25362.

Kumar, S., Tamura, K., and Nei, M. (2004). MEGA3: Integrated software for molecular evolutionary genetics analysis and sequence alignment. *Brief. Bioinformatics* **5**, 150–163.

Lamb, J. R., Tugendreich, S., and Hieter, P. (1995). Tetratrico peptide repeat interactions: To TPR or not to TPR? *Trends Biochem. Sci.* **20**, 257–259.

LeMoual, H., and Koshland, D. E. (1996). Molecular evolution of the C-terminal cytoplasmic domain of a superfamily of bacterial receptors involved in taxis. *J. Mol. Biol.* **261**, 568–585.

Li, M. S., and Hazelbauer, G. L. (2005). Adaptational assistance in clusters of bacterial chemoreceptors. *Mol. Microbiol.* **56**, 1617–1626.

Letunic, I., Copley, R. R., Pils, B., Pinkert, S., Schultz, J., and Bork, P. (2006). SMART 5.0: Domains in the context of genomes and networks. *Nucleic Acids Res.* **32**, D142–D144.

Martin, A. C., Wadhams, G. H., and Armitage, J. P. (2001). The roles of the multiple CheW and CheA homologues in chemotaxis and in chemoreceptor localization in *Rhodobacter sphaeroides*. *Mol. Microbiol.* **40**, 1261–1272.

McEvoy, M. M., Hausrath, A. C., Randolph, G. B., Remington, S. J., and Dahlquist, R. M. (1998). Two binding modes reveal flexibility in kinase/response regulator interactions in the bacterial chemotaxis pathway. *Proc. Natl. Acad. Sci. USA* **95**, 7333–7338.

McGinnis, S., and Madden, T. L. (2004). BLAST: At the core of a powerful and diverse set of sequence analysis tools. *Nucleic Acids Res.* **32**, W20–W25.

Motaleb, M. A., Miller, M. R., Li, C., Bakker, R. G., Goldstein, S. F., Silversmith, R. E., Bourret, R. B., and Charon, N. W. (2005). CheX is a phosphorylated CheY phosphatase essential for *Borrelia burgdorferi* chemotaxis. *J. Bacteriol.* **187**, 7963–7969.

Notredame, C., Higgins, D. G., and Heninga, J. (2000). T-Coffee: A novel method for fast and accurate multiple sequence alignment. *J. Mol. Biol.* **302**, 205–217.

Overbeek, R., Fonstein, M., D'Souza, M., Pusch, G. D., and Maltsev, N. (1999). The use of gene clusters to infer functional coupling. *Proc. Natl. Acad. Sci. USA* **96,** 2896–2901.

Park, S. Y., Beel, B. D., Simon, M. I., Bilwes, A. M., and Crane, R. B. (2004a). In different organisms, the mode of interaction between two signaling proteins is not necessarily conserved. *Proc. Natl. Acad. Sci. USA* **101,** 11646–11651.

Park, S. Y., Chao, X., Gonzalez-Bonet, G., Beel, B. D., Bilwes, A. M., and Crane, B. R. (2004b). Structure and function of an unusual family of protein phosphatases: The bacterial chemotaxis proteins CheC and CheX. *Mol. Cell.* **16,** 563–574.

Pei, J. M., Sadreyev, R., and Grishin, N. V. (2003). PCMA: Fast and accurate multiple sequence alignment based on profile consistency. *Bioinformatics* **19,** 427–428.

Pittman, M. S., Goodwin, M., and Kelly, D. J. (2001). Chemotaxis in the human gastric pathogen *Helicobacter pylori*: Different roles for CheW and the three CheV paralogues, and evidence for CheV2 phosphorylation. *Microbiology* **147,** 2493–2504.

Porter, S. L., and Armitage, J. P. (2004). Chemotaxis in *Rhodobacter sphaeroides* requires an atypical histidine protein kinase. *J. Biol. Chem.* **279,** 54573–54580.

Porter, S. L., Warren, A. V., Martin, A. C., and Armitage, J. P. (2002). The third chemotaxis locus of *Rhodobacter sphaeroides* is essential for chemotaxis. *Mol. Microbiol.* **46,** 1081–1094.

Rebbapragada, A., Johnson, M. S., Harding, G. P., Zuccarelli, A. J., Fletcher, H. M., Zhulin, I. B., and Taylor, B. L. (1997). The Aer protein and the serine chemoreceptor Tsr independently sense intracellular energy levels and transduce oxygen, redox, and energy signals for *Escherichia coli* behavior. *Proc. Natl. Acad. Sci. USA* **94,** 10541–10546.

Reinelt, S., Hofmann, E., Gerharz, T., Bott, M., and Madden, D. R. (2003). The structure of the periplasmic ligand-binding domain of the sensor kinase CitA reveals the first extracellular PAS domain. *J. Biol. Chem.* **278,** 39189–39196.

Rosario, M. M., Fredrick, K. L., Ordal, G. W., and Helmann, J. D. (1994). Chemotaxis in *Bacillus subtilis* requires either of two functionally redundant CheW homologs. *J. Bacteriol.* **176,** 2736–2739.

Schwede, T., Kopp, J., Guex, N., and Peitsch, M. C. (2003). SWISS-MODEL: An automated protein homology-modeling server. *Nucleic Acids Res.* **31,** 3381–3385.

Shiomi, D., Zhulin, I. B., Homma, M., and Kawagishi, I. (2002). Dual recognition of the bacterial chemoreceptor by chemotaxis-specific domains of the CheR methyltransferase. *J. Biol. Chem.* **277,** 42325–42333.

Shu, C. J., Ulrich, L. E., and Zhulin, I. B. (2003). The NIT domain: A predicted nitrate-responsive module in bacterial sensory receptors. *Trends Biochem. Sci.* **28,** 121–124.

Sim, J. H., Shi, W., and Lux, R. (2005). Protein-protein interactions in the chemotaxis signaling pathway of *Treponema denticola*. *Microbiology* **151,** 1801–1807.

Stock, A. M., Robinson, V. L., and Goudreau, P. N. (2000). Two-component signal transduction. *Annu. Rev. Biochem.* **69,** 183–215.

Sun, H., Zusman, D. R., and Shi, W. (2000). Type IV pilus of *Myxococcus xanthus* is a motility apparatus controlled by the frz chemosensory system. *Curr. Biol.* **10,** 1143–1146.

Szurmant, H., Muff, T. J., and Ordal, G. W. (2004). *Bacillus subtilis* CheC and FliY are members of a novel class of CheY-P-hydrolyzing proteins in the chemotactic signal transduction cascade. *J. Biol. Chem.* **279,** 21787–21792.

Szurmant, L., and Ordal, G. W. (2004). Diversity in chemotaxis mechanisms among the bacteria and archaea. *Microbiol. Mol. Biol. Rev.* **68,** 301–319.

Taylor, B. L., and Zhulin, I. B. (1999). PAS domains: Internal sensors of oxygen, redox potential and light. *Microbiol. Mol. Biol. Rev.* **63,** 479–506.

Terry, K., Go, A. C., and Ottemann, K. M. (2006). Proteomic mapping of a suppressor of non-chemotactic *cheW* mutants reveals that *Helicobacter pylori* contains a new chemotaxis protein. *Mol. Microbiol.* **61**, 871–882.

Ulrich, L. E., Koonin, E. V., and Zhulin, I. B. (2005). One-component systems dominate signal transduction in prokaryotes. *Trends Microbiol.* **13**, 52–56.

Ulrich, L. E., and Zhulin, I. B. (2005). Four-helix bundle: A ubiquitous sensory module in prokaryotic signal transduction. *Bioinformatics.* **21**(Suppl. 3), iii45–iii48.

Ulrich, I. E., and Zhulin, I. B. (2007). MiST: The Microbial Signal Transduction database. *Nucleic Acids Res.* **35**, D386–D390.

Wadhams, G. H., and Armitage, J. P. (2004). Making sense of it all: Bacterial chemotaxis. *Nat. Rev. Mol. Cell Biol.* **5**, 1024–1037.

West, A. H., and Stock, A. M. (2001). Histidine kinases and response regulator proteins in two-component signaling systems. *Trends Biochem. Sci.* **26**, 369–376.

Wheeler, D. L., Barrett, T., Benson, D. A., Bryant, S. H., Canese, K., Chetvernin, V., Church, D. M., DiCuccio, M., Edgar, R., Federhen, S., Geer, L. Y., Helmberg, W., *et al.* (2006). Database resources of the National Center for Biotechnology Information. *Nucleic Acids Res.* **34**, D173–D180.

Whitchurch, C. B., Leech, A. J., Young, M. D., Kennedy, D., Sargent, J. L., Bertrand, J. J., Semmler, A. B., Mellick, A. S., Martin, P. R., Alm, R. A., Hobbs, M., Beatson, S. A., *et al.* (2004). Characterization of a complex chemosensory signal transduction system which controls twitching motility in *Pseudomonas aeruginosa. Mol. Microbiol.* **52**, 873–893.

Williams, S. B., and Stewart, V. (1999). Functional similarities among two-component sensors and methyl-accepting chemotaxis proteins suggest a role for linker region amphipathic helices in transmembrane signal transduction. *Mol. Microbiol.* **33**, 1093–1102.

Wuichet, K., and Zhulin, I. B. (2003). Molecular evolution of sensory domains in cyanobacterial chemoreceptors. *Trends Microbiol.* **11**, 200–203.

Yoshihara, S., Geng, X., and Ikeuchi, M. (2002). pilG Gene cluster and split pilL genes involved in pilus biogenesis, motility and genetic transformation in the cyanobacterium *Synechocystis* sp. PCC 6803. *Plant Cell Physiol.* **43**, 513–521.

Zhang, W., and Phillips, G. N. (2003). Structure of the oxygen sensor in *Bacillus subtilis*: Signal transduction of chemotaxis by control of symmetry. *Structure* **11**, 1097–1110.

Zhao, R., Collins, E. J., Bourret, R. B., and Silversmith, R. E. (2002). Structure and catalytic mechanism of the *E. coli* chemotaxis phosphatase CheZ. *Nat. Struct. Biol.* **9**, 570–575.

Zhulin, I. B. (2001). The superfamily of chemotaxis transducers: From physiology to genomics and back. *Adv. Microb. Physiol.* **45**, 157–198.

Zhulin, I. B., Nikolskaya, A. N., and Galperin, M. Y. (2003). Common extracellular sensory domains in transmembrane receptors for diverse signal transduction pathways in bacteria and archaea. *J. Bacteriol.* **185**, 285–294.

[2] Two-Component Systems in Microbial Communities: Approaches and Resources for Generating and Analyzing Metagenomic Data Sets

By MIRCEA PODAR

Abstract

Two-component signal transduction represents the main mechanism by which bacterial cells interact with their environment. The functional diversity of two-component systems and their relative importance in the different taxonomic groups and ecotypes of bacteria has become evident with the availability of several hundred genomic sequences. The vast majority of bacteria, including many high rank taxonomic units, while being components of complex microbial communities remain uncultured (i.e., have not been isolated or grown in the laboratory). Environmental genomic data from such communities are becoming available, and in addition to its profound impact on microbial ecology it will propel molecular biological disciplines beyond the traditional model organisms. This chapter describes the general approaches used in generating environmental genomic data and how that data can be used to advance the study of two component-systems and signal transduction in general.

Introduction

Single cell organisms sense, react, and adapt to environmental variables through a complex network of signal transduction systems. Traditionally, two-component systems (TCS) have been considered as the main signaling cascade in bacteria. In a typical system, a linear intermolecular phosphorelay connects sensing of an extracellular signal to a cellular response (e.g., movement, enzymatic reaction, or gene expression). Functional diversification of TCSs appears to be driven primarily by global duplication of individual pathways followed by coevolution of their components (Koretke *et al.*, 2000). There is good evidence (Koretke *et al.*, 2000) that archaeal and eukaryal TCS originated in bacteria; therefore, lateral gene transfer may also play an important role in the acquisition of new sensing capabilities by members of environmental communities. A significant number of two-component systems integrate the sensory function in a single protein together with histidine kinase-response regulatory relay components (Dutta *et al.*, 1999; Galperin, 2006a; Ulrich and Zhulin, 2005). The complexity of some of these hybrid systems can be staggering and, based on the

METHODS IN ENZYMOLOGY, VOL. 422   0076-6879/07 $35.00
DOI: 10.1016/S0076-6879(06)22002-0

variability of their architecture, it appears that they evolve by both domain recruitment and duplication. One component systems, a more recently recognized category of signal transduction proteins (Ulrich et al., 2005), incorporate both sensory and effector functions but lack the phosphorelay component. These systems appear to be even more abundant than the TCS in microbial genomes and likely predated the TCS.

Several two-component systems, such as those involved in chemotaxis and sporulation, have been studied extensively at genetic, biochemical, structural, and genomic levels in model organisms such as *Escherichia coli* and *Bacillus subtilis* (e.g., Bilwes et al., 1999; Kobayashi et al., 2001; Yamamoto et al., 2005). Complete genome sequences have allowed system-atic investigations of the function of all two-component systems in a handful of other species (Murata and Suzuki, 2006; Skerker et al., 2005; Throup et al., 2000; Zhang et al., 2006). Outside of the few model systems, however, the specificity determinants and the biological functions of multiple TCS present in many organisms are unknown. Large-scale comparative genomic studies have shown that the number, type, and complexity of the signal transduction systems in bacterial genomes are highly variable and dependent on genome size, environment, and lifestyle (Galperin, 2005). Species that live in dynamic environments or complex communities tend to have a much higher number of signal transduction genes than those occupying habitats that maintain constant conditions. The fraction and the distribution of the different categories of signal transduction genes in bacterial genomes have therefore been proposed to be indicators of species adaptability and the organisms' potential to interact with the environment and other mem-bers of the community (Galperin, 2005). Consequently, computational and experimental studies of signal transduction pathways are increasingly important as we aim to understand the physiology and ecology of microbes in their natural settings (Wuichet and Zhulin, 2003; Zhulin, 2001).

Most microbes are members of complex communities, with diversities including hundreds or thousands of distinct taxa. A wide range of approaches, collectively described as environmental genomics ("metagenomics"), have been developed to study such communities without culturing individual organisms (Handelsman, 2004). Most notably, large-scale microbial com-munity shotgun sequencing has become feasible and, combined with envi-ronmental gene expression and proteomics, is poised to revolutionize microbial ecology. Large-scale community genomic projects have already generated a wealth of information, from snapshots of the metabolic poten-tial of a variety of complex communities (Delong et al., 2006; Tringe et al., 2005; Venter et al., 2004) to nearly complete metabolic reconstruction and individual genome assemblies from relatively simple consortia (Martin et al., 2006; Tyson et al., 2004; Woyke et al., 2006). Members of a community are often times metabolically interdependent or utilize specific resources, which limit their cultivability in isolation. Environmental sequencing has already uncovered such metabolic networks and specializations (Hallam, 2004; Martin *et al.*, 2006; Tyson *et al.*, 2004; Woyke *et al.*, 2006). The study of the Iron Mine biofilm, for example, has provided genomic information leading to the cultivation of a bacterial species from a phylum with no prior cultured representatives (Tyson *et al.*, 2005).

Signal transduction pathways play critical roles in the formation and dynamics of microbial communities. Characterization of these pathways and their integration in the context of community metabolism inferred from metagenomic sequence data is an important step in understanding community ecophysiology and may reveal biological features of specific organisms (e.g., chemotaxis) that can facilitate their isolation and cultivation. As there have been several other chapters (Galperin, 2005, 2006a), including some in this volume (see Galperin and Nikolskaya, 2007; Wuichet *et al.*, 2007), that detail methods of searching for and analyzing signal transduction genes in microbial genomes, this chapter focuses on current approaches and resources for generating and analyzing environmental sequence data.

Generating Metagenomic Data

Since the first use of ribosomal RNA sequences to characterize the diversity of bacteria and archaea in environmental samples, molecular approaches have become ubiquitous in microbial ecology. While rRNA sequencing led to the discovery of numerous microbial phyla, it provides little if any information about the metabolic potential of the organism. A wide range of different approaches have been designed for tapping into the genetic potential of uncultured microbes, all starting with an environmental sample containing the target community (reviewed by Handelsman, 2004). Figure 1 presents some of the more common steps and various alternatives for generating environmental sequence data.

Depending on the specific goals of the project, various steps of the sample collection and processing require special considerations. While there are no published guidelines so far, there are efforts in the environmental genomic community to establish a list of "meta-data" associated with microbial community sample. Some of these include location, time of collection, physicochemical characteristics, associated biotic factors, sample handling, storage, and processing. Such information is important not only to ensure reproducibility and allow comparative analyses but also to provide the foundation for establishing correlations between genomic inferences and determined features of the environment/community. Relevant to studying

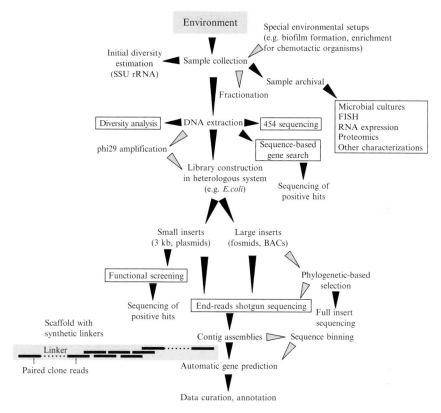

FIG. 1. Network diagram of most commonly used strategies and steps for obtaining environmental genomic data. Dark arrowheads indicate major steps, associated with typical projects. Gray arrowheads indicate optional steps, which are applicable to a subset of projects. Boxes pinpoint distinctive strategies that generate different types of data.

signal transduction and chemotaxis in microbial communities, special experimental setups ("biotraps") can be designed or are available commercially (Microbial Insights Inc., www.microbe.com) to enrich for members of the community that have the potential to sense and be attracted to specific physical or chemical cues (Geyer *et al.*, 2005).

While a description of techniques for DNA isolation and environmental library construction is beyond the scope of this chapter, it should be emphasized that when dealing with complex microbial communities in their natural environmental niches, many factors (e.g., cell walls, organic and inorganic compounds in the environment) can interfere with DNA isolation and result in poor recovery or loss of original community structure.

Specially designed commercially available kits for environmental DNA isolation (e.g., Epicentre, www.epibio.com) or previously tested and published procedures for specific types of communities are therefore recommended. A more recent technique that has been used when the microbial density is extremely low and does not allow sufficient DNA isolation is whole genome amplification (Abulencia *et al.*, 2006). While the resulting amplified DNA can be used for metagenomic sequencing, the amplification is biased and the original community structure is altered.

Following isolation, genomic DNA is usually cloned into various types of vectors (plasmids, fosmids, BACs) after mechanical or enzymatic fragmentation and fractionation. When large-scale sequencing is involved, the sequencing centers have specific protocols and vector systems for generating libraries in order to ensure quality and compatibility with their operation (e.g., www.jgi.doe.gov).

Cloning of relatively small genomic fragments (up to several tens of kilobases) in bacterial expression vectors allows the identification of specific enzymes or metabolic pathways using functional selection or screening techniques. Occasionally, phylogenetically informative gene sequences are found that allow taxonomic inferences for the originating organism and connections with the ecology of the community. Prior information about specific families of genes allows the design of oligonucleotides targeting conserved motifs. These can be used to extract novel members of a known gene family by library hybridization techniques or polymerase chain reaction (PCR). These approaches have been used extensively and have significantly expanded the sequence space for several enzyme families (Robertson *et al.*, 2004).

Identifying an environmental gene involved in a signal transduction pathway represents the first step in characterizing that pathway in the context of its host organism. Because pathway genes are often clustered in operons, analyzing the genomic neighborhood of the identified gene can lead to the discovery of additional members of the pathway. If the initial gene discovery was based on PCR or sequencing of small insert library clones, probing large insert libraries (e.g., fosmids, BACs) can pinpoint a relatively large fraction of the genome of the host organism. BAC sequencing is both efficient for the discovery of complete pathways and may contain phylogenetically informative genes that allow taxonomic identification of the organism (Beja *et al.*, 2000).

The approach that generates the largest amount of sequence information from a community genomic library (small or large inserts) is shotgun-end sequencing. When large insert clones are sequenced, the identification of genes of interest based on the end reads provides a means for selecting clones for complete sequencing. A newer sequencing technology developed by 454 Life Sciences (www.454.com) allows even higher throughput and bypasses cloning but so far there have been few reports on its use in generating metagenomic data (Angly *et al.*, 2006). Overall, the amount of shotgun sequencing will depend on a number of factors, including community structure and diversity, scope (genome closure or broad survey), and, not the least, available resources.

Assembly of Environmental Sequence Data

By far the most complex aspect of large-scale environmental genomics involves the computational steps following the generation of raw sequence data. Shotgun metagenomic sequencing is a rather young approach, with the first analysis of a microbial community having been published in 2004 (Tyson *et al.*, 2004). Consequently, the bioinformatics of metagenomic sequence analysis is not yet standardized and as robust as genomic analysis of individual organisms. As of the time of this writing, the number of completed and published microbial metagenomic analyses has increased to 12 with several dozen other projects under way (Genomes OnLine Database; Liolios *et al.*, 2006). There is a sustained effort to develop better approaches to deal with the difficulties inherent in environmental sequence analysis (Chen and Pachter, 2005; Foerstner *et al.*, 2006). A team from the Joint Genome Institute has performed systematic step-by-step comparisons of various methods used to assemble and analyze metagenomic data (Mavromatis *et al.*, 2007). As these studies provide details on the methods and links to the various software and databases, only some of the differences between the computational aspects of single genomes versus environmental data are highlighted here.

Two fundamental differences between a cultured microbe and an environmental microbial sample have to be considered when analyzing sequence data. First, in the case of cultured organisms, the cells used for DNA isolation represent a clonal population and will have the same genomic sequence. In environmental samples, even for organisms that represent the same "species," there are many independent lineages that result in varying degrees of sequence variation (polymorphisms) within that species' population. That heterogeneity has a significant impact on the assessment of sequence quality, especially on sequence assemblies. Current assemblers (PHRAP, JAZZ, Arachne) are designed to handle single, homogeneous genomes and generate various errors when presented with environmental sequences. A detailed analysis of the performance of the three assemblers using simulated and real metagenomic data indicates that Arachne generates the most accurate assemblies, whereas PHRAP assembled the largest number of sequences but with lower precision, with JAZZ being in between (Mavromatis *et al.*, 2007).

Depending on the complexity of the population structure, individual whole genome assemblies for the different lineages may not be feasible even with sufficient sequence coverage. Instead, assembled scaffolds are binned into a population "pan genome" that encodes the metabolic core of the species. This provides valuable genetic data for understanding the evolutionary ecological processes that affect the structure and dynamics of the population (Whitaker and Banfield, 2006).

The second fundamental difference between cultured organisms and environmental samples is that natural microbial communities are usually highly diverse at multiple taxonomic levels. While several of the recently analyzed communities were simple and allowed genomic and metabolic reconstruction of most of their species members, others inferred diversities in the thousands of species, without containing dominant ones (Tringe *et al.*, 2005). A whole continuum of species diversities and abundance can exist in between. What this means in terms of assembly is that the more complex a community is, the less chance there is of getting larger contigs of any single represented genomes. For example, in the case of a 150-Mbp of soil metagenomic data, which represented an estimated 3000 species with no dominant ones, the largest scaffold was less than 10 kb and over 99% of the sequence reads did not assemble into contigs (Tringe *et al.*, 2005).

When the sequence reads do not assemble, a useful approach is to link each clone's pair reads through an artificial linker, after reverse complementing one of the sequences. That way, the genomic relationship between the two sequences is maintained and, if, for example, one of the reads forms a contig with a read from a different clone, a larger scaffold connected through artificial linkers can be assembled (Fig. 1). The linker could have stop codons in each reading frame (5′TAANTAANTAATTANTTANTTA) that would prevent open reading frame (ORF) read through from one sequence into the other with the resulting formation of a chimeric gene during ORF prediction (M. Podar, unpublished result).

Because of the many variables involved in large-scale metagenomic data processing, the overall goals of the project have to be balanced with the diversity and structure of the community, allowing an estimation of the necessary amount of sequence to reach a desired assembly depth (Chen *et al.*, 2005).

Gene Prediction in Environmental Sequence Data

Gene finding, the next important step in environmental sequence analysis, is also laden with challenges when compared to gene prediction in completed microbial genomes. Most bacterial and archaeal gene prediction algorithms (e.g., GLIMMER, Fgenesb) use intrinsic properties of the genome (oligonucleotide composition, codon usage bias) to derive gene models, which are then applied to finding the coding regions. Other methods (e.g., NCBI ORF Finder) do not use a priori models but identify all reading frames and then estimate each one's potential to represent a gene based on a variety of sequence comparisons (Blast, HMM search, amino acid composition) with genes in existing databases. Because environmental genomic data usually consist of complex mixtures of sequences from various genomes, model-based approaches for gene prediction have low accuracy. In addition, those methods usually fail when sequences are limited to individual reads (~700 bp) and contain only short fragments of genes, with no start and/or stop codons. In such cases blastx comparisons can be used to identify coding regions based on sequence similarity with existing genes in the databases. Care needs to be exercised when predicting genes based on such sequences, as not every potential ORF may code for a protein. A comparative analysis of real and simulated data sets indicates that Fgenesb (Softberry) is more reliable than the Glimmer/Critica approach in identifying genes in metagenomic sequences (Mavromatis et al., 2007). A number of novel gene prediction approaches have been published, tailored to the specific features of the metagenomic data, including one that can handle short sequences generated using the 454 technology (Krause et al., 2006; Noguchi et al., 2006). These software packages are freely available and can be installed on standard computers, although their performance has not been independently tested yet on large data sets.

In general the analysis and annotation of metagenomic data sets involve grouping (binning) of sequences using sequence characteristics (e.g., GC content, codon, or oligonucleotide frequencies) or inferred taxonomic affiliation (Teeling et al., 2004). Binning is useful when the diversity is relatively low and can lead to a separation of the various types of cellular metabolisms present in the community. The best example of the successful use of binning to study such communities comes from work on the acid mine biofilms and symbiotic bacteria from the gut of the Olavius algarvensis worm (Tyson et al., 2004; Woyke et al., 2006). A comparison of the various types of binning approaches is part of the study by the JGI group (Mavromatis et al., 2007).

Analysis of Two-Component System Genes in Environmental Sequence Data

Once metagenomic sequences have been assembled, binned, and coding regions (genes and gene fragments) identified, searching for two-component systems can be performed either at the level of an individual metagenomic data set or as part of comparative analysis across multiple metagenomes and microbial genomes. Currently, there are a large number of public databases that store annotated microbial genomes and allow various types of searches and sequence analyses (Galperin, 2006b). So far, however, the number of data repositories for environmental sequences is limited. While GenBank serves as the main repository for all public sequences, the level of annotation for environmental data and the options for comparative analyses are limited. The most advanced management and analysis system for metagenomic data available, IMG/M (http://img.jgi.doe.gov/m), has been developed at the Joint Genome Institute and the Lawrence Berkeley National Laboratory (Markowitz *et al.*, 2006). IMG/M is a part of the Integrated Microbial Genome, IMG (http://img.jgi.doe.gov), a larger platform that includes all publicly available microbial and phage genomes and a few unicellular eukaryotes (Markowitz *et al.*, 2006). The system allows for a wide range of data access, analysis, and customizable input/retrieval of information and represents the best starting point in searching for signal transduction genes in deposited environmental genomic data. A database specifically dedicated for storage and analysis of environmental signal transduction information is not yet available. However, the Microbial Signal Transduction database (MiST) (Ulrich and Zhulin, 2007) will eventually incorporate functions for searching and classification of signal transduction components in user's query sequences (I. Zhulin, personal communication).

Searching for TCS Genes in Environmental Sequences

There are multiple starting points in searching environmental sequences for the presence of TCS genes, depending on the level of curation and annotation of data. In the case of raw, nonannotated sequences, for which gene prediction has not been applied or for short, unassembled reads, sequence similarity-based searches (blastx) against the NCBI or the IMG databases can identify coding regions with significant hits to known signal transduction proteins. Once potential genes of interest are identified, the potential start and stop codons for the gene can be ascertained by analyzing the open reading frames with online tools (e.g., NCBI ORF Finder) or local software (e.g., Vector NTI, Invitrogen). Evidence for the presence of a complete gene can be derived by finding different genes (or gene fragments) upstream and downstream of the TCS gene. Because the architecture is modular for many TCS genes, one would not be able to conclude that the gene is complete unless such flanking genes are identified.

Once an amino acid sequence for a TCS gene has been obtained, a number of different analysis tools can be used for more in-depth study. The various conserved domains present in TCS genes have been encoded as

Hidden Markov Models (HMMs), which can be downloaded from the PFAM database (http://pfam.janelia.org/) and used in local analyses using the HMMER package (http://hmmer.janelia.org/). Alternatively, the domain architecture of the environmental TCS genes can be analyzed online using the PFAM, SMART (http://smart.embl-heidelberg.de/), or IMG databases or the CDART tool at NCBI and, based on the known relatives, a functional classification of the sequence can be derived. Further functional classification and the place of the identified TCS gene in a specific signal transduction pathway can be inferred using the STRING database (http://string.embl.de/), which allows identification of functionally interacting proteins, the KEGG database (http://www.genome.jp/kegg/), or the Cluster of Orthologous Groups (COG, http://www.ncbi.nlb.nih.gov/COG/), available through the IMG/M database. Additional information regarding specific searches for the different type of TCS genes is available elsewhere in this volume (Galperin and Nikolskaya, 2007; Wuichet et al., 2007) or has been published previously (Galperin, 2005, 2006a).

Once the close homologs of the environmental sequence have been identified and the domain architecture uncovered, phylogenetic analyses can provide additional insights into the specific function of the TCS system, as well as identifying the possible taxonomic level of the host organism. This is especially the case when additional genes involved in signal transduction are physically linked on the same contig and may therefore be part of a coregulated operon. Phylogenetic analysis begins with construction of a protein sequence alignment. In the case of multidomain TCS genes, it is best to analyze a single domain at a time (e.g., the histidine kinase domain, the receiver domain). For publicly available sequences, the individual domain sequences can be retrieved from the SMART database. Alternatively, aligning the sequences in HMMAlign (part of HMMER) will only result in alignment of the specific HMM domain used as input, the rest of the protein will remain unaligned, and the approximate domain boundaries will be evident. Numerous other software packages for protein sequence alignments are available (Wallace et al., 2005), some of them using available three-dimensional structure information to guide the process (O'Sullivan et al., 2004). Following alignment, it is advised to inspect the result using a sequence alignment editor (e.g., BioEdit, http://mbio.ncsu.edu/BioEdit/bioedit.html) and manually correct minor mistakes as well as mask out regions of the alignment that are not aligned properly, are highly variable, or contain numerous insertions or deletions in some proteins. After a confident alignment is obtained, phylogenetic analysis can be performed, usually a distance/neighbor joining (for very large numbers of sequences and low computing power) or maximum likelihood. Numerous publications describe and review the various options for phylogenetic analyses, the potential pitfalls associated with the different methods, the interpretation of results, and the statistical confidence (Felsenstein, 2004). One of the most widely used software package, which incorporates programs for parsimony, distance, and maximum likelihood analyses, as well as obtaining statistical support, is Joe Felsenstein's PHYLIP (http://evolution.genetics. washington.edu/phylip.html). A more recent program that allows fast maximum likelihood analyses under complex models of evolution, as well as bootstrapping, is PHYML (http://atgc.lirmm.fr/phyml/).

Large-Scale Comparative Analyses of Two-Component Systems

The amount of available genomic information from complete microbial genomes, as well as environmental genomic projects, is increasing exponentially. Comparative analyses of the types, abundance, and distribution of two-component system genes across genomes and metagenomes can reveal important contributions of various signal transduction systems to specific organisms and communities and allow correlations with environmental parameters.

The IMG/M database allows rapid retrieval and comparative analyses of signal transduction genes from all deposited environmental genomic data, as well as from completed genomes. Using the COG classification, one can select all gene families involved in signal transduction at the start of the analysis or focus only on a specific group of proteins, for example, histidine kinases. There are 14 different COGs for various types of histidine kinases, and searching for them retrieves a table with direct links to the individual genes in every selected metagenome or genome. A normalization of the number of genes retrieved based on the database size allows the identification of COG categories, which are over- or underrepresented. Results for such analysis are shown in Table I. Based on this it appears, for example, that chemotaxis (CheA) is overrepresented in the Whale Fall 1 metagenome, whereas kinases that regulate osmosensitive channels are at high abundance in a human gut and in the soil metagenome. Such results can then be followed-up with additional abundance analyses for other components of the pathways and provide direct access to the genes and the gene neighborhood on the encoding contigs.

Metagenomic profiling of microbial communities is still in its infancy, and the statistical significance of analyzing a small fraction of the genomic information of complex communities is not fully understood. Nevertheless, mining such data sets for specific genes and gene families involved in signal transduction allows access to the great microbial diversity, which, until recently, was available only in the form of ribosomal RNA sequences. Undoubtedly, there are many aspects of signal transduction still hidden in the vast uncultured microbial world waiting to be discovered.

TABLE I

ANALYSIS OF THE DISTRIBUTION OF HISTIDINE KINASE FAMILIES (COG CATEGORIES) IN SEVERAL METAGENOMIC DATA SETS IN IMG/M[a]

COG ID	COG name	Acid mine	Human gut #7	Sludge (U.S)	Soil (MN)	Whale Fall 1	Whale Fall 2	Whale Fall 3
COG0642	Signal transduction His kinase	0	0	0.43	0.26	0.75	0.18	0.06
COG0643	CheA and related kinases	0	0	0.53	0.01	1.34	0.05	0.01
COG2205	Osmosensitive K+ channel His kinase	0	1.61	0.37	1.58	0.79	0.22	0.22
COG3290	His kinase regulating citrate/malate metabolism	0	0.91	0	0	0	0	0
COG3850	His kinase nitrate/nitrite specific	0	0	0.01	0.69	0.1	0	0
COG3851	His kinase glucose-6-phosphate specific	0	0	1.22	0.08	0	0	0
COG3852	His kinase nitrogen specific	0	0	0.61	0	0.97	1.41	0.24
COG3920	Signal transduction His kinase	0	0	0	0	0.21	0.21	0.16
COG4191	His kinase regulating C4-dicarboxylate transport system	0	0	0.67	0.2	1.33	0.1	0
COG4192	His kinase regulating phosphoglycerate transport system	0	0	0	0.3	0.35	0.34	0
COG4564	Signal transduction His kinase	0	0	0.57	0.9	2.39	0.21	0.86
COG4585	Signal transduction His kinase	0	0.81	1.82	1.05	0	0	0.14
COG5000	His kinase involved in nitrogen fixation and metabolism regulation	0.38	0	1.24	0.8	0.74	1.41	0
COG5002	Signal transduction His kinase	0	0.34	0.14	0.26	0	0	0

[a] Results are normalized relative to the size of the individual metagenomes (z score). Values above 0.5 are shaded and are suggested to represent high abundance of that gene family in the respective microbial community. The metagenomic projects that describe the various data sets have been published elsewhere (Gill et al., 2006; Martin et al., 2006; Tringe et al., 2005; Tyson et al., 2004).

Acknowledgments

I thank Professor M. Simon for stimulating discussions, support, and encouragement and I. Zhulin, N. Kiripides, and the developers of IMG for collaboration and sharing unpublished data.

References

Abulencia, C. B., Wyborski, D. L., Garcia, J. A., Podar, M., Chen, W., Chang, S. H., Chang, H. W., Watson, D., Brodie, E. L., Hazen, T. C., and Keller, M. (2006). Environmental whole-genome amplification to access microbial populations in contaminated sediments. *Appl. Environ. Microbiol.* **72,** 3291–3301.

Angly, F. E., Felts, B., Breitbart, M., Salamon, P., Edwards, R. A., Carlson, C., Chan, A. M., Haynes, M., Kelley, S., Liu, H., Mahaffy, J. M., Mueller, J. E., *et al.* (2006). The marine viromes of four oceanic regions. *PLoS. Biol.* **4,** e368.

Beja, O., Suzuki, M. T., Koonin, E. V., Aravind, L., Hadd, A., Nguyen, L. P., Villacorta, R., Amjadi, M., Garrigues, C., Jovanovich, S. B., Feldman, R. A., and Delong, E. F. (2000). Construction and analysis of bacterial artificial chromosome libraries from a marine microbial assemblage. *Environ. Microbiol.* **2,** 516–529.

Bilwes, A. M., Alex, L. A., Crane, B. R., and Simon, M. I. (1999). Structure of CheA, a signal-transducing histidine kinase. *Cell* **96,** 131–141.

Chen, K., and Pachter, L. (2005). Bioinformatics for whole-genome shotgun sequencing of microbial communities. *PLoS. Comput. Biol.* **1,** 106–112.

Delong, E. F., Preston, C. M., Mincer, T., Rich, V., Hallam, S. J., Frigaard, N. U., Martinez, A., Sullivan, M. B., Edwards, R., Brito, B. R., Chisholm, S. W., and Karl, D. M. (2006). Community genomics among stratified microbial assemblages in the ocean's interior. *Science* **311,** 496–503.

Dutta, R., Qin, L., and Inouye, M. (1999). Histidine kinases: Diversity of domain organization. *Mol. Microbiol.* **34,** 633–640.

Felsenstein, J. (2004). "Inferring Phylogenies." Sinauer, Sunderland, MA.

Foerstner, K. U., von Mering, C., and Bork, P. (2006). Comparative analysis of environmental sequences: Potential and challenges. *Philos. Trans. R. Soc. Lond. B Biol. Sci.* **361,** 519–523.

Galperin, M. Y. (2005). A census of membrane-bound and intracellular signal transduction proteins in bacteria: Bacterial IQ, extroverts and introverts. *BMC Microbiol.* **5,** 35.

Galperin, M. Y. (2006a). Structural classification of bacterial response regulators: Diversity of output domains and domain combinations. *J. Bacteriol.* **188,** 4169–4182.

Galperin, M. Y. (2006b). The Molecular Biology Database Collection: 2006 update. *Nucleic Acids Res.* **34,** D3–D5.

Galperin, M. Y., and Nikolskaya, A. N. (2007). Identification of sensory and signal-transducing domains in two-component signaling systems. *Methods Enzymol.* **244**(3), (this volume).

Geyer, R., Peacock, A. D., Miltner, A., Richnow, H. H., White, D. C., Sublette, K. L., and Kastner, M. (2005). *In situ* assessment of biodegradation potential using biotraps amended with 13C-labeled benzene or toluene. *Environ. Sci. Technol.* **39,** 4983–4989.

Gill, S. R., Pop, M., Deboy, R. T., Eckburg, P. B., Turnbaugh, P. J., Samuel, B. S., Gordon, J. I., Relman, D. A., Fraser-Liggett, C. M., and Nelson, K. E. (2006). Metagenomic analysis of the human distal gut microbiome. *Science* **312,** 1355–1359.

Handelsman, J. (2004). Metagenomics: Application of genomics to uncultured microorganisms. *Microbiol. Mol. Biol. Rev.* **68,** 669–685.

Kobayashi, K., Ogura, M., Yamaguchi, H., Yoshida, K., Ogasawara, N., Tanaka, T., and Fujita, Y. (2001). Comprehensive DNA microarray analysis of *Bacillus subtilis* two-component regulatory systems. *J. Bacteriol.* **183,** 7365–7370.

Koretke, K. K., Lupas, A. N., Warren, P. V., Rosenberg, M., and Brown, J. R. (2000). Evolution of two-component signal transduction. *Mol. Biol. Evol.* **17,** 1956–1970.

Krause, L., Diaz, N. N., Bartels, D., Edwards, R. A., Puhler, A., Rohwer, F., Meyer, F., and Stoye, J. (2006). Finding novel genes in bacterial communities isolated from the environment. *Bioinformatics* **22,** e281–e289.

Liolios, K., Tavernarakis, N., Hugenholtz, P., and Kyrpides, N. C. (2006). The Genomes On Line Database (GOLD) v.2: A monitor of genome projects worldwide. *Nucleic Acids Res.* **34,** D332–D334.

Markowitz, V. M., Ivanova, N., Palaniappan, K., Szeto, E., Korzeniewski, F., Lykidis, A., Anderson, I., Mavromatis, K., Kunin, V., Garcia, M. H., Dubchak, I., Hugenholtz, P., *et al.* (2006). An experimental metagenome data management and analysis system. *Bioinformatics* **22,** e359–e367.

Markowitz, V. M., Korzeniewski, F., Palaniappan, K., Szeto, E., Werner, G., Padki, A., Zhao, X., Dubchak, I., Hugenholtz, P., Anderson, I., Lykidis, A., Mavromatis, K., *et al.* (2006). The integrated microbial genomes (IMG) system. *Nucleic Acids Res.* **34,** D344–D348.

Martin, H. G., Ivanova, N., Kunin, V., Warnecke, F., Barry, K. W., McHardy, A. C., Yeates, C., He, S., Salamov, A. A., Szeto, E., Dalin, E., Putnam, N. H., *et al.* (2006). Metagenomic analysis of two enhanced biological phosphorus removal (EBPR) sludge communities. *Nat. Biotechnol.* **24,** 1263–1269.

Mavromatis, K., Ivanova, N., Barry, K., Shapiro, H. J., Goltsman, E., McHardy, A. C., Rigoutsos, I., Salamov, A. A., Korzeniewski, F., Land, M., Lapidus, A., Grigoriev, I., *et al.* (2007). On the the fidelity of processing metagenomic sequencing using simulated datasets. Submitted for publication.

Murata, N., and Suzuki, I. (2006). Exploitation of genomic sequences in a systematic analysis to access how cyanobacteria sense environmental stress. *J. Exp. Bot.* **57,** 235–247.

Noguchi, H., Park, J., and Takagi, T. (2006). MetaGene: Prokaryotic gene finding from environmental genome shotgun sequences. *Nucleic Acids Res.* **34,** 5623–5630.

O'Sullivan, O., Suhre, K., Abergel, C., Higgins, D. G., and Notredame, C. (2004). 3D Coffee: Combining protein sequences and structures within multiple sequence alignments. *J. Mol. Biol.* **340,** 385–395.

Robertson, D. E., Chaplin, J. A., DeSantis, G., Podar, M., Madden, M., Chi, E., Richardson, T., Milan, A., Miller, M., Weiner, D. P., Wong, K., McQuaid, J., *et al.* (2004). Exploring nitrilase sequence space for enantioselective catalysis. *Appl. Environ. Microbiol.* **70,** 2429–2436.

Skerker, J. M., Prasol, M. S., Perchuk, B. S., Biondi, E. G., and Laub, M. T. (2005). Two-component signal transduction pathways regulating growth and cell cycle progression in a bacterium: A system-level analysis. *PLoS Biol.* **3,** e334.

Teeling, H., Waldmann, J., Lombardot, T., Bauer, M., and Glockner, F. O. (2004). TETRA: A web-service and a stand-alone program for the analysis and comparison of tetranucleotide usage patterns in DNA sequences. *BMC Bioinformatics* **5,** 163.

Throup, J. P., Koretke, K. K., Bryant, A. P., Ingraham, K. A., Chalker, A. F., Ge, Y., Marra, A., Wallis, N. G., Brown, J. R., Holmes, D. J., Rosenberg, M., and Burnham, M. K. (2000). A genomic analysis of two-component signal transduction in *Streptococcus pneumoniae*. *Mol. Microbiol.* **35,** 566–576.

Tringe, S. G., von Mering, C., Kobayashi, A., Salamov, A. A., Chen, K., Chang, H. W., Podar, M., Short, J. M., Mathur, E. J., Detter, J. C., Bork, P., Hugenholtz, P., *et al.* (2005). Comparative metagenomics of microbial communities. *Science* **308,** 554–557.

Tyson, G. W., Chapman, J., Hugenholtz, P., Allen, E. E., Ram, R. J., Richardson, P. M., Solovyev, V. V., Rubin, E. M., Rokhsar, D., and Banfield, J. F. (2004). Community structure and metabolism through reconstruction of microbial genomes from the environment. *Nature* **428,** 37–43.

Tyson, G. W., Lo, I., Baker, B. J., Allen, E. E., Hugenholtz, P., and Banfield, J. F. (2005). Genome-directed isolation of the key nitrogen fixer *Leptospirillum ferrodiazotrophum* sp. nov. from an acidophilic microbial community. *Appl. Environ. Microbiol.* **71,** 6319–6324.

Ulrich, L. E., Koonin, E. V., and Zhulin, I. B. (2005). One-component systems dominate signal transduction in prokaryotes. *Trends Microbiol.* **13,** 52–56.

Ulrich, L. E., and Zhulin, I. B. (2007). MiST: The Microbial Signal Transduction database. *Nucleic Acids Res.* **35,** D386–D390.

Venter, J. C., Remington, K., Heidelberg, J. F., Halpern, A. L., Rusch, D., Eisen, J. A., Wu, D., Paulsen, I., Nelson, K. E., Nelson, W., Fouts, D. E., Levy, S., *et al.* (2004). Environmental genome shotgun sequencing of the Sargasso Sea. *Science* **304,** 66–74.

Wallace, I. M., Blackshields, G., and Higgins, D. G. (2005). Multiple sequence alignments. *Curr. Opin. Struct. Biol.* **15,** 261–266.

Whitaker, R. J., and Banfield, J. F. (2006). Population genomics in natural microbial communities. *Trends Ecol. Evol.* **21,** 508–516.

Woyke, T., Teeling, H., Ivanova, N. N., Huntemann, M., Richter, M., Gloeckner, F. O., Boffelli, D., Anderson, I. J., Barry, K. W., Shapiro, H. J., Szeto, E., Kyrpides, N. C., *et al.* (2006). Symbiosis insights through metagenomic analysis of a microbial consortium. *Nature* **443,** 950–955.

Wuichet, K., and Zhulin, I. B. (2003). Molecular evolution of sensory domains in cyanobacterial chemoreceptors. *Trends Microbiol.* **11,** 200–203.

Wuichet, K., Alexander, R. P., and Zhulin, I. B. (2007). Comparative genomic and protein sequence analyses of a complex system controlling bacterial chemotaxis. *Methods Enzymol.* **244**(1), (this volume).

Yamamoto, K., Hirao, K., Oshima, T., Aiba, H., Utsumi, R., and Ishihama, A. (2005). Functional characterization *in vitro* of all two-component signal transduction systems from *Escherichia coli. J. Biol. Chem.* **280,** 1448–1456.

Zhang, W., Culley, D. E., Wu, G., and Brockman, F. J. (2006). Two-component signal transduction systems of *Desulfovibrio vulgaris*: Structural and phylogenetic analysis and deduction of putative cognate pairs. *J. Mol. Evol.* **62,** 473–487.

Zhulin, I. B. (2001). The superfamily of chemotaxis transducers: From physiology to genomics and back. *Adv. Microb. Physiol.* **45,** 157–198.

[3]   Identification of Sensory and Signal-Transducing Domains in Two-Component Signaling Systems

By MICHAEL Y. GALPERIN and ANASTASIA N. NIKOLSKAYA

Abstract

The availability of complete genome sequences of diverse bacteria and archaea makes comparative sequence analysis a powerful tool for analyzing signal transduction systems encoded in these genomes. However, most signal transduction proteins consist of two or more individual protein domains, which significantly complicates their functional annotation and makes automated annotation of these proteins in the course of large-scale genome sequencing projects particularly unreliable. This chapter describes certain common-sense protocols for sequence analysis of two-component histidine kinases and response regulators, as well as other components of the prokaryotic signal transduction machinery: Ser/Thr/Tyr protein kinases and protein phosphatases, adenylate and diguanylate cyclases, and c-di-GMP phosphodiesterases. These protocols rely on publicly available computational tools and databases and can be utilized by anyone with Internet access.

Introduction

Sequence analysis of regulatory proteins played a key role in the discovery of two-component signal transduction. Indeed, the sequence alignments of the chemotaxis response regulator CheY and transcriptional regulators OmpR and ArcA from *Escherichia coli* with *Bacillus subtilis* sporulation proteins Spo0F and Spo0A by James Hoch and colleagues (Trach *et al.*, 1985) and with the N-terminal fragment of the chemotaxis methylesterase CheB by Ann and Jeffry Stock and Daniel Koshland (Stock *et al.*, 1985) convinced them that all these protein fragments were homologous. This homology, in turn, suggested "an evolutionary and functional relationship between the chemotaxis system and systems that are thought to regulate gene expression in response to changing environmental conditions" (Stock *et al.*, 1985). This prescient conclusion has been verified in subsequent studies that described phosphorylation of these proteins and identified their common CheY-like receiver (REC) domain as an evolutionarily stable compact structural unit (Stock *et al.*, 1989, 1993; Volz and Matsumura, 1991) that undergoes a distinctive change upon phosphorylation (Kern *et al.*, 1999; Lee *et al.*, 2001).

METHODS IN ENZYMOLOGY, VOL. 422   0076-6879/07 $35.00
                    DOI: 10.1016/S0076-6879(06)22003-2

Identification of the receiver domain was followed by sequence analysis of histidine kinases, most importantly by Parkinson and Kofoid (1992), who described five conserved sequence motifs (H, N, G1, F, and G2 boxes), and by Grebe and Stock (1999), who classified histidine kinases into 11 families based on sequence similarity in their kinase domains (http://www.uni-kl.de/FB-Biologie/AG-Hakenbeck/TGrebe/HPK/Table4.htm).[1] These papers provided a solid basis for recognition of histidine kinases in genomic sequences and analysis of the diversity in their domain organization (Dutta *et al.*, 1999).

The importance of sequence analysis in studies of bacterial and archaeal signal transduction systems has received an additional boost from genome sequencing projects, which provided virtually unlimited material for comparative studies. However, these studies revealed a stunning complexity and diversity of signal transduction systems in various microorganisms. The total number of sensory histidine kinases encoded in the genomes of *E. coli* K12 and *B. subtilis*, 30 and 36, respectively, proved to be quite modest compared to the sets of histidine kinases encoded by such environmental organisms as *Pseudomonas aeruginosa* (62 proteins), *Streptomyces coelicolor* (95 proteins), or *Myxococcus xanthus* (138 proteins); see http://www.ncbi.nlm.nih.gov/Complete_Genomes/SignalCensus.html (Galperin, 2005). Furthermore, the list of microbial environmental receptors has been expanded and now, in addition to histidine kinases and methyl-accepting chemotaxis proteins, includes Ser/Thr protein kinases and protein phosphatases, as well as adenylate and diguanylate cyclases and c-di-GMP phosphodiesterases (Galperin, 2004, 2005; Kennelly, 2002; Kennelly and Potts, 1996; Römling *et al.*, 2005). All these environmental receptors share a pool of sensory domains, which can be extracytoplasmic (periplasmic or, in gram-positive bacteria, extracellular), membrane-embedded, or cytoplasmic (Galperin *et al.*, 2001; Nikolskaya *et al.*, 2003; Zhulin *et al.*, 2003); see reviews by Taylor and Zhulin (1999) and Galperin (2004). Another important development was characterization of a complex system of "one-component" intracellular signaling proteins (Galperin, 2004; Ulrich *et al.*, 2005), such as the anaerobic nitric oxide reductase transcription regulator NorR, which combines a sensor GAF domain with an enhancer-binding ATPase and a DNA-binding domain (Gardner *et al.*, 2003; Pohlmann *et al.*, 2000). To complicate the picture even further, certain receptors contain more than one sensory domain and/or more than one output domain and participate in the cross-talk between different signal transduction pathways (Galperin, 2004). However, this very complexity

---

[1] All URLs and database references in this chapter were correct at the time of writing (October 2006). We apologize for any confusion that might arise from subsequent changes in database content, sequence, and/or annotation updates.

makes case-by-case sequence analysis of signal transduction proteins so effective. The following paragraphs discuss the computational tools and databases used most commonly in sequence analysis of sensory and signal transduction proteins and describe analytical methods used for recognizing histidine kinases, response regulators, and other bacterial signaling components in genomic sequences and for delineating their constituent domains.

Computational Tools for Domain Identification

Identification of the CheY-like receiver (REC) domain (Stock *et al.*, 1985; Trach *et al.*, 1985) as a common phosphoacceptor domain in various two-component systems demonstrated the power of comparative sequence analysis in studies of prokaryotic signal transduction systems. In subsequent studies, many other conserved protein domains involved in signal transduction were identified and included in public domain databases, such as Pfam, SMART, InterPro, and CDD (Table I). Each of these databases comes with a search tool that allows comparing any given protein sequence against the domain library to identify the (known) domains that this protein consists of. In addition, these databases contain precomputed profiles for previously sequenced proteins and provide graphical views of their domain architectures. As new genomic data are added, deduced proteins are automatically analyzed for domain content. Therefore, unless the protein to be analyzed is a newly sequenced one that is still absent from the NCBI protein database and/or UniProt, its domain architecture should be available in protein domain databases. Importantly, position-specific scoring matrices (PSSMs) and hidden Markov models (HMMs) used for sequence searches in domain databases already reflect sequence divergence within each protein family. This makes comparing a protein sequence against a domain database (a library of PSSMs or HMMs) much more sensitive than any pairwise comparisons, used, for example, in the BLAST algorithm. However, domain recognition and functional annotation of multidomain proteins are tedious processes that cannot be automated readily (see later). Furthermore, the standard methodology of assigning protein function based on the function of its closest experimentally characterized homolog is not readily applicable to signal transduction components, as proteins with very similar sequences (e.g., *B. subtilis* response regulators PhoP and ResD) may have dramatically different biological functions. As a result, many signal transduction proteins have incomplete, biased, or even erroneous annotation. Given that most protein annotations these days are made in an automated high-throughput fashion, it would be unrealistic to put too much trust into these annotations, especially when planning long-term experimental research. For many experimentally uncharacterized proteins, an imprecise annotation,

TABLE I

COMPUTATIONAL RESOURCES FOR SEQUENCE ANALYSIS OF SIGNAL TRANSDUCTION PROTEINS

Name	URL	Comment	References
Specialized databases of signal transduction proteins			
KEGG	http://www.genome.ad. jp/kegg/ pathway/ko/ ko02020.html	Graphical representation of two-component systems in bacteria with sequenced genomes	Kanehisa *et al.* (2004)
MiST	http://genomics.ornl. gov/mist/	Predicted signal transduction proteins from all completely sequenced prokaryotic genomes	Ulrich and Zhulin (2007)
Sentra	http://compbio.mcs. anl.gov/sentra/	Predicted signal transduction proteins from all completely sequenced prokaryotic genomes	D'Souza *et al.* (2007)
Signaling Census	http://www.ncbi.nlm. nih.gov/Complete_ Genomes/ SignalCensus.html	Total counts of signal transduction proteins in completely sequenced prokaryotic genomes	Galperin (2005)
KinG	http://hodgkin.mbu. iisc.ernet.in/~king/	Kinases in Genomes, a listing of Ser/Thr/Tyr-specific protein kinases encoded in complete genomes of prokaryotes and eukaryotes	Krupa *et al.* (2004)
Sequence motif databases			
PROSITE	http://www.expasy.org/ prosite/	Protein sequence patterns and profiles that define protein domains	Hulo *et al.* (2006)
BLOCKS	http://blocks.fhcrc.org/	Protein sequence motifs represented as Blocks, multiply aligned ungapped sequence segments	Henikoff *et al.* (2000)
PRINTS	http://www.bioinf.man. ac.uk/ dbbrowser/ PRINTS/	Protein fingerprints groups of conserved motifs used to characterize each protein family	Attwood *et al.* (2003)
Protein domain databases			
Pfam	http://www.sanger.ac. uk/Software/Pfam/	An extensive collection of protein domains, including those with unknown functions	Finn *et al.* (2006)
SMART	http://smart.embl.de/	Prokaryotic and eukaryotic signaling domains and domain architectures	Letunic *et al.* (2006)

(*continued*)

TABLE I (*continued*)

Name	URL	Comment	References
ProDom	http://prodom.prabi.fr/	An automatically generated listing of protein domains	Servant *et al.* (2002)
CDD	http://www.ncbi.nlm. nih.gov/Structure/ cdd/cdd.shtml	Conserved domains with curated alignments based on available three-dimensional structures	Marchler-Bauer *et al.* (2005)
Protein family databases (full-length proteins)			
COG	http://www.ncbi.nlm. nih.gov/COG/	Clusters of orthologous groups that represent either whole proteins or individual domains	Tatusov *et al.* (2000)
PIRSF	http://pir.georgetown. edu/pirsf/	Families of proteins with shared domain architecture and full-length similarity	Wu *et al.* (2004)
Integrated motif, domain and family database			
InterPro	http://www.ebi.ac.uk/ interpro/	An umbrella database combining results from several of the aforementioned databases	Mulder *et al.* (2005)
Protein structure database			
PDB	http://www.rcsb.org/ pdb	Three-dimensional structures of proteins and other molecules	Berman *et al.* (2000)

such as "response regulator, OmpR type" (http://pir.georgetown.edu/cgi-bin/ipcSF?id=PIRSF003173) in the PIRSF protein family classification system (Nikolskaya *et al.*, 2006) or "COG0745: Response regulators consisting of a CheY-like receiver domain and a winged-helix DNA-binding domain" in the COG database (Tatusov *et al.*, 2000) would actually be far more accurate than a more precise, but likely erroneous, functional assignment. Protein annotations in specialized databases, dedicated to signal transduction, such as Sentra and MiST, are much more reliable but even these might need to be verified.

As a starting point in sequence analysis of a putative signal transduction protein, it is often useful to compare it against the several (or even all) domain databases listed in Table I. Because each of these databases uses its own search tool, the results are likely to be nonuniform, both in terms of domain recognition and in terms of domain boundaries for the same domain. A careful analysis of all meaningful annotations from these different sources, taking into account the similarity scores, the underlying experimental evidence, and the available references, is the best way to avoid costly mistakes. This chapter provides several examples of such an analysis.

Sequence Analysis of Histidine Kinases

Overview

A typical sensory histidine kinase consists of at least three distinct domains: a sensor (signal input) domain, a His-containing phosphoacceptor (dimerization) HisKA domain, and an ATP-binding HATPase domain (Dutta *et al.*, 1999; Grebe and Stock, 1999; Hoch, 2000; Stock *et al.*, 2000). There are numerous variations on this common theme. Sensor domains can be periplasmic, membrane-embedded, or cytoplasmic, and a single histidine kinase can contain two or more sensory domains. Extracytoplasmic sensor domains are connected to the intracellular HisKA domains by one or more transmembrane segments and, sometimes, the cytoplasmic helical linker (HAMP) domain (Aravind and Ponting, 1999; Williams and Stewart, 1999). In addition, certain histidine kinases contain C-terminal CheY-like phosphoacceptor (receiver, REC) domains (these enzymes are commonly referred to as hybrid histidine kinases) and/or histidine phosphotransfer (HPt) domains (Dutta *et al.*, 1999; Matsushika and Mizuno, 1998; Mizuno, 1997).

The diversity of histidine kinases makes recognizing them in genomic sequences a formidable task. In 1992, Parkinson and Kofoid described conserved sequence motifs (H, N, G1, F, and G2 boxes), common for most histidine kinases. These motifs were subsequently used in numerous papers, most significantly in the histidine kinase classification by Grebe and Stock (1999) (http://www.uni-kl.de/FB-Biologie/AG-Hakenbeck/TGrebe/HPK/Table6.htm). However, proteins that lack one or more such motifs can still function as histidine kinases. Examples include proteins that belong to the HPK8 and HPK10 families in the classification by Grebe and Stock (1999)(COG2972 and COG3275 in the COG database), such as *Pseudomonas aeruginosa* sensor protein FimS (Yu *et al.*, 1997) and many others. While these motifs can be captured by such databases as BLOCKS, PRINTS, or PROSITE (Table I), in the past several years the motif-based approach to identification of histidine kinases has been largely replaced by an approach based on domain analysis. Sensory domains are by far the most diverse ones of the three core domains in histidine kinases; many of them are unique or have a narrow phylogenetic representation. Phosphoacceptor (dimerization) HisKA domains, which contain the H box with the phosphoryl-accepting His residue, are less diverse and have similar three-dimensional structures, consisting of long α-helices (Stock *et al.*, 2000; Wolanin *et al.*, 2002). Still, recognizing these domains through sequence similarity alone may be complicated. In the latest version of the Pfam database, HisKA domains are divided into four separate domain families: HisKA (PF00512), HisKA_2 (PF07568), HisKA_3 (PF07730), and HWE_HK (PF07536), which are unified into the His Kinase A (phosphoacceptor) domain clan (Finn *et al.*, 2006). It should be noted that the four HisKA domains currently in Pfam do not cover the full diversity of these domains: some experimentally characterized sensory histidine kinases, such as *Clostridium perfringens* VirS (Cheung and Rood, 2000), as well as many archaeal histidine kinases, contain dimerization domains that are not recognized by either of the Pfam profiles, at least at standard confidence levels ($E < 10^{-3}$). The ATPase domain of histidine kinases, referred to as HATPase_c domain (PF02518) in the Pfam database, contains the N, G1, F, and G2 boxes of Parkinson and Kofoid (1992). It is by far the most conserved domain in histidine kinases and the easiest one to recognize in sequence similarity searches. However, very similar ATPase domains of the GHKL family can be found in a stand-alone form in the DNA gyrase (*gyrB* gene product) and DNA repair protein MutL, or as a component of the heat-shock protein HSP90 (Ban and Yang, 1998; Dutta and Inouye, 2000). Therefore, recognition of a histidine kinase by sequence analysis relies on finding a (usually C-terminal) ATPase domain of the GKHL superfamily that does not belong to the GyrB, MutL, or HtpG family. This domain should be preceded by a histidine kinase A (phosphoacceptor) domain: either one of the HisKA domains listed in Pfam or a poorly conserved domain of ~60 amino acid residues that consists of predicted α-helices and contains an invariant His residue. Finally, there should be an N-terminal fragment, corresponding to a sensory domain, which may or may not have close homologs in the existing protein databases. The presence of these three domains would qualify the protein in question as a two-component sensory histidine kinase. Determining its exact substrate (ligand) specificity would require further analysis and may be impossible without additional experimental data. A significant fraction of histidine kinases are encoded in conserved operons with their cognate response regulators that can be easily recognized by their highly conserved REC domain. Although not a universal trait, presence of an adjacent gene encoding a response regulator could strengthen the case for the analyzed protein being a histidine kinase. Thus, sequence analysis of potential sensory kinases should always include examination of their gene neighborhoods.

Identification of MA3481 as a Sensory Histidine Kinase

1. Find the entry for the *Methanosarcina acetivorans* protein MA3481 in GenBank (accession No. AE010299) or directly in one of the protein databases: UniProtKB (Accession No. Q8TKC7) or the NCBI protein database (AAM06847 or gi|20092292).

2. Inspect the annotation of MA3481 and its constituent domains in each of these databases. Note that both the NCBI protein database (a noncurated database) and the UniProt (a curated database) annotate MA3481 as a

"hypothetical protein," in keeping with its original annotation by the scientists at the Whitehead Institute Center for Genome Research (Cambridge, MA). Nevertheless, both NCBI and UniProt entries include the list of the domains that are recognized in the MA3481 sequence by various tools, which provide numerous hints that MA3481 might be a histidine kinase.

a. In the NCBI entry, these domains come from the CDD databases and are linked to the CDD entries for the PAS domain and the HATPase_c domain. The exact borders of each domain vary depending on the source of the domain entry: in the COG database the PAS domain (COG2202) occupies amino acid residues 243 to 446, whereas the CDD's own cd00130 entry recognizes a much smaller region, 339 to 439, as the PAS domain. Likewise, COG3920 "Signal transduction histidine kinase" covers amino acid residues 436 to 702, whereas the SMART entry SM00387 recognizes only the C-terminal HATPase_c domain (amino acid residues 557 to 702) with a somewhat unreliable expectation value of 0.004. The complete domain organization can be viewed on the NCBI web site by clicking the "Conserved Domains" link or by entering the link http://www.ncbi.nlm.nih.gov/Structure/cdd/wrpsb.cgi?INPUT_TYPE=precalc&SEQUENCE=19917534 where the last eight digits correspond to the gi number. When the protein in question is not in the database, one would need to compare its sequence against the CDD using RPS-BLAST (see later).

b. The UniProt entry for MA3481 contains even more hints that MA3481 is a histidine kinase. For example, this entry contains links to the Gene Ontology (GO) "Molecular function: two-component sensor activity (GO: 0000155)" and "Biological process: two-component signal transduction system (GO: 0000160)." Still, these annotations should be treated with caution. For example, one of them says "Biological process: regulation of transcription, DNA-dependent (GO:0006355), which is probably not true for MA3481, as DNA-binding response regulators are very rare in archaea (Galperin, 2006) and, like most archaeal histidine kinases, MA3481 does not appear to regulate transcription. The UniProt entry for MA3481 also contains links to the PROSITE database and several domain databases, such as InterPro, Pfam, SMART, and TIGRFAMs. A very useful link to the "Graphical view of the domain structure" in InterPro (http://www.ebi.ac.uk/interpro/ISpy?mode=single&ac=Q8TKC7) allows one to take a birds-eye view at all the domains recognized by individual tools used in InterPro. Again, all of them recognize the PAS (or PAS/PAC) domain in the 320–449 region of the protein and the ATPase domain in the 550–706 region of the protein. Two InterPro tools also cover the intermediate 450–550 region: Pfam recognizes it as the HisKA_2 (PF07568) domain (corresponding to the InterPro entry IPR011495), whereas PROSITE unifies it with the

ATPase domain into the single HIS_KIN entry spanning the entire C-terminal half of MA3481 from Glu-459 to its C terminus. Pfam graphical view, which can be obtained by following a link from the UniProt entry or by going directly to http://www.sanger.ac.uk/cgi-bin/Pfam/swisspfamget.pl?name=Q8TKC7 will show the presence of all three required domains, i.e., the sensory PAS domain, the phosphorylation/dimerization HisKA_2 domain, and the C-terminal HATPase_c domain. In addition, Pfam shows that the N terminus of MA3481 consists of seven transmembrane segments, which might form an additional sensory domain (see later).

3. Although most sequence analysis tools recognize the C-terminal region of MA3481 as the HATPase domain, it is necessary to check conservation of the key (active site) residues to make sure that this protein actually can function as an ATPase (or kinase). The easiest way to do that is to use the CDD tool that compares the sequence in question against the consensus sequence for the given domain. In the current version this can be done by clicking the "plus" sign in the CDD output. Although the current CDD entry does not have any information about the active site residues, this alignment allows one to recognize the G1 box and the less obvious N and G2 boxes (the F box is poorly conserved in the MA3481 sequence). Therefore, it is necessary to verify conservation of the key residues in the HATPase_c domain of MA3481 by comparing it against the active site residues of a well-characterized ATPase domain, for example, with the nucleotide-binding domain of *Thermotoga maritima* CheA (TM0702, gi|15643465), whose structure (PDB entry: 1I5A and others) has been solved in a complex with an ATP analog (Bilwes *et al.*, 1999, 2001).

4. The quickest way to compare two closely related sequences is by aligning them using the Blast 2 sequences tool (Tatusova and Madden, 1999), available on the NCBI BLAST web page http://www.ncbi.nlm.nih.gov/BLAST/ in the "Special" category. (Other on-line tools for aligning two sequences, such as EMBOSS, http://www.ebi.ac.uk/emboss/align/, or LALIGN, http://fasta.bioch.virginia.edu/fasta_www2/fasta_www.cgi?rm=lalign, can be used as well.) Clicking on "bl2sec" opens two sequence windows; paste the gi number of MA3481, 20092292, into one of them and the gi number of the *T. maritima* CheA, 15643465, into the other, change the program to be run from "blastn" to "blastp," and press the "Align" key. Although the resulting alignment shows only a relatively low sequence similarity (21% identity in the 279 amino acid overlap; expect value E = 0.71), comparing it with the article by Bilwes and colleagues (2001) shows that the key nucleotide-binding residues of CheA, Asn-409, His-413, and Gly-506 are all conserved in MA3481. Hence, MA3481 appears to have a functional ATPase domain.

5. According to the Pfam database, MA3481 contains a dimerization and phosphoacceptor domain HisKA_2 (PF07568). The only thing that needs to be checked here is the presence of the phosphoryl-accepting His residue. A BLAST search using the MA3481 residues 430–550 as a query reveals a large number of sequences with a conserved (HNQDR)HR motif, characteristic of the recently described HWE family of histidine kinases, where the conserved His residue serves as the phosphorylation site (Karniol and Vierstra, 2004). A similar result can be obtained by comparing MA3481 with *Agrobacterium tumefaciens* protein AGR_C_3927 (gi| 15157306), which was experimentally characterized by Karniol and Vierstra (2004). These comparisons show that MA3481 indeed contains a functional dimerization/phosphoacceptor domain with a phosphoryl-accepting His residue.

6. Summing up, the aforementioned analysis shows that MA3481 contains a C-terminal HATPase domain, preceded by a HisKA domain and a PAS domain. In addition, it contains an N-terminal 7TM region, which could be another sensor domain (see later). Although the MA3481 gene neighborhood does not contain a gene for a potential response regulator, MA3481 satisfies all the key criteria listed previously and can be confidently annotated as sensory histidine kinase.

Analysis of Sensory Domains in Histidine Kinases

The sensory domains of histidine kinases are extremely diverse, reflecting the diversity of the signals they perceive. However, as far as we can judge, members of the same domain family typically recognize the same (or very close) substrates. Therefore, functional characterization of a sensory domain in one organism often serves as a basis for functional annotation of all proteins that contain the same domain. Still, functions of many periplasmic, membrane-embedded, or intracellular sensory domains are still unknown. Furthermore, not every N-terminal domain in every histidine kinase necessarily serves as a sensor. For example, the N-terminal membrane domain of the *E. coli* UhpB, a histidine kinase that regulates transport and metabolism of glucose-6-phosphate and related sugars, does not appear to work as a sensor. Instead, its function appears to be limited to the interaction with UhpC, another membrane protein that is evolutionarily related to sugar phosphate transport proteins but actually works as a sensor of sugar phosphate in the UhpB–UhpC complex (Island and Kadner, 1993). Therefore, describing new sensory domains requires certain caution. Still, in most cases one can safely assume that the (periplasmic) N-terminal region of a histidine kinase is its sensory domain. This assumption is definitely justified for those domains that are found in combination with more than one type of membrane sensors, for example, histidine kinases and chemotaxis methyl-accepting proteins, adenylate or diguanylate cyclases (Galperin *et al.*, 2001; Nikolskaya *et al.*, 2003; Zhulin *et al.*, 2003).

Analysis of the Putative Sensory Domain of MA3481

1. Extract the sequence of MA3481 from UniProt (Accession No. Q8TKC7) or the NCBI protein database (Accession No. AAM06847 or gi|20092292). Note that its N-terminal region is not covered by any known domain in CDD, whereas in Pfam it is represented by seven predicted transmembrane segments.

2. Select the first 240 amino acid residues of the MA3481 sequence, copy them and paste into the PSI-BLAST window on the NCBI web site (http://www.ncbi.nlm.nih.gov/BLAST/) and press the "Run BLAST" and "Format" keys.

3. Upon receiving the results, press "Run PSI-BLAST iteration 2" and then "Format" keys again. Continue this procedure until convergence, that is, until PSI-BLAST reports "No new sequences were found above the 0.005 threshold!" The search should converge after six or seven iterations, resulting in a list of ~90 database hits.

4. Visually inspect the degree of sequence conservation by scrolling down from the highest-scoring to the lowest-scoring proteins. Note that none of them has been experimentally characterized so far. In addition, although most proteins in the hit list are annotated as "sensory transduction histidine kinase," several of them (including *T. maritima* protein TM0972 and *Thiobacillus denitrificans* protein Tbd_2578) are annotated as "GGDEF domain" or "diguanylate cyclase," whereas others are annotated as "Protein phosphatase 2C-like" (*Clostridium thermocellum* protein EAM46991, gi|67851426) or "Stage II sporulation E" (*Alkaliphilus metalliredigenes* protein EAO80373, gi|77637946). This diversity of annotations deserves a further investigation to check if the N-terminal region of MA3481 can be found in signaling proteins other than histidine kinases.

5. By pressing the "Taxonomy reports" link on top of the BLAST output, generate the listing of database hits, sorted by their taxonomic representation, and save to your disk in HTML and/or plain text format.

6. Using formatting options for the BLAST results, remove the low-scoring hits by changing the "Expect value range" parameter to the range from 0 to 1e-10 and press the "Format" key. Save the resulting BLAST output file to your disk in HTML and/or plain text format.

7. Manually inspect each of the nonhistidine kinase hits in the output by following the link to the source protein and then checking out the "Conserved domains" link. This should confirm that the N-terminal region

of MA3481, used in the PSI-BLAST search, indeed can be fused to a variety of signal output domains and, hence, comprises a novel sensory domain. By analogy to the previously described *membrane-associated sensory* domains of unknown function, MASE1 and MASE2 (Nikolskaya *et al.*, 2003), we can tentatively name it MASE3.

8. To create a multiple alignment of the MASE3 domain, return to formatting options and change the "Alignment view" from "Pairwise" to "flat-query-anchored without identities." In addition, unselect the Graphical Overview, Linkout, and Sequence Retrieval boxes in the "Show" menu and again press "Format." Save the resulting multiple alignment to your disk in HTML and/or plain text format (the plain text file can also be obtained by selecting the "Alignment in Plain text" option in the "Show" menu). After manual curation (trimming, removal of the unjustified gaps, etc.), such an alignment can be colored using Microsoft Word or a dedicated software program such as GeneDoc (http://www.psc.edu/biomed/genedoc/), BoxShade (http://www.ch.embnet.org/software/BOX_form.html), Cinema (http://aig.cs.man.ac.uk/research/utopia/utopia.php), or Protein Colourer (http://www.ebi.ac.uk/proteincol/index.html) and used for publication.

9. In certain cases, it might make sense to remove sequences coming from the unfinished genome sequences from the alignment. This can be done using the "Limit results by entrez query" option and selecting the limit "srcdb_refseq_provisional[prop]" or a similar one.

10. The most conserved residues in the domain can be visualized using the WebLogo tool (Crooks *et al.*, 2004). For a first look, change the BLAST results formatting options to the "query-anchored without identities" alignment view and save the resulting alignment as a text file. Remove the unaligned amino acid residues by deleting all lines that do not start with a gi number, fill the empty spaces with dots or dashes, and submit the resulting alignment to http://weblogo.berkeley.edu/. The formatted logo (Fig. 1) shows several well-conserved residues and groups of positively charged residues that could be used to determine the membrane topology of the domain using the "positive-inside" rule (von Heijne, 1992). For a publishable sequence logo, use the manually curated alignment from step 8.

11. It would be helpful to check whether all identified instances of MASE3 domain indeed consist of seven transmembrane segments. This could be done using a variety of software tools, listed in Table II. As always, it is recommended that at least three different methods are used, their results compared, and any discrepancies in the outcomes analyzed on a case-by-case basis.

12. In conclusion, the described procedure identified a new integral membrane sensory domain (MASE3) found in histidine kinases,

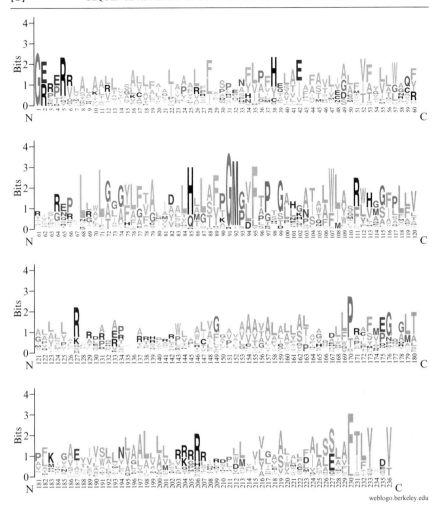

FIG. 1. Sequence logo of the MASE3 domain generated using the WebLogo (http://weblogo.berkeley.edu) tool from a multiple alignment of 35 different sequences of the MASE3 domain aligned to the *Methanosarcina acetivorans* histidine kinase MA3481. Residue numbering starts from Gly-4 of the MA3481 sequence. (See color insert.)

diguanylate cyclases, c-di-GMP phosphodiesterases, and PP2C-type protein phosphatases from a variety of bacteria (α-, β-, γ-, and δ-proteobacteria, firmicutes, and *Thermotoga* spp.) and archaea. The signal sensed by this domain is currently unknown and can only be identified through experimental studies of the respective signaling proteins.

TABLE II
COMPUTATIONAL RESOURCES FOR PREDICTION OF SIGNAL PEPTIDES AND TRANSMEMBRANE
SEGMENTS IN PROTEINS

Name	URL	No. of sequences	References
Prediction of signal peptides			
SignalP	http://www.cbs.dtu.dk/services/ SignalP	≤2000	Bendtsen *et al.* (2004)
LipoP	http://www.cbs.dtu.dk/services/ LipoP/	≤4000	Juncker *et al.* (2003)
PSORT	http://www.psort.org/	≤2000[a]	Gardy *et al.* (2005)
SOSUI signal	http://bp.nuap.nagoya-u.ac.jp/ sosui/sosuisignal/	1	Gomi *et al.* (2004)
Prediction of transmembrane segments			
TMHMM	http://www.cbs.dtu.dk/services/ TMHMM/	≤4000	Krogh *et al.* (2001)
ConPred	http://bioinfo.si.hirosaki-u.ac.jp/ %7EConPred2/	≤100	Arai *et al.* (2004)
DAS	http://mendel.imp.ac.at/sat/DAS/ DAS.html	≤50	Cserzo *et al.* (2004)
HMMTOP	http://www.enzim.hu/hmmtop/	N/A[b]	Tusnady and Simon (2001)
MEMSAT	http://bioinf.cs.ucl.ac.uk/psipred/	1	Jones *et al.* (1994)
Phobius	http://phobius.cgb.ki.se/	1	Kall *et al.* (2004)
PSORT	http://www.psort.org/	≤2000[a]	Gardy *et al.* (2005)
SOSUI	http://bp.nuap.nagoya-u.ac.jp/ sosui/	N/A[b]	Hirokawa *et al.* (1998)
SVMtm	http://ccb.imb.uq.edu.au/svmtm/	≤25[a]	Yuan *et al.* (2004)
TMpred	http://www.ch.embnet.org/ software/TMPRED_form.html	1	Hofmann and Stoffel (1993)
TMAP	http://bioweb.pasteur.fr/seqanal/ interfaces/tmap.html	1	Persson and Argos (1997)
TopPred	http://bioweb.pasteur.fr/seqanal/ interfaces/toppred.html	1	Claros and von Heijne (1994)

[a] The actual limitation for PSORT is 600,000 characters, for SVMtm it is 10 KB.
[b] The server accepts multiple sequences but the exact limit is not specified.

Sequence Analysis of Response Regulators

Overview

All response regulators of the two-component signal transduction system contain the CheY-like phosphoacceptor (receiver, REC) domain (Stock *et al.*, 2000; West and Stock, 2001), either in a stand-alone form (e.g., the chemotaxis response regulator CheY or the sporulation regulator

SpoOF) or fused to an effector, or output, domain, which is usually located at the C terminus of the polypeptide chain (Grebe and Stock, 1999; Stock et al., 2000). Two-domain response regulators are typically thought of as transcriptional regulators that combine the REC domain with a DNA-binding output domain. Indeed, studies have shown that the great majority of output domains are involved in DNA binding (Galperin, 2006; Ulrich et al., 2005). However, a substantial fraction of response regulators have RNA-binding, enzymatic, or ligand-binding (noncatalytic) output domains, or uncharacterized output domains whose function is unknown (Galperin, 2006; Ulrich et al., 2005). Phylogenetic analysis of the receiver and output domains in various response regulators has shown that receiver domains typically coevolve with the corresponding effector domains, although some of them show signs of relatively recent domain shuffling (Pao and Saier, 1995).

The mechanism of two-component signal transduction includes phosphoryl transfer from the His residue in the HisKA domain of the sensor histidine kinase to an Asp residue in the REC domain of its cognate response regulator (Stock et al., 2000; West and Stock, 2001). Phosphorylation induces conformational changes in the REC domain (Kern et al., 1999), which affects its binding properties, including its association with the output domain (if any). In different response regulators, there appear to be several different mechanisms of signal transmission. These include dimerization of the REC domain (in the OmpR/PhoB family and potentially in all DNA-binding response regulators [Toro-Roman et al., 2005a,b]), direct protein–protein interaction with a variety of target proteins (in stand-alone receiver domains, such as CheY or SpoOF), and a relief-of-inhibition mechanism (and potentially also a stimulatory effect on catalysis) in case of enzymatically active output domains (Anand and Stock, 2002).

The most common response regulators are the DNA-binding transcriptional regulators that belong to the two largest families: (1) the OmpR/PhoB family regulators that contain winged helix-turn-helix (HTH) output domains (Martinez-Hackert and Stock, 1997a,b) and (2) the NarL/FixJ family regulators that contain a single HTH motif in the middle of a four-helix bundle (Baikalov et al., 1996). Other, less common, DNA-binding response regulators contain DNA-binding output domains of the Fis type (NtrC and PrrA/ActR families), AraC type (YesN family), LytTR type (AgrA/LytR family), SpoOA type (SpoOA family), and several others (Galperin, 2006). Within each family of response regulators, the signaling specificity is determined by minute details of the interactions of the REC domains with the cognate histidine kinases and of the HTH domains with the target sites on the DNA. As a result, response regulators within each particular family typically show high levels of sequence similarity, even when they regulate dramatically different biological processes. This circumstance makes it almost impossible to assign function to newly sequenced response regulators based solely on sequence similarity. Therefore, the goals of sequence analysis have to be far more modest: (1) identification of protein in question as a response regulator, based on the presence of the REC domain; (2) identification of the output domain of the given response regulator (if any) and its function (if known); and (3) assignment of this response regulator to a particular family, followed by assignment of a general function, such as DNA binding, RNA binding, small molecule ligand binding, or an enzymatic activity.

Identification of Spo0A as a Response Regulator

By definition, almost any protein containing the receiver (REC) domain can be considered a response regulator. Exceptions include hybrid histidine kinases that contain a C-terminal REC domain and other multidomain signal transduction proteins that combine the REC domain with various sensory and/or output domains (see, e.g., Fig. 2 in Galperin [2006]). The relatively high sequence conservation of the REC domain makes its identification relatively straightforward. Comparing the protein in question against any of the domain databases, such as CDD, Pfam, InterPro, or ProDom, using their default parameters, is usually sufficient to find out whether this protein contains the REC domain and, if it does, what are the domain boundaries. The output of this comparison will also show if this protein also contains a recognized output domain (see Galperin [2006] for a recently compiled listing). Consider the following example of the well-characterized DNA-binding response regulator Spo0A, which controls the initiation of sporulation in gram-positive bacteria (Stephenson and Lewis, 2005).

1. Retrieve the sequence of *Bacillus anthracis* Spo0A (BA_4394) from UniProt (SP0A_BACAN, Accession No. P52928), the NCBI protein database (Accession No. AAP28110; gi|30258892), or TIGR web site (http://www.tigr.org/tigr-scripts/CMR2/GenePage.spl?locus=BA_4394). Inspect the annotation of BA_4394 in each of these databases. Note that this protein is uniformly annotated as "Stage 0 sporulation protein A."

2. Inspect the domain architecture of BA_4394 as represented in Pfam. To do this, go to the Pfam search page at http://www.sanger.ac.uk/Software/Pfam/search.shtml and enter the UniProt name, accession number, or the protein sequence in the appropriate windows. Results will show at http://www.sanger.ac.uk/cgi-bin/Pfam/swisspfamget.pl?name=P52928. Also, look at the domain representation in CDD (click the "Conserved Domains" link from the NCBI protein entry or go to http://www.ncbi.nlm.

nih.gov/Structure/cdd/cdd.shtml and then enter the NCBI accession number, gi number, or sequence). Results will also show at http://www.ncbi.nlm. nih.gov/Structure/cdd/wrpsb.cgi?INPUT_TYPE=precalc&SEQUENCE= 30258892. Both CDD and Pfam recognize an N-terminal REC domain in the BA_4394 sequence with convincing similarity scores. Furthermore, inspection of domain alignment shows conservation of the phosphoacceptor Asp residue, confirming that BA_4394 contains a functional REC domain and, hence, is a genuine response regulator.

Identification of the Output Domain of Spo0A

In some cases, including the one just given, comparing a sequence of a response regulator against protein domain databases shows that (1) the REC domain is the only one recognized in the given sequence and (2) it occupies only a certain part of the protein, leaving 50 or even more amino acid residues not assigned to any domain. Given that some protein domains can be as short as 25 amino acid residues (e.g., ATP-hook, [Aravind and Landsman, 1998]) and many HTH domains are not much longer (Aravind *et al.*, 2005), these unassigned regions could well belong to as yet unrecognized protein domains. However, some of such unassigned regions appear to lack any (predicted) secondary structure and therefore should not be considered separate domains. The following example continues sequence analysis of the *B. anthracis* protein Spo0A (BA_4394).

1. Note that the C-terminal region of BA_4394 (amino acid residues 130–264) is not covered by any known domain in CDD, whereas in Pfam it is represented by the putative domain Pfam-B_5095 (Pfam-B is the section of Pfam that contains unannotated and putative domains).

2. Copy the 134 amino acid C-terminal sequence fragment of BA_4394 (residues 131–264) into the PSI-BLAST search page at the NCBI web site, http://www.ncbi.nlm.nih.gov/BLAST/. Run PSI-BLAST until convergence using the default parameters on the web page. The search should converge after three or four iterations, resulting in a list of ~90 database hits. Visually inspect the degree of sequence conservation by scrolling down from the highest-scoring to the lowest-scoring proteins to confirm that all the hits are genuine homologs.

3. By pressing the "Taxonomy reports" link on top of the BLAST output, generate and save the listing of database hits, sorted by their taxonomic representation. You will see that, with a single exception, all high-scoring hits (bit score of >133, which corresponds to the expectation value $E < 2 \times 10^{-30}$) belong to the Firmicutes (low G+C gram-positive bacteria). The only exception is *Symbiobacterium thermophilum*, which,

because of its relatively high G+C content, has been initially assigned to the phylum *Actinobacteria* but obviously belongs to the Firmicutes.

4. In the BLAST output, find hits to the known three-dimensional structures (marked by red squares with the letter "S"). Follow the links to Protein Data Base (PDB) entry 1FC3 to view and see the detailed description of this domain (Lewis *et al.*, 2000) and then the link "Structure" to view its six-helix structure. Alternatively, follow the link to PDB entry 1LQ1 to see this domain bound to the DNA (Zhao *et al.*, 2002). Note that the C-terminal DNA-binding fragment of Spo0A forms a compact and stable three-dimensional structure and thus comprises a separate well-defined protein domain.

Sequence Analysis of Prokaryotic Signal Transducers

Overview

Analysis of the rapidly accumulating genome sequences from diverse bacteria and archaea revealed the great variety of sensory proteins. The characteristic architecture of histidine kinases and MCPs, which include a periplasmic sensory domain, a transmembrane segment with one or more transmembrane helices, and a cytoplasmically located output domain, was predicted for many proteins encoded in the newly sequenced genomes (Galperin, 2004; Galperin *et al.*, 2001). However, while their N-terminal sensory domains were shared with histidine kinases and/or MCPs (Zhulin *et al.*, 2003), their C-terminal output domains could be adenylate cyclases, diguanylate cyclases, or c-di-GMP phosphodiesterases of EAL or HD-GYP type, as well as Ser/Thr protein kinases and protein phosphatases (in Ser/Thr protein kinases, the protein kinase domain is typically at the N terminus and the sensory domains at the C terminus). The computational predictions were followed by experimental data detailing participation of these new (predicted) receptors in bacterial signal transduc- tion. Thus, activities of cyanobacterial membrane-bound adenylate cyclase and *Rhodobacter sphaeroides* bacteriophytochrome BphG1 (diguanylate cyclase/phosphodiesterase) were shown to respond to the red and blue light (Ohmori and Okamoto, 2004; Tarutina *et al.*, 2006). Activities of many other bacterial receptor-type proteins appear to be regulated by environmental factors as well; however, the nature of these factors still remains obscure (Galperin, 2004, 2005; Kennelly, 2002; Lory *et al.*, 2004; Römling *et al.*, 2005; Zhang and Shi, 2004). Still, recognition of a potential receptor protein in a given piece of DNA sequence could be an important step toward understanding the function of that protein and/or its neighbors. We briefly describe here the domain-based approaches to the identification of diguanylate cyclases (the GGDEF domain), c-di-GMP-specific phosphodiesterases (EAL and HD-GYP domains), class III adenylate cyclases (the CyaA domain), eukaryotic-type Ser/Thr/Tyr-specific protein kinases (the STYK domain), and PP2C-family Ser/Thr/Tyr-specific protein phosphatases (the PP2C-SIG domain) and mention several caveats of such searches that may lead to false-positive hits.

Identification of Diguanylate Cyclases (the GGDEF Domain)

Diguanylate cyclase activity, synthesis of the bacterial second messenger c-di-GMP from two molecules of GTP, is a property of the so-called GGDEF domain, named after its most conserved sequence motif, Gly-Gly-Asp/Glu-Glu-Phe (Galperin *et al.*, 2001; Jenal and Malone, 2006; Römling *et al.*, 2005). The GGDEF domain is structurally related to the eukaryotic adenylate cyclase domains (Chan *et al.*, 2004) and has a number of well-conserved residues (see the sequence logo of this domain at http://www.blackwell-synergy.com/doi/suppl/10.1111/j.1365–2958.2005.04697.x/suppl_file/FigS1A.zip), which makes its identification fairly straightforward. The key problem with its sequence analysis is recognition of inactivated and/or truncated GGDEF domains, which has to be done by meticulously checking the active site residues (Christen *et al.*, 2006; Paul *et al.*, 2004) and deciding whether any given mutation allows correct folding of the protein and is compatible with its activity. Unfortunately, very few residues outside the GGDEF loop and the allosteric I-site (Christen *et al.*, 2006) have been mutated so far, and their contribution to activity remains unknown. The listings of all GGDEF-containing proteins encoded in completely sequenced bacterial genomes are available on the SignalCensus web site and in Sentra and MiST databases.

*Identification of Type I c-di-GMP-Specific Phosphodiesterases
(the EAL Domain)*

Just like the GGDEF domain, the type I c-di-GMP-specific phos-phodiesterase (the EAL domain) is very well conserved and easily iden-tified through comparison with any of the protein domain databases. In addition to the conserved EAL motif, two acidic residues required for the activity have been identified (see the sequence logo of this domain at http://www.blackwell-synergy.com/doi/suppl/10.1111/j.1365–2958.2005.04697.x/suppl_file/FigS1B.zip). An alignment of active and inactive EAL domains (Schmidt *et al.*, 2005) could provide further clues to which amino acid residues are required for activity.

Identification of Type II c-di-GMP-Specific Phosphodiesterases (the HD-GYP Domain)

The HD-GYP domain (Galperin *et al.*, 1999), recently proven to function as a c-di-GMP-specific phosphodiesterase (Ryan *et al.*, 2006), is an extended variant of the widespread HD-type phosphohydrolase domain (Aravind and Koonin, 1998) that contains extra conserved residues at its C terminus. Because of that, Pfam and SMART databases do not recognize HD-GYP as a separate domain and list its N-terminal 110 amino acid residues as a HD catalytic domain. In contrast, PIRSF and COGs list HD-GYP as a separate domain, while InterPro has a separate entry for RpfG-like response regulators that combine REC and HD-GYP domains (Ryan *et al.*, 2006). In sequence similarity searches of HD-GYP domains, generic HD domains often show up with higher similarity scores than genuine HD-GYP domains. Here, the listing of HD-GYP-containing proteins on the SignalCensus web site could be used as a guide.

Identification of Adenylate Cyclases

Several unrelated (analogous) forms (classes) of the adenylate cyclase (EC 4.6.1.1) have been described, but only one of them, usually referred to as class III, is widespread in bacteria and has been shown to function as an environmental sensor (Lory *et al.*, 2004; Ohmori and Okamoto, 2004). The class III adenylate cyclase domain (designated the Guanylate_cyc domain in Pfam and PROSITE and A/G cyclase in InterPro) is well conserved and can be easily recognized through protein sequence comparison against any of the domain databases. It should be noted that the eukaryotic form of this domain can use both ATP and GTP as substrates, producing, respectively, cAMP and cGMP, while in most bacteria it appears to be specific for ATP and have little, if any, guanylate cyclase activity (Baker and Kelly, 2004; Shenoy and Visweswariah, 2004). Therefore, for bacteria, the name "adenylate cyclase domain" appears to be more appropriate than any other. Many class III adenylate cyclases are cytoplasmic enzymes and only a small fraction of them are membrane bound and can be considered genuine environmental sensors.

Identification of Ser/Thr/Tyr-Specific Protein Kinases

Curiously, the vast majority of prokaryotic Ser/Thr/Tyr-specific protein kinases belong to the so-called eukaryotic-type protein kinase superfamily (Kennelly and Potts, 1996). This superfamily includes several other kinase families, such as choline kinases, lipopolysaccharide kinases, and aminoglycoside 3′-phosphotransferases. This fact makes correct identification of Ser/Thr/

Tyr protein kinases fairly complicated, particularly because Ser/Thr protein kinases from different families sometimes show less similarity to each other than to 3-deoxy-D-manno-octulosonic acid (KDO) kinase (the product of the *waaP* gene, assigned in Pfam to a separate domain PF06293). In addition, there is a long-standing controversy regarding the functions of the proteins of ABC1/AarF/UbiB family, which are required for ubiquinone biosynthesis (Poon *et al.*, 2000). These widespread proteins, which belong to the Pfam family PF03109 and COG0661 and most likely function as protein kinases that regulate the ubiquinone biosynthesis pathway, are sometimes misannotated as ABC transporters or even as ABC transporter substrate-binding proteins. Identification of an unknown protein as a Ser/Thr/Tyr protein kinase should take into account domain assignments in several domain databases, presence or absence of additional—sensory or signal output—domains, and genomic context, for example, the operon structure of the adjacent genes. Functional assignments for completely sequenced genomes are available on MiST, Sentra, and SignalCensus web sites, as well as in KinG, a database dedicated specifically to Ser/Thr/Tyr protein kinases (Krupa *et al.*, 2004).

Identification of Ser/Thr/Tyr Protein Phosphatases

Prokaryotic Ser/Thr/Tyr-specific protein phosphatases of the PP2C family are reasonably well conserved and can be easily recognized through protein sequence comparison against any of the domain databases. However, only a small fraction of these enzymes have an attached sensory domain and can be considered genuine environmental sensors.

Functional Annotation of Multidomain Proteins

The complexity of microbial signal transduction machinery and the paucity of experimentally characterized proteins make annotating signaling proteins even in well-studied organisms an arduous task. For example, of the 30 histidine kinases encoded by *E. coli* K12, functions of five (RstB, YehU, YpdA, YfhK, YedV) are unknown and several others have poorly defined substrates. For (predicted) signal transduction proteins encoded in the newly sequenced genomes this task becomes even more daunting. Although assigning the signal transduction protein to a general class, such as "histidine kinase," "response regulator," "methyl-accepting chemotaxis protein," or "diguanylate cyclase," is usually easy (see earlier discussion), it is practically impossible to identify the signal that this protein responds to. Indeed, standard methods of protein sequence analysis, where the function of a newly sequenced protein is assigned based on the function of its experimentally characterized (close) homolog, are rarely applicable to signal transduction proteins. Thus, the degree of sequence similarity of the response regulators of the NtrC family is determined more by their central ATPase domains than by their N-terminal REC or C-terminal DNA-binding Fis-like domains, which are in fact responsible for the specificity of these transcriptional regulators. As a result, proteins that show the highest similarity scores are very likely to regulate entirely different biological processes. Therefore, functional annotation of newly sequenced signal transduction proteins should rely on the following simple rules.

1. If at all possible, the protein should be classified as signaling, signal transduction, or signal output protein.

2. For signaling proteins, the enzymatic activity of the signal transduction domain (histidine kinase, Ser/Thr/Tyr kinase, protein phosphatase, adenylate cyclase, diguanylate cyclase, c-di-GMP phosphodiesterase) can be used as a basis for protein annotation.

3. Any protein that contains more than one enzymatic output domain should be annotated as "multidomain signaling protein containing such and such domain." The only exceptions to this rule are the proteins combining the GGDEF and EAL domains, which, unless one of the domains is known (or predicted) to be inactive, can be annotated as "diguanylate cyclase/ phosphodiesterase."

4. It is also useful to identify the transmembrane segments, if any, and decide whether the respective signaling protein contains a periplasmic (extracytoplasmic) or membrane-embedded sensory domain, or it is sensing the cytoplasmic milieu.

5. If the signaling protein contains a known sensory domain, its name (or, better yet, ligand specificity) should be included in the annotation, for example, "Citrate-sensing histidine kinase," "Histidine kinase with two PAS and one GAF sensory domains," "Adenylate cyclase with a PAS sensory domain," or "Diguanylate cyclase with MASE2 sensory domain."

6. Any protein containing the REC domain is annotated as a response regulator, as a hybrid histidine kinase, or as a multidomain signal transduction protein.

7. Response regulators are classified into families and named according to the specificities of their output domains, for example, "DNA-binding response regulator, OmpR family" or a "Transcriptional regulator, OmpR/ PhoB family." In the rare cases when the transmitted signal is known, this information has to be reflected in the name, whereas the family designation can be omitted, for example, NarL can be annotated either as "Nitrate/nitrite response regulator NarL" or as "Transcriptional regulator of nitrate/nitrite response, NarL family."

8. Response regulators with enzymatic output domains can have family-based or domain-based names. For consistency, it would be appropriate to annotate them based on their enzymatic activities, just as it is being done for signaling proteins. For example, the response regulator of the REC-GGDEF domain architecture can be called a "Response regulator, WspR family" or a "Response regulator with a diguanylate cyclase output domain." Finally, response regulators with ligand-binding output domains have to be named based on their domain architectures, for example, "Response regulator with a PAS output domain, REC-PAS."

9. Finally, other components of the signal transduction machinery, unless their function has been established experimentally, should be annotated based on their domain composition. For example, we would recommend that uncharacterized homologs of the nitric oxide reductase transcription regulator NorR be annotated as "Transcriptional regulator containing GAF, AAA-type ATPase, and Fis-like DNA binding domains, NorR/HyfR family," as some of them may turn out to respond to other signals than nitric oxide.

10. Most importantly, all signal transduction proteins should be annotated based on their total domain composition, not just based on the database hits that might not cover the whole protein.

In conclusion, it is worth noting that complete automation of the sequence analyses and functional annotations described in this chapter is still impossible. Although analysis of protein domain organization is usually fairly straightforward, proper utilization of proper databases and software tools requires a human with certain biological education. While this may be considered a nuisance by some, this reflects a more general principle that to recognize novelty, one has to be aware of the state of the art. If the latest developments are any indication, signal transduction proteins with complex domain architectures would not be amenable to fully automated analysis for years to come.

References

Anand, G. S., and Stock, A. M. (2002). Kinetic basis for the stimulatory effect of phosphorylation on the methylesterase activity of CheB. *Biochemistry* **41,** 6752–6760.

Arai, M., Mitsuke, H., Ikeda, M., Xia, J. X., Kikuchi, T., Satake, M., and Shimizu, T. (2004). ConPred II: A consensus prediction method for obtaining transmembrane topology models with high reliability. *Nucleic Acids Res.* **32,** W390–W393.

Aravind, L., Anantharaman, V., Balaji, S., Babu, M. M., and Iyer, L. M. (2005). The many faces of the helix-turn-helix domain: Transcription regulation and beyond. *FEMS Microbiol. Rev.* **29,** 231–262.

Aravind, L., and Koonin, E. V. (1998). The HD domain defines a new superfamily of metal-dependent phosphohydrolases. *Trends Biochem. Sci.* **23,** 469–472.

Aravind, L., and Landsman, D. (1998). AT-hook motifs identified in a wide variety of DNA-binding proteins. *Nucleic Acids Res.* **26,** 4413–4421.

Aravind, L., and Ponting, C. P. (1999). The cytoplasmic helical linker domain of receptor histidine kinase and methyl-accepting proteins is common to many prokaryotic signalling proteins. *FEMS Microbiol. Lett.* **176,** 111–116.

Attwood, T. K., Bradley, P., Flower, D. R., Gaulton, A., Maudling, N., Mitchell, A. L., Moulton, G., Nordle, A., Paine, K., Taylor, P., Uddin, A., and Zygouri, C. (2003). PRINTS and its automatic supplement, prePRINTS. *Nucleic Acids Res.* **31,** 400–402.

Baikalov, I., Schroder, I., Kaczor-Grzeskowiak, M., Grzeskowiak, K., Gunsalus, R. P., and Dickerson, R. E. (1996). Structure of the *Escherichia coli* response regulator NarL. *Biochemistry* **35,** 11053–11061.

Baker, D. A., and Kelly, J. M. (2004). Structure, function and evolution of microbial adenylyl and guanylyl cyclases. *Mol. Microbiol.* **52,** 1229–1242.

Ban, C., and Yang, W. (1998). Crystal structure and ATPase activity of MutL: Implications for DNA repair and mutagenesis. *Cell* **95,** 541–552.

Bendtsen, J. D., Nielsen, H., von Heijne, G., and Brunak, S. (2004). Improved prediction of signal peptides: SignalP 3.0. *J. Mol. Biol.* **340,** 783–795.

Berman, H. M., Westbrook, J., Feng, Z., Gilliland, G., Bhat, T. N., Weissig, H., Shindyalov, I. N., and Bourne, P. E. (2000). The Protein Data Bank. *Nucleic Acids Res.* **28,** 235–242.

Bilwes, A. M., Alex, L. A., Crane, B. R., and Simon, M. I. (1999). Structure of CheA, a signal-transducing histidine kinase. *Cell* **96,** 131–141.

Bilwes, A. M., Quezada, C. M., Croal, L. R., Crane, B. R., and Simon, M. I. (2001). Nucleotide binding by the histidine kinase CheA. *Nat. Struct. Biol.* **8,** 353–360.

Chan, C., Paul, R., Samoray, D., Amiot, N. C., Giese, B., Jenal, U., and Schirmer, T. (2004). Structural basis of activity and allosteric control of diguanylate cyclase. *Proc. Natl. Acad. Sci. USA* **101,** 17084–17089.

Cheung, J. K., and Rood, J. I. (2000). Glutamate residues in the putative transmembrane region are required for the function of the VirS sensor histidine kinase from *Clostridium perfringens*. *Microbiology* **146,** 517–525.

Christen, B., Christen, M., Paul, R., Schmid, F., Folcher, M., Jenoe, P., Meuwly, M., and Jenal, U. (2006). Allosteric control of cyclic di-GMP signaling. *J. Biol. Chem.* **281,** 32015–32024.

Claros, M. G., and von Heijne, G. (1994). TopPred II: An improved software for membrane protein structure predictions. *Comput. Appl. Biosci.* **10,** 685–686.

Crooks, G. E., Hon, G., Chandonia, J. M., and Brenner, S. E. (2004). WebLogo: A sequence logo generator. *Genome Res.* **14,** 1188–1190.

Cserzo, M., Eisenhaber, F., Eisenhaber, B., and Simon, I. (2004). TM or not TM: Transmembrane protein prediction with low false positive rate using DAS-TMfilter. *Bioinformatics* **20,** 136–137.

D'Souza, M., Glass, E. M., Syed, M. H., Zhang, Y., Rodriguez, A., Maltsev, N., and Galperin, M. Y. (2007). Sentra, a database of signal transduction proteins for comparative genome analysis. *Nucleic Acids Res.* **35,** D271–D273.

Dutta, R., and Inouye, M. (2000). GHKL, an emergent ATPase/kinase superfamily. *Trends Biochem. Sci.* **25,** 24–28.

Dutta, R., Qin, L., and Inouye, M. (1999). Histidine kinases: Diversity of domain organization. *Mol. Microbiol.* **34,** 633–640.

Finn, R. D., Mistry, J., Schuster-Bockler, B., Griffiths-Jones, S., Hollich, V., Lassmann, T., Moxon, S., Marshall, M., Khanna, A., Durbin, R., Eddy, S. R., Sonnhammer, E. L., *et al.* (2006). Pfam: Clans, web tools and services. *Nucleic Acids Res.* **34,** D247–D251.

Galperin, M. Y. (2004). Bacterial signal transduction network in a genomic perspective. *Environ. Microbiol.* **6,** 552–567.

Galperin, M. Y. (2005). A census of membrane-bound and intracellular signal transduction proteins in bacteria: Bacterial IQ, extroverts and introverts. *BMC Microbiol.* **5,** 35.

Galperin, M. Y. (2006). Structural classification of bacterial response regulators: Diversity of output domains and domain combinations. *J. Bacteriol.* **188,** 4169–4182.

Galperin, M. Y., Gaidenko, T. A., Mulkidjanian, A. Y., Nakano, M., and Price, C. W. (2001). MHYT, a new integral membrane sensor domain. *FEMS Microbiol. Lett.* **205,** 17–23.

Galperin, M. Y., Natale, D. A., Aravind, L., and Koonin, E. V. (1999). A specialized version of the HD hydrolase domain implicated in signal transduction. *J. Mol. Microbiol. Biotechnol.* **1,** 303–305.

Galperin, M. Y., Nikolskaya, A. N., and Koonin, E. V. (2001). Novel domains of the prokaryotic two-component signal transduction systems. *FEMS Microbiol. Lett.* **203,** 11–21.

Gardner, A. M., Gessner, C. R., and Gardner, P. R. (2003). Regulation of the nitric oxide reduction operon (*norRVW*) in *Escherichia coli*: Role of NorR and Σ^{54} in the nitric oxide stress response. *J. Biol. Chem.* **278,** 10081–10086.

Gardy, J. L., Laird, M. R., Chen, F., Rey, S., Walsh, C. J., Ester, M., and Brinkman, F. S. (2005). PSORTb v.2.0: Expanded prediction of bacterial protein subcellular localization and insights gained from comparative proteome analysis. *Bioinformatics* **21,** 617–623.

Gomi, M., Sonoyama, M., and Mitaku, S. (2004). High performance system for signal peptide prediction: SOSUIsignal. *Chem-Bio. Info. J.* **4,** 142–147.

Grebe, T. W., and Stock, J. B. (1999). The histidine protein kinase superfamily. *Adv. Microb. Physiol.* **41,** 139–227.

Henikoff, J. G., Greene, E. A., Pietrokovski, S., and Henikoff, S. (2000). Increased coverage of protein families with the blocks database servers. *Nucleic Acids Res.* **28,** 228–230.

Hirokawa, T., Boon-Chieng, S., and Mitaku, S. (1998). SOSUI: Classification and secondary structure prediction system for membrane proteins. *Bioinformatics* **14,** 378–379.

Hoch, J. A. (2000). Two-component and phosphorelay signal transduction. *Curr. Opin. Microbiol.* **3,** 165–170.

Hofmann, K., and Stoffel, W. (1993). TMbase: A database of membrane spanning proteins segments. *Biol. Chem. Hoppe-Seyler* **374,** 166.

Hulo, N., Bairoch, A., Bulliard, V., Cerutti, L., De Castro, E., Langendijk-Genevaux, P. S., Pagni, M., and Sigrist, C. J. (2006). The PROSITE database. *Nucleic Acids Res.* **34,** D227–D230.

Island, M. D., and Kadner, R. J. (1993). Interplay between the membrane-associated UhpB and UhpC regulatory proteins. *J. Bacteriol.* **175,** 5028–5034.

Jenal, U., and Malone, J. (2006). Mechanisms of cyclic-di-GMP signaling in bacteria. *Annu. Rev. Genet.* **40,** 385–407.

Jones, D. T., Taylor, W. R., and Thornton, J. M. (1994). A model recognition approach to the prediction of all-helical membrane protein structure and topology. *Biochemistry* **33,** 3038–3049.

Juncker, A. S., Willenbrock, H., Von Heijne, G., Brunak, S., Nielsen, H., and Krogh, A. (2003). Prediction of lipoprotein signal peptides in Gram-negative bacteria. *Protein Sci.* **12,** 1652–1662.

Kall, L., Krogh, A., and Sonnhammer, E. L. (2004). A combined transmembrane topology and signal peptide prediction method. *J. Mol. Biol.* **338,** 1027–1036.

Kanehisa, M., Goto, S., Kawashima, S., Okuno, Y., and Hattori, M. (2004). The KEGG resource for deciphering the genome. *Nucleic Acids Res.* **32,** D277–D280.

Karniol, B., and Vierstra, R. D. (2004). The HWE histidine kinases, a new family of bacterial two-component sensor kinases with potentially diverse roles in environmental signaling. *J. Bacteriol.* **186,** 445–453.

Kennelly, P. J., and Potts, M. (1996). Fancy meeting you here! A fresh look at "prokaryotic" protein phosphorylation. *J. Bacteriol.* **178**, 4759–4764.

Kennelly, P. J. (2002). Protein kinases and protein phosphatases in prokaryotes: A genomic perspective. *FEMS Microbiol. Lett.* **206**, 1–8.

Kern, D., Volkman, B. F., Luginbuhl, P., Nohaile, M. J., Kustu, S., and Wemmer, D. E. (1999). Structure of a transiently phosphorylated switch in bacterial signal transduction. *Nature* **402**, 894–898.

Krogh, A., Larsson, B., von Heijne, G., and Sonnhammer, E. L. (2001). Predicting transmembrane protein topology with a hidden Markov model: Application to complete genomes. *J. Mol. Biol.* **305**, 567–580.

Krupa, A., Abhinandan, K. R., and Srinivasan, N. (2004). KinG: A database of protein kinases in genomes. *Nucleic Acids Res.* **32**, D153–D155.

Lee, S. Y., Cho, H. S., Pelton, J. G., Yan, D., Henderson, R. K., King, D. S., Huang, L., Kustu, S., Berry, E. A., and Wemmer, D. E. (2001). Crystal structure of an activated response regulator bound to its target. *Nat. Struct. Biol.* **8**, 52–56.

Letunic, I., Copley, R. R., Pils, B., Pinkert, S., Schultz, J., and Bork, P. (2006). SMART 5: Domains in the context of genomes and networks. *Nucleic Acids Res.* **34**, D257–D260.

Lewis, R. J., Krzywda, S., Brannigan, J. A., Turkenburg, J. P., Muchova, K., Dodson, E. J., Barak, I., and Wilkinson, A. J. (2000). The trans-activation domain of the sporulation response regulator Spo0A revealed by X-ray crystallography. *Mol. Microbiol.* **38**, 198–212.

Lory, S., Wolfgang, M., Lee, V., and Smith, R. (2004). The multi-talented bacterial adenylate cyclases. *Int. J. Med. Microbiol.* **293**, 479–482.

Marchler-Bauer, A., Anderson, J. B., Cherukuri, P. F., DeWeese-Scott, C., Geer, L. Y., Gwadz, M., He, S., Hurwitz, D. I., Jackson, J. D., Ke, Z., Lanczycki, C. J., Liebert, C. A., *et al.* (2005). CDD: A Conserved Domain Database for protein classification. *Nucleic Acids Res.* **33**, D192–D196.

Martinez-Hackert, E., and Stock, A. M. (1997a). The DNA-binding domain of OmpR: Crystal structures of a winged helix transcription factor. *Structure* **5**, 109–124.

Martinez-Hackert, E., and Stock, A. M. (1997b). Structural relationships in the OmpR family of winged-helix transcription factors. *J. Mol. Biol.* **269**, 301–312.

Matsushika, A., and Mizuno, T. (1998). The structure and function of the histidine-containing phosphotransfer (HPt) signaling domain of the *Escherichia coli* ArcB sensor. *J. Biochem. (Tokyo)* **124**, 440–445.

Mizuno, T. (1997). Compilation of all genes encoding two-component phosphotransfer signal transducers in the genome of *Escherichia coli*. *DNA Res.* **4**, 161–168.

Mulder, N. J., Apweiler, R., Attwood, T. K., Bairoch, A., Bateman, A., Binns, D., Bradley, P., Bork, P., Bucher, P., Cerutti, L., Copley, R., Courcelle, E., *et al.* (2005). InterPro, progress and status in 2005. *Nucleic Acids Res.* **33**, D201–D205.

Nikolskaya, A. N., Arighi, C. N., Huang, H., Barker, W. C., and Wu, C. H. (2006). PIRSF family classification system for protein functional and evolutionary analysis. *Evolutionary Bioinformatics Online* **2**, 209–221.

Nikolskaya, A. N., Mulkidjanian, A. Y., Beech, I. B., and Galperin, M. Y. (2003). MASE1 and MASE2: Two novel integral membrane sensory domains. *J. Mol. Microbiol. Biotechnol.* **5**, 11–16.

Ohmori, M., and Okamoto, S. (2004). Photoresponsive cAMP signal transduction in cyanobacteria. *Photochem. Photobiol. Sci.* **3**, 503–511.

Pao, G. M., and Saier, M. H., Jr. (1995). Response regulators of bacterial signal transduction systems: Selective domain shuffling during evolution. *J. Mol. Evol.* **40**, 136–154.

Parkinson, J. S., and Kofoid, E. C. (1992). Communication modules in bacterial signaling proteins. *Annu. Rev. Genet.* **26,** 71–112.

Paul, R., Weiser, S., Amiot, N. C., Chan, C., Schirmer, T., Giese, B., and Jenal, U. (2004). Cell cycle-dependent dynamic localization of a bacterial response regulator with a novel di-guanylate cyclase output domain. *Genes Dev.* **18,** 715–727.

Persson, B., and Argos, P. (1997). Prediction of membrane protein topology utilizing multiple sequence alignments. *J. Protein Chem.* **16,** 453–457.

Pohlmann, A., Cramm, R., Schmelz, K., and Friedrich, B. (2000). A novel NO-responding regulator controls the reduction of nitric oxide in *Ralstonia eutropha. Mol. Microbiol.* **38,** 626–638.

Poon, W. W., Davis, D. E., Ha, H. T., Jonassen, T., Rather, P. N., and Clarke, C. F. (2000). Identification of *Escherichia coli ubiB*, a gene required for the first monooxygenase step in ubiquinone biosynthesis. *J. Bacteriol.* **182,** 5139–5146.

Römling, U., Gomelsky, M., and Galperin, M. Y. (2005). C-di-GMP: The dawning of a novel bacterial signalling system. *Mol. Microbiol.* **57,** 629–639.

Ryan, R. P., Fouhy, Y., Lucey, J. F., Crossman, L. C., Spiro, S., He, Y. W., Zhang, L. H., Heeb, S., Camara, M., Williams, P., and Dow, J. M. (2006). Cell-cell signaling in *Xanthomonas campestris* involves an HD-GYP domain protein that functions in cyclic di-GMP turnover. *Proc. Natl. Acad. Sci. USA* **103,** 6712–6717.

Schmidt, A. J., Ryjenkov, D. A., and Gomelsky, M. (2005). Ubiquitous protein domain EAL encodes cyclic diguanylate-specific phosphodiesterase: Enzymatically active and inactive EAL domains. *J. Bacteriol.* **187,** 4774–4781.

Servant, F., Bru, C., Carrere, S., Courcelle, E., Gouzy, J., Peyruc, D., and Kahn, D. (2002). ProDom: Automated clustering of homologous domains. *Brief Bioinform.* **3,** 246–251.

Shenoy, A. R., and Visweswariah, S. S. (2004). Class III nucleotide cyclases in bacteria and archaebacteria: Lineage-specific expansion of adenylyl cyclases and a dearth of guanylyl cyclases. *FEBS Lett.* **561,** 11–21.

Stephenson, K., and Lewis, R. J. (2005). Molecular insights into the initiation of sporulation in Gram-positive bacteria: New technologies for an old phenomenon. *FEMS Microbiol. Rev.* **29,** 281–301.

Stock, A., Koshland, D. E., Jr., and Stock, J. (1985). Homologies between the *Salmonella typhimurium* CheY protein and proteins involved in the regulation of chemotaxis, membrane protein synthesis, and sporulation. *Proc. Natl. Acad. Sci. USA* **82,** 7989–7993.

Stock, A. M., Martinez-Hackert, E., Rasmussen, B. F., West, A. H., Stock, J. B., Ringe, D., and Petsko, G. A. (1993). Structure of the Mg^{2+}-bound form of CheY and mechanism of phosphoryl transfer in bacterial chemotaxis. *Biochemistry* **32,** 13375–13380.

Stock, A. M., Mottonen, J. M., Stock, J. B., and Schutt, C. E. (1989). Three-dimensional structure of CheY, the response regulator of bacterial chemotaxis. *Nature* **337,** 745–749.

Stock, A. M., Robinson, V. L., and Goudreau, P. N. (2000). Two-component signal transduction. *Annu. Rev. Biochem.* **69,** 183–215.

Tarutina, M., Ryjenkov, D. A., and Gomelsky, M. (2006). An unorthodox bacteriophytochrome from *Rhodobacter sphaeroides* involved in turnover of the second messenger c-di-GMP. *J. Biol. Chem.* **281,** 34751–34758.

Tatusov, R. L., Galperin, M. Y., Natale, D. A., and Koonin, E. V. (2000). The COG database: A tool for genome-scale analysis of protein functions and evolution. *Nucleic Acids Res.* **28,** 33–36.

Tatusova, T. A., and Madden, T. L. (1999). BLAST 2 Sequences, a new tool for comparing protein and nucleotide sequences. *FEMS Microbiol. Lett.* **174,** 247–250.

Taylor, B. L., and Zhulin, I. B. (1999). PAS domains: Internal sensors of oxygen, redox potential, and light. *Microbiol. Mol. Biol. Rev.* **63,** 479–506.

Toro-Roman, A., Mack, T. R., and Stock, A. M. (2005a). Structural analysis and solution studies of the activated regulatory domain of the response regulator ArcA: A symmetric dimer mediated by the a4-b5-a5 face. *J. Mol. Biol.* **349,** 11–26.

Toro-Roman, A., Wu, T., and Stock, A. M. (2005b). A common dimerization interface in bacterial response regulators KdpE and TorR. *Protein Sci.* **14,** 3077–3088.

Trach, K. A., Chapman, J. W., Piggot, P. J., and Hoch, J. A. (1985). Deduced product of the stage 0 sporulation gene *spo0F* shares homology with the Spo0A, OmpR, and SfrA proteins. *Proc. Natl. Acad. Sci. USA* **82,** 7260–7264.

Tusnady, G. E., and Simon, I. (2001). The HMMTOP transmembrane topology prediction server. *Bioinformatics* **17,** 849–850.

Ulrich, L. E., Koonin, E. V., and Zhulin, I. B. (2005). One-component systems dominate signal transduction in prokaryotes. *Trends Microbiol.* **13,** 52–56.

Ulrich, L. E., and Zhulin, I. B. (2007). MiST: A microbial signal transduction database. *Nucleic Acids Res.* **35,** D386–D390.

Volz, K., and Matsumura, P. (1991). Crystal structure of *Escherichia coli* CheY refined at 1.7-Å resolution. *J. Biol. Chem.* **266,** 15511–15519.

von Heijne, G. (1992). Membrane protein structure prediction: Hydrophobicity analysis and the positive-inside rule. *J. Mol. Biol.* **225,** 487–494.

West, A. H., and Stock, A. M. (2001). Histidine kinases and response regulator proteins in two-component signaling systems. *Trends Biochem. Sci.* **26,** 369–376.

Williams, S. B., and Stewart, V. (1999). Functional similarities among two-component sensors and methyl-accepting chemotaxis proteins suggest a role for linker region amphipathic helices in transmembrane signal transduction. *Mol. Microbiol.* **33,** 1093–1102.

Wolanin, P. M., Thomason, P. A., and Stock, J. B. (2002). Histidine protein kinases: Key signal transducers outside the animal kingdom. *Genome Biol.* **3,** REVIEWS3013.

Wu, C. H., Nikolskaya, A., Huang, H., Yeh, L. S., Natale, D. A., Vinayaka, C. R., Hu, Z. Z., Mazumder, R., Kumar, S., Kourtesis, P., Ledley, R. S., Suzek, B. E., *et al.* (2004). PIRSF: Family classification system at the Protein Information Resource. *Nucleic Acids Res.* **32,** D112–D114.

Yu, H., Mudd, M., Boucher, J. C., Schurr, M. J., and Deretic, V. (1997). Identification of the *algZ* gene upstream of the response regulator *algR* and its participation in control of alginate production in *Pseudomonas aeruginosa. J. Bacteriol.* **179,** 187–193.

Yuan, Z., Mattick, J. S., and Teasdale, R. D. (2004). SVMtm: Support vector machines to predict transmembrane segments. *J. Comput. Chem.* **25,** 632–636.

Zhang, W., and Shi, L. (2004). Evolution of the PPM-family protein phosphatases in *Streptomyces*: Duplication of catalytic domain and lateral recruitment of additional sensory domains. *Microbiology* **150,** 4189–4197.

Zhao, H., Msadek, T., Zapf, J., Madhusudan Hoch, J. A., and Varughese, K. I. (2002). DNA complexed structure of the key transcription factor initiating development in sporulating bacteria. *Structure* **10,** 1041–1050.

Zhulin, I. B., Nikolskaya, A. N., and Galperin, M. Y. (2003). Common extracellular sensory domains in transmembrane receptors for diverse signal transduction pathways in bacteria and archaea. *J. Bacteriol.* **185,** 285–294.

[4] Features of Protein–Protein Interactions in Two-Component Signaling Deduced from Genomic Libraries

By ROBERT A. WHITE, HENDRIK SZURMANT, JAMES A. HOCH, and TERENCE HWA

Abstract

As more and more sequence data become available, new approaches for extracting information from these data become feasible. This chapter reports on one such method that has been applied to elucidate protein–protein interactions in bacterial two-component signaling pathways. The method identifies residues involved in the interaction through an analysis of over 2500 functionally coupled proteins and a precise determination of the substitutional constraints placed on one protein by its signaling mate. Once identified, a simple log-likelihood scoring procedure is applied to these residues to build a predictive tool for assigning signaling mates. The ability to apply this method is based on a proliferation of related domains within multiple organisms. Paralogous evolution through gene duplication and divergence of two-component systems has commonly resulted in tens of closely related interacting pairs within one organism with a roughly one-to-one correspondence between signal and response. This provides us with roughly an order of magnitude more protein pairs than there are unique, fully sequenced bacterial species. Consequently, this chapter serves as both a detailed exposition of the method that has provided more depth to our knowledge of bacterial signaling and a look ahead to what would be possible on a more widespread scale, that is, to protein–protein interactions that have only one example per genome, as the number of genomes increases by a factor of 10.

Introduction

Since the mid-1990s the number of complete and published bacterial genomes has grown from just a handful to approximately 300 today with ~700 ongoing bacterial sequencing projects (Liolios *et al.*, 2006). Each significant increase in the number of sequences allows new methods to bridge the gap between data and information. This chapter reports on a method that can transform sequence data of a large number of interacting proteins into detailed spatial information about the interaction between the proteins. This information can then be used to predict specific interaction partners.

METHODS IN ENZYMOLOGY, VOL. 422   0076-6879/07 $35.00
Copyright 2007, Elsevier Inc. All rights reserved.   DOI: 10.1016/S0076-6879(06)22004-4

The underlying premise of the method is that residues that interact in real space are coupled in sequence space. When this coupling is between positions that correspond to variable columns of a multiple sequence alignment (MSA), the coupling can be detected through mutual constraint of the amino acid substitutions in the two columns. This idea began in the study of RNA secondary structure in the mid-1980s where canonical Watson–Crick pairing of bases was used in conjunction with multiple RNA sequences to elucidate the secondary structure of ribosomal RNA (Winker *et al.*, 1990; Woese *et al.*, 1983). In proteins, the multiplicity of interactions is much larger due to the increased alphabet diversity, that is, 20 amino acids vs 4 bases in RNA. Furthermore, there does not appear to be any Watson–Crick-like rule for amino acid interactions. Consequently, the covariance between positions in proteins is significantly more "fuzzy" than Watson–Crick pairing and requires many more sequences to distinguish signal from the many sources of noise.

Various unique measures have been introduced to investigate covariance between amino acid positions in proteins (Altschuh *et al.*, 1987; Atchley *et al.*, 2000; Göbel *et al.*, 1994; Kass and Horovitz, 2002; Lockless and Ranganathan, 1999; Suel *et al.*, 2003). These methods have been largely applied to single proteins in attempts to provide insight into tertiary structure and, in one case, to identify couplings between interacting proteins (Kass and Horovitz, 2002). Different approaches have been taken to identify interacting pairs of proteins that rely on detecting functional coupling through symmetries in the phylogenetic trees of coupled proteins (Goh *et al.*, 2000; Koretke *et al.*, 2000; Pazos and Valencia, 2001; Ramani and Marcotte, 2003) or more gross genetic structural considerations that track the relative locations of interacting orthologues in different organisms (Enright *et al.*, 1999; Marcotte *et al.*, 1999; Overbeek *et al.*, 1999).

This chapter presents a coupled approach that blends the identification of structural determinants of protein pair interactions with protein–protein interaction prediction. We first employ a mutual information measure to identify coupled positions between known interacting proteins and then apply a log-likelihood scoring procedure to the coupled positions to potentially interacting protein pairs.

The choice of the mutual information-based measure is motivated by its simplicity and our ability to explicitly identify and correct for sample-size effects, that is, the effect that the number of distinct functionally coupled proteins available in genomic databases has on the reliability of the measure. In a comparative analysis of various measures of covariance, Fodor and Aldrich (2004) studied the extent to which correlation between positions is coupled to the column entropy of each position for various covariance metrics. In their study, measured mutual information (MI) is shown to associate "significant" covariance preferentially between highly variable columns with very little power to predict residues that are in close contact in the tertiary structure. While it is universally appreciated that covariance analysis between columns in proteins requires large numbers of sequences, the important question of how many sequences are required to trust the results has not been systematically addressed. We show that the preferential assignment of high correlation values between highly disordered columns of the measured MI is largely due to sample size effects and diminishes as the sample size increases. We propose a natural sample-size correction to the measured MI and describe how it varies with the number of unique proteins in the database. This corrected MI value is much more reliable as it becomes sample-size independent from 400 sequences and up.

The bioinformatic method developed here will be applied to analyze interactions between partners of the bacterial two-component signaling (TCS) proteins. The two components involved in TCS pathways are a sensor kinase (SK) and a response regulator (RR). The proto-typical path of two-component signal transduction begins with a ligand-controlled auto-kinase reaction of the membrane-bound SK, which phosphorylates a conserved histidine on a HisKA domain. The HisKA-P is transferred to a conserved aspartate on the receiver domain (Rec) of the cytoplasmic RR. The phosphorylation state of the RR is coupled to its activity as a transcription factor and the path from signal to response is complete. In short, the ultimate association between signal and response is achieved through a transient phosphotransfer interaction between the HisKA and RR proteins (Hoch, 2000; Ninfa et al., 1988; Parkinson, 1993; Stock et al., 1989, 2000). The features of TCS systems that are ideally suited to test our methods are as follows.

1. Bacteria typically possess tens of distinct two-component systems, and mated TCS genes are often chromosomally adjacent (e.g., over 90% of TCS proteins in *Escherichia coli* and *Bacillus subtilis* are chromosomally adjacent). Therefore, *in silico* genomic screening will produce more than 10 times as many TCS proteins as there are sequenced bacteria, which leads to an unprecedented number of putatively interacting protein pairs.

2. A cocrystal has been obtained showing the interaction between structural homologs of prototypical TCS pairs that can be compared to our results (Zapf et al., 2000).

3. Numerous experimental studies support the idea that the signaling networks are predominantly organized in a one-to-one manner (Bijlsma and Groisman, 2003; Howell et al., 2006; Skerker et al., 2005; Verhamme et al., 2002), meaning that interactions between nonmated pairs are rarely found in physiologically relevant conditions—even when two mated pairs

may have close phylogenetic relationships with each other. Therefore, if we understand which amino acid pairs determine the interaction specificity we should be able to predict pairing propensity among all the TCS proteins within a particular organism.

Our goal in this chapter is to detail the procedure for identifying coupled pairs of columns and to illustrate how these pairs are used in the generation of a protein-pair scoring function. We provide the motivations for the choices we have made in developing this approach. We have made a great deal of effort to make our procedure transparent and free of *hidden optimizations* so that this procedure can be applied to other protein interactions as the number of examples of such interactions occurring in databases grows by an order of magnitude. Consequently, we have attempted to describe the method in general terms and refer to the TCS system only as an application in sections following the description of the method.

This chapter is structured as follows: The section titled "Identifying Coupled Columns" consists of four steps that include alignment, functional association, determination, and sample size correction of the MI measure. This section concludes with an overview of the application of this part of the method to identify structural features of the phosphotransfer interaction of TCS proteins. The next section, entitled "Predicting Protein–Protein Interaction," consists of two steps that include the construction of a simple log-likelihood scoring procedure and the subsequent profiling of potential mates within a certain environment, for example, with one organism. Again this section concludes with an application: The genomic profiling of the TCS complement of proteins in *E. coli*. We have attempted to present the procedure as a recipe that can be implemented, with little effort, by nonspecialists in bioinformatics. Perl scripts for all procedures are available upon request.

Identifying Coupled Columns

To study column couplings in a common protein–protein interaction, one must first have a procedure for obtaining a widely inclusive set of homologs and producing multiple sequence alignments for each domain of interest. There are a handful of freely available tools that will aid in this step of the procedure. One semiautomated method for identifying diverse sets of homologs is position-specific iterative BLAST (PSI-BLAST) (Altschul *et al.*, 1997). PSI-BLAST begins with an initial Blastp query that produces a number of hits based on pairwise sequence alignment, which is, itself, determined from generic protein-scoring matrices. A number of these hits—above a user-defined threshold of sequence similarity—can then be used to build a protein-specific scoring matrix. The position-specific scoring matrix can then be used to scan the database again, build a more precise profile, and repeat. This iterative procedure broadens the net of "homology" to include potentially distant relations likely to share the same structure but may not be identified through pair-wise homology searches. When successive iterations fail to produce unique hits, the set of proteins can then be compiled and then processed to produce a MSA with freely available tools (Edgar, 2004; Katoh *et al.*, 2005; Notredame *et al.*, 2000; Thompson *et al.*, 1994). For a large family of commonly studied proteins (including the TCS proteins), it is likely that a hidden Markov model (HMM) for the domain of interest is already available in the Pfam database (Bateman *et al.*, 2002), obtained from methods similar in spirit to the PSI-BLAST method mentioned earlier. Such HMMs can be used to both query a database for homologs of the model and align homologs to the model.

Given the freely available Hmmer software package (Eddy, 1998), an HMM obtained from Pfam (for example) and a protein database, search and alignment of homologous domains consist of the following steps.

1. If "HisKA.hmm" is the name of the HMM used to query the database called "refseq.faa," the following line (in courier font) at the unix command prompt produces a list of hits that meet the threshold criteria "E" in the file "outfile.txt."

```
hmmsearch-EHisKA.hmmrefseq.faa>outfile.txt
```

2. The contents of "outfile.txt" will have an entry for each domain that fits the profile above a user-defined score.
3. The file "outfile.txt" can be converted to a FASTA alignment format using any scripting language.
4. The output of this procedure will be a list of protein names and the associated sequence in an aligned format to be used to detect correlations between columns.

The Pfam accession numbers for the "HisKA" and "Response_reg" HMMs are PF00512 and PF00072, respectively. We henceforth refer to these as the HisKA domain and Rec domain, respectively. For this study, we queried the refseq database release 19 (Pruitt and Maglott, 2001), which contains the sequences from 300 complete bacterial genomes.

Independent Measures of Functional Association

A necessary requirement in assessing coupling between columns in a large library of interacting proteins is the possession of a reliable method for identifying interacting pairs. There is a balance between reliability and the number of functional pairs that can be identified. Ideally, direct experimental conformation of functional pairing in physiological conditions for each pair would be the foundation of library construction. However, current high-throughput methods of large-scale interaction determination are impractical for this purpose due to factors, including cost and the limited number of organisms that can be manipulated in the laboratory.

Sophisticated genomic methods of large-scale determination of functional association exist (Marcotte et al., 1999; von Mering et al., 2005) and seem to be steadily improving as experimental data and growing genomic databases produce an ever more finely resolved knowledge of protein–protein interactions. A discussion of the techniques of generating large libraries of interacting domains is outside the scope of this chapter. It is important to point out, however, that the potential of our method outside of TCS is directly coupled to accurate methods for identifying a large number of interacting proteins.

There are two extremes in which the generation of an interacting pair library is relatively straightforward. Consider, for example, known protein interactions in widely utilized metabolic pathways. If it is known that two proteins interact within an organism and they each have one gene per genome, then functional coupling of these proteins in other organisms may be inferred by virtue of their coexistence in the genomes of those organisms. If, however, there are multiple homologs of each of the interacting proteins (e.g., isozymes), coexistence of these genes in a genome does not yield much information on who interacts with whom. In this case, knowledge of which genes are placed in the same operon would be much more informative. This latter scenario applies to TCS proteins and will be exploited to provide the putatively functionally coupled protein pairs we analyze.

Measurement of Column Coupling by Constraint

We identify substitutional constraint between MSA columns using a commonly employed measure in information theory called mutual information (Shannon and Weaver, 1949). For a schematic example of how MI is used, consider the following. Suppose there are two random variables, i and j, that each can take on any one of two discrete values (e.g., A = {0,1}) with a certain probability. As an illustration consider that the probability mass function (pmf) for the possible values of i, that is, $P_i(A_i)$, is given by $P_i(0) = 0.8$ and $P_i(1) = 0.2$ while the pmf for j, $P_j(A)$, is $P_j(0) = 0.4$ and $P_j(1) = 0.6$. Suppose we are given a set of pairs of i and j values; the number of possible i-j pairs is 4, that is, 1-1; 1-0; 0-1 and 0-0. If the variables i and j are independent in the pairs we were given, then we expect the observed pair distribution $P_{i,j}^{obs}(A_i, A_j)$ to be given by the null joint pmf

$P_{i,j}^{null}(A_i, A_j) = P_i(A_i) \cdot P_j(A_j)$ given a sufficient number of pairs. If variables i and j are *not* independent in the given set of pairs, the *observed* joint pmf will differ from the *null* joint pmf. Heuristically if the observed differs from the null, knowledge of one variable in a pair constrains the pmf of the other member of the pair. The MI is a measurement of this constraint, whose value computed from observed data is

$$M_{i,j}^{obs} = \sum_{\substack{\text{all pairs} \\ A_i, A_j}} P_{i,j}^{obs}(A_i, A_j)\log[P_{i,j}^{obs}(A_i, A_j)/P_i(A_i) \cdot P_j(A_j)]. \qquad (1)$$

Theoretically, $M_{i,j}^{obs} = 0$ if the two variables i and j are independent in the pair set and $M_{i,j}^{obs} > 0$ for all possible relationships among the pairs. The more constraint placed by one variable on the other—or the more *information* one learns about i by knowing j—the higher the value of M. Adapting the aforementioned notation to columns of MSA, i now refers to the ith column of one MSA (e.g., HisKA) and j refers to the jth column of the other MSA (e.g., RR), while A_i refers to the amino acid at the ith column.

To use the MI measure to identify coupling between columns of MSA for interacting protein domains, a feature of MI deserves note. MI measurements between absolutely conserved columns yield a zero value since for those columns $P_{i,j}^{obs} = P_i \cdot P_j = 1$. Although sequence conservation, in itself, is not a statistically relevant method for prediction of protein–protein interactions (Caffrey *et al.*, 2004), conserved patches may couple with additional data to provide insight into how two proteins interact. As a consequence, some meshing of information contained in the MI with conserved columns may provide an additional layer to improve interaction predictions. We do not report on such a meshing in this work.

Accounting for Sample-Size Effects on MI. In the example given earlier we assumed that the true pmfs were known. In practice we have only a finite amount of data to estimate the probability distribution. It is known that MI yields spuriously large values for small sample sizes (Li, 1990). That is, a measurement of MI for a finite number of pairs that are *constructed* to be independent turns out to have values larger than what is expected (i.e., zero). The smaller the sample set, the larger the apparent MI value between independent variables, hence more likely the false interpretation that the variables are mutually correlated. Therefore, sample size effects can have profound influences on the interpretation of MI values near the "small" sample size limit.

In applying MI measurements to proteins—even if we have 1000 unique proteins—we are teetering near the small sample size limit: Each MSA column has a maximum of 21 possible values (20 amino acids and 1 gap).

With such a large potential alphabet size, the sample size effects will depend on the number of amino acids observed in the particular columns being measured. The extent to which the MI measurements are inflated between two columns due to sample size effects will be most pronounced between two columns that possess diverse sets of potential amino acids (only ~5 counts per position if the amino acids are evenly distributed) and less important in columns that are moderately conserved (Li, 1990). Such effects may introduce a structure in the column coupling that stems only from the level of conservation observed in the individual columns and confound the interpretation of the raw MI values (Fodor and Aldrich, 2004).

We write the observed MI, M^{obs}, as the "true" MI value, M^{true}, plus some value, δ, that is generally a function of the sample size in the true probability distributions and goes to zero as the sample size increases. The "true" value is what we are interested in and is defined as the value we would obtain if we had an infinitely large and perfectly representative sample size. Thus,

$$M_{i,k}^{true} = M_{i,j}^{obs} - \delta(P_i, P_j, P_{i,j}^{true}, N, \Upsilon), \tag{2}$$

where P^{true} refers to the true joint distribution, N is the sample size, and Υ represents everything else that we have ignored, such as inhomogeneities in the sample distribution and correlations due to shared phylogeny. Alternatively, the probability distributions can be written in terms of their measured values and their fluctuations, which are dependent on N. We do not know the true distributions and we cannot directly generate fluctuations in the observed distributions because we do not know the "rules" for recreating another sample size with exactly the same characteristics as the one we are studying.

To estimate the effect of small sample size to the observed MI value on a column-by-column basis, we compute the MI from data generated by a null model with no correlation between columns. Keeping the entropy,

$$\sigma_i = - \sum_{all\ A_i} P_i(A_i) \log P_i(A_i), \tag{3}$$

of each individual column fixed, we generate random alignments with the same number of sequences as our sample set. This can be achieved by assuming that the sample set culled from the database defines the true column probability distributions and using these distributions to randomly generate an aligned set of random amino acid sequences. One can alternatively generate random alignments using the HMMs used to generate the database, with the hmmemit program in the HMMER suite (http://hmmer.janelia.org/). Because neither of these methods introduces any correlation between columns, $M^{true} = 0$ for each of the column pairs. Consequently, by measuring the MI between the columns in the null alignments we are directly determining the sample size error for the null model, that is, $M_{i,k}^{null} = \delta(P_i, P_j, P_i \cdot P_j, N, \Upsilon^{null})$, where Υ^{null} encodes the method for generating the null. Because the MI is always positive, the sample size error must also be positive. We assume the positive definiteness of the sample size correction persists for our correlated set and that its value approximates the magnitude of the sample size error. The simplified form for the true value of the MI, which we will henceforth refer to as $S_{i,j}$, is then

$$S_{i,j} \equiv M_{i,j}^{true} \approx M_{i,j}^{obs} - \langle M_{i,j}^{null} \rangle, \tag{4}$$

where brackets denote an average over many (e.g., 200) null alignments containing the same number of sequences as the sample set, and column indices have been reintroduced to stress that the null value is unique from column pair to column pair.

The effect of this correction is to down weight the MI measured between columns with high sequence entropy. It has been shown (Fodor and Aldrich, 2004) that $M_{i,j}^{obs}$ itself performed poorly in isolating contact interactions within proteins when compared to alternative methods. When we discuss the application of the $S_{i,j}$ measure, it will be made clear that small sample size measurements (roughly less than 400 sequences) of the uncorrected MI are essentially equivalent to that observed in null alignments. We will show that as the sample size increases to roughly 2500 proteins the system size correction is reduced to around 10% of the corrected value, and the $S_{i,j}$ values obtained for small sample sizes down to roughly 200 sequences bear a strong overall resemblance to the $M_{i,j}^{obs}$ values obtained for large sample sizes.

Determination of "Significant" Coupling: Self-Consistant
  Threshold Determination

Here we want to capture constraints between columns to identify functional couplings. The ideal data set to use would be one that uniformly fills the sequence space "allowed" by the interaction. In the case of the phosphotransfer interaction in TCS, such an ideal data set would contain the set of all allowed amino acid pairings between the proteins that result in phosphotransfer between the proteins. Furthermore, each of the possible "solutions" to the interaction should be represented evenly throughout the data set. That is, each solution should appear an equal number of times in the functionally coupled set of proteins.

There are fundamental and practical barriers to obtaining a representative data set. One fundamental problem is because the sequence space of allowed solutions is not known; it is difficult to determine the coverage of allowed solutions in the data set. Another fundamental issue of biological importance is the various environments in which the proteins are operating may render a perfectly functional solution in one organism useless in another. Of course the overriding barrier is a practical one: even though the details of all possible allowed solutions are unknown, we can reasonably assume that there are many more allowed solutions than observed solutions. Such is the state of our understanding of the underlying features of a functional interaction that we shall not hope to touch the surface of all possible solutions through current methods. Rather, we content ourselves with issues involving the uniformity of the coverage in the space of solutions defined by the set of functionally coupled protein pairs.

It is unreasonable to assume homogeneous representation, even within the subset of solutions present in the set of functionally coupled proteins. Some protein pairs will uniquely define a particular solution while some solutions will be present in many protein pairs. This may be due to the fact that a particular solution is more resistant to evolutionary change than another. A less interesting but more likely source of inhomogeneous representation is sampling bias in the sequencing of genomes (e.g., there are currently over two times as many sequenced proteobacterial genomes as there are firmicutes). Whatever the source, such heterogeneities will cause overrepresented solutions to appear more stable than underrepresented solutions. How can we isolate the columns that are coupled due to functional constraints from those that arise out of inhomogeneities in the sampling?

One of the main effects of inhomogeneities in the sample set is a global increase in the MI measured between all column pairs. Suppose that in a data set there are multiple copies of the same pair due to redundancy in the database or sequencing of closely related organisms (in our application of the method we have omitted redundant matches). Because the amino acid pairs observed in these copies will be identical and multiply counted they will tend to increase the MI between all variable columns. This inflation is a sampling artifact and has no bearing on the "real" constraint between columns. Of course, the presence of this artifact is not limited to the presence of exact copies. Clusters of sequences with global sequence similarity over a range of scales will contribute to the in global inflation of $S_{i,j}$.

Our strategy to separating out the global artifactual inflation of $S_{i,j}$ is based on the premise that functional sources of coupling between columns result in an increase in the $S_{i,j}$ value *in addition to* the spuriously high values originating from sampling inhomogeneities. Therefore, we view the determination of significant values of $S_{i,j}$ as a signal detection problem in a noisy background. The background "noise" is a supposedly smooth distribution of systematically high, although functionally indeterminate, $S_{i,j}$ values largely determined by heterogeneities in sampling the space of observed

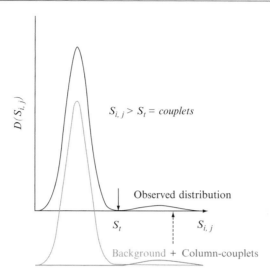

FIG. 1. Schematic diagram of threshold determination. Couplets are defined to be column pairs that have corrected MI values, $S_{i,j}$, above the onset of an anomaly, S_t, in an assumed background distribution. (See color insert.)

solutions. The "signal" consists of the $S_{i,j}$ values that lie significantly above the background distribution.

Most residues in the interacting proteins have little to do with the interaction except to maintain functions of the protein not directly related to the interaction. Therefore, most of the positive $S_{i,j}$ scores are due to sampling artifacts and should not be interpreted on a continuum of relation to the functional coupling between the proteins. The challenge is finding a threshold feature in the distribution of scores to provide a natural determination of a cutoff, above which the scores can reasonably be considered significant. Anomalous counts of large $S_{i,j}$ values in the tail of the background distribution will then correspond to functionally significant column couplings, henceforth referred to as "couplets"; see Fig. 1 for illustration.

Couplet Interaction Networks, Clustering, and "Best Friend"
(BF) Transformation

Couplet networks are defined by a set of nodes and edges. Nodes represent MSA columns. Edges exist between nodes if the two columns form a couplet. This definition of the network is completely independent of the spatial location of the correlated positions. However, the general spatial constraints on the regions involved in an interaction will have consequences on the expected structure of a couplet network. A particular protein–protein interaction is not likely to rely on one couplet. Rather, the interaction will be mediated by a number of spatially localized positions near an interaction patch. How can spatial localization relate to generic character of the couplet network?

To answer this question we first discuss the notion of "correlation chains." Suppose there are two positions in protein A labeled a1, a2 and two positions in protein B labeled b1, b2. Let a1 be in close physical proximity to b1 and a2 be in close proximity to both b1 and b2. In this scenario we would expect an edge set of the network to be a1-b1, a2-b1, and a2-b2. However, we are likely to find a1-b2 also in the network because the fact that a1 is coupled to b1, b1 is coupled to a2, and a2 is coupled to b2 induces an *indirect correlation* between a1 and b2 through a correlation chain. Such correlation chains between residues in spatially localized patches will generally result in more edges among the nodes involved in the patch than are expected through direct contact between residues.

The increased connectivity of regions in couplet networks that correspond to distinct contact patches introduces in principle the need for another layer of analysis, namely clustering of the couplet network to isolate distinct coupled regions of the proteins mediating the interaction. In our study of the TCS proteins, there turn out to be two clusters that are completely disconnected, that is, no edges exist between the clusters. Therefore, no sophisticated clustering techniques were required. For networks that do not break into completely disconnected clusters, methods for determining optimal bipartions of a graph (network) may be used in future studies of other interactions (Sören and Michael Malmros, 1993).

The indirect edges can muddle the interpretation of the results when mapped to structures. To lessen this effect, we assume that indirect correlations will generally be weaker than direct correlations. This assumption allows us to visualize the direct interactions by transforming the couplet network into a directed "best friend" network as follows. For each node "a" in the network, find the coupled node "b" with the highest value of $S_{i,j}$. Make that edge directional by adding an arrow from "a" to "b". Each node will then have one outgoing arrow but the relationships are generally not symmetrical, that is, the BF of node "a" may be node "b" but the BF of "b" may be node "c" and not node "a". The remaining edges in this graph are not directed. This transformation adds depth to the couplet network without changing the connectivity of the graph.

Results for TCS Phosphotransfer Pairs

Across TCS in bacteria the phosphotransfer domain of the histidine kinase (HK) has been categorized into 11 subfamilies based on sequence features (Grebe and Stock, 1999). HK families 1–4, which are all described by one hidden Markov model (designated HisKA), are the most represented out of the 11 subfamilies, composing around 70% of known HK domains. HisKA includes PhoR, EnvZ, NtrB, FixL, and all known examples of eukaryotic histidine kinases. The receiver domains (Rec) also break up into phylogenetic subfamilies that, in large part, correspond to their signaling mate HK family definition (Koretke *et al.*, 2000). In contrast to the HK superfamily, all families of Rec domain are described by one HMM. By searching whole genomes with two hidden Markov models we can, at once, identify and align both a large fraction of the HisKA domains and virtually all of the Rec domains. This convenience, coupled with the fact that many TCS phosphotransfer mates have genes that are adjacent to each other on the chromosome, yields over 2700 pairs of HisKA/Rec domain pairs implied to be functionally mated through chromosomal adjacency. With an interacting-pair library built in this way, we expect to miss many functional pairs but we hope to minimize the number of false-positive associations.

Sample Size Studies of Libraries of TCS Pairs: How Many Is "Enough?". Systematic study of subalignments selected randomly from the 2700 total TCS pairs gives two significant results.

1. Figure 2A shows an overlay of the sample size correction, $M_{i,j}^{null}$ (gray dots), and the corrected MI values, $S_{i,j}$ (blue dots), plotted as a function of increasing average entropy between columns, that is, $x = (\sigma_i + \sigma_j)/2$, in a series of increasingly large subsets of the set of functionally coupled TCS protein pairs (total size 2790). For sample sizes less than 400 sequences, the sample size correction calculated from the null alignments (i.e., $M_{i,j}^{null}$) is comparable to the signal ($S_{i,j}$)(Fig. 2A). Therefore, use of the raw MI signal for sample sizes under 400 is not advised unless one is trying to isolate pairs of columns that have average entropy. At 2700 sequences, the sample size contribution to the raw MI signal diminishes to roughly 20% of the corrected value for the strongly coupled pairs with $S_{i,j} > 0.3$. It is interesting to note that plots of sample size corrections versus mean column entropy (Fig. 2A) are quite similar in character to a previous comparison of covariance methods (Fodor and Aldrich, 2004) using the raw MI, suggesting that their analysis was performed on alignments dominated by sample-size effects.

2. As any useful sample-size correction should be, the corrected values of MI (blue dots) are roughly invariant to sample size changes. Visually inspecting the corrected values in Fig. 2A reveals no overall changes from 200 sequences up to 2700 sequences in scale or overall profile of the $S_{i,j}$ vs x plots. For a more detailed measure, we consider the difference between the particular $S_{i,j}$ value of the superset containing all 2700 pairs and subsets selected randomly from the superset (Fig. 2B). Fluctuations of the derived

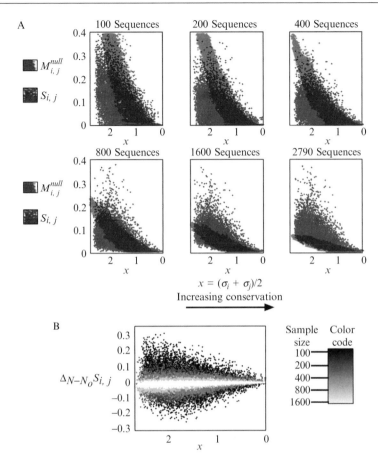

$$x = (\sigma_i + \sigma_j)/2$$

Increasing conservation

FIG. 2. Sample size analysis of corrected MI. (A) Overlaid plots of the corrected MI value $S_{i,j}(x)$ (blue) and the sample-size correction $M_{i,j}^{null}(x)$ (grey) for increasingly large subsets of the set of functionally coupled domains, where x is the average entropy between the columns i and j. For alignments consisting of 100 sequences, the sample-size correction is greater than the corrected values with larger contributions between columns with large entropy, that is, highly variable. The relative invariance of the corrected MI value over the range of sample sizes, even when the correction varies widely, suggests a useful correction. (B) Direct comparison of each value between the largest sample size available and the smaller sets $\Delta_{N-N_0}S_{i,j} \equiv S_{i,j}(N,x) - S_{i,j}(N_0,x)$, where N is the sample size and $N_0 = 2790$. If possible, a perfect correction would show no change between $S_{i,j}$ values for different sample sizes. The fact that the fluctuations around $\Delta_{N-N_0}S_{i,j} = 0$ are roughly symmetric for sample sizes greater than 200 suggests that the deviations from zero are not caused by a systematic problem with the correction. Rather, the fluctuations are likely to be caused by other sources, such as sampling of the different areas of the phylogenetic landscape being represented by the random selection process of the subsets. (See color insert.)

$S_{i,j}$ value around the expected ideal value of zero for 200, 400, 800, and 1600 sequence sets are conservatively ±40, ±20, ±10, and ±4%, respectively, for sample-size corrected values above $S_{i,j} > 0.3$. The fact that these fluctuations are not significantly skewed away from zero suggests that there is no major systemic problem with the correction. Instead the fluctuations are likely an indicator of heterogeneities in the sample set that are more apparent for smaller samples.

Overall, we conclude that the raw MI is a horrible indicator of column coupling for sample sizes less than 400; it remains poor for sizes from 400 to 3000 and is probably just fine for alignments containing more than 3000 sequences. In contrast, the system size-corrected MI appears to give very similar results for alignments containing 400 on up. The quality of the results is discussed in the following sections. We expect that the corrected MI score (i.e., $S_{i,j}$) to perform as well as other covariance-based methods for alignments with less than 400 sequences.

Threshold Location in TCS Pairs and Couplet Network Clustering. As shown in Fig. 3, the distribution of $S_{i,j}$ values between HisKA and Rec pairs exhibits an exponential background (the dashed line) over 2 decades

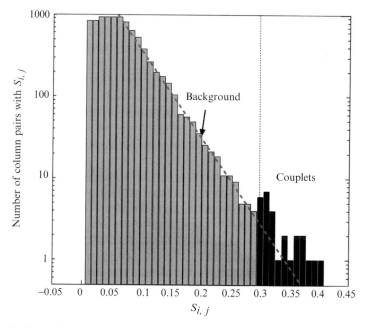

FIG. 3. Determination of the threshold for couplets in TCS proteins. There are more column pairs with $S_{i,j} > 0.3$ than would be expected from the systematic, apparently exponential, background established over two orders of magnitude of counts.

terminated by a "bump" in the tail of the distribution for $S_{i,j} > S_o \approx 0.3$, followed by more high values than would be indicated by the apparent exponential background. There are 20 column couplets ($\sim$0.2% of all possible column pairings) involving 11 unique residue positions in the Rec domains ($\sim$10% of the Rec positions) and 13 unique positions in the HisKA domains ($\sim$20% of the HisKA positions).

The network of couplets in TCS pairs falls naturally into two separated clusters. The first cluster (Fig. 4A) consists of the following five Rec positions (using the residue numbers of Spo0F in circles): 14, 15, 18, 21, 22, and 8 HisKA positions (using the residue numbers of HK853 in squares): 267, 268, 271, 272, 275, 291, 294, 298. The second cluster (Fig. 4B) consists of five Rec positions 56, 84, 87, 90, 99 and four HisKA positions 251, 252, 257, 264.

Mapping Couplets to an Exemplary Structure: Spo0B/Spo0F Is an Exemplary Structure for TCS Pairs. The gold standard for covariance analysis has historically been the degree to which the covarying residues are in close proximity to each other when mapped to an exemplary structure (Atchley *et al.*, 2000; Clarke, 1995; Fodor and Aldrich, 2004; Kass and Horovitz, 2002; Pazos *et al.*, 1997). The current state of knowledge of TCS proteins places us in a unique situation in this regard. There is no known structure of a complex between a SK domain that is described by the HisKA HMM and a domain described by the Rec HMM. However, there *is* a cocrystal structure of a Rec domain (Spo0F) and an HisKA structural homolog (Spo0B) (Zapf *et al.*, 2000). The hypothesis is that the interaction observed in the cocrystal structure is exemplary of the interactions between HisKA and Rec domains (which have numerous structural examples independently). To emulate how this method would be used if no complex structure exists, we will first map the couplets to the individual structures to investigate the possibility of creating a structural model for the generic HisKA/Rec interactions. We then take the positions in the Spo0B structural homolog that are reliably aligned to the HisKA HMM and map the corresponding couplets to the Spo0B/Spo0F structure.

The recently obtained crystal structure of the HisKA domain HK853 found in *Thermatoga maritima* provides the best example structure in the HisKA family (Marina *et al.*, 2005). Numerous structures of the receiver domain have been obtained, including Spo0F, which exists as a single domain that plays the role of a phosphotransferase between the HisKA domains KinA, KinB, KinC, KinD, and KinE and the phosphotransferase Spo0B in the sporulation phospho relay in *Bacillus subtilis*. If we take HK853 and Spo0F as the representatives of their respective structures, align their sequences with their respective HMMs, and map the positions that correspond to the couplets obtained in steps 1–4 to the respective structures, we obtain the result shown in Fig. 4C. The colorings of the

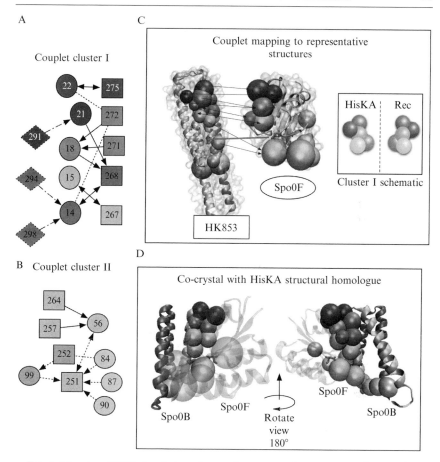

FIG. 4. Mapping of TCS phosphotransfer couplets to exemplary structures. (A and B) The two distinct clusters formed in the couplet network. The assignment of arrows (or lack thereof) in the networks is achieved by the "best friend" transformation described in the text. Circles correspond to positions in the Rec domain and are numbered according to the SpoOF protein. Rectangles represent positions in the HisKA domain and are numbered according to the HK853 kinase (Marina *et al.*, 2005). Diamond shapes denote HisKA residues that cannot be reliably mapped to the SpoOB phosphotransferase structure through the HisKA HMM. Solid edges between nodes denote residue pairs that have a minimum distance d < 8 Å when mapped to the SpoOB/SpoOF complex (Zapf *et al.*, 2000), dotted edges d > 8 Å, and dash dotted edges are unknown. Coloring for the nodes is achieved by selecting best friends of the HisKA nodes; if they share a best friend they are colored identically. (C) Mapping of couplet network to HK853 and SpoOF with residue colorings (in bead format using VMD 1.8.5 [Humphrey *et al.*, 1996]) derived from node coupling coloring. Edges correspond to edges in the couplet diagram, and the thick red edge connects the His-Asp phosphotransfer residues. (Inset) Schematic of the mirror symmetry of couplets. (D) Front and back views of couplets mapped to the phosphotransfer complex of phosphotransferases SpoOB/SpoOF. In both C and D the His-Asp residues where the phosphotransfer takes place are shown in red "licorice" representation. (See color insert.)

nodes of the BF network (Fig. 4A and B) are determined by tracing the "best friend" of each HisKA position (those in boxes). The color scheme matches the residues shown in bead representation in Fig. 4C and D.

As suggested by Fig. 4C, residues in cluster I on the Rec domain form a pattern in apparent registry with the cluster I residues on the HisKA domain. A model for the interaction suggested by these results would be that the mirror-symmetrical cluster I regions (inset of Fig. 4C) represent the docking site of the two proteins and define the interaction orientation. This model is consistent with the close proximity of the conserved His and Asp residues (in "licorice") required for phosphotransfer.

To test this model of interaction, we consider the SpoOF/SpoOB complex. The overall alignment of SpoOB with the HisKA HMM is poor. However, in the region near the conserved histidine residue (from -10 to $+15$ where the conserved His lies at position 0), the sequence aligns well to the HisKA HMM. If we align the residues near the SpoOB histidine to the HisKA HMM and map the coupled residues to the SpoOB/SpoOF structure, we find that the model suggested by visual inspection of the mapping to exemplary structures SpoOF and HK853 is represented with striking clarity in the SpoOB/SpoOF complex (Fig. 4D) as the similarly colored positions are, indeed, in close contact.

Of the 16 total couplets that can be mapped to the SpoOB/SpoOF complex, 9 have minimum distances between side chains less than 8 Å. The 7 couplets associated with two HisKA positions (251 and 252) in cluster II cannot be understood through direct contact interaction as they are coupled to Rec positions that are up to ~30 Å apart in the SpoOB/SpoOF complex. Furthermore, these residues in the HK853 structure have side chains involved in the homodimer core of the HK853 molecule. The significance of these long-distance couplings shall be discussed elsewhere (White $et\ al.$, 2006).

Predicting Protein–Protein Interaction

The procedure for identifying couplets presented in steps 1–4 is descriptive on the column-to-column level in families of interacting proteins. Results from this approach illuminate structural features of a shared mode of interaction common to these families. Another approach is needed to address the question of whether two specific members of this family interact. As a first attempt, we apply the following simple log-likelihood scoring procedure between specific amino acid pairs in the couplets identified earlier.

The score $\zeta_{i,j}(A_i, A_j)$ for a particular amino acid pair (A_i, A_j) in the couplet compares observed frequency of these pairs in the set of functionally coupled protein pairs to the null assumption of statistical independence given by

$$\zeta_{i,j}(A_i, A_j) = \log\left[\frac{P_{i,j}(A_i, A_j)}{P_i(A_i) \cdot P_j(A_j)}\right] \tag{5}$$

where probabilities P are as defined previously. Functional coupling between the amino acid pair (A_i, A_j) would be manifested in a positive correlation in their occurrences, hence a positive score.

In the actual applications of this scoring function, it is possible that a particular pair of amino acids not present in the functionally coupled library would be observed in a pair of proteins being scored, causing a singularity in score value. This may be significant if the amino acids are each common in their respective columns. We account for this problem with two ad-hoc bounds. The first is a lower bound on the joint probability distribution, $P_{i,j}(A_i, A_j) \geq \frac{1}{n}$ where n is the number of functionally coupled pairs in the library. In essence, this is a kind of pseudo count (Durbin *et al.*, 1998) for each possible pairing. This bound sets a limit on the score that can be contributed by a pairing that is not observed in the library but maintains the obviously significant relationship between nonobservance of amino acid pairing between amino acids that are nonetheless common in their respective columns. To reduce the effect of overvaluing rare occurrences of amino acids within a column, we set a second bound on which amino acids will be scored, by bounding the individual PMFs to $P_i(A_i) \geq \frac{1}{\sqrt{n}}$. As a result, both amino acids (A_i, A_j) must occur in the MSA $\sim$50 times or more in our 2700 member library in order for the couplet ij to be scored; otherwise, the couplet receives score zero, that is, it does not contribute to interaction. To maintain transparency, we have not attempted to optimize these bounds for a particular output.

To obtain a score between two proteins X and Y, we make the simplifying assumption that all residue pairings between couplets are independent and equally important. The total score is then obtained by summing individual scores from all couplets to obtain the total protein pair score $\psi(X,Y)$:

$$\psi(X,Y) = \sum_{(i,j) \subset C} \zeta_{i,j}(A_i^X, A_j^Y), \tag{6}$$

where C is the set of all couplets, that is, column pairings that have an above threshold MI score, $S_{i,j} > S_0$ (see Fig. 3). In HisKA/Rec pairs, this sum will have 20 terms corresponding to the edges in the couplet network shown in Fig. 4A and B.

It should be noted that the scoring procedure is "raw" in the sense that it does not account for sampling effects as the couplet identification procedure does. We believe that by isolating the couplets through a procedure that does incorporate such effects, we have captured the essence of what can be described with this simple method. In the future, efforts to incorporate approaches to account for sampling effects will, no doubt, further improve the descriptive abilities of the scoring function.

Coenvironment Interaction Profiling

Considering that the scoring procedure takes no account of biomolecular details of the interactions, we expect the score to have, at best, a coarse-grained correspondence to association constants between the various possible pairs. Therefore, instead of attempting to map the scores to interaction energetics, we determine interaction propensity through comparison with all possible interacting partners within a shared environment.

The procedure for this comparison is simple and qualitative. Suppose there are 10 members of protein family X (e.g., HisKA) and 10 members of protein family Y (e.g, Rec) within one genome. For each member of X, compute the scores between X and all members of the Y family, generating $\psi(X,Y_1)$, $\psi(X,Y_2)$, ..., $\psi(X,Y_{10})$. Pairs with low-ranking scores are unlikely to interact as they have a significant number of amino acid pairings that are more rare in the set of functionally coupled pairs than would be expected from the null hypothesis. Pairs that have the highest score with a significant separation to the next highest score are likely to be monogamous functional mates. Proteins that have a small separation between the highest and the next highest scores are more likely to be promiscuous.

This genomic profiling method assumes that all possible pairs are present at the same time and have access to each other in the cell with one exception: Response regulator domains that lie on the same gene as an HisKA domain (so-called "hybrids") are not included in the profiling procedure. Given that most HisKA domains are membrane bound, Rec domains that are members of hybrids are not likely to have physical access to other SKs in the cell. The next section reports on the scores between cohybrid mates but omits Rec domains from consideration when determining which Rec domains are mated with other HisKA domains. There are, of course, other means of achieving interaction specificity that are not captured by the scoring function, such as nonoverlapping periods of protein presence in the cell and localization constraints, which are beyond the scope of this work.

Example of Bioinformatic Profiling in Escherichia coli

As a test case we have performed the bioinformatic profiling procedure to the TCS systems in the gram-negative bacterial representative *E. coli*. In the *E. coli* K12 strain there are 20 domains described by the HisKA HMM (5 of these 20 are members of hybrids) and 32 Rec domains that are not members of hybrids. Because there are kinases that are not described by the HisKA HMM, most notably NarX and NarQ, we profile the TCS interactions by scoring all Rec domains within the organisms against each HisKA domain.

The top 9 Rec domain scores with each HisKA domain are shown in Fig. 5. Of the 15 nonhybrid kinases, 14 have, as their top scoring Rec domain, their known functional mate (identified through associations in the manually drawn pathways in the KEGG database [Kanehisa, 1997; Kanehisa et al., 2006]). The graphical overview (Fig. 6) of all the scores normalized by the maximum score for each kinase shows that most of the 32 possible scores are negative. The average separation between the highest score and the next highest score in the 14 known mates with the top score is 49%. These results suggest that the scoring function is accurately reflecting both mating sensitivity and specificity.

Results for the hybrid kinases are not as striking but they remain consistent with the notion that the scoring function is capturing essential features of the interaction. The five hybrid kinases are ArcB, BarA, EvgS, RcsC, and TorS with the sign $\pm$ and rank x, out of the possible 32 cytoplasmic Rec domains shown as ($x/32 \pm$). Results for four of the five known pairs are as follows: ArcB/ArcA ($8/32$ +), BarA/UvrY ($4/32$ +), EvgS/EvgA ($7/32$ +), TorS/TorR ($14/32$ −). The phosphotransfer pair of the hybrid kinase RcsC is a phosphotransferase RcsD, which is not described by the HisKA domain HMM and is therefore beyond the purview of this method. RcsC is part of a proposed phospho relay, which presumably involves the transfer of phosphate through a RcsC–RcsD–RcsB pathway (Majdalani et al., 2005; Takeda et al., 2001). Interestingly, the RcsC/RcsB is the top-ranked score for profiling of the HisKA domain of RcsC, suggesting that phosphate can be transferred directly from RcsC to RcsB (i.e., a feedforward loop) or that the molecular determinants of phosphotransfer specificity are transferred along the His-Asp-His-Asp phospho-relay chain (analogous to the correlation chains discussed earlier). RcsB is recently conjectured to be phosphorylated by other kinases. These interactions make the Rcs signaling pathway rather complex.

We have also scored Rec domains that lie on the hybrids with their same-gene counter part and dub these scores "self-scores." The self-scores of the five hybrids are all positive and are ranked ($x/(32 + 1)$) as follows: ArcB ($6/33$ +), BarA ($3/33$ +), EvgS ($9/33$ +), RcsC ($8/33$ +), TorS ($9/33$ +). It is generally known that overexpression of nonmated phosphotransfer pairs can induce phosphotransfer (Bijlsma and Groisman, 2003). The ability to overcome physiologically tuned specificity by increasing concentration suggests that in cases of high local concentration—as is the case between cohybrid phosphotransfer domains—the constraints for maintenance of a

Genomic profiling of *E. coli*

Top SK-RR pairing scores (un-normalized)

SK	atoS	Ψ	baeS	Ψ	basS	Ψ	cpxA	Ψ	creC	Ψ
	atoC	10.5	baeR	12.8	basR	14.4	cpxR	6.5	creB	3.7
	zraR	8.4	ompR	8.3	ygix	11.3	rstA	5.2	kdpE	−0.6
	glnG	8	cpxR	1.5	phoP	7.7	ompR	3.7	phoP	−0.9
Response reg.	yfhA	5.7	kdpE	1	narP	0.1	yedW	2	fimZ	−1.2
	citB	5.1	rstA	0.6	creB	−0.9	baeR	1.5	cheY	−2.2
	cheY	0	phoB	−1.2	kdpE	−1	creB	0.7	hnr	−2.4
	dcuR	−0.9	phoP	−1.6	rcsB	−1	torR	−0.5	yfhA	−2.6
	ypdB	−1.1	citB	−2.2	uvrY	−1.3	phoP	−1.7	evgA	−2.6
	yehT	−3	creB	−2.3	torR	−3.7	phoB	−2.8	basR	−2.7

SK	cusS	Ψ	envZ	Ψ	glnL	Ψ	kdpD	Ψ	phoR	Ψ
	cusR	18.1	ompR	8.2	glnG	5.6	kdpE	5.6	phoB	12
	yedW	4.3	rstA	3.9	zraR	−0.5	baeR	3.5	kdpE	4.9
	ygiX	0.4	baeR	3.2	torR	−0.6	yedW	2.9	citB	1.7
Response reg.	basR	−1.2	cpxR	2.3	citB	−0.7	creB	2.7	ompR	1.7
	phoP	−1.5	yedW	0.3	arcA	−1.1	ygiX	1.8	rstA	1.4
	baeR	−1.7	phoB	−4.6	hnr	−1.8	ompR	1.4	cpxR	1.3
	rcsB	−3	creB	−4.7	yfhA	−1.9	phoP	1.2	creB	0.4
	narL	−4.7	dcuR	−5.9	phoB	−2.2	cusR	1.1	baeR	−0.4
	uhpA	−4.8	cusR	−6	cheY	−3	phoB	0.4	phoP	−2.6

SK	rstB	Ψ	yedV	Ψ	yfhK	Ψ	ygiY	Ψ	zraS	Ψ
	rstA	9.3	yedW	7.7	phoP	2.8	ygiX	15.5	zraR	10.7
	ompR	4.2	cusR	3.5	cpxR	2.2	basR	11.5	atoC	10.6
	baeR	2	hnr	−1	ompR	1.6	phoP	8.3	glnG	8.5
Response reg.	cpxR	1.3	dcuR	−1.9	kdpE	0.4	cusR	1.5	yfhA	4.3
	dcuR	−0.3	ypdB	−2.2	fimZ	−0.1	kdpE	−0.5	ypdB	2.5
	hnr	−2.2	narL	−2.4	phoB	−0.3	torR	−1.2	cheY	1.9
	yfhA	−3.4	ygix	−2.6	basR	−0.6	rcsB	−1.4	citB	1.2
	yedW	−3.7	uvrY	−2.7	cheY	−0.8	narL	−1.4	yehT	0.7
	citB	−3.9	rcsB	−3.2	hnr	−1.2	creB	−1.4	hnr	−2.1

SK/RR Hybrids

SK	arcB	Ψ	barA	Ψ	evgS	Ψ	rcsC	Ψ	torS	Ψ
	uhpA	7.8	uhpA	5.1	cheY	3	rcsB	10.2	uhpA	9.1
	rcsB	6.4	hnr	4.9	uvrY	2.7	uhpA	8.1	rcsB	8.7
	uvrY	5.6	barA*	4.8	rcsB	2.6	uvrY	5	uvrY	6
Response reg.	narP	3.9	rcsB	3.6	uhpA	2.2	fimZ	2.7	narP	5
	cheY	3.8	uvrY	3	cheB	1.2	cheY	2.7	cheY	4.1
	arcB*	3.6	evgA	2.4	hnr	0.9	narP	2.6	dcuR	2.5
	cheB	2.7	cheB	2.2	evgA	0.8	cheB	0.9	evgA	1.3
	evgA	2.5	narP	1.8	dcuR	0.7	rcsC*	0.5	narL	1.1
	arcA	2.3	cheY	1.4	evgS*	0.5	evgA	0.3	torS*	0.8

FIG. 5. Top nine Rec domain scores of all HisKA domains in *E. coli*. Raw scores are listed in ranked order with the highest score on the top for each kinase described by the HisKA HMM. Response regulator mates of kinases (known through KEGG associations) are shown in bold. Bold genes with asterisks denote self-scores of hybrids (see text). Genes yfhK and torS are the only kinases that do not have their known matches in the top nine scores.

functional phosphotransfer pair are not as stringent as those between diffusive partners. This release of constraint for cohybrid phosphotransfer domains would result in generally lower scores but not "forbidden" scores, that is, negative scores. The *E. coli* hybrid self-scores are consistent with this

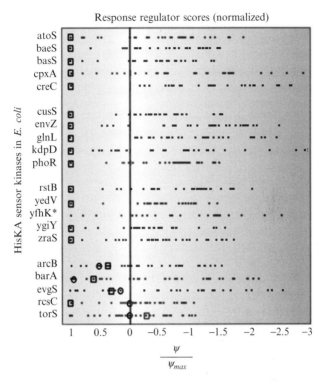

FIG. 6. Graphic representations of all Rec domain scores with each HisKA domain in *E. coli*. For each HisKA domain, scores are normalized by the maximum score for that domain. The blue (red) region represents all positive (negative) scores. Each dot is a Rec domain score. Dots with surrounding squares represent the known mate. Dots with circles surrounding are self-scores of the hybrid TCS proteins. The known mate of yfhK is the most negative score off the scale of this graph. (See color insert.)

rational, as they are all positive, but not the top score, as is found in the nonhybrid HisKA profiles.

The scoring function appears to be in reasonable accord with what is known about the TCS phosphotransfer landscape of *E. coli*. In 23 of the 25 known or presumed HisKA/Rec domain pairings (including self-scores), the score of the known functional mate is positive and, in most cases, the top score. Overall we believe that the scoring function can be used as a tool to help guide experimentalists in the elucidation of TCS signaling networks in organisms whose phosphotransfer mates are not identified easily through chromosomal adjacency, such as *Myxococcus xanthus* and *Nostoc puncti- forme*, and perhaps uncover additional connections in more thoroughly characterized organisms.

Summary

We have described a method that leverages the gross knowledge of interaction to provide structural details of the interaction, such as the docking orientation and contact sites. This knowledge is then used to create a scoring procedure that allows prediction of unexpected interactions, as well as expected negative interactions. The requirements of this method include the ability to reliably align the sequences of the proteins involved in the interaction and a large sequence library of unique interacting pairs.

As an example of the utility of the method, we applied it to TCS proteins and gained valuable insight into this ubiquitous and well-studied mechanism. We have provided substantial support for the notion that the interaction mode captured in the Spo0B/Spo0F structure is exemplary of the phosphotransfer interaction across a large fraction of TCS proteins by identifying couplets from a large collection of TCS mates that map directly to the Spo0B/Spo0F interaction region. Furthermore, when these couplets are scored with a simple, log-likelihood procedure we are able to reflect the known phosphotransfer landscape of *E. coli*. With the overall correspondence between our approach and experimental data—both structural and biochemical—we have reduced the question of interaction pair specificity in TCS proteins to 20 couplets involving 22 positions across the interacting proteins. In the future, as more data become available, we believe this approach will prove a straightforward and inexpensive guide to the experimental elucidation of other protein–protein interactions.

Acknowledgments

This work is supported by the NIH through an Admin Supplement for the Study of Complex Biological Systems to Grant GM19416 (awarded to J.A.H.) and the National Academies/Keck Futures Initiative Award (to T.H.). R.A.W. acknowledges a NSF postdoctoral fellowship award DBI-0532925 and the hospitality of the NSF PFC-sponsored Center for Theoretical Biological Physics (Grants PHY-0216576 and PHY-0225630).

References

Altschuh, D., Lesk, A. M., Bloomer, A. C., and Klug, A. (1987). Correlation of co-ordinated amino acid substitutions with function in viruses related to tobacco mosaic virus. *J. Mol. Biol.* **193**, 693–707.

Altschul, S. F., Madden, T. L., Schaffer, A. A., Zhang, J., Zhang, Z., Miller, W., and Lipman, D. J. (1997). Gapped BLAST and PSI-BLAST: A new generation of protein database search programs. *Nucleic Acids Res.* **25**, 3389–3402.

Atchley, W. R., Wollenberg, K. R., Fitch, W. M., Terhalle, W., and Dress, A. W. (2000). Correlations among amino acid sites in bHLH protein domains: An information theoretic analysis. *Mol. Biol. Evol.* **17**, 164–178.

Bateman, A., Birney, E., Cerruti, L., Durbin, R., Etwiller, L., Eddy, S. R., Griffiths-Jones, S., Howe, K. L., Marshall, M., and Sonnhammer, E. L. (2002). The Pfam protein families database. *Nucleic Acids Res.* **30,** 276–280.

Bijlsma, J. J., and Groisman, E. A. (2003). Making informed decisions: Regulatory interactions between two-component systems. *Trends Microbiol.* **11,** 359–366.

Caffrey, D. R., Somaroo, S., Hughes, J. D., Mintseris, J., and Huang, E. S. (2004). Are protein-protein interfaces more conserved in sequence than the rest of the protein surface? *Protein Sci.* **13,** 190–202.

Clarke, N. D. (1995). Covariation of residues in the homeodomain sequence family. *Protein Sci.* **4,** 2269–2278.

Durbin, R., Eddy, S. R., Krogh, A., and Mitchison, G. (1998). "Biological Sequence Analysis: Probabilistic Models of Proteins and Nucleic Acids." Cambridge University Press.

Eddy, S. R. (1998). Profile hidden Markov models. *Bioinformatics* **14,** 755–763.

Edgar, R. (2004). MUSCLE: A multiple sequence alignment method with reduced time and space complexity. *BMC Bioinformatics* **5,** 113.

Enright, A. J., Iliopoulos, I., Kyrpides, N. C., and Ouzounis, C. A. (1999). Protein interaction maps for complete genomes based on gene fusion events. *Nature* **402,** 86–90.

Fodor, A. A., and Aldrich, R. W. (2004). Influence of conservation on calculations of amino acid covariance in multiple sequence alignments. *Proteins Structure Function Bioinformatics* **56,** 211–221.

Göbel, U., Sander, C., Schneider, R., and Valencia, A. (1994). Correlated mutations and residue contacts in proteins. *Proteins Structure Function Genetics* **18,** 309–317.

Goh, C.-S., Bogan, A. A., Joachimiak, M., Walther, D., and Cohen, F. E. (2000). Co-evolution of proteins with their interaction partners. *J. Mol. Biol.* **299,** 283–293.

Grebe, T. W., and Stock, J. B. (1999). The histidine protein kinase superfamily. *Adv. Microb. Physiol.* **41,** 139–227.

Hoch, J. A. (2000). Two-component and phosphorelay signal transduction. *Curr. Opin. Microbiol.* **3,** 165–170.

Howell, A., Dubrac, S., Noone, D., Varughese, K. I., and Devine, K. (2006). Interactions between the YycFG and PhoPR two-component systems in *Bacillus subtilis*: The PhoR kinase phosphorylates the non-cognate YycF response regulator upon phosphate limitation. *Mol. Microbiol.* **60,** 535.

Kanehisa, M. (1997). A database for post-genome analysis. *Trends Genet.* **13,** 375–376.

Kanehisa, M., Goto, S., Hattori, M., Aoki-Kinoshita, K. F., Itoh, M., Kawashima, S., Katayama, T., Araki, M., and Hirakawa, M. (2006). From genomics to chemical genomics: New developments in KEGG. *Nucleic Acids Res.* **34,** D354–D357.

Kass, I., and Horovitz, A. (2002). Mapping pathways of allosteric communication in GroEL by analysis of correlated mutations. *Proteins Struct. Funct. Genet.* **48,** 611–617.

Katoh, K., Kuma, K.-i., Toh, H., and Miyata, T. (2005). MAFFT version 5: Improvement in accuracy of multiple sequence alignment. *Nucleic Acids Res.* **33,** 511–518.

Koretke, K. K., Lupas, A. N., Warren, P. V., Rosenberg, M., and Brown, J. R. (2000). Evolution of two-component signal transduction. *Mol. Biol. Evol.* **17,** 1956–1970.

Li, W. (1990). Mutual information functions versus correlation functions. *J. Stat. Phys.* **V60,** 823–837.

Liolios, K., Tavernarakis, N., Hugenholtz, P., and Kyrpides, N. C. (2006). The Genomes On Line Database (GOLD) v.2: A monitor of genome projects worldwide. *Nucleic Acids Res.* **34,** D332–D334.

Lockless, S. W., and Ranganathan, R. (1999). Evolutionarily conserved pathways of energetic connectivity in protein families. *Science* **286,** 295–299.

Majdalani, N., Heck, M., Stout, V., and Gottesman, S. (2005). Role of RcsF in signaling to the Rcs phosphorelay pathway in *Escherichia coli*. *J. Bacteriol.* **187,** 6770–6778.

Marcotte, E. M., Pellegrini, M., Ng, H.-L., Rice, D. W., Yeates, T. O., and Eisenberg, D. (1999). Detecting protein function and protein-protein interactions from genome sequences. *Science* **285,** 751–753.

Marina, A., Waldburger, C. D., and Hendrickson, W. A. (2005). Structure of the entire cytoplasmic portion of a sensor histidine-kinase protein. *EMBO J.* **24,** 4247–4259.

Ninfa, A. J., Ninfa, E. G., Lupas, A. N., Stock, A., Magasanik, B., and Stock, J. (1988). Crosstalk between bacterial chemotaxis signal transduction proteins and regulators of transcription of the Ntr regulon: Evidence that nitrogen assimilation and chemotaxis are controlled by a common phosphotransfer mechanism. *Proc. Natl. Acad. Sci. USA* **85,** 5492–5496.

Notredame, C., Higgins, D. G., and Heringa, J. (2000). T-coffee: A novel method for fast and accurate multiple sequence alignment. *J. Mol. Biol.* **302,** 205–217.

Overbeek, R., Fonstein, M., D'Souza, M., Pusch, G. D., and Maltsev, N. (1999). The use of gene clusters to infer functional coupling. *Proc. Natl. Acad. Sci. USA* **96,** 2896–2901.

Parkinson, J. S. (1993). Signal transduction schemes of bacteria. *Cell* **73,** 857–871.

Pazos, F., Helmer-Citterich, M., Ausiello, G., and Valencia, A. (1997). Correlated mutations contain information about protein-protein interaction. *J. Mol. Biol.* **271,** 511–523.

Pazos, F., and Valencia, A. (2001). Similarity of phylogenetic trees as indicator of protein-protein interaction. *Protein Eng.* **14,** 609–614.

Pruitt, K. D., and Maglott, D. R. (2001). RefSeq and LocusLink: NCBI gene-centered resources. *Nucleic Acids Res.* **29,** 137–140.

Ramani, A. K., and Marcotte, E. M. (2003). Exploiting the co-evolution of interacting proteins to discover interaction specificity. *J. Mol. Biol.* **327,** 273–284.

Shannon, C. E., and Weaver, W. (1949). "The Mathematical Theory of Communication." University of Illinois Press.

Skerker, J. M., Prasol, M. S., Perchuk, B. S., Biondi, E. G., and Laub, M. T. (2005). Two-component signal transduction pathways regulating growth and cell cycle progression in a bacterium: A system-level analysis. *PLoS Biol.* **3,** e334.

Sören, H., and Michael Malmros, S. (1993). The optimal graph partitioning problem. *OR Spectrum* **V15,** 1–8.

Stock, A. M., Robinson, V. L., and Goudreau, P. N. (2000). Two-component signal transduction. *Annu. Rev. Biochem.* **69,** 183–215.

Stock, J. B., Ninfa, A. J., and Stock, A. M. (1989). Protein phosphorylation and regulation of adaptive responses in bacteria. *Microbiol. Rev.* **53,** 450–490.

Suel, G. M., Lockless, S. W., Wall, M. A., and Ranganathan, R. (2003). Evolutionarily conserved networks of residues mediate allosteric communication in proteins. *Nat. Struct. Biol.* **10,** 59–69.

Takeda, S.-i., Fujisawa, Y., Matsubara, M., Aiba, H., and Mizuno, T. (2001). A novel feature of the multistep phosphorelay in *Escherichia coli*: A revised model of the RcsC→YojN→RcsB signalling pathway implicated in capsular synthesis and swarming behaviour. *Mol. Microbiol.* **40,** 440–450.

Thompson, J. D., Higgins, D. G., and Gibson, T. J. (1994). CLUSTAL W: Improving the sensitivity of progressive multiple sequence alignment through sequence weighting, position-specific gap penalties and weight matrix choice. *Nucleic Acids Res.* **22,** 4673–4680.

Verhamme, D. T., Arents, J. C., Postma, P. W., Crielaard, W., and Hellingwerf, K. J. (2002). Investigation of *in vivo* cross-talk between key two-component systems of *Escherichia coli*. *Microbiology* **148,** 69–78.

von Mering, C., Jensen, L. J., Snel, B., Hooper, S. D., Krupp, M., Foglierini, M., Jouffre, N., Huynen, M. A., and Bork, P. (2005). STRING: Known and predicted protein-protein associations, integrated and transferred across organisms. *Nucleic Acids Res.* **33,** D433–D437.

Winker, S., Overbeek, R., Woese, C. R., Oisen, G. J., and Pfluger, N. (1990). Structure detection through automated covariance search. *Comput. Appl. Biosci.* **6,** 365–371.

Woese, C. R., Gutell, R., Gupta, R., and Noller, H. F. (1983). Detailed analysis of the higher-order structure of 16S-like ribosomal ribonucleic acids. *Microbiol. Mol. Biol. Rev.* **47,** 621–669.

Zapf, J., Sen, U., Madhusudan Hoch, J. A., and Varughese, K. I. (2000). A transient interaction between two phosphorelay proteins trapped in a crystal lattice reveals the mechanism of molecular recognition and phosphotransfer in signal transduction. *Structure* **8,** 851–862.

[5] Sporulation Phosphorelay Proteins and Their Complexes: Crystallographic Characterization

By KOTTAYIL I. VARUGHESE, HAIYAN ZHAO,
VIDYA HARINI VELDORE, and JAMES ZAPF

Abstract

Bacteria use two-component systems to adapt to changes in environmental conditions. In response to deteriorating conditions of growth, certain types of bacteria form spores instead of proceeding with cell division. The formation of spores is controlled by an expanded version of two-component systems called the phosphorelay. The phosphorelay comprises a primary kinase that receives the signal/stimulus and undergoes autophosphorylation, followed by two intermediate messengers that regulate the flow of the phosphoryl group to the ultimate response regulator/transcription factor. Sporulation is initiated when the level of phosphorylation of the transcription factor reaches a critical point. This chapter describes efforts to understand the mechanism of initiation of sporulation at the molecular level using X-ray crystallography as a tool. Structural analyses of individual members, as well as their complexes, provide insight into the mechanism of phosphoryl transfer and the origin of specificity in signal transduction.

Introduction

Bacteria monitor the surroundings and adapt to fluctuations in environmental and growth conditions using two-component/phosphorelay systems for signal detection and response (Hoch and Silhavy, 1995; Parkinson and Kofoid, 1992). We have been studying the signaling pathway that controls the onset of sporulation in *Bacillus subtilis*. When conditions for growth turn unfavorable, *B. subtilis* forms spores (Fig. 1). The initiation of sporulation is controlled by an expanded version of the two-component system called phosphorelay (Fig. 2) (Burbulys *et al.*, 1991), consisting of four main components: a histidine kinase, a secondary messenger (Spo0F), a phosphotransferase (Spo0B), and a transcription factor (Spo0A). In the sporulation pathway, signals are processed by five different histidine kinases, but the bulk of the signal input is through the primary kinase, Kin-A (Jiang *et al.*, 2000). The activities of kinases are modulated by a variety of specific signals, but the nature of the signals is unknown.

METHODS IN ENZYMOLOGY, VOL. 422
Copyright 2007, Elsevier Inc. All rights reserved.
0076-6879/07 $35.00
DOI: 10.1016/S0076-6879(06)22005-6

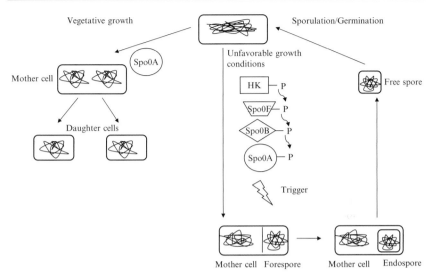

FIG. 1. Vegetative growth and sporulation. In the sporulation phosphorelay of *Bacillus*, the histidine kinases pass the phosphoryl group to an intermediate response regulator, Spo0F, and subsequently the phosphotransferase, Spo0B, transfers it to the transcription factor, Spo0A. When the intracellular concentration of Spo0A-P reaches a critical value, sporulation is initiated. Otherwise the bacterium proceeds with vegetative growth.

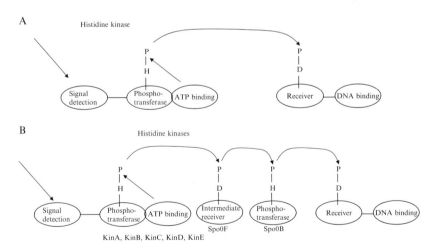

FIG. 2. Domain organization and signal transmission in two-component/phosphorelay systems. (A) A prototypical two-component system. (B) Phosphorelay that controls the sporulation in *B. subtilis*. Signals are processed by five different kinases, KinA, KinB, KinC, KinD, and KinE, in this phosphorelay.

Activation of the signaling pathway involves four phosphotransfer reactions, and the first one is autophosphorylation of a conserved histidine residue in the kinase, dependent on the conversion of a bound ATP to ADP. The kinase, then, transfers the phosphoryl group to an aspartate on Spo0F, which transfers it to a histidine residue on Spo0B. Subsequently, Spo0B transfers it to an aspartate on Spo0A. Therefore, the order of the phosphoryl flow is ATP→His→Asp→His→Asp. The initiation of sporulation is dictated by the degree of phosphorylation of the transcription factor Spo0A (Ferrari *et al.*, 1985; Fujita and Losick, 2005).

Sensor kinases are modular in nature and comprise two parts: an N-terminal stimulus detection domain and an autokinase domain consisting of a phosphotransferase subdomain containing an active site histidine residue and an ATP-binding subdomain (Fig. 2). The autokinase domains are of similar length and show many conserved amino acid motifs, whereas the signal detection domains show considerable variations in size and amino acid sequence reflecting the diversity of signals. A prevalent feature of the signal detection domain is the presence of one or more domain(s) with close homology to the mammalian Per-Arnt-Sim class of proteins, also named clock proteins, which are involved in circadian rhythms. Most of the response regulators are transcription factors consisting of two domains. The N-terminal domain receives the phosphoryl group from the kinase and regulates the DNA-binding activity of the C-terminal effector domain.

We are involved in an effort to explore the mechanism of phosphoryl transfer and signaling through structural characterization of the components of the phosphorelay. We have determined the crystal structures of Spo0F, Spo0B, the molecular complex of Spo0F with Spo0B, and the C-terminal domain of Spo0A (Spo0AC) bound to a DNA duplex containing a specific recognition sequence called the 0A box. The methods employed for preparing the protein, crystallizing the protein, and determining the crystal structure are described. The second half of the chapter shows how structural data were used to gain insights into the mechanisms of molecular recognition and phosphoryl transfer.

Methods

Expression Systems and Expression Conditions

All the analyses described in this chapter are performed using recombinant protein samples. All the genes except Spo0AC are cloned into the pET20b+ expression vector (Novagen) in the *Nde*I and *Bam*HI sites. The inserts are polymerase chain reaction (PCR) amplified with gene-specific forward and reverse primers, using the genomic DNA of *B. subtilis* (strain: JH12908) as a template. *Nde*I and *Bam*HI sites are introduced in the forward and reverse primers, respectively. The Spo0AC gene is cloned in the pET16b vector in the *Nco*I and *Bam*HI sites.

The recombinant plasmids are transformed in the BL21(DE3) strain of *E. coli*, and the transformants are grown in LB broth supplemented with 100 μg/ml ampicillin at 37° (Spo0AC, being an exception, is grown at 30°). Selenomethionine containing Spo0AC is expressed in the methionine auxotrophic strain B834(DE3) of *E. coli*. B834 cells are grown in defined media (Boles *et al.*, 1991) containing 40 mg/l of seleno-DL-methionine. After the optical density (600 nm) reaches 1.0, the culture is induced with 2 mM isopropyl-β-D-thiogalactoside and is further grown for 5 hours. The bacteria are then harvested by centrifugation at 6000 rpm for 10 min at 4°. All the recombinant proteins are obtained in the soluble fraction as analyzed from their expression profiles by SDS–PAGE.

All the proteins crystallized are expressed without His tags; we had less success with proteins expressed with His tags. Within days of purification, His-tagged Spo0 proteins show signs of aggregation; high molecular weight peaks appeared on size exclusion chromatography. Addition of additives such as excess EDTA to eliminate metals that may be acting as nucleating centers for His tags failed to suppress aggregation, and our conjecture is that the presence of aggregation, not the presence of a His tag, obstructed crystallization.

Purification

The recombinant proteins are purified using conventional procedures. The choice of buffers used in purification procedures was decided based on their theoretical pI value and solubility properties (Table I).

Cells are lysed by sonication, the cell debris is separated by centrifugation, and the recombinant proteins are purified from the supernatants. Spo0F is purified using a three-step protocol: Trisacryl-DEAE followed by hydroxylapatite and subsequently with size exclusion chromatography. Spo0B is purified using a four-step procedure. The Spo0AC protein is purified using S-cation exchange followed by heparin agarose and size exclusion chromatography. Table II describes the various buffer conditions

TABLE I
PROTEIN DATA

Name	Chain length (amino acids)	Molecular mass (Da)	pI
Spo0F	124	14227.68	4.89
Spo0B	192	22542.54	5.13
Spo0A-C(1,2,149–267)	121	13539.72	9.36

TABLE II
PURIFICATION CHART

Protein	Resuspension buffer	Column 1 (Tris-acryl-DEAE) elution buffer	Buffer change	Column 2 (hydroxylapatite) elution buffer	Buffer exchange	Column 3 (Tris-acryl-DEAE) elution buffer	Column 4 (Sepharose S-100) equilibration buffer	Crystallization buffer
SpoOB (Zhou et al., 1997)	50 mM Tris-HCl, pH 8.0, 50 mM KCl, 1 mM EDTA, 1 mM PMSF, 1 mM DTT	0-0.3 M KCl, 50 mM Tris-HCl, pH 8.0, 50 mM KCl, 1 mM EDTA, 1 mM PMSF, 1 mM DTT	50 mM Tris-HCl, pH 7.4, 1 mM PMSF, 1 mM DTT	0.1 M potassium phosphate, 50 mM Tris-HCl, pH 7.4, 1 mM PMSF, 1 mM DTT	50 mM Tris-HCl, pH 8.0, 50 mM KCl, 1 mM EDTA, 1 mM PMSF, 1 mM DTT	0.1-0.3 M KCl, 50 mM Tris-HCl, pH 8.0, 50 mM KCl, 1 mM EDTA, 1 mM PMSF, 1 mM DTT	25 mM Tris-HCl, pH 7.4, 1 mM EDTA, 1 mM DTT, 75 mM KCl	25 mM Tris-HCl, pH 7.4, 1 mM DTT
SpoOF (Madhusudan et al., 1996b)	25 mM Tris-HCl, pH 7.8, 10 mM KCl, 1 mM MgCl$_2$, 5 mM βME, 0.5 mM PMSF, 1 leupeptin, 1 μM pepstatin	10-210 mM KCl, 25 mM Tris-HCl, pH 7.8, 10 mM KCl, 1 mM MgCl$_2$, 5 mM βME, 0.5 mM PMSF, 1 μM leupeptin, 1 μM pepstatin	25 mM Bis-Tris, pH 7.3, 50 mM KCl, 2 mM MgCl$_2$	800 mM MgCl$_2$, 25 mM Bis-Tris, pH 7.3, 50 mM KCl, 2 mM MgCl$_2$	—	—	25 mM Bis-Tris, pH 7.3, 50 mM KCl	25 mM Bis-Tris, pH 7.3, 50 mM KCl, 2 mM MgCl$_2$
SpoOAC (Zhao et al., 2002)	20 mM phosphate buffer, pH 6.8, 150 mM NaCl, 1 mM EDTA, 1 mM PMSF	S-cation exchange 250-800 mM NaCl, 20 mM sodium phosphate buffer, pH 6.8, 1 mM EDTA	20 mM phosphate buffer, pH 6.8, 150 mM NaCl, 1 mM EDTA	Heparin agarose 250-800 mM NaCl, 20 mM sodium phosphate buffer, pH 6.8, 1 mM EDTA	Superdex 75 gel filtration 20 mM phosphate buffer, pH 6.8, 150 mM NaCl, 1 mM EDTA	20 mM phosphate buffer, pH 6.8, 150 mM NaCl, 1 mM EDTA	20 mM phosphate buffer, pH 6.8, 150 mM NaCl	10 mM Tris-HCl, pH 7.5, 10 mM KCl, 2 mM MgCl$_2$

used for resuspension of the cells, as well as binding and elution, used during different stages of purification for each of the proteins. The choice of buffer conditions was found to be critical for concentrating proteins to 10 to 20 mg/ml used in crystallization experiments. A solubility screen was devised using the hanging drop method with no added precipitants. Typically, unconcentrated proteins ($\sim$1 mg/ml) are mixed with equal volumes of test buffer solution (2 μl) on a microscope slide, and the drops are inverted over the reservoir solution in a Linbro plate. A range of buffers, pH values, salts, metal ions, and salt concentrations was tested. Lack of conspicuous precipitation indicates that a particular condition may be suitable for concentrating proteins. For crystallization, proteins are concentrated using Centricons with the appropriate molecular weight cutoff. Procedures employed for forming the DNA complex of Spo0AC are listed later.

The DNA oligomers are purified by ion exchange followed by HPLC and lyophilized. The lyophilized powder is dissolved in 10 mM Tris-HCl, pH 7.5, 5 mM KCl, and 2 mM MgCl$_2$. Duplex formation is achieved by slow cooling after heating the sample to 100°. The purified Spo0AC is then mixed with the oligonucleotide sample in 2:1 stoichiometric ratio and then further concentrated to 18 mg/ml using the Centricon. The concentration is verified by the Bradford assay before proceeding with the crystallization experiments.

In order to produce diffraction quality crystals, we use oligonucleotides of varying lengths 11 to 21, containing a 0A box (TGTCGAA) to form complexes with the C-terminal domain of Spo0A. The best crystals are obtained with 16 base oligonucleotides, which form a 15-bp duplex with one base overhanging at both 5′ ends, as shown in Fig. 3A.

```
A   ′-TTCGTGTCGAATTTTG-3′
    3′-AGCACAGCTTAAAACA-5′


B           〈--0A--〉  〈--0A--〉
    5′ -  TTCGTGTCGAATTTTGTCGTGTCGAATTTTGTTCGTG......
    3′ -  AGCACAGCTTAAAACAGCACAGCTTAAAACAAGCAC......
                    〈--0A--〉  〈--0A--〉


C   〈--0A--〉  〈--0A--〉
    5′ TGTCGAATAATGACGAA
    3′ ACAGCTTATTACTGCTT
```

Fig. 3. Elongation of the DNA duplex and formation of 0A boxes in the lattice. (A) Synthetic duplex. (B) Formation of the elongated helix in the crystal lattice. (C) In the AbrB promoter, two 0A boxes are separated by 3 bp.

In the crystal lattice, the 15-bp duplexes are arranged in a head-to-tail fashion to create a continuous helix. This mode of association gives rise to two additional Spo0A-binding sites at the junctions spaced 3 bp from the intrinsic sites (Fig. 3B). The association produces AbrB promoter sites in the crystal lattice (Fig. 3C). AbrB is a global pleiotropic regulator of several genes in *B. subtilis*.

Crystallization

We crystallized the wild-type Spo0F, calcium, and manganese complexes of the Y13S mutant of Spo0F, Spo0B, the molecular complex of Spo0B and Spo0F, beryllofluoride Spo0F in complex with Spo0B, and the DNA complex of the Spo0A effector domain. All crystals are grown by the vapor diffusion method by hanging drop techniques. Equal volumes of protein solution (typically 1–3 μl) and reservoir solutions are mixed and equilibrated using a Linbro plate. The precipitants used for each protein are listed in Table III.

Structure Determination

Structure solutions are carried out using well-established methods: isomorphous replacement, molecular replacement, and multiwavelength anomalous dispersion techniques (Hendrickson, 1991; Rossmann and Arnold, 2001).

Y13S Mutant Spo0F and Wild-Type Spo0F. Native crystals of the Y13S mutant of Spo0F diffracted to 3 Å on irradiation with monochromatic Cu Kα radiation from a rotating anode RU200. The structure is solved by single isomorphous replacement and anomalous dispersion techniques using a uranium derivative. To produce the heavy atom derivative, the crystals are soaked in 1 mM uranium oxynitrate for 2 weeks. The uranium position is determined using isomorphous difference Patterson techniques, and the electron density maps are computed with single isomorphous and anomalous signals using the package PHASES (Furey and Swaminathan, 1997). Model building is carried out using the programs Frodo and O (Jones *et al.*, 1991). For higher resolution data collection at a synchrotron beam line, the crystals are flash frozen on a liquid nitrogen stream, after dipping into a cryo solvent containing 25% PEG300, 25% glycerol, and 150 mM calcium chloride in 75 mM sodium acetate buffer, pH 4.5; it produces significantly enhanced diffraction to 2.0 Å, and this data set is used for final refinements of the coordinates. Crystals of the wild-type Spo0F diffract to 2.45-Å resolution, and the structure is solved by molecular replacement using the mutant Spo0F as a search model.

TABLE III
CRYSTALLIZATION CONDITIONS

Protein	Stoichiometry/ concentration (mg/ml)	Protein buffer	Precipitant
Y13S SpoOF Ca complex Mn complex	25	25 mM Bis-Tris, pH 7.3, 50 mM KCl	6–7.5% PEG3350, 10% glycerol, 200 mM CaCl$_2$ in 100 mM sodium acetate buffer, pH 4.5
Wild-type SpoOF metal free	15	25 mM Bis-Tris, pH 7.3, 50 mM KCl	25% PEG4500, 15% ethanol in 100 mM sodium potassium phosphate buffer, pH 7.8
SpoOB	75	25 mM Tris-HCl, pH 7.4, 1 mM DTT	2M ammonium sulfate in 100 mM Tris-HCl buffer, pH 8.5
SpoOAC: DNA	2:1/18 (protein)	10 mM Tris-HCl, pH 7.5, 10 mM KCl, 2 mM MgCl$_2$	12% PEG3350, in 100 mM Tris-HCl buffer, pH 7.0, 100 mM NaCl, 2 mM CaCl$_2$, 2% isopropanol as a precipitant
SpoOF:SpoOB	1.3:1/9:10.3	25 mM Bis-Tris, pH 7.3, 50 mM KCl, 25 mM Tris-HCl pH 7.4, 1 mM DTT, 2 mM MgCl$_2$, 10 mM AlCl$_3$, 30 mM NaF	0.5 M KCl, 24% PEG 2000 in 100 mM Tris-HCl buffer, pH 8.1
SpoOF:BeF^{3-}: SpoOB	1.3:1/9:10.3	25 mM Tris-HCl buffer, pH 7.5, 1 mM DTT, 10 mM Bis-Tris buffer, pH 7.3, 50 mM KCl, 5.3 mM BeCl$_3$, 35 mM NaF, 7 mM MgCl$_2$	0.5 M KCl, 25% PEG2000, 100 mM Tris-HCl, pH 8.1

SpoOB. Crystals are dipped in a cryo solvent prepared by mixing 200 mM ammonium sulfate and 48% PEG400 in 100 mM Tris-HCl buffer at pH 8.5 before freezing in a nitrogen stream. Native crystals diffract to a resolution of 2.2 Å, and diffraction data are collected at the Stanford Synchrotron Radiation Laboratory (SSRL). The structure is solved by multiple isomorphous replacement using three heavy atom derivatives: (i) KauCl$_3$ (soak time: 1 day), (ii) K$_2$PtCl$_4$ (soak time: 5 days), and (iii) CH$_3$HgCOOH (soak time: ½ day). The heavy atom phases are obtained from the difference pattern maps calculated using the PHASES package. Model building is carried out using the program O.

Spo0F:Spo0B Complexes. X-ray intensity data for the crystals of both the apo- and the beryllofluoride-bound complex of Spo0F with Spo0B are collected at SSRL. The flash frozen crystals diffract to a resolution of 3 Å. A mixture of 40% PEG400, 5% PEG3350, 200 mM KCl, 2 mM MgCl$_2$, and 100 mM Tris-HCl, pH 8.1, is used as a cryo solvent. The structure is solved using molecular replacement methods with the atomic coordinates of Spo0F and Spo0B as search models.

Spo0AC:DNA Complex. Crystals of selenomethionine Spo0AC bound to DNA are used for data collection on a tunable synchrotron beam line. The crystals are frozen using a mixture of 200 mM NaCl, 2% isopropanol, 20 mM CaCl$_2$, and 8% PEG3350 in 100 mM Tris-HCl, pH 7.5. With the goal of estimating the phases of reflections, intensity data are collected at the inflection point (0.9798 Å), at the peak (0.97949 Å), and at a remote wavelength (0.92526 Å). The selenium positions are determined using the program SOLVE (Terwilliger and Berendzen, 1999), and an electron density map is computed. Crystals of the Spo0AC–DNA complex with four consecutive thymines changed to iodo-uracils (5′-TTCGTGTCGAA*TT-TTG*-3′) are grown and intensity data are collected. A difference electron density map reveals all iodine positions. Phases are recomputed incorporating data from the iodinated crystal. Model building is carried out with a solvent-flattened electron density map computed using RESOLVE (Terwilliger, 2000). Using the iodine positions, the base pairs of the DNA duplex are unambiguously assigned during model building.

All the computations for structure refinements in our study are carried out using the X-PLOR/CNS (Brunger, 1992; Brunger *et al.*, 1998) suite of programs.

Insights from Structural Analysis

The structure of the intermediate response regulator Spo0F represents an α/β fold with a central parallel β sheet made of five strands with helices on either side (Fig. 4A) (Madhusudan *et al.*, 1996a, 1997; Mukhopadhyay *et al.*, 2004). The structure was found to be similar to CheY. The receiver domain three-dimensional structures of a number of response regulators are known so far, and all of them have the same fold as CheY and Spo0F. The active site is located on top of the β sheet and contains three aspartate residues (Asp10, Asp11, and Asp54), a threonine (Thr 82), and a lysine (Lys104). All these residues are highly conserved in response regulators. The site of phosphorylation, Asp54, is located at the bottom of the active site. The structures of mutant (Y13S) and wild-type Spo0F are very similar.

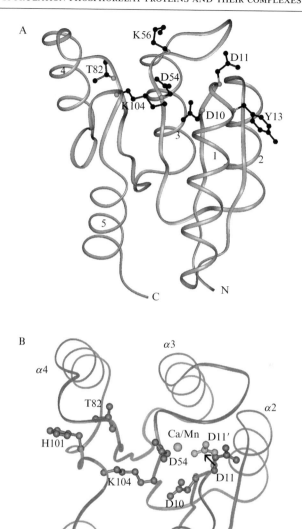

FIG. 4. Spo0F structure. (A) Ribbon representation of the structure of Spo0F. The central β sheet consists of five parallel β strands. There are five α-helices, α1 to α5 (labeled as 1 to 5). N and C termini are labeled as N and C. The five catalytic residues, Asp10, Asp11, Asp54,

The side chain of Tyr13 is positioned on the surface of the molecule, where it can interact with Rap phosphatases (Fig. 4A).

The crystal structures of metal-free and metal-bound forms are similar. The most significant difference between the two states is that, in the metal-free structure, the metal cavity is not fully formed and the metal induces formation of the cavity. Figure 4B shows that the side chain of Asp11, which points away from the active site, reorients to coordinate with the metal. Using similar crystallization conditions, the structure of mutant SpoOF was also solved in the presence of manganese, as manganese ions are known to play an important role in sporulation.

Autophosphatase Activities Are Customized for Specific Roles

Response regulators are activated by phosphorylation, and the stability of the phosphorylated state appears to be tailored to their roles. SpoOF~P is one of the most "stable" phosphorylated response regulators (the half-life of SpoOF~P is on the order of hours), whereas the chemotaxis response regulator CheY~P is one of the least stable. The half-life of CheY~P is on the order of seconds (Lukat *et al.*, 1991), which is about the same amount of time bacteria take to change the direction of swimming. Sporulation in *B. subtilis* is initiated over a period of hours, similar to the rate of SpoOF autodephosphorylation. We wondered if we could correlate this large difference in the autodephosphorylation rate with any specific characteristics of the active sites. An inspection of the active site of SpoOF shows that Lys56 is located on the edge of the active site covering Asp54, providing a partial shield from external water molecules and enhancing the stability of Asp-P (Fig. 4A). The Lys56Met mutant has the same autophosphatase activity as the wild type, whereas substitution by Ala increases the autophosphatase activity 3-fold. The corresponding residue in CheY is Asn, and substitution of Lys56 by Asn results in a 23-fold increase in autophosphatase activity in SpoOF. We hypothesized that Asn could position a water molecule suitable for nucleophilic attack at the phosphate atom, causing an increase in the autophosphatase activity. The autophosphatase activity of response regulators appears to depend on the shape of the active site, the shielding of the phosphoryl group, which restricts access of water molecules, and the interactions in the active site that stabilize the phosphoryl group. Thus residues

Thr82, and Lys104, are shown as a ball-and-stick model. Lys56, which plays a critical role in determining autophosphatase activity, is also shown. The side chain of the residue Y13 is solvent accessible. (B) A view from the top of the molecule showing the difference between metal-free and metal-bound conformations. Upon metal binding, the side chain of Asp11 turns toward the metal atom to coordinate with it, shown as D11'.

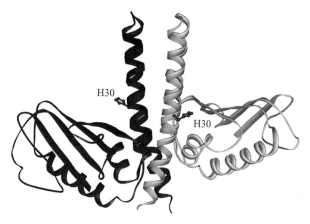

FIG. 5. Ribbon representation of the Spo0B dimer. One protomer is shown in a dark shade. His30, the site of phosphorylation, is shown as a ball and stick.

surrounding the Asp~P residue attenuate the rate of dephosphorylation to match the biological function of the response regulator. Divalent cations such as magnesium play an important role in phosphotransfer reactions, and it has been shown that the presence of magnesium ions increases the autophosphatase rate by a factor of 10 (Zapf et al., 1998).

Our structure of Spo0B provided a first view of the structural assembly for histidine kinases (Varughese et al., 1998). In the crystal lattice, Spo0B exists as a dimer, which is also physiologically relevant. Each monomer contributes a helical hairpin to the dimer interface and, in so doing, forms a four-helical bundle (Fig. 5). The site of phosphorylation is His30 and its side chain is solvent exposed, making it accessible to Spo0F and Spo0A. The dimerization domain is followed by a second domain, similar in fold to ATP-binding domains. However, the ATP-binding domain in Spo0B is devoid of any consensus sequence signatures, such as Walker A or B or Q loop motifs, which could bind the ATP molecule.

Interactions of Spo0F and Spo0B. The crystal structures of the apo- and beryllofluoride-bound forms of Spo0F in complex with Spo0B showed how the two proteins associate to exchange the phosphoryl group (Fig. 6A and B). Our analysis indicates that the apo form represents a transition state intermediate, whereas the beryllofluoride complex represents a pretransition state (Varughese et al., 2006; Zapf et al., 2000). Differences between the two complexes are localized to the active site pockets in the individual proteins, while there is no drastic change in the mode of interaction. The association of the two molecules brings the active histidine of Spo0B and the active

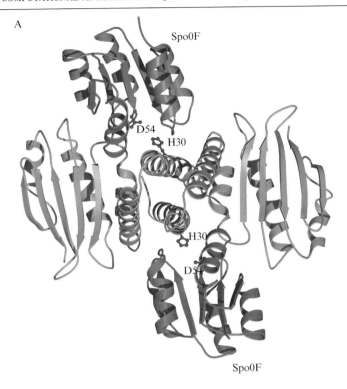

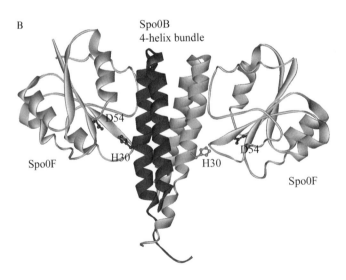

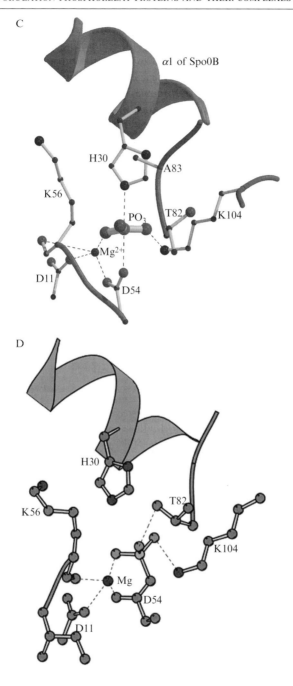

aspartate of Spo0F in close proximity and creates the correct geometry for phosphoryl transfer. Figure 6C shows a model for the transition state intermediate generated using the crystal structure of the apo complex of Spo0F with Spo0B. The model shows that the aspartate and the histidine are ideally oriented in the crystal lattice for phosphotransfer. The negative charge of the phosphoryl group is neutralized through the interactions of Mg^{2+} and Lys104. The active site shields the phosphoryl group from external water molecules to prevent hydrolysis. Therefore the association of Spo0F and Spo0B creates an environment for phosphotransfer. Figure 6D shows the interactions of beryllofluride Spo0F with Spo0B and represents the pretransition state interactions of the "phosphorylated" Spo0F with Spo0B. It is important to point out that Spo0F and Spo0B have very similar conformations in the two crystal structures. The main difference is in the orientation of the side chain Thr82 of Spo0F. Therefore the complex can go from a pretransition state to a transition state with a small change in the conformation of the loop carrying Thr82.

   Spo0A:DNA Interactions. The effector domain fold consists of six helices with two of them forming a helix-turn-helix motif, commonly observed in DNA-binding transcription factor proteins. The domain forms head-to-tail tandem dimers orienting the recognition helices separated by about 34 Å corresponding to the helical pitch of the DNA. The recognition helix interacts with the major groove of DNA (Fig. 7) (Zhao *et al.*, 2002).

Specificity in Signal Transduction

   A significant portion of bacterial genomes (up to 1.5%) is composed of two-component pairs used to detect unique environmental signals and elicit unique responses (Alm *et al.*, 2006). In particular, *B. subtilis* has at least 35 two-component systems (Kunst *et al.*, 1997). Hence, it is intriguing how a particular histidine protein kinase specifically recognizes its partner and activates it to produce the correct response when there are a number of response regulators with very similar three-dimensional structures. Using

---

FIG. 6. Association of Spo0F with Spo0B and phosphory transfer. (A) A view of the Spo0B:Spo0F complex down the axis of the four-helix bundle. (B) A view of the Spo0F:Spo0B complex perpendicular to the four-helix bundle. For clarity, the C-terminal domains of Spo0B are omitted. Residues His30 of Spo0B and Asp54 of Spo0F are in close proximity for phosphoryl transfer. (C) A model for the transition state intermediate, created by placing a phosphoryl group between His30 and Asp54. The phosphorus atom forms partial covalent bonds with O^{δ} of Asp and N^{ε} of His and is in a penta-coordinated state. Negative charges on the phosphoryl oxygens are compensated through interactions with Mg^{2+} and Lys104 (reproduced with permission from Zapf *et al.*, 2000). (D) Active site interactions in the crystal structure of beryllofluoride Spo0F with Spo0B (reproduced with permission from Varughese *et al.*, 2006). (See color insert.)

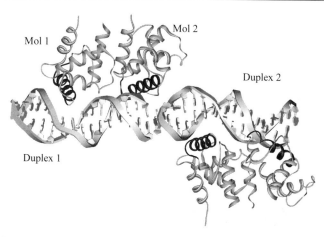

FIG. 7. Interactions of the Spo0AC dimer with DNA. In the asymmetric unit of the crystal, two DNA duplexes (1 and 2) associate in a head-to-tail manner to form a extended duplex. The recognition helix in each monomer is highlighted as a dark shade.

the Spo0F:Spo0B structure as a prototype for interactions between histidine protein kinase and response regulators, we carried out a sequence comparison of the interacting residues in all the response regulators and kinases in *B. subtilis* (Hoch and Varughese, 2001; Varughese, 2002; Zapf *et al.*, 2000). Sequence comparisons of the response regulators show that a hydrophobic core of the contact surface is well conserved, whereas interacting residues at the periphery are variable. A very similar pattern was also noticed for the interacting surface on the four-helix bundle (Fig. 8). Therefore, we conclude that binding is initiated through the hydrophobic core and that discrimination is affected through the peripheral variable residues (Mukhopadhyay and Varughese, 2005).

Cross Talk

Although each two-component system appears to detect and respond to a specific signal(s), they do not always act independently of each other. For example, regulatory links exist between the PhoPR two-component system that participates in the cellular response to phosphate limitation and the essential YycFG two-component system in *B. subtilis* (Howell *et al.*, 2006). The PhoR sensor kinase can activate the YycF response regulator during a phosphate limitation-induced stationary phase, while the reciprocal cross-phosphorylation does not occur. Could crystallographic studies provide an explanation for this unusual observation?

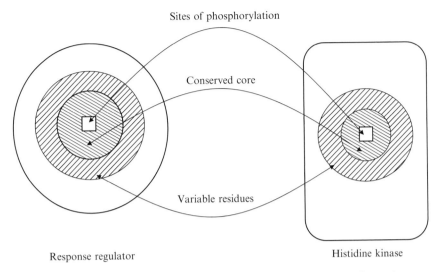

Sites of phosphorylation

Conserved core

Variable residues

Response regulator   Histidine kinase

FIG. 8. Recognition surfaces of kinases and response regulators. The binding surfaces are composed of a central core of conserved residues. Outside this core, the residues are more variable.

The recognition specificity between proteins lies on their interacting surfaces. The Spo0F:Spo0B crystal structure (Zapf *et al.*, 2000) is a model for the association of the response regulators with histidine kinases (Fig. 9). The site of phosphorylation of a histidine kinase is located on the four-helix bundle; consequently, most of the significant interactions of the response regulator during phosphoryl transfer are with the four-helix bundle. In this context, it is interesting to examine those amino acids of the four-helical bundles of YycG and PhoR that interact with their respective response regulators. The most relevant for comparison between YycG and PhoR are positions −4, −3, −1, +1, +3, +4, +7, +8, +10, +11, and +12 shown in Table IV (numbered relative to the phosphorylated His residue) (Howell *et al.*, 2006). Figure 9 shows that these residues are suitably positioned to interact with the response regulators. In YycG and PhoR, the residues at positions −4, −3, −1, +1, +4, and +7 are identical. The amino acids at +3, +8, +10, and +12, although different, are similar in nature (Arg-Thr-Arg-Tyr versus Lys-Ser-Lys-Phe). There is a very conspicuous difference in position 11 where YycG has a Ser, whereas PhoR has a Gly, suggesting that this position plays a role in YycG, discriminating against PhoP. Gly is the smallest amino acid. Therefore, the presence of Gly in PhoR creates an additional space on the surface of the four-helix bundle for the response

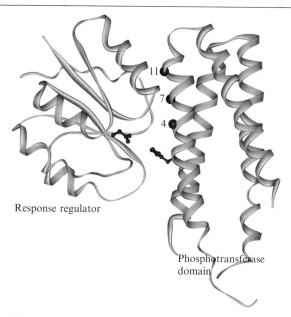

FIG. 9. A model for response regulator interactions with histidine kinases. The C^α positions of three residues of helix $\alpha 1$ facing the response regulator are labeled based on their relative positions with respect to the site of phosphorylation.

TABLE IV

COMPARISON OF SEQUENCES AROUND THE SITE OF PHOSPHORYLATION IN THE YycG AND PhoR[a]

```
                    *
YycG    ANVSHELRTPLTTMRSYLEALA
PhoR    ANVSHELKTPITSIKGFTETLL
```

[a] The site of phosphorylation is denoted by an asterisk.

regulator to approach. YycG has a Ser at this position, which complements YycF but most likely causes steric repulsion of PhoP. Additionally, the residues in YycG at positions +3, +8, +10, and +12 (Arg-Thr-Arg-Tyr) are larger than the corresponding residues in PhoR (Lys-Ser-Lys-Phe), again implying possible steric repulsion of PhoP by YycG. Therefore, interactions observed in the Spo0F:Spo0B complex provide a structural basis for the observed cross talk.

Conclusion

Our crystallographic studies on proteins in the sporulation phosphorelay provide a picture of the molecular architecture used by two-component systems for phosphoryl transfer and signal transduction. The mechanism of phosphotransfer and specificity between histidine protein kinase/response regulator pairs was delineated by the SpoOF/SpoOB complex structure. These structures have proved to be useful models for explaining the properties of other two-component systems, for example, the cross talk occurring between the YycG histidine protein kinase and the PhoR response regulator. Studies also show that the phosphorylated and unphosphorylated response regulators associate with the phosphotransferase domain in a similar manner. It has also been shown how the transcription factor SpoOA specifically recognizes its DNA motif, the OA box. This series of crystal structures reveal the structural features that underlie the adaptive signal responses carried by two-component systems.

Structural characterization of sporulation phosphorelay proteins was helpful in delineating the mechanism of phosphotransfer and the origin of specificity in two-component signaling. Additionally, the crystallographic studies were helpful in explaining the rare occurrence of cross-communication.

Acknowledgments

Most of the studies described in this chapter were performed in collaboration with Dr. James Hoch. The analysis of "cross talk" was done in collaboration with Dr. Kevine Devine. Main support for the original research came from NIH. Preparation of this chapter was supported by the Arkansas Biosciences Institute.

References

Alm, E., Huang, K., and Arkin, K. (2006). The evolution of two-component systems in bacteria reveals different strategies for niche adaptation. *PLoS Comput. Biol.* **2,** e143.

Boles, J. O., Cisneros, R. J., Weir, M. S., Odom, J. D., Villafranca, J. E., and Dunlap, R. B. (1991). Purification and characterization of selenomethionyl thymidylate synthase from *Escherichia coli*: Comparison with the wild-type enzyme. *Biochemistry* **30,** 11073–11080.

Brunger, A. T. (1992). X-PLOR Version 3.1. Yale University, New Haven, CT.

Brunger, A. T., Adams, P. D., Clore, G. M., DeLano, W. L., Gros, P., Grosse-Kunstleve, R. W., Jiang, J. S., Kuszewski, J., Nilges, M., and Pannu, N. S. (1998). Crystallography and NMR system: A new software suite for macromolecular structure determination. *Acta Crystallogr.* **D54,** 905–921.

Burbulys, D., Trach, K. A., and Hoch, J. A. (1991). The initiation of sporulation in *Bacillus subtilis* is controlled by a multicomponent phosphorelay. *Cell* **64,** 545–552.

Ferrari, F. A., Trach, K., LeCoq, D., Spence, J., Ferrari, E., and Hoch, J. A. (1985). Characterization of the *spo0A* locus and its deduced product. *Proc. Natl. Acad. Sci. USA* **82**, 2647–2651.

Fujita, M., and Losick, R. (2005). Evidence that entry into sporulation in *Bacillus subtilis* is governed by a gradual increase in the level and activity of the master regulator Spo0A. *Genes Dev.* **19**, 2236–2244.

Furey, W., and Swaminathan, S. (1997). PHASES-95: A program package for the processing and analysis of diffraction data from macromolecules. *Methods Enzymol.* **277**, 590–620.

Hendrickson, W. A. (1991). Determination of macromolecular structures from anomalous difraction of synchrotron radiation. *Science* **254**, 51–58.

Hoch, J. A., and Silhavy, T. J. (1995). "Two-Component Signal Transduction." American Society for Microbiology Press, Washington, DC.

Hoch, J. A., and Varughese, K. I. (2001). Keeping signals straight in phosphorelay signal transduction. *J. Bacteriol.* **183**, 4941–4949.

Howell, A., Dubrac, S., Noone, D., Varughese, K. I., and Devine, K. (2006). Interactions between the YycFG and PhoPR two-component systems in *Bacillus subtilis*: The PhoR kinase phosphorylates the non-cognate YycF response regulator upon phosphate limitation. *Mol. Microbiol.* **59**, 1199–1215.

Jiang, M., Shao, W., Perego, M., and Hoch, J. A. (2000). Multiple histidine kinases regulate entry into stationary phase and sporulation in *Bacillus subtilis*. *Mol. Microbiol.* **38**, 535–542.

Jones, T. A., Zou, J. Y., Cowan, S. W., and Kjeldgaard, M. (1991). Improved methods for binding protein models in electron density maps and the location of errors in these models. *Acta Crystallogr.* **A47**, 110–119.

Kunst, F., Ogasawara, N., Moszer, I., Albertini, A. M., Alloni, G., Azevedo, V., Bertero, M. G., Bessieres, P., Bolotin, A., Borchert, S., Borriss, R., Boursier, L., *et al.* (1997). The complete genome sequence of the gram-positive bacterium *Bacillus subtilis*. *Nature* **390**, 237–238.

Lukat, G. S., Lee, B. H., Mottonen, J. M., Stock, A. M., and Stock, J. B. (1991). Roles of the highly conserved aspartate and lysine residues in the response regulator of bacterial chemotaxis. *J. Biol. Chem.* **266**, 8348–8354.

Rossmann, M. G., and Arnold, E. (eds.) (2001). "International Tables for Crystallography," Vol. F. Kluwer Dordrecht.

Madhusudan, M., Zapf, J., Whiteley, J. M., Hoch, J. A., Xuong, N. H., and Varughese, K. I. (1996a). Crystal structure of a phosphatase-resistant mutant of sporulation response regulator Spo0F from *Bacillus subtilis*. *Structure* **4**, 679–690.

Madhusudan, M., Zapf, J., Whiteley, J. M., Hoch, J. A., Xuong, N. H., and Varughese, K. I. (1996b). Crystallization and preliminary X-ray analysis of a Y13S mutant of Spo0F from *Bacillus subtilis*. *Acta Crystallogr.* **D52**, 589–590.

Madhusudan, M., Zapf, J., Hoch, J. A., Whiteley, J. M., Xuong, N. H., and Varughese, K. I. (1997). A response regulatory protein with the site of phosphorylation blocked by an arginine interaction: Crystal structure of Spo0F from *Bacillus subtilis*. *Biochemistry* **36**, 12739–12745.

Mukhopadhyay, D., Sen, U., Zapf, J., and Varughese, K. I. (2004). Metals in the sporulation phosphorelay: Manganese binding by the response regulator Spo0F. *Acta Crystallogr.* **D60**, 638–645.

Mukhopadhyay, D., and Varughese, K. I. (2005). A computational analysis on the specificity of interactions between histidine kinases and response regulators. *J. Biomol. Struct. Dyn.* **22**, 555–562.

Parkinson, J. S., and Kofoid, E. C. (1992). Communication modules in bacterial signaling proteins. *Annu. Rev. Genet.* **26**, 71–112.

Terwilliger, T. C. (2000). Maximum-likelihood density modification. *Acta Crystallogr.* **D56,** 965–972.

Terwilliger, T. C., and Berendzen, J. (1999). Automated MAD and MIR structure solution. *Acta Crystallogr.* **D55,** 849–861.

Varughese, K. I. (2002). Molecular recognition of bacterial phosphorelay proteins. *Curr. Opin. Microbiol.* **5,** 142–148.

Varughese, K. I., Madhusudan, M., Zhou, X.-Z., Whiteley, J. M., and Hoch, J. A. (1998). Formation of a novel four-helix bundle and molecular recognition sites by dimerization of a response regulator phosphotransferase. *Mol. Cell* **2,** 485–493.

Varughese, K. I., Tsigelny, I., and Zhao, H. (2006). The crystal structure of beryllofluoride Spo0F in complex with the phosphotransferase Spo0B represents a phosphotransfer pretransition state. *J. Bacteriol.* **188,** 4970–4977.

Zapf, J., Sen, U., Madhusudan, M., Hoch, J. A., and Varughese, K. I. (2000). A transient interaction between two phosphorelay proteins trapped in a crystal lattice reveals the mechanism of molecular recognition and phosphotransfer in signal transduction. *Structure* **8,** 851–862.

Zapf, J. W., Madhusudan, M., Grimshaw, C. E., Hoch, J. A., Varughese, K. I., and Whiteley, J. M. (1998). A source of response regulator autophosphatase activity: The critical role of a residue adjacent to the Spo0F autophosphorylation active site. *Biochemistry* **37,** 7725–7732.

Zhao, H., Msadek, T., Zapf, J., Madhusudan Hoch, J. A., and Varughese, K. I. (2002). DNA complexed structure of the key transcription factor initiating development in sporulating bacteria. *Structure* **10,** 1041–1050.

Zhou, X. Z., Madhusudan, Whiteley, J. M., Hoch, J. A., and Varughese, K. I. (1997). Purification and preliminary crystallographic studies on the sporulation response regulatory phosphotransferase protein, spo0B, from *Bacillus subtilis. Proteins* **27,** 597–600.

[6]   Control Analysis of Bacterial Chemotaxis Signaling

By Tau-Mu Yi, Burton W. Andrews, and Pablo A. Iglesias

Abstract

Bacteria such as *Escherichia coli* demonstrate the remarkable ability to migrate up gradients of attractants and down gradients of repellents in a rapid and sensitive fashion. They employ a temporal sensing strategy in which they estimate the concentration of ligand at different time points and continue moving in the same direction if the concentration is increasing in time, and randomly reorient if the concentration is decreasing in time. The key to success is accurate sensing of ligand levels in the presence of extra-cellular and intracellular disturbances. Research from a control theory perspective has begun to characterize the robustness of the bacterial che-motaxis signal transduction system to these perturbations. Modeling and theory can describe the optimal performance of such a sensor and how it can be achieved, thereby illuminating the design of the network. This chapter describes some basic principles of control theory relevant to the analysis of this sensing system, including sensitivity analysis, Bode plots, integral feedback control, and noise filters (i.e., Kalman filters).

Introduction

Chemotaxis, movement toward chemical attractants and away from chemical repellants, is a universal behavior in living organisms (Berg, 1993). Motile single-cell organisms such as bacteria are proficient at che-motaxis. The bacterium *Escherichia coli* is a typical example; it performs a biased random walk in which straight runs are interspersed with random changes in direction termed tumbles. When going up a chemoattractant gradient, the bacteria tend to run and thus continue in the same direction. When going down the gradient they tend to tumble and thus reorient their direction of movement.

This strategy is termed temporal sensing because the bacteria estimate the changes in chemoattractant concentration over time. If the concen-tration is increasing in time (i.e., moving up a gradient), then the bacteria are more likely to run; if the concentration is decreasing in time, then the bacteria are more likely to tumble. To a theorist interested in understanding the regulation and robustness of this sensing system, there are many potential challenges to temporal sensing. In particular, the cell must distinguish between random fluctuations in chemoattractant levels caused

METHODS IN ENZYMOLOGY, VOL. 422
Copyright 2007, Elsevier Inc. All rights reserved.
0076-6879/07 $35.00
DOI: 10.1016/S0076-6879(06)22006-8

by external noise and meaningful changes in chemoattractant levels caused by movement through the gradient.

To the biologist, the bacterial chemotaxis signal transduction pathway is a canonical example of a two-component signaling system (Stock *et al.*, 1991). Bacteria can sense the levels of attractant or repellant via a receptor-mediated signaling pathway that connects to the flagellar motor. More specifically, the receptor complex, which consists of receptor, the histidine kinase CheA, and the adaptor protein CheW, phosphorylates the response regulator CheY. Phosphorylated CheY interacts with the flagellar motor to induce clockwise rotation and tumbling behavior. Attractant inhibits the receptor complex, resulting in counterclockwise flagellar rotation and straight runs. Receptor complex activity is regulated by methylation, which mediates adaptation. Methylation by CheR increases receptor activity; demethylation by CheB decreases activity.

A central question of interest to both the theorists and the biologists is how does this system achieve accurate sensing in the presence of internal and external disturbances? Biological systems are constantly subjected to many different types of perturbations and yet they are able to function in a remarkably robust fashion. One example is the robustness of perfect adaptation of bacterial chemotaxis signaling to the chemoattractant aspartate (Alon *et al.*, 1999; Barkai and Leibler, 1997). To address this issue, mathematical modeling is necessary. These models can then be analyzed using techniques from control theory.

The application of control theory to biological systems originated with Wiener (1961) and flourished in the late 1960s (Milsum, 1966; Savageau, 1971), but only recently has it caught the attention of the wider molecular biology audience. Interestingly the bacterial chemotaxis system has been a focal point of research into biological control and robustness (Baker *et al.*, 2006; Kollmann *et al.*, 2005), and the experimentalists in this field have been influenced by this new perspective. Control theory covers many topics, and here we introduce basic concepts relevant to the investigation of biological complex systems. We start with the notion of sensitivity to perturbations (i.e., nonrobustness) and then generalize this idea to the sensitivity function $S(\omega)$, which is best described using a Bode plot, showing the sensitivity over a range of frequencies. We also describe how two control strategies, integral feedback control and noise filtering, reduce the sensitivity to steady-state and high-frequency disturbances.

Basic Concepts in Dynamics and Mathematical Modeling

Biological systems are quite dynamic; the properties are changing in time. Yet most qualitative descriptions of biological phenomena fail to depict dynamical information accurately. For example, an arrow diagram does not distinguish between fast and slow processes, nor does it describe the timing of events. A mathematical representation contains this information.

A dynamical variable $x(t)$ is a quantity that is evolving in time. A dynamical system is composed of a set of dynamical variables or state variables denoted by the vector $\mathbf{x(t)} = [x_1(t)\,x_2(t)...\,x_n(t)]$ that captures the "state" of the system. For biological systems, the state variables are typically the concentrations of various components in the cell (e.g., proteins, small molecules, etc.). The rate of change of a dynamical variable $x(t)$ is given by an ordinary differential equation (ODE): $dx/dt = f(x)$. For biological processes, the right-hand side of the equation represents the biochemical reaction kinetics that underlie biological functioning. Typically, these dynamics follow chemical rate laws specified by mass-action or enzymatic kinetics. We can write the dynamics of a system of species in vector form as $d\mathbf{x}/dt = \mathbf{f(x)}$. Most systems take inputs $\mathbf{u(t)}$ and transform them into outputs $\mathbf{y(t)}$. Mathematically, we describe a dynamical system with inputs and outputs using the notation $d\mathbf{x}/dt = \mathbf{f(x, u)}$; $\mathbf{y} = \mathbf{g(x, u)}$.

This chapter uses a simple model of chemoreceptor dynamics (Iglesias and Levchenko, 2001). The state variables of this system are the concentrations of unmethylated receptor ([R]), unmethylated receptor bound to ligand ([RL]), methylated receptor ([M]), and methylated receptor bound to ligand ([ML]). The variables are made dimensionless (normalized) by dividing by the total receptor concentration. These four species comprise the state of the system: $\mathbf{x} = [x_1\,x_2\,x_3\,x_4]$, where $x_1 = [R], x_2 = [RL], x_3 = [M]$, and $x_4 = [ML]$. The input u to the system is ligand concentration, $u = [L]$ (normalized by dividing by 1 μM), and the output y is the level of receptor activity, $y = Act$. The following mathematical model describes the dynamics and input–output behavior of the system:

$$\frac{d[R]}{dt} = -k_a[L][R] + k_d[RL] - k_1 CR + k_{-1}a_1[M]CB$$

$$\frac{d[RL]}{dt} = k_a[L][R] - k_d[RL] - k_1 CR + k_{-1}a_2[ML]CB$$

$$\frac{d[M]}{dt} = -k_a[L][M] + k_d[ML] + k_1 CR - k_{-1}a_1[M]CB$$

$$\frac{d[ML]}{dt} = k_a[L][M] - k_d[ML] + k_1 CR - k_{-1}a_2[ML]CB$$

$$Act = a_1[M] + a_2[ML]$$

Because this model is for illustrative purposes, we set the parameters $k_a = k_d = k_1 = k_{-1} = 1$ s^{-1} (note that because the variables and input are dimensionless and have no units, the parameters have units of time only).

In addition, the levels of CheR (CR) and CheB (CB) are set to be 1. Finally, the parameters $a_1 = 1$ and $a_2 = 0.5$ represent the fraction of methylated unbound receptors and methylated bound receptors that are active, respectively. Thus, receptor activity, the output of the system, is given by the expression $Act = a_1[M] + a_2[ML]$. The initial conditions are $[R]_0 = 1$, $[RL]_0 = [M]_0 = [ML]_0 = 0$.

The following reactions are represented in this model: receptor–ligand association ($k_a[L][R]$ and $k_a[L][M]$), receptor–ligand dissociation ($k_d[RL]$ and $k_d[ML]$), receptor methylation (k_1CR), and receptor demethylation ($k_{-1}a_1[M]CB$ and $k_{-1}a_2[ML]CB$). Note that the demethylation rate depends on the number of active receptors. This feedback mechanism is responsible for robust perfect adaptation, which is described in the section on integral feedback control. The reaction dynamics of the model are shown schematically in Fig. 1A.

Given a system of ODEs and the values of the state variables at the initial time point (initial conditions), one can solve for $\mathbf{x(t)}$ to obtain a time history using the computer to integrate the differential equations. For a small time step Δt, the computer calculates each of the derivatives and determines the change $\Delta\mathbf{x(t)}$ in the state variables. In this fashion, the trajectories of the state variables in time are determined. Figure 1B plots results of a simulation of receptor activity (Act) versus time for two different ligand concentrations.

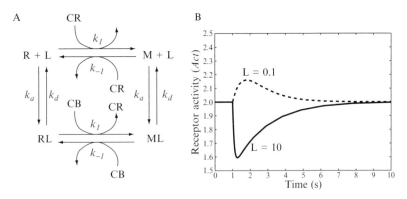

FIG. 1. Chemoreceptor model and simulation. (A) Schematic diagram of model showing reaction dynamics. Receptor–ligand binding and receptor methylation/demethylation are represented. L, ligand; R, unmethylated receptor; M, methylated receptor; RL, unmethylated receptor bound to ligand; ML, methylated receptor bound to ligand; CR, CheR; and CB, CheB. (B) Model simulation showing the change in receptor activity (Act) in response to a change in ligand concentration. Ligand levels were increased from 1 to 10 μM (solid line) or decreased from 1 to 0.1 μM (dashed line) at $t = 1$ s.

A central concept in dynamics is that of the steady state. We are often interested in the state of the system at long timescales. For a stable system, the state variables will no longer change, and the derivatives will go to 0 as $t \to \infty$. The steady-state values of the variables can be determined by running the simulation for a long time. The transient behavior refers to short timescales when the state variables are still changing.

Finally, this model is deterministic; there are no stochastic variables. Randomness does not enter into the model dynamics. Every simulation starting with the same initial conditions will produce exactly the same time course plots. The section on noise filtering introduces stochastic variables representing external noise that perturbs the system. One must appreciate that a deterministic model is an approximation of the true system dynamics in which stochastic fluctuations play an important role (Rao et al., 2002).

There are many software packages for constructing, simulating, and analyzing mathematical models of biological systems. A broad cross section of these programs can be found on the home page of the Systems Biology Markup Language (SBML) web site (http://sbml.org). For model construction using a graphical notation, we recommend the program Cell Designer (http://celldesigner.org). This chapter uses MATLAB, a large suite of programs to facilitate technical computing, which possesses a broad following in the science and engineering communities. The big advantage of MATLAB has been the development of specialized "toolboxes" by MATLAB users for particular application domains. For example, we use the Systems Biology Toolbox to calculate the steady-state parameter sensitivities in the next section. There are many excellent primers and tutorials on MATLAB available on the Internet or in books.

Critical to the success of systems biology is that mathematical models such as the one described earlier be accessible in a standardized computer readable format to other researchers so that one does not have to transcribe mathematical equations from a paper into a MATLAB file. The accepted model format standard is the Systems Biology Markup Language (SBML), and we recommend that all systems biology practitioners save a copy of their model in SBML.

Robustness and Steady-State Sensitivity Analysis

The sensitivity of the state variables and outputs to changes in the inputs and parameters is a key concept in control theory (Saltelli et al., 2005; Stelling et al., 2004). Typically one selects a particular output or variable (controlled variable) and then measures the sensitivity of the quantity of interest to particular disturbances (e.g., variations in parameters or inputs). If the sensitivity is low, then the controlled variable is robust to the particular disturbance. A robust system is one in which the key controlled variables are not sensitive to a set of common disturbances. In general, one can classify the disturbances as being internal or external to the system (e.g., intracellular versus extracellular perturbations).

This section focuses on steady-state sensitivities or the sensitivity of the steady-state value of the controlled variable y_{ss} to a disturbance. The perturbation will be a step change in the value of an input or parameter p from its nominal value to a perturbed value. The sensitivity is the slope of the change in the steady-state value of y_{ss} to the change in p: $S = dy_{ss}/dp$. This absolute sensitivity can be converted into a normalized sensitivity by determining the fractional change in y_{ss} over the fractional change in p: $S_n = \frac{dy_{ss}/y_{ss}}{dp/p}$. Here we focus on normalized sensitivities because they do not depend on the absolute magnitudes of y_{ss} or p.

Typically, sensitivities are determined using the computer (Saltelli et al., 2005). One can compute the sensitivity as the local derivative of the variable of interest, y_{ss}, and the perturbed parameter p. One calculates the steady-state value y_{ss} for a given set of nominal values for the inputs and parameters. Then, one calculates the perturbed steady-state y_{ss} value after p has been perturbed. This calculation takes the form of the finite-difference approximation of dy_{ss}/dp:

$$S = \frac{dy_{ss}}{dp} = \frac{y_{ss}(p + \Delta p) - y_{ss}(p)}{2\Delta p} + \frac{y_{ss}(p) - y_{ss}(p - \Delta p)}{2\Delta p}$$
$$= \frac{y_{ss}(p + \Delta p) - y_{ss}(p - \Delta p)}{2\Delta p}$$

To compute the normalized sensitivities one multiplies S by p/y_{ss}: $S_n = \frac{p}{y_{ss}} S$.

It is straightforward to perform this calculation using existing software. Here we use the Systems Biology Toolbox (http://www.sbtoolbox.org/) (Schmidt and Jirstrand, 2006), which is a MATLAB toolbox; another good choice is the BioSens program developed at UCSB. Both pieces of software can be downloaded for free from their respective websites and are relatively easy to install. What follows is a step-by-step sequence of commands to carry out the steady-state sensitivity analysis on our chemoreceptor model. The MATLAB commands are written in bold after the command prompt ($\gg$), and comments are placed after the percent sign in italics.

1. $\gg$ **model = SBmodel('ChemoReceptor.xml')** % *Import SBML version of chemoreceptor model into the Systems Biology Toolbox*

2. $\gg$ **output = SBsensdatastat(model)** % *Compute steady-state values*

3. $\gg$ **SBsensstat(output)** % *Calculate and plot sensitivities*

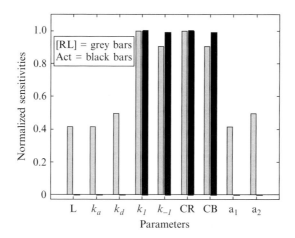

FIG. 2. Steady-state sensitivities of [RL] and Act. A bar graph depicts the normalized steady-state sensitivities (S_n) of unmethylated receptor–ligand complex, ([RL]) gray bars, and receptor activity, (*Act*) black bars, for the input [L] and the eight parameters listed on the *x* axis. For receptor activity, $S_n = 0$ for some parameters and no bar is shown.

We determined the normalized sensitivities of the variable [RL] (unmethylated receptor bound to ligand) and the output *Act* (receptor activity) to the input [L] and the eight parameters, k_a, k_d, k_1, k_{-1}, CR, CB, a_1, a_2 (Fig. 2). [RL] was sensitive to step perturbations in all of the parameters, from those involved in receptor–ligand binding (k_a, k_d, and [L]) to those involved in receptor modification. However, receptor activity was completely insensitive to step changes in ligand levels, which indicated that the model exhibits perfect adaptation. Receptor activity was also robust to changes in k_a, k_d, a_1, and a_2, but was sensitive to those parameters involved in receptor methylation and demethylation, which influence the feedback control loop that regulates receptor activity.

Thus, steady-state sensitivity analysis is easy to perform and furnishes a valuable description of the robustness of specific variables and outputs to step changes in specific parameters and inputs. Finally, it is possible to determine the sensitivities analytically by starting with a linear system (or by linearizing a nonlinear system) and using formulas derived from Metabolic Control Analysis (Heinrich and Schuster, 1996).

Constructing and Interpreting a Bode Plot

The previous section determined the steady-state sensitivity of an output *y* to a step perturbation *p*. How sensitive are the transient properties of *y* to a perturbation that is changing in time? Many disturbances are quite dynamic and the early or transient response of the system can be important. In order to generalize the notion of sensitivity to all timescales (and hence all frequencies), we need to introduce the concepts of the Fourier transform, the frequency response, and the Bode plot.

A signal $f(t)$, any quantity changing in time, can be decomposed into sinusoids (sine and cosine functions). The decomposition of a signal $f(t)$ into a combination of sinusoids is the foundation of Fourier analysis. The Fourier transform $F(\omega)$ indicates the contribution of sinusoids of frequency ω to the overall signal $f(t)$. Thus, $F(\omega)$ is the "frequency domain" representation of the "time domain" signal $f(t)$. Low frequencies represent longer timescales (steady state is when $\omega = 0$ and $t \rightarrow \infty$); high frequencies represent shorter timescales (i.e., rapidly changing signals).

Given an input signal $u(t)$, one is often interested in calculating the output $y(t)$ produced by the system. For a linear system, there is a simple relationship between the Fourier transform of the input, $U(\omega)$, and the Fourier transform of the output, $Y(\omega)$: $Y(\omega) = H(\omega)U(\omega)$, in which $H(\omega)$ is termed the frequency response of the system. One can interpret $H(\omega)$ as converting each input sinusoid component of frequency ω into an output sinusoid component of the same frequency ω but possibly different phase and magnitude depending on the phase and magnitude of $H(\omega) = |H(\omega)|e^{i\angle H(\omega)}$.

The frequency response provides a complete characterization of a linear system: knowing $H(\omega)$ for all ω means knowing the response of the system to all frequency components of the input and hence the response to the entire input. It should be noted that a nonlinear system does not possess a frequency response. However, one can linearize the nonlinear system and then determine the frequency response of the linearized approximation.

It is convenient to describe the frequency response $H(\omega)$ graphically. Visualization of $H(\omega)$ is known as a Bode plot with magnitude $|H(\omega)|$ and phase $\angle H(\omega)$ plotted separately as a logarithmic function of ω. The magnitude or gain is often measured in decibels, defined by x dB $= 20\log_{10} x$. Bode plots are useful because they provide a broad range of information on both the low- and the high-frequency properties of a system that can be compared to an extensive catalog of reference plots from a variety of canonical systems.

In the context of sensitivity analysis, we are interested in the frequency response of some output that is perturbed by some disturbance input. This frequency response, termed the sensitivity function $S(\omega)$, describes the sensitivity of an output at all frequencies to a varying parameter or input and generalizes the notion of the steady-state sensitivity, $S(\omega = 0)$. For our receptor model, we wish to calculate the sensitivity function for receptor activity to changes in chemoattractant concentration.

Because our model is nonlinear, we first need to linearize the model and then determine the Bode plot. Within MATLAB, both tasks are straightforward.

1. Construct a Simulink simulation file of the model in question. Simulink is a simulation environment in MATLAB. Briefly, one needs to make some minor adjustments to the MATLAB model file to convert it into an S-function. Then, within Simulink one can attach an input port (In1) and an output port (Out1) to the model S-function.
2. Linearize the model.

   a. ≫ [t, x, y] = sim('SimuChemoReceptor', 20, [], [0 u0]);
   % *Simulate the system for an extended period of time (20 s) subjected to an input u0.*

   b. ≫ Xss = x(end, :); % *Extract the steady-state values for state variables at end of simulation.*

   c. ≫ [A, B, C, D] = linmod('SimuChemoReceptor', Xss, u0); % *Linearize system around steady-state point and determine state-space matrices A, B, C, D*

   d. ≫ sys = ss(A, B, C, D); % *Convert state-space matrices into MATLAB system variable*

3. ≫ bode(sys); % *Construct Bode plot of system*

Both the magnitude and the phase of the frequency response will appear on the Bode plot. Each contains valuable information and they should be viewed together for maximum benefit. However, here we focus exclusively on the magnitude plot because we are most interested in the magnitude of the sensitivity.

Figure 3A displays the Bode plot of the sensitivity function of receptor activity to ligand levels for the simple chemoreceptor model. At steady state, the sensitivity is 0, and the log magnitude of $S(\omega)$ drops to $-\infty$ at $\omega = 0$. At lower frequencies, the plot is a line of positive slope, which indicates the existence of a differentiator. In other words, the system is differentiating the ligand concentration with respect to time, which is expected for the model to perform temporal sensing. At a specific break-point frequency, the plot levels off. Beyond this break-point frequency, the system no longer acts as a differentiator. At even higher frequencies the plot "rolls off" and the magnitude of the sensitivity function decreases. This roll-off indicates that higher frequency components in the input are filtered out or that the system is not fast enough to respond to these fast changes.

For comparison, we altered some of the parameters of the receptor modification reactions. Increasing the receptor methylation rate by letting CR = 2 instead of 1 resulted in a higher gain (magnitude) but did not significantly change the break point in the Bode plot (Fig. 3A). In the

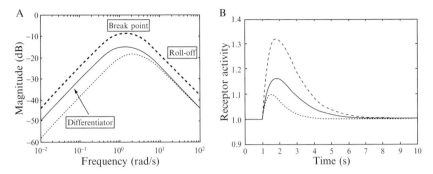

FIG. 3. Bode plot of sensitivity function compared to time-course plot of simulation. (A) Bode plot of the sensitivity function of receptor activity to ligand concentration for different values of CR and CB. CR = 1, CB = 1 (solid line); CR = 2, CB = 1 (dashed line); and CR = 2, CB = 2 (dotted line). The main features of these plots include the line of positive slope at lower frequencies (differentiator), the plateau (break point), and the decline at higher frequencies (roll-off). Altering the methylation/demethylation dynamics affects the gain (magnitude or height of plot) and break-point frequency. (B) Time-course simulations of receptor activity for different values of CR and CB. Receptor activities were normalized to 1 for direct comparison of simulations. See (A) for definition of lines. Ligand levels were dropped from 1 to 0.1 μM at $t = 1$ s. Comparing the two plots shows the relationship between peak height (time domain) and gain (frequency domain) and between adaptation time and break-point frequency.

time domain, this change corresponded to an elevated peak height, but a relatively similar adaptation time (i.e., time it takes for receptor activity to settle back to its steady-state value) when the chemoattractant concentration was dropped from 1 to 0.1 μM (Fig. 3B). Letting CB = 2 and CR = 2 shifted the break-point frequency to the right and produced a modest reduction in the gain in the Bode plot. The time course plot showed a modestly reduced peak height and a significantly reduced adaptation time. Through such comparisons, one can elucidate the inverse relationship between the break-point frequency and the adaptation time. Indeed, it makes sense because the break-point frequency is the frequency domain correlate of the integration time (i.e., time interval over which the system is measuring the attractant concentration before comparing to the next time interval), and the integration time depends on the adaptation time (Andrews et al., 2006).

Primer on Integral Feedback Control

A goal of control systems is to reduce the sensitivity of the most critical outputs to the most common disturbances (Skogestad and Postlethwaite, 1996). One of the fundamental tenets of control theory is that robustness is achieved through feedback control. In such a feedback system, the controller measures the difference between the current output (controlled variable) and the desired output and, based on this error, takes some control action that reduces the error. For example, the speed of a car is regulated by a controller (i.e., driver) that compares the current speed to the desired speed and then takes a control action (i.e., presses the accelerator or brake) to reduce the error. This control system is robust to a variety of internal (e.g., car performance) and external (e.g., wind) perturbations. For biological systems, feedback control is necessary to achieve the homeostatic regulation of the various biological components and activities.

There are three basic types of feedback control that are classified according to the mathematical operation used to convert the error into a control action (Franklin et al., 2005). In proportional control, the error term is multiplied by a constant before being fed back. In integral control, the error is integrated over time and then fed back. In derivative control, the error is differentiated. Proportional control reduces both transient and steady-state errors; integral control eliminates steady-state errors; and derivative control is especially good at decreasing transient errors.

Here we focus on integral control, which is shown schematically in Fig. 4A. This type of feedback structure ensures that the output will approach its desired level despite step changes in the parameters and inputs. Thus, there is robust tracking of a specific output steady-state value. One can understand this property by examining the feedback diagram. A generic system with gain k takes an input u and produces the output y, which here represents the deviation from the ideal output (i.e., error). This output error is integrated over time and is then fed back into the system, thereby producing the control loop. The key to integral control is that the feedback term $x = \int y \, dt$, so that $\frac{dx}{dt} = y$. Thus, if the system is stable, at steady state all of the derivatives approach 0, and so $y \to 0$ as $t \to \infty$ for all values of k and u. We refer to the equation $\frac{dx}{dt} = y$ as the integral control equation, which is characteristic of integral feedback.

From the aforementioned analysis, it is clear that integral feedback control is sufficient to produce robust perfect adaptation. The steady-state output (e.g., flagellar motion) of an integral feedback system will be the same despite step changes in the input (e.g., chemoattractant levels) or in the parameters. More importantly, a key result from control theory states that integral feedback is not only sufficient, but also necessary for robust perfect regulation against step perturbations (Yi et al., 2000). Thus, the fact that the bacterial chemotaxis signaling system exhibits robust perfect adaptation toward the chemoattractant aspartate (Alon et al., 1999) argues that integral feedback control must exist in this signaling system.

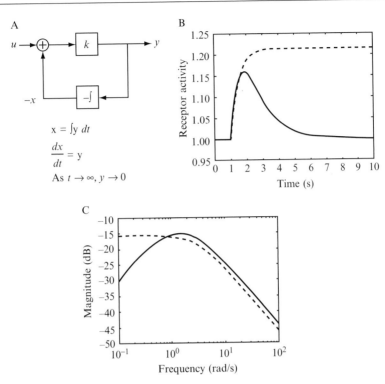

FIG. 4. Properties of integral feedback control. (A) Block diagram of integral feedback. The variable u is the input, k is the gain of the system, y is the deviation from the desired output (i.e., output error), and x is the feedback term. (B) Time-course simulations of integral control model (solid line) and modified model lacking integral control (dashed line). Receptor activities were normalized to 1 for direct comparison of simulations. Ligand levels were decreased from 1 to 0.1 μM at $t = 0$ s. The modified model no longer exhibits perfect adaptation. (C) Bode plot of integral feedback model (solid line) and model in which integral feedback is disrupted (dashed line). The plot of the modified model no longer exhibits the characteristic line of positive slope representing a differentiator.

Given a complex model, how does one know whether a particular output is under integral feedback control? One approach is to calculate the steady-state sensitivity of the output to step perturbations in the input and parameters. In Fig. 2, we observe that receptor activity in the chemoreceptor model is robust to changes in five of the parameters. Receptor activity is sensitive only to those parameters directly involved in the integral feedback regulation mechanism mediated by receptor methylation/demethylation. Thus, sensitivity analysis can provide strong evidence for integral feedback.

A more conclusive approach would be to identify the integral control equation in the model. The goal is to manipulate the differential equations so that on the right-hand side there is some expression that represents the difference between the output and some set point or steady-state value. At steady state for a stable system, the left-hand side containing the derivatives of state variables will go to 0, thus driving the right-hand side, which represents the output error to 0. What follows is the derivation of the integral control equation for the simple chemoreceptor model:

$$
\begin{aligned}
\frac{d([M] + [ML])}{dt} &= 2k_1 CR - k_{-1} a_1 [M]CB - k_{-1} a_2 [ML]CB \\
&= 2k_1 CR - k_{-1} CB(a_1 [M] + a_2 [ML]) \\
&= -k_{-1} CB(Act - Act_0)
\end{aligned}
$$

where the steady-state receptor activity $Act_0 = \frac{2k_1 CR}{k_{-1} CB}$. Here we see that the steady-state receptor activity depends on the parameters k_1, k_{-1}, CR, and CB, all of which are involved in receptor methylation and demethylation as noted previously. Not surprisingly the analysis in Fig. 2 revealed that receptor activity was sensitive to only these four parameters.

A key insight from this derivation is that the integral feedback loop depends on the fact that the receptor demethylation rate is a function of receptor activity. One can remove this dependence by modifying the receptor demethylation terms so that they are a function of methylated receptor and not receptor activity ($k_{-1}[M]CB$ and $k_{-1}[ML]CB$). In the time course plot of the modified model, one can observe that perfect adaptation is lost; the steady-state receptor activity level changes when the chemoattractant level changes (Fig. 4B). More revealingly, in the Bode plot of the sensitivity function of receptor activity to ligand levels for the modified model, we observe that the system no longer acts as a differentiator of ligand levels; the line of positive slope is gone and replaced by a line of constant gain (Fig. 4C). Thus, it is the integral feedback that gives rise to the differentiator, and indeed one can show analytically that any integral feedback control system will differentiate the input.

Noise Filtering and the Kalman Filter

The previous section considered step disturbances to the input and parameters. What about higher frequency disturbances? One would expect stochastic fluctuations in the measurement of chemoattractant concentration because a discrete number of ligand molecules are randomly diffusing into the local neighborhood of the cell and then binding receptor (Berg and Purcell, 1977). These fluctuations could interfere with the ability of the bacteria to sense real changes in the level of ligand as they move through a gradient. How can the cell determine the signal of true variations in chemoattractant levels from the noise of the stochastic fluctuations?

In engineering, filtering refers to the extraction of a signal from an observed variable that is corrupted by noise. In many cases, it is assumed that the latter is an additive stochastic process. In general, the signal tends to be of lower frequency and the noise tends to be of higher frequency. When there is an overlap in the frequency distributions of the noise and signal, precise separation is not possible, so that determining which filter is best has to be considered.

Perhaps the simplest noise-filtering strategy is to average the noisy input signal over a longer time interval. We examine a simple receptor–ligand binding reaction, $R + L \leftrightarrow RL$, in which the level of ligand $[L] = x + n$ is the sum of a deterministic component x and a stochastic noise term n. Let k_a and k_d be the association and dissociation rate constants for binding ligand. We shall vary these values while maintaining a constant equilibrium dissociation constant $K_d = k_d/k_a = 1$. For receptor–ligand complex ([RL]), when the dissociation and association rates are slower, then the noise is "averaged out" because slower binding kinetics result in longer binding time intervals (Fig. 5A). In the frequency domain, when we plot the sensitivity function of [RL] to the input [L], we can see that the slower binding rate constants change the position of the roll-off so that it occurs at lower frequencies, thereby filtering out additional noise (Fig. 5B). Deciding on the optimal cutoff frequency for the filter depends on the frequency characteristics of both signal and noise and is the subject of optimal filtering.

One popular form of optimal filter, known as the Kalman filter (Kalman, 1960), is based on slightly different assumptions of how the signals are generated. In this case, the signal to be estimated is the state $x(t)$ of a linear Markov process; that is, one in which knowledge of the entire past of the system ($x(t')$, for $t' < t$) is embedded in the state at time t so that $dx/dt = Ax + v$, where v is known as the process noise and A is a matrix describing the evolution of the state x. The estimate is based on some corrupted observation of the state: $y = Cx + n$, where C is a matrix and n is the observation noise. In this case, the Kalman filter is a recursive algorithm, which provides an estimate $\hat{x}$ of x based on y.

We illustrate some concepts of Kalman filtering by using a simple example from bacterial chemotaxis (Andrews *et al.*, 2006):

$$\frac{dx}{dt} = v$$

$$y = x + n$$

The variable x represents the level of ligand and y represents the cell measurement of ligand that is corrupted by noise n. Real changes in x are caused by the movement of the bacteria, which is influenced by rotational

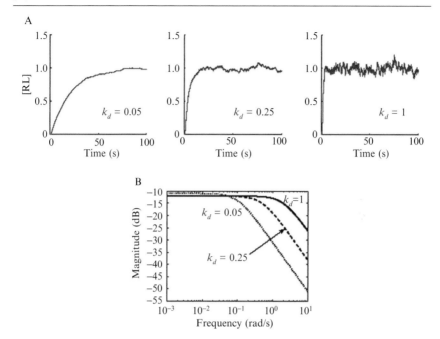

FIG. 5. Filtering of noise in ligand levels by changing the binding rate constants. (A) The dynamics of receptor–ligand complex, [RL], was simulated in response to a unit step ligand input that was corrupted by additive Gaussian noise with zero mean and 0.5 variance. The K_d was kept constant ($K_d = k_d/k_a = 1$) while the values of k_d (and k_a) were varied from 1 to 0.05. For the slower binding kinetics the amount of noise in [RL] levels was reduced because the noise in ligand levels was averaged over a longer time period. (B) Bode plot of sensitivity function of [RL] to the input [L]. Reducing k_d shifts the position of the roll-off to lower frequencies: $k_d = 1$ (solid line), $k_d = 0.25$ (dashed line), and $k_d = 0.05$ (dotted line). This change in the Bode plot explains the enhanced filtering caused by the slower binding dynamics. (See color insert.)

diffusion and is represented by the stochastic random variable v. We treat the variables v and n as zero-mean white noise processes. In optimal filtering the goal is to obtain the estimate $\hat{x}$ that minimizes the expected square error $E[(x(t) - \hat{x}(t))^2]$. One can determine the optimal steady-state filter for this system by computing the Kalman filter. In this example, it is a first-order, low-pass filter ($\omega_{cf}/(s + \omega_{cf})$) with the cutoff frequency related to the ratio of signal-to-noise covariances: $\omega_{cf} = \sqrt{E[v^2]/E[n^2]}$. Intuitively, we can think of the cutoff frequency as a compromise to the tradeoff of signal detection and noise attenuation. At frequencies below ω_{cf}, the power of the signal is greater than that of the noise; at frequencies above ω_{cf}, the power of the noise is greater.

Briefly, one can perform the matrix computations to solve for the Kalman filter in MATLAB. Given matrices A and C describing the system $dx/dt = Ax + v; y = Cx + n$, where the process ($v$) and measurement ($n$) noises have covariances V and N, respectively. One would calculate the optimal state estimator $\hat{x}$ by solving the following differential equation $\frac{d\hat{x}}{dt} = A\hat{x} + L(y - C\hat{x})$, where $L = PC^T N^{-1}$ and P is the solution to the algebraic Riccati (matrix) equation: $0 = AP + PA^T - PC^T N^{-1} CP + V$. The Riccati equation can be solved using the **care.m** function in MATLAB. In our example just given, $A = 0$ and $C = 1$.

Future Perspectives and Further Information

This chapter introduced basic concepts of control theory using a simple model of chemoreceptor dynamics. All of the exercises described previously could be performed with a few modifications on more complex models. In particular we emphasized the use of Bode plots to furnish a global view of the response or sensitivity of the system to changes in the input or parameters. How would the additional dynamics in a bigger model change the Bode plot? One would expect that the general shape of the Bode plot shown in Fig. 3A, which is characteristic of a system that performs temporal sensing, would be preserved. What would change would be the quantitative specifics, such as the gain of the system (i.e., magnitude of the plateau of the plot), the frequency of the break point, and the frequency of the roll-off. These important features are critical to the robust performance of the system and depend on the details of the system and its environment.

Beyond the basic topics covered here are a range of deeper questions that both the experimentalist and the theorist must address: (1) What are the most relevant perturbations that affect this system? (2) What sort of performance objectives is the system trying to achieve? (3) What control strategies is the system employing? (4) How optimal are these control strategies with respect to (1) and (2)? Powerful techniques from robust control theory allow one to tackle these issues in a rigorous fashion. The goal of robust control theory is to take into explicit consideration the multitude of internal and external uncertainties when assessing the performance of a system and its controller.

Finally, for those interested in learning more about dynamical systems, we recommend the text by Strogatz (1994). For those interested in an insightful introduction to mathematical modeling, we recommend the book by Gershenfeld (1999). Åström and Murray (2006) have written an excellent introduction to control theory that is currently available only on the Internet, and a more advanced treatment that includes robust control theory is provided by Skogestad and Postlethwaite (1996). We also recommend the following systems biology books (Bower and Bolouri, 2001; Fall *et al.*, 2002; Szallasi *et al.*, 2006).

References

Alon, U., Surette, M. G., Barkai, N., and Leibler, S. (1999). Robustness in bacterial chemotaxis. *Nature* **397**, 168–171.

Andrews, B. W., Yi, T.-M., and Iglesias, P. A. (2006). Optimal noise filtering in the chemotactic response of *E. coli. PLOS Comput. Biol.* **2**, 1407–1418.

Åström, K. J., and Murray, R. M. (2006). "Feedback Systems: An Introduction for Scientists and Engineers."

Baker, M. D., Wolanin, P. M., and Stock, J. B. (2006). Systems biology of bacterial chemotaxis. *Curr. Opin. Microbiol.* **9**, 187–192.

Barkai, N., and Leibler, S. (1997). Robustness in simple biochemical networks. *Nature* **387**, 913–917.

Berg, H. C. (1993). "Random Walks in Biology." Princeton University Press, Princeton, NJ.

Berg, H. C., and Purcell, E. M. (1977). Physics of chemoreception. *Biophys. J.* **20**, 193–219.

Bower, J. M., and Bolouri, H. (2001). "Computational Modeling of Genetic and Biochemical Networks." MIT Press, Cambridge, MA.

Fall, C. P., Marland, E. S., Wagner, J. M., and Tyson, J. J. (2002). "Computational Cell Biology." Springer-Verlag, New York.

Franklin, G. F., Powell, J. D., and Emami-Naeini, A. (2005). "Feedback Control of Dynamic Systems." Prentice Hall.

Gershenfeld, N. (1999). "The Nature of Mathematical Modeling." Cambridge University Press, Cambridge, UK.

Heinrich, R., and Schuster, S. (1996). "The Regulation of Cellular Systems." Chapman & Hall, New York.

Iglesias, P. A., and Levchenko, A (2001). A general framework for achieving integral control in chemotactic biological signaling mechanisms. *In* "IEEE Conference on Decision and Control," pp. 843–848. Orlando, FL.

Kalman, R. E. (1960). A new approach to linear filtering and prediction problems. *Trans. ASME J. Basic Eng. Series D* **82**, 35–46.

Kollmann, M., Lovdok, L., Bartholome, K., Timmer, J., and Sourjik, V. (2005). Design principles of a bacterial signalling network. *Nature* **438**, 504–507.

Milsum, J. H. (1966). "Biological Control Systems Analysis." McGraw-Hill, New York.

Rao, C. V., Wolf, D. M., and Arkin, A. P. (2002). Control, exploitation and tolerance of intracellular noise. *Nature* **420**, 231–237.

Saltelli, A., Ratto, M., Tarantola, S., and Campolongo, F. (2005). Sensitivity analysis for chemical models. *Chem. Rev.* **105**, 2811–2827.

Savageau, M. A. (1971). Parameter sensitivity as a criterion for evaluating and comparing the performance of biochemical systems. *Nature* **229**, 542–544.

Schmidt, H., and Jirstrand, M. (2006). Systems Biology Toolbox for MATLAB: A computational platform for research in systems biology. *Bioinformatics* **22**, 514–515.

Skogestad, S., and Postlethwaite, I. (1996). "Multivariable Feedback Control: Analysis and Design." Wiley, New York.

Stelling, J., Gilles, E. D., and Doyle, F. J., 3rd (2004). Robustness properties of circadian clock architectures. *Proc. Natl. Acad. Sci. USA* **101,** 13210–13215.

Stock, J. B., Lukat, G. S., and Stock, A. M. (1991). Bacterial chemotaxis and the molecular logic of intracellular signal transduction networks. *Annu. Rev. Biophys. Biophys. Chem.* **20,** 109–136.

Strogatz, S. H. (1994). "Nonlinear Dynamics and Chaos." Addison-Wesley, Reading, MA.

Szallasi, Z., Stelling, J., and Periwal, V. (2006). "System Modeling in Cellular Biology." MIT Press, Cambridge, MA.

Wiener, N. (1961). "Cybernetics." MIT Press, New York.

Yi, T.-M., Huang, Y., Simon, M. I., and Doyle, J. (2000). Robust perfect adaptation in bacterial chemotaxis through integral feedback control. *Proc. Natl. Acad. Sci. USA* **97,** 4649–4653.

[7]   Classification of Response Regulators Based on
Their Surface Properties

By Douglas J. Kojetin, Daniel M. Sullivan,
Richele J. Thompson, and John Cavanagh

Abstract

The two-component signal transduction system is a ubiquitous signaling module present in most prokaryotic and some eukaryotic systems. Two conserved components, a histidine protein kinase (HPK) protein and a response regulator (RR) protein, function as a biological switch, sensing and responding to changes in the environment, thereby eliciting a specific response. Extensive studies have classified the HPK and RR proteins using primary sequence characteristics, domain identity, domain organization, and biological function. We propose that structural analysis of the surface properties of the highly conserved receiver domain of RRs can be used to build on previous classification methods. Our studies of the OmpR subfamily RRs in *Bacillus subtilis* and *Escherichia coli* reveal a notable correlation between the RR receiver domain surface classification and previous classification of cognate HPK proteins. We have extended these studies to analyze the receiver domains of all predicted RR proteins in the marine-dwelling bacterium *Vibrio vulnificus*.

Introduction

Biologists have long used classification as a means of organizing and understanding the complex nature within and between organisms. With the advent of large-scale genomic sequencing, it becomes possible for the experimentalist to infer the function of a protein found in a less-studied organism based on information available from the extensive studies of other model organisms, such as *Bacillus subtilis* and *Escherichia coli*—the so-called *structure–function* relationship (Ouzounis *et al.*, 2003). Classification by protein *structure* broadly describes the use of the primary sequence (amino acid composition) and structural folds (three-dimensional structures of individual protein domains) to suggest possible relationships between distinct sets of proteins—in essence, the *chemical* makeup of proteins. However, classification based on protein *function* refers to the use of *biological* information, such as the role of the protein or domain within the organism. When combined, these methods of classification can

METHODS IN ENZYMOLOGY, VOL. 422   0076-6879/07 $35.00
                         DOI: 10.1016/S0076-6879(06)22007-X

provide hypotheses that allow the experimentalist to focus and direct future *in vitro* and *in vivo* studies in a specific direction.

Two-component signal transduction systems have been a subject of classification on both structural and functional levels. The first formal classification of these proteins included 348 histidine protein kinase (HPK) proteins and 298 response regulator (RR) proteins (Grebe and Stock, 1999). The analysis utilized features such as primary sequence alignments (conserved amino acid residues), tertiary structure (fold and domain organization), protein function (enzymatic, DNA binding, ligand binding, etc.), and biological function (involved in sporulation, chemotaxis, etc.). The study revealed that distinct subgroups of HPKs exist, and a strong correlation between HPK and RR subfamilies was noted, suggesting that subclasses of two-component systems have evolved independently. In a subsequent study, the HPKs in *B. subtilis* were analyzed and classified into groups as a basis for determining specificity of the RR for its cognate HPK (Fabret *et al.*, 1999). These studies provided a detailed classification of HPKs based on features such as homology around the conserved phospho-histidine residue and the lengths of intracellular and extracellular loops and domains. However, only a broad classification of RRs was provided from these previous studies, focusing primarily on sequence similarity of the receiver domain and function of the output domain.

Classification of the Receiver Domain of RRs Using Protein Interaction Surfaces

Information derived from a SpoOF–SpoOB cocrystal structure (Zapf *et al.*, 2000) has been used previously to analyze sequence alignments of the receiver domains of the OmpR subfamily RRs (Hoch and Varughese, 2001). A correlation was observed between residues that comprise the interaction surface between the receiver domain of RR proteins and the four-helix bundle motif of HPKs and phosphotransferase proteins. These residues consist of three general types of amino acids: (i) essential invariant catalytic residues directly involved in the phosphotransfer mechanism, (ii) anchor residues that establish broad orientational contacts for catalysis, and (iii) recognition residues that ensure the correct two proteins come together. This patch of residues on the RR protein, suggested to be important for recognition specificity, is found on the surface containing α-helix 1, α-helix 5, and the β4-α4 loop. Subsequent cocrystal structures and biochemical analyses have confirmed the role of this receiver domain surface in the interaction between the RR and its cognate four-helix bundle motif (Tzeng and Hoch, 1997; Xu *et al.*, 2003; Zhao *et al.*, 2002). Examples of this interaction are shown in Fig. 1. This was further confirmed by another

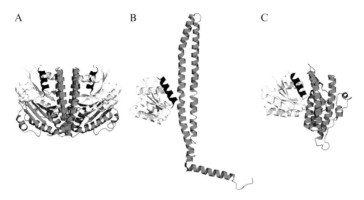

FIG. 1. Response regulator/four-helix bundle cocrystal structures. The receiver domain of the RR protein is shown in white with α-helix 1 highlighted in black. Four-helix bundle proteins are shown in gray. (A) Spo0F (RR) and Spo0B (four-helix bundle); (B) CheY (RR) and CheZ (four-helix bundle); (C) Sln1 (RR) and Ypd1 (four-helix bundle).

study that noted the four-helix bundle motif contains recognition specificity for its cognate RR (Ohta and Newton, 2003). All of these observations lead to the notion that the interaction surface on the RR is conserved, not only within a single protein among a variety of related organisms, but perhaps among all RR proteins.

Hoch and Varughese (2001) noted that the interaction surface residues on the RR, which they described as a mosaic pattern of surface residues, resists evolutionary change among orthologous RRs. However, when comparing a related group of proteins, such as those in the OmpR subfamily, the differences in conservation are dependent on the role of the amino acid. Catalytic residues are almost strictly conserved. Anchor residues are also conserved, although perhaps to a lesser degree. Recognition residues, however, display a moderate degree of variability and likely provide the molecular foundation for protein recognition and specificity. To ensure that unproductive interactions between the HPK and the incorrect RR are prevented, a means of discrimination must be discernible. This discrimination most likely comes in the form of subtle surface variability for the so-called recognition residues. If this is the case, then it is plausible that a specific region on the receiver domain surface that shows heightened variability across the family may act as the primary contributor in HisKA/Hpt domain interactions. Additionally, there may be similar surface patterns of amino acids that have been evolutionarily conserved or modified between smaller groups of proteins.

Based on the aforementioned studies, an interesting question arises: can biological information, such as the evolutionary-conserved interaction surface on RR receiver domains, as detailed by Hoch and Varughese (2001), be used in conjunction with sequential and structural information to further subclassify members of this related subfamily? In other words, can we use the similarities and differences of the residues on the receiver domain of the RR that comprise the HisKA/Hpt domain interaction surface to further subclassify these protein domains? This chapter outlines our experience in performing a subclassification analysis of the receiver domain of RRs in the OmpR subfamily from *B. subtilis* and *E. coli* using a comparative modeling approach (Kojetin *et al.*, 2003). Additionally, we describe an extension of this analysis to all predicted receiver domains in the pathogenic marine bacterium *Vibrio vulnificus*.

Modeling and Subclassification of Receiver Domains of OmpR Subfamily RRs in *B. subtilis* and *E. coli*

We used the OmpR subfamily RR proteins from *B. subtilis* and *E. coli* to test this method of classification. All of these proteins fall within the previously classified group R_A1 (Grebe and Stock, 1999) and have a conserved domain architecture, including an N-terminal receiver domain followed by a C-terminal winged-helix DNA-binding domain. Figure 2 shows an example of a full-length OmpR subfamily protein structure, DrrB from *T. maritime* (Robinson *et al.*, 2003). Within the OmpR subfamily, the receiver and DNA-binding domains display a large degree of structural similarity, although a few exceptions exist. There is modest structural variability in α-helix 4 in the

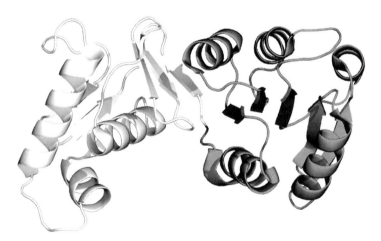

Fig. 2. A structural example of an OmpR subfamily RR: DrrB from *T. maritima*. The N-terminal receiver domain is shown in gray and the C-terminal DNA-binding output domain in white.

receiver domain and the $\alpha 2$-$\beta 2$ loop in the DNA-binding domain that may help guide domain–domain interactions and target specificity.

The best, and preferred, method to map biological and functional information onto a three-dimensional structure of a protein would be to use one that has been experimentally determined using a technique such as nuclear magnetic resonance (NMR) spectroscopy or X-ray crystallography. At present, only a handful of experimentally determined receiver domain structures exist for this group of proteins. However, the receiver domain of RRs has a structural fold that is highly conserved within and between different organisms. If a protein sequence of unknown structure (the *target*) possesses a high degree of sequence similarity to a protein with a experimentally determined structure (the *template*), it is possible to use comparative "homology" modeling to approximate the three-dimensional structure of the protein target based on the actual structure of the protein template. The following outlines the general strategy used for the development of OmpR subfamily RR receiver domain comparative models.

Target and Template Identification

The target sequences for the OmpR subfamily RR receiver domains from *B. subtilis* and *E. coli* used in this study are shown in Table I. The sequences were categorized using the Protein Information Resource (Wu *et al.*, 2002) and Kyoto Encyclopedia of Genes and Genomes (Kanehisa *et al.*, 2002) databases and obtained from the SWISS-PROT/TrEMBL database (Bairoch and Apweiler, 2000). A PSI-BLAST (Altschul *et al.*, 1997) search against protein structures present in the Protein Data Bank deposited before April 2002 (Berman *et al.*, 2000) was used to rank template structures (three-dimensional structures used as templates to create models of the target sequences) using the E-value.

Target–Template Alignment and Comparative Model Building

Sequence alignments between the target sequence and template structures were derived using the SALIGN and ALIGN2D commands in MODELLER 6v2 (Marti-Renom *et al.*, 2000). Alignments were inspected visually to assure the quality of the alignment based on the known conserved and active site residues, as well as conserved secondary structure elements found within the receiver domains of RRs. Fifty models per target were calculated using default MODELLER parameters, with one exception—the degree of refinement was set to very fast MD annealing 'refine 1'. Sequences of the four most similar structures, determined based on an assay described later for ArcA from *E. coli*, were used to generate structural models of the

TABLE I

BACILLUS SUBTILIS AND ESCHERICHIA COLI OmpR SUBFAMILY RR PROTEINS USED AS TARGET SEQUENCES FOR MODELING

	B. subtilis				E. coli		
Protein	Sequence[a]	Entry name[b]	Accession[c]	Protein	Sequence[a]	Entry name[b]	Accession[c]
CssR	1–120	CSSR_BACSU	O32192	ArcA	1–118	ARCA_ECOLI	P03026
PhoP	1–118	PHOP_BACSU	P13792	BaeR	12–125	BAER_ECOLI	P30846
ResD	8–121	RESD_BACSU	P35163	BasR	1–116	BASR_ECOLI	P30843
SpaR	1–115	SPAR-BACSU	P33112	CpxR	1–115	CPXR_ECOLI	P16244
YbdJ	1–116	N/A	O31432	CreB	1–119	CREB_ECOLI	P08368
YcbL	1–112	YCBL_BACSU	P42244	CusR	1–116	CUSR_ECOLI	P77380
YccH	1–119	YCCH_BACSU	P70955	KdpE	1–116	KDPE_ECOLI	P21866
YclJ	1–118	YCLJ_BACSU	P94413	OmpR	1–120	OMPR_ECOLI	P03025
YkoG	1–122	YKOG_BACSU	O34903	PcoR	1–117	PCOR_ECOLI	Q47456
YrkP	1–116	YRKP_BACSU	P54443	PhoB	1–120	PHOB_ECOLI	P08402
YtsA	1–119	N/A	O34951	PhoP	1–116	PHOP_ECOLI	P23836
YvcP	1–119	N/A	O06978	QseB	1–116	QSEB_ECOLI	P52076
YvrH	1–120	N/A	P94504	RstA	6–119	RSTA_ECOLI	P52108
YxdJ	1–116	YXDJ_BACSU	P42421	TorR	1–117	TORR_ECOLI	P38684
YycF	1–117	YYCF_BACSU	P37478	YedW	1–115	YEDW_ECOLI	P76340

[a] Sequence range was taken from the "Features" section of the profile in the Swiss-Prot/ExPASY database if available or estimated from a sequence alignment if the sequence range was unavailable from the profile at the time of modeling.

[b] Swiss-Prot/ExPASY database entry name, if available.

[c] Swiss-Prot/ExPASY database primary accession number.

template sequences. Nearly all aspects of model generation and analysis were semiautomated using perl scripts written in-house.

Evaluation of Comparative Models

Several methods were used to evaluate the quality of the models. Models were initially ranked based on their MODELLER defined molecular probability density function, or objective function, which encompasses the overall energy constraint violation (Fiser *et al.*, 2002). PROCHECK-NMR (Laskowski *et al.*, 1996) was used to analyze Ramachandran stereochemical parameters of the models. ERRAT (Colovos and Yeates, 1993) was used to analyze interactions between different atom types, which identify regions in the structure that are acceptable by means of a confidence level value that encompasses the percentage of residues in acceptable conformations. VERIFY3D (Eisenberg *et al.*, 1997) was used to analyze the side-chain environment based on the solvent accessibility of the side chain and the fraction of the side chain covered by polar atoms. Cα RMSD between the model and the solved structure of PhoB (PDB: 1B00) from *E. coli* (Sola *et al.*, 1999), as it generally had the most sequence similarity to the OmpR subfamily receiver domains in the study, was calculated using the SUPERPOSE command in MODELLER. As a final measure of quality, models were inspected visually to ensure that the secondary structure was consistent with that of known RR structures and conserved residues were in expected regions of the structure. In summary, final models were generally chosen based on their overall quality, defined by the following criteria: low objective function values, low disallowed Ramachandran regions, and highest favored/additionally allowed Ramachandran regions, lowest Cα RMSD values, highest ERRAT, and VERIFY3D values.

Model Visualization

Models constructed were visualized using the program PYMOL (DeLano, 2002). Hydrophobic plots were produced by highlighting the side-chain atoms of the following residues (one-letter code): I, L, V, G, A, F, C, M, S, T, W, Y, P, and H. In order to visualize subtle changes in surface side-chain hydrophobicity, a color-coded hydrophobic scale was utilized that highlighted side chains based on a relative hydrophobicity scale derived from averaged physicochemical properties of the amino acid side chains (Kyte and Doolittle, 1982; Wolfenden *et al.*, 1981). Hydrophobic rankings were assigned and color coded as a gradient, from high to low as follows: ILV (red), GAF (orange), CM (yellow), ST (green), and WYPH (blue).

Determination of Optimal Modeling Parameters

The optimal parameters for use in constructing the models were determined using an assay modified from a previous study (Kirton *et al.*, 2002). The receiver domain of ArcA from *E. coli* was chosen to perform the assay.

The sequence of the ArcA receiver domain, consisting of residues 1–118, was compared to three-dimensional structures found in the PDB by means of a PSI-BLAST search against the sequences in PDB and ranked based on their sequence similarity by evaluation of the expectation value, or E-value (Table II). The E-value describes the statistical significance of each alignment in terms of the number of times a hit would be expected to get a false relationship with a similar score (Jones and Swindells, 2002)—generally, the lower the E-value, the more significant the score. Proteins with abnormal structural elements compared to conventional, monomeric RR domains, such as Spo0A (PDB: 1DZ3), which has a domain swapped α-helix 5 (Lewis *et al.*, 2000), were not used as templates.

In the Kirton study, models of cytochrome P450 2C5 (473 residues) were generated using either a single template or multiple templates with varied methods of alignment. The optimal modeling method involved the use of five templates, and alignments were based on the secondary structure of the templates with the predicted structure of the target sequence. Unlike the target of the Kirton study, the receiver domains of RR proteins are shorter in sequence (approximately 120 residues) and have a higher degree of sequence similarity. Thus, the quality of the alignment between the target and templates should be greater in comparison to the Kirton study.

Table III outlines the setup for the determination of optimal modeling parameters in terms of (i) the number of templates used to model the target and (ii) the degree of MD refinement. The identity and the number of the templates used were adapted from the Kirton study and used the PSI-BLAST rankings shown in Table II. The degree of refinement used in the assay included no refinement, very fast MD annealing ('refine 1'), or fast MD annealing ('refine 2'). Other slower methods of MD annealing were available but required considerably more computational time and did not show an improvement in the calculated models (data not shown) and, therefore, were left out of the assay. For each set of parameters, 20 models were constructed and the lowest energy model was analyzed based on aforementioned criteria outlined for the evaluation of comparative models.

Results of the parameter assay shown in Table III revealed that the use of one template produced models with low Cα RMSD to the solved structure of PhoB, but higher disallowed Ramachandran values compared to models produced from multiple templates. The evaluation of a model built from one template will vary and is dependent on the degree of sequence similarity between the target and the template. The more similar the target sequence is to the template, the lower the Cα RMSD of the target–template superimposition. The increase in Cα RMSD between models calculated from a single template versus multiple templates can be easily understood, as using multiple templates allows a wider sampling of structure space, and thus a potential for an increase in Cα RMSD as compared to a single

TABLE II
PSI-BLAST Results for the ArcA Receiver Domain from *E. coli*

PDB	Protein	Length	Score	Bits	E-value	Identities	%	Positives	%	Gaps	%
1B00	PhoB	120	139	350	1.00E-34	43	35.8	69	57.5	3	2.5
1DC7	NtrC	112	137	347	2.00E-34	25	22.3	54	48.2	1	0.9
1NAT	Spo0F	119	128	322	1.00E-31	32	26.9	61	51.3	1	0.8
3CHY	CheY	117	125	314	1.00E-30	29	24.8	53	45.3	4	3.4
1TMY	CheY-T	118	124	312	3.00E-30	35	29.7	56	47.5	2	1.7
1DBW	FixJ	115	120	301	5.00E-29	26	22.6	50	43.5	1	0.9
1DCF	ETR1	112	109	273	8.00E-26	22	19.6	50	44.6	7	6.3
1A2O	CheB	109	107	268	3.00E-25	29	26.6	53	48.6	6	5.5
1A04	NarL	115	98.6	245	1.00E-22	31	27	61	53	3	2.6

TABLE III

OPTIMAL MODELING PARAMETER ASSAY FOR THE ArcA RECEIVER DOMAIN FROM *E. COLI*

Templates	Refinement[a]	$(\phi,\psi)_M{}^b$	$(\phi,\psi)_A{}^b$	$(\phi,\psi)_G{}^b$	$(\phi,\psi)_D{}^b$	ERRAT[c]	RMSD[d]
1	none	91.33	6.09	1.59	1.05	94.3	0.43
1	refine_1	91.24	6.68	1.18	1	93	0.44
4	none	93.57	5.16	0.93	0.39	88.9	1.1
4	refine_1	94.37	4.82	0.78	0.05	90.4	1.1
4	refine_2	94.66	4.83	0.5	0.05	91	1.11
5	none	94.57	4.73	0.49	0.25	91.5	1.13
5	refine_1	94.52	4.77	0.64	0.1	88.5	1.14
5	refine_2	94.9	4.21	0.84	0.1	90.2	1.15
6	none	96.33	3.35	0.15	0.2	91.7	1.11
6	refine_1	95.85	3.64	0.35	0.2	90.9	1.11
6	refine_2	95.9	3.63	0.25	0.25	90.8	1.11
8	none	94.7	4.78	0.35	0.2	88.1	1.14
8	refine_1	93.75	4.97	0.88	0.45	85.7	1.15

[a] Degree of refinement defined in MODELLER.
[b] Ramachandran values: M, most favored; A, additionally allowed; G, generously allowed; D, disallowed.
[c] Value derived from ERRAT analysis.
[d] Cα RMSD to PhoB (PDB: 1B00).

template. As a result, we decided to utilize more than one template to calculate the models.

In most cases, the use of no MD refinement resulted in lower quality models. Fast MD annealing ('refine 2'), as compared to very fast MD annealing ('refine 1'), did not improve the quality of the models and, in some cases, resulted in a decrease in some quality criteria. Increasing the number of templates from one to four resulted in an increase in the quality of the templates, but going from four to eight templates resulted in decreased template quality, most notably in disallowed Ramachandran values, ERRAT analysis, and Cα RMSD superimposition to the PhoB structure. This quality decrease is likely due to the introduction of templates with lower homology, thus increasing the gap size of the alignment between the target and the average template. From the results shown in Table III, it was determined that using four templates and very fast MD annealing (refine 1), termed model4-refine1, would produce adequate models for this study.

Model vs Experimentally Determined Structure: PhoB from E. coli

As a second test for optimal parameters, PhoB from *E. coli* was modeled using a similar, abbreviated methodology. In this case, the solved structure of PhoB (PDB: 1B00) was used neither as a template nor for Cα RMSD

TABLE IV

OPTIMAL MODELING PARAMETER ASSAY FOR THE PhoB RECEIVER DOMAIN FROM *E. COLI*

Templates	Refinement[a]	$(\phi,\psi)_M$[b]	$(\phi,\psi)_A$[b]	$(\phi,\psi)_G$[b]	$(\phi,\psi)_D$[b]	ERRAT[c]	RMSD[d]
1	none	94.26	5.05	0.33	0.38	86.36	1.41
1	refine_1	94.36	5.01	0.45	0.2	87.56	1.4
4	none	94.67	4.65	0.23	0.48	91	1.2
4	refine_1	95.33	4.44	0.2	0.05	91.7	1.19
5	none	93.9	5.46	0.33	0.32	90	1.18
5	refine_1	94.51	5.21	0	0.3	92.06	1.17

[a] Degree of refinement defined in MODELLER.
[b] Ramachandran values: M, most favored; A, additionally allowed; G, generously allowed; D, disallowed.
[c] Value derived from ERRAT analysis.
[d] $C\alpha$ RMSD to PhoB (PDB: 1B00).

superimposition so as to not bias modeling run. Instead, the structure with the highest degree of structural similarity was used for the calculation of $C\alpha$ RMSD. Based on the results, shown in Table IV, it was determined that the rationale for determining modeling parameters and our choice of the model4-refine1 strategy were both satisfactory.

Figure 3 shows a comparison of the *E. coli* PhoB model and *E. coli* PhoB solved structure. The model and the solved structure are very similar. Additionally, the hydrophobic surface characteristics between the model and the solved structure reveal similar trends in amino acid surface composition, particularly within the surface composed of α-helix 1, α-helix 5, and the $\beta4$-$\alpha4$ loop. There are differences in the surface composition in the surface defined by α-helix 4. This helix has been noted to display a high degree of structural deviation within experimentally determined receiver domain RR structures (Birck *et al.*, 2003). Accordingly, this deviation in helix positioning makes it difficult to obtain a correct topology for the α-helix 4 in the models generated.

Modeled Receiver Domains of the OmpR Subfamily of RRs from B. subtilis *and* E. coli

Thirty receiver domain models of the OmpR subfamily RRs were calculated and analyzed, 15 each from *B. subtilis* and *E. coli*. Table V shows a summary of the quality evaluation parameters of the final models. Figure 4 shows the superimposition of models generated from *B. subtilis* or *E. coli* and reveals that the backbone traces of the models are extremely similar. The backbone similarity is an artifact of the comparative modeling process. When the structures of these proteins are solved using traditional methods, such as

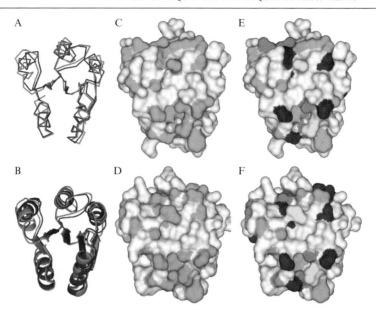

FIG. 3. Modeled vs experimentally determined structure of PhoB from *E. coli*. Cα align-ment between the modeled (blue) and the solved (red) structure of PhoB shown in (A) ribbon and (B) cartoon diagrams. Comparison of the hydrophobic surfaces of the model (C, E) and experimentally determined structure (D, F) of PhoB using a (C, D) single color for all hydrophobic residues or (E, F) a color-coded hydrophobic scale as described in the text. (See color insert.)

X-ray crystallography or NMR, these proteins are unlikely to have such similar backbone structures. As mentioned previously, a structure deviation in the models is observed in the positioning of α-helix 4, which also has been observed in experimentally solved three-dimensional RR structures (Birck *et al.*, 2003). However, the emphasis of this work is on the more structurally conserved region consisting of the α-helix 1 and the α-helix 1/α-helix 5 interface, which shows a very high degree of similarity with respect to the surface characteristics of receiver domains with solved structures.

The analysis provided by Hoch and Varughese (2001) is a good platform from which to develop and test a method for subclassifying the receiver domain of RRs. In their study, a sequence alignment-based approach was utilized to group residues into categories based on their suggested role in protein recognition. We build on this by using a three-dimensional approach to provide a more visual method for discerning differences in the HisKA/Hpt interaction surfaces among the receiver domains of the OmpR subfam-ily RRs in *B. subtilis* and *E. coli*. The hydrophobic surface characteristics of the modeled receiver domains of the OmpR subfamily of RRs from

TABLE V

STATISTICS FOR MODELS OF OmpR SUBFAMILY RECEIVER DOMAINS IN *B. SUBTILIS* AND *E. COLI*

Protein	PDF[a]	$(\phi,\psi)_M$[b]	$(\phi,\psi)_A$[b]	$(\phi,\psi)_G$[b]	$(\phi,\psi)_D$[b]	ERRAT[c]	RMSD[d]	VERIFY3D[e]
				B. subtilis				
CssR	5934.719	94.30%	4.70%	0.90%	0.00%	97.3	1.055	50.92
PhoP	6060.0322	93.40%	6.60%	0.00%	0.00%	95.5	1.1694	41.05
ResD	5886.9604	95.00%	5.00%	0.00%	0.00%	99.1	1.0841	47.26
SpaR	5890.122	90.00%	9.00%	1.00%	0.00%	96.3	0.7885	39.12
YbdJ	6055.074	91.00%	6.00%	3.00%	0.00%	86.1	1.1721	45.41
YcbL	5181.8891	95.90%	3.10%	1.00%	0.00%	97.1	1.021	42.33
YccH	6804.8242	93.60%	6.40%	0.00%	0.00%	90.1	1.1753	34.89
YclJ	5920.5512	91.50%	5.70%	2.80%	0.00%	92.7	0.7271	46.8
YkoG	5970.409	95.30%	4.70%	0.00%	0.00%	98.2	1.2321	54.3
YrkP	5821.97	93.20%	5.80%	1.00%	0.00%	97.2	1.1569	50.32
YtsA	6143.8984	93.50%	6.50%	0.00%	0.00%	95.5	0.9415	48.83
YvcP	5965.3872	90.70%	9.30%	0.00%	0.00%	89.2	0.8896	48.31
YvrH	5872.4931	90.70%	9.30%	0.00%	0.00%	94.6	0.879	50.42
YxdJ	5488.5961	95.20%	3.80%	1.00%	0.00%	81.5	0.7809	46.7
YycF	5618.9492	93.30%	5.70%	1.00%	0.00%	98.2	0.8776	56.54

(continued)

TABLE V (*continued*)

Protein	PDF[a]	$(\phi,\psi)_M$[b]	$(\phi,\psi)_A$[b]	$(\phi,\psi)_G$[b]	$(\phi,\psi)_D$[b]	ERRAT[c]	RMSD[d]	VERIFY3D[e]
				E. coli				
ArcA	5939.2714	94.30%	5.70%	0.00%	0.00%	93.6	1.1401	52.6
BaeR	5584.657	91.80%	8.20%	0.00%	0.00%	97.2	0.7545	50.96
BasR	5946.1103	91.30%	8.70%	0.00%	0.00%	90.7	1.012	54.15
CpxR	5575.602	95.10%	4.90%	0.00%	0.00%	99.1	0.6516	52.45
CreB	5885.5844	91.30%	7.80%	1.00%	0.00%	89.2	1.1174	46.32
CusR	5473.1123	91.00%	8.00%	1.00%	0.00%	96.3	1.1455	55.65
KdpE	5529.2871	94.10%	5.90%	0.00%	0.00%	97.2	0.9648	59.55
OmpR	5635.6743	95.30%	4.70%	0.00%	0.00%	92.9	0.7747	57.78
PcoR	5842.973	95.00%	4.00%	1.00%	0.00%	97.2	1.0511	45.73
PhoB	5665.297	97.10%	2.90%	0.00%	0.00%	92.9	1.1542	49.36
PhoP	5570.972	95.30%	3.80%	0.90%	0.00%	97.2	1.0595	45
RstA	5553.8344	94.00%	6.00%	0.00%	0.00%	91.5	1.0776	44.76
TorR	5644.8134	94.20%	4.90%	1.00%	0.00%	95.4	0.8336	63.27
YedW	5840.814	89.20%	9.80%	1.00%	0.00%	95.3	1.0631	55.42
YgiX	5929.464	96.00%	4.00%	0.00%	0.00%	97.2	1.1188	46.92

[a] Molecular probability density function as defined by MODELLER.
[b] Ramachandran values: M, most favored; A, additionally allowed; G, generously allowed; D, disallowed.
[c] Value derived from ERRAT analysis.
[d] Cα RMSD to PhoB (PDB: 1B00).
[e] Value derived from VERIFY3D analysis.

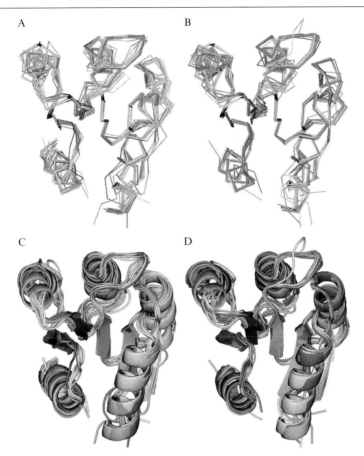

FIG. 4. Superimposition of modeled regulatory domains. Cα superimpositions displayed as PyMOL ribbon (A, B) or cartoon (C, D) diagrams for the OmpR subclass regulatory domains from *B. subtilis* (A, C) and *E. coli* (B, D).

B. subtilis and *E. coli*, respectively, are shown in Figs. 5 and 6. Hydrophobic residues are shown in "conventional" single-color format in order to visually detect hydrophobic surface trends between these protein domains. Comparison of the surface characteristics reveals a conserved patch of hydrophobic residues consisting of the α-helix 1/α-helix 5 interface. This observation is consistent with the notion that interactions between RR and HisKA/Hpt domains are principally hydrophobic in nature (Hoch and Varughese, 2001; Zapf *et al.*, 2000).

In order to detect subtle changes in surface side-chain hydrophobicity, a color-gradient scheme, based on hydrophobic characteristics of the amino

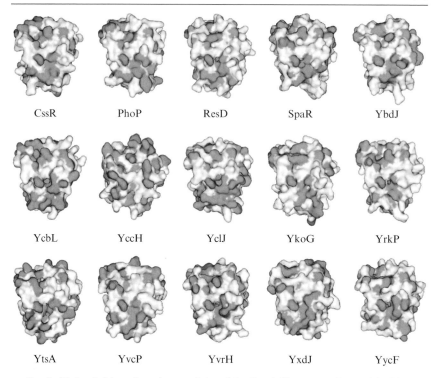

FIG. 5. Hydrophobic surface characteristics of the *B. subtilis* comparative models. The perspective is such that the region comprising the α-helix 1 and α-helix 1/α-helix 5 interface is visible.

acid side chains, was used to represent different levels of hydrophobicity rather than the simpler single color plots.

As mentioned previously, two available hydrophobic scales were combined and weighted to provide a color-gradient scheme to allow the relative hydrophobic strengths and composition to be compared and contrasted quickly. The general subclassification procedure consists of several stages. An example of this scoring procedure is shown in Fig. 7 for the receiver domains of PhoP, ResD, and YycF from *B. subtilis*. Initially, a surface square of approximately 225 $\mathring{A}^2$ (15 × 15 $\mathring{A}$) is centered about the α-helix 1/α-helix 5 interface. This area encompasses the HisKA/Hpt domain recognition residues in α-helix 1 and α-helix 5. This square was divided into three vertical strips (~5 $\mathring{A}$ each wide), roughly corresponding to α-helix 5 and the β4-α4 loop (strip 1), α-helix 1 (strip 3), and the interface between these two regions (strip 2). Within each strip, the overall hydrophobic content was evaluated for each amino acid type using the previously described sliding scale from 5 (red) to 1 (blue), more hydrophobic to less hydrophobic. Averaged

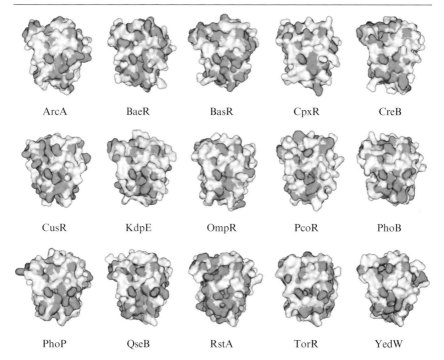

| ArcA | BaeR | BasR | CpxR | CreB |

| CusR | KdpE | OmpR | PcoR | PhoB |

| PhoP | QseB | RstA | TorR | YedW |

FIG. 6. Hydrophobic surface characteristics of the *E. coli* comparative models. The perspective is such that the region comprising the α-helix 1 and α-helix 1/α-helix 5 interface is visible.

hydrophobic scores were calculated for each strip, and this represents the initial classification step. Further subclassification is achieved by considering the characteristics of the individual amino acid, which includes side-chain types, clustering of residues, and relative amino acid positions.

Figures 8 and 9 show the modeled receiver domains of OmpR subfamily RRs from *B. subtilis* and *E. coli*, respectively, and are grouped as subclasses that display similarity in the α-helix 1 and α-helix 1/α-helix 5 interface surface regions. Table VI shows a breakdown of the subclasses developed within this study, in addition to the previously determined HPK classifications (Fabret *et al.*, 1999; Grebe and Stock, 1999). There is a significant correlation between the receiver domain subclasses generated from the comparative surface analysis and that of the HPK subclasses grouped using multiple sequence alignments and domain architecture (Fabret *et al.*, 1999; Grebe and Stock, 1999). In the case of the *B. subtilis* OmpR subfamily, the receiver domain groupings correlate to greater than 60% to the group IIIA HPK subclasses that are classified according to domain architecture and configuration (Fabret *et al.*, 1999). The HPK groups

Strip analysis
 –strip 1: $(1 + 1)/2 = 1$
 –strip 2: $(5 + 4 + 5)/3 = 4.7$
 –strip 3: $(2 + 2 + 1)/3 = 1.7$

$\alpha 1/\alpha 5$ interface hydrophobic content
 –R: 2, O: 1, Y: 0, B: 2, G: 3

PhoP

Strip analysis
 –strip 1: $(1 + 1)/2 = 1$
 –strip 2: $(5 + 4 + 5)/3 = 4.7$
 –strip 3: $(3)/1 = 3$

$\alpha 1/\alpha 5$ interface hydrophobic content
 –R: 2, O: 1, Y: 1, B: 0, G: 2

ResD

Strip analysis
 –strip 1: $(1 + 2)/2 = 1.5$
 –strip 2: $(5 + 4 + 5)/3 = 4.7$
 –strip 3: $(1 + 4)/2 = 2.5$

$\alpha 1/\alpha 5$ interface hydrophobic content
 –R: 2, O: 2, Y: 0, B: 1, G: 2

YycF

FIG. 7. Example of color-coded hydrophobic surface classification. Strip plots of the receiver domains of PhoP, ResD, and YycF from *B. subtilis* are shown. Details about the breakdown during the strip analysis and α-helix1/α-helix5 interface hydrophobic content are provided for each protein. (See color insert.)

developed for the *E. coli* OmpR subfamily based on primary sequence alignment show significant correlation to the RR receiver domain groupings as well (Grebe and Stock, 1999).

Modeling and Subclassification of the Receiver Domain of RRs in *V. vulnificus*

Brief Outline of Methods Used for Modeling

Modeling of the receiver domains of *V. vulnificus* RR proteins was performed in a similar fashion as described for OmpR subfamily RRs, with the following caveats. In the target and template identification, a list of predicted RR receiver domains of *V. vulnificus* YJ016 (Table VII) were obtained using the SUPERFAMILY (Gough *et al.*, 2001) database, and the respective GENBANK (Benson *et al.*, 2005) entry was used to approximate the receiver domain sequence. A different version of MODELLER (8v2) was used for the sequence alignment and modeling procedures, and 10 models per target were calculated. Models generated were not subjected to the robust analysis described for the OmpR subfamily RRs.

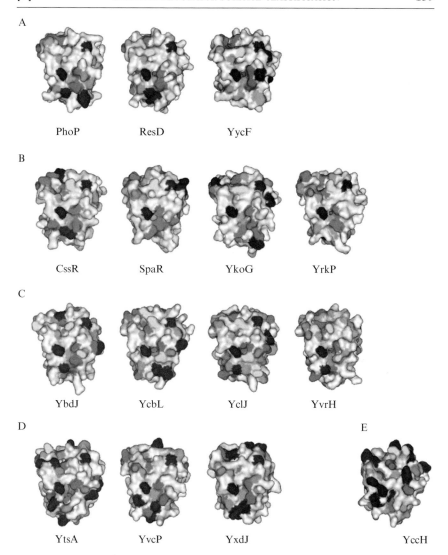

FIG. 8. Color-gradient hydrophobic surface characteristics of *B. subtilis* comparative models. Models are categorized into the subclasses listed in Table VI (A–E). The perspective is such that the region comprising the α-helix 1 and α-helix 1/α-helix 5 interface is visible. (See color insert.)

The aforementioned analysis performed on OmpR subfamily proteins is biased due to the degree of structural similarity between the target and templates, between all templates used, as well as the refinement performed by the program (A. Fiser, personal communication). Instead, the structural

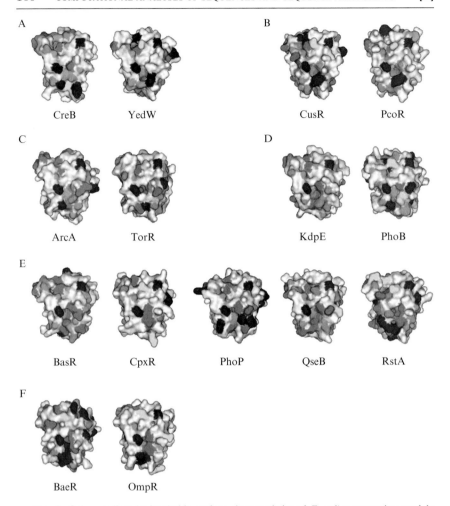

Fig. 9. Color-gradient hydrophobic surface characteristics of *E. coli* comparative models. Models are categorized into the subclasses listed in Table VI (A–F). The perspective is such that the region comprising the α-helix 1 and α-helix 1/α-helix 5 interface is visible. (See color insert.)

quality of the lowest energy models was inspected visually, in addition to stereochemical analysis of Ramachandran values using PROCHECK-NMR (Laskowski *et al.*, 1996), to choose the best model for analysis.

The BioPython package (Chapman and Chang, 2000) was used for semiautomation of the sequence analysis and modeling procedures. The initial goal was to utilize the accompanying built-in packages to (i) obtain *V. vulnificus* RR protein sequences from the GenBank database, (ii) use

TABLE VI
RECEIVER DOMAIN SUBCLASSES WITHIN THE OmpR SUBFAMILY OF RRs FROM *B. SUBTILIS* AND *E. COLI*

Protein	Subclass	1^a	2^a	3^a	R^b	O^b	Y^b	G^b	B^b	RR class[c]	HPK	HPK class[c]	HPK class[d]
							B. subtilis						
PhoP	A	1	4.7	1.7	2	1	0	2	3	R A1	PhoR	HPK 1a	4
ResD	A	1	4.7	3	2	1	1	0	3	R A1	ResE	HPK 1a	4
YycF	A	1.5	4.7	2.5	2	2	0	1	2	R A1	YycG	HPK 1a	4
CssR	B	1	4.7	0	3	1	0	0	2		CssS		3
SpaR	B	1	5	0	3	1	0	0	1	R A1	SpaK	HPK 3c	
YkoG	B	1	4.8	3	4	1	0	0	2	R A1	YkoH	HPK 1a	3
YrkP	B	1	4.7	0	2	1	0	0	1		YrkQ	HPK 4	3
YbdJ	C	1	4.7	3	2	1	1	0	2		YbdK		2
YcbL	C	2	3.6	2	1	1	1	1	2		YcbM		1
YclJ	C	1	3.5	2.5	1	1	1	2	2	R A1	YclK	HPK 1a	3
YvrH	C	1	4.3	4	2	3	1	0	2	R A1	YvrG	HPK 1a	3
YtsA	D	2.5	5	2	2	2	2	1	1	R A1	YtsB	HPK 3i	2
YvcP	D	3	4.7	0	3	2	0	0	2	R A1	YvcQ	HPK 3i	2
YxdJ	D	1	3.8	0	3	1	0	1	2	R A1	YxdK	HPK 3i	2
YccH	E	1	3	0	1	1	0	0	3		YccG		5
CreB	A	1	3.5	1	1	1	1	1	3	R A1	CreC	HPK 3c	

(continued)

TABLE VI (*continued*)

Protein	Subclass	1[a]	2[a]	3[a]	R[b]	O[b]	Y[b]	G[b]	B[b]	RR class[c]	HPK	HPK class[c]	HPK class[d]
							E. coli						
YedW	A	1	3.8	1	1	2	0	2	2		YedV		
CusR	B	1	3	0	1	1	0	1	2		CusS		
PcoR	B	1	3	0	1	1	0	1	2	R A1	PcoS	HPK 2a	
ArcA	C	1	3.3	4	3	1	0	3	2	R A1	ArcB	HPK 1b, hybrid	
TorR	C	1	3	3.5	3	1	0	2	2	R A1	TorS	HPK 1b, hybrid	
KdpE	D	1	4.5	3	2	0	0	1	1	R A1	KdpD	HPK 1a	
PhoB	D	1	4.3	2.5	2	3	1	0	3	R A1	PhoR	HPK 1a	
BasR	E	1	4.8	4	5	2	0	1	1	R A1	BasS	HPK 2a	
CpxR	E	1	5	3	2	0	1	0	1	R A1	CpxA	HPK 2b	
PhoP	E	1	4.7	5	4	1	0	0	1	R A1	PhoQ	HPK 3a	
QseB	E	1	4.7	3.5	4	1	0	1	1	R A1	QseC	HPK 2a	
RstA	E	1.5	5	4	3	1	0	1	1	R A1	RstB	HPK 2b	
BaeR	F	1	4.7	0	2	1	0	0	3	R A1	BaeS	HPK 1a	
OmpR	F	1	5	0	2	1	0	0	2	R A1	EnvZ	HPK 2b	

[a] Average strip position hydrophobic content.
[b] Total gradient hydrophobic content on the α-helix 1/α-helix 5 surface.
[c] HPK superfamily subclassification (Grebe and Stock, 1999).
[d] *B. subtilis* OmpR subfamily subclassification (Fabret *et al.*, 1999).

the GenBank sequence records to narrow down the receiver domain sequence ranges, and (iii) automatically feed this information for subsequent external analyses. During initial stages of implementation, it was apparent that the GenBank sequence record describing the receiver domain (Region; region_name="REC") was often inaccurate or missing, which is likely a result of incomplete genome annotation, particularly on newly sequenced genomes. Accordingly, full-length sequences were obtained for each *V. vulnificus* RR, and the sequence range for each receiver domain was obtained manually using information provided by the SUPERFAMILY database, an initial PSI-BLAST search, and visual inspection of the alignments, focusing on conserved active site residues and the length of secondary structural elements, α-helix 5 in particular. PISCES (Wang and Dunbrack, 2003) was used to restrict the PSI-BLAST search to a 90% nonredundant sequence search using the database pdbaanr.

Model Visualization

As described earlier, models were visualized using the molecular visualization program PYMOL (DeLano, 2002). The side-chain hydrophobicity of surface-accessible residues was monitored utilizing the aforementioned color-coded scale: ILV (red), GAF (orange), CM (yellow), ST (green), and WYPH (blue).

Modeled Receiver Domains of RRs from V. vulnificus

Of the 92 domains predicted to be CheY like by the SUPERFAMILY database, 81 were modeled successfully. Following the procedure outlined earlier for the OmpR subfamily of RRs in *B. subtilis* and *E. coli*, a 15 × 15-Å square was centered about the α-helix 1/α-helix 5 interface and was subsequently divided into three vertical strips, each approximately 5 Å wide. Again, this division corresponds roughly to α-helix 5 and the β4-α4 loop (strip 1), the α-helix 1/α-helix 5 interface (strip 2), and to α-helix 1 (strip 3). The average hydrophobic content based on amino acid composition was calculated for each strip, which was then combined with the hydrophobic content totals for the 225-Å^2 region in the initial step in model classification. Further subclassification was conducted through the visual inspection of the model surfaces, studying the amino acid compositions and their positions within each strip, and by the clustering of models given commonalities of hydrophobic patch characteristics. This visual inspection allows for models that might share similar hydrophobic scores to be placed in different classes, given different amino acid content and positioning. In this manner, the 81 *V. vulnificus* RR receiver domains modeled successfully were subcategorized into the 16 classes described in Table VII, an example of which (class 1)

TABLE VII

PREDICTED RR RECEIVER DOMAINS IN *V. VULNIFICUS* USED AS TARGET SEQUENCES FOR MODELING

GenBank accession ID	Sequence	Subclass	1^a	2^a	3^a	R^b	O^b	Y^b	G^b	B^b
37675987	1–124	1	2.0	3.5	4.7	4	1	1	1	1
37676307	988–1111	1	1.5	5.0	3.7	4	0	0	1	2
37676847	1–124	1	1.0	5.0	4.0	3	0	1	0	1
37678875	1–124	1	1.0	5.0	4.3	4	2	0	0	1
37679852	446–569	1	1.0	5.0	3.3	5	0	1	1	1
37676294	727–850	1	0.0	3.3	4.0	3	2	0	1	1
37675985	1–124	1	1.7	3.8	3.5	3	1	0	3	2
37676244	694–816	1	1.0	5.0	5.0	4	0	0	0	2
37675988	15–128	2	4.5	3.7	3.0	4	1	0	0	2
37677080	589–712	2	3.3	4.3	4.0	5	3	0	0	2
37680448	597–720	2	3.3	4.7	3.0	4	2	0	2	1
37677345	2–125	2	4.5	4.7	3.3	5	2	0	1	1
37677132	987–1110	2	3.0	5.0	4.5	5	1	0	0	1
37678325	466–585	2	3.0	5.0	5.0	5	0	0	0	1
37679417	1–124	2	3.0	5.0	5.0	6	2	0	0	1
37675989	1017–1140	3	3.3	3.0	4.0	3	1	1	0	2
37676593	1114–1237	3	5.0	4.0	5.0	4	0	0	1	0
37676469	13–136	3	3.0	4.0	5.0	3	4	0	1	1
37676880	731–850	4	1.0	4.7	3.0	2	2	0	1	1
37681007	661–784	4	1.0	5.0	4.0	3	2	0	0	1
37680917	721–844	4	1.0	4.3	3.5	2	2	0	1	1
37679896	29–153	4	1.0	4.2	4.3	3	4	1	0	2
37680253	17–139	4	1.0	5.0	4.7	4	1	0	0	1
37680589	31–154	4	1.0	4.7	5.0	3	1	0	0	1
37678380	1–120	5	1.0	4.8	2.0	3	1	0	1	1
37681297	1–124	5	1.5	4.8	0.0	3	1	0	1	1
37679342	1–124	5	2.3	4.3	2.0	2	1	1	2	1

37676377	11–134	6	1.3	3.3	1.0	2	0	0	2	4
37676245	1–124	6	1.3	3.0	2.0	1	1	0	4	3
37679274	1–124	6	1.0	3.2	0.0	2	1	0	0	3
37675679	1–124	7	2.7	3.0	3.0	3	0	1	1	3
37676982	1–115	7	1.5	2.8	4.5	2	2	0	1	2
37677138	50–172	7	1.0	3.5	5.0	2	2	0	0	3
37678420	1–124	7	1.0	4.0	4.0	3	1	0	0	3
37679811	1–124	7	2.5	3.4	3.0	4	1	0	2	2
37680601	19–142	7	2.5	3.0	4.5	2	3	0	1	3
37680766	133–256	7	1.0	3.2	5.0	3	1	0	0	3
37676273	1–124	7	1.0	3.8	4.5	3	2	0	0	3
37680487	1–124	8	1.0	3.0	4.0	3	3	0	2	2
37677144	1–120	8	3.3	3.2	0.0	2	2	0	1	2
37679788	1–124	9	2.0	3.0	3.0	2	2	1	0	4
37676309	13–126	9	1.0	3.3	4.0	3	1	1	2	1
37676477	4–128	9	3.0	4.0	4.0	2	1	1	0	1
37676593	991–1113	9	1.0	4.0	3.0	1	3	1	1	1
37680408	1–124	9	2.5	3.8	3.0	2	2	1	2	1
37675668	510–633	9	1.0	4.3	5.0	4	0	1	0	1
37679812	486–609	9	1.5	3.3	3.7	2	2	2	1	1
37680265	449–571	9	2.3	4.0	4.0	2	1	2	0	1
37681313	1036–1156	9	1.0	3.6	5.0	3	1	1	1	2
37675909	1–124	10	1.0	3.0	0.0	1	0	1	0	2
37676244	541–660	10	2.5	3.8	1.0	1	2	2	0	2
37679248	1–121	10	2.5	3.0	0.0	1	2	1	1	2
37679606	1–120	10	3.0	4.0	0.0	2	1	1	0	1
37677352	1–119	10	1.0	4.0	1.7	1	1	1	2	2
37676593	1259–1381	11	1.5	3.0	3.8	2	1	2	4	1
37676721	4–127	11	1.5	4.0	5.0	3	1	0	1	1

(continued)

TABLE VII (continued)

GenBank accession ID	Sequence	Subclass	1^a	2^a	3^a	R^b	O^b	Y^b	G^b	B^b
37679475	181–309	11	3.0	4.0	3.0	1	2	0	1	1
37680289	1–124	11	3.0	4.7	4.0	3	2	1	1	0
37680644	1–124	11	0.0	4.7	4.0	2	2	0	0	0
37677133	5–131	12	2.3	3.3	4.0	1	4	0	2	2
37678240	1–124	12	1.0	3.8	3.0	1	3	0	2	1
37679379	13–127	12	1.0	3.4	4.0	1	4	0	2	1
37680766	12–133	12	3.3	2.5	4.0	1	2	1	1	1
37676889	754–877	13	2.3	4.5	3.0	1	2	1	1	1
37678907	1–120	13	1.3	4.3	3.0	3	1	1	1	3
37677310	1–124	13	1.5	3.8	1.0	1	2	0	3	4
37679284	1–123	14	2.0	3.7	0.0	1	2	0	2	2
37679910	22–145	14	1.5	3.3	4.5	2	3	1	3	1
37680667	1–124	14	3.0	3.2	4.0	1	5	0	1	2
37675773	1–124	15	3.0	3.8	3.5	3	2	1	2	1
37678828	523–646	15	2.5	5.0	3.0	4	1	0	2	1
37679341	513–636	15	2.0	5.0	2.0	3	1	0	2	2
37676235	189–327	15	3.0	4.0	3.0	3	2	0	2	1
37678830	1–124	15	1.0	3.7	3.5	2	1	0	2	2
37676308	1–124	15	1.5	3.7	4.0	3	1	0	3	1
37676845	1–120	16	2.5	3.0	3.5	1	2	0	2	1
37677342	1–124	16	1.0	4.0	2.0	1	0	2	1	1
37679142	178–301	16	4.0	4.0	5.0	1	3	0	0	2
37676089	170–299	16	4.0	4.0	3.5	1	6	0	0	0
37676596	5–133	16	1.0	3.4	4.0	2	1	1	2	1

[a] Average strip position hydrophobic content.
[b] Total gradient hydrophobic content on the α-helix 1/α-helix 5 surface.

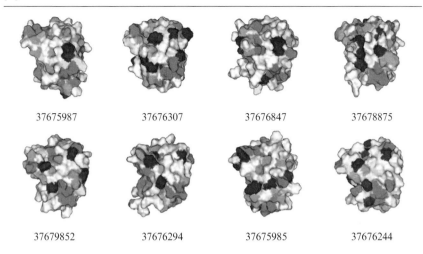

| 37675987 | 37676307 | 37676847 | 37678875 |

| 37679852 | 37676294 | 37675985 | 37676244 |

Fig. 10. Examples of *V. vulnificus* subclass 1 models. The perspective is such that the region comprising the α-helix 1 and α-helix 1/α-helix 5 interface is visible. (See color insert.)

is illustrated in Fig. 10. No attempts were made to correlate these 16 classes to the 6 classes developed for the OmpR subfamily members in *B. subtilis* and *E. coli*. Rather, this analysis was meant entirely as a subclassification of RR domains within *V. vulnificus*.

Practical Aspects and Future Directions of This Approach

Due simply to the scale of this analysis, a deficiency in this approach becomes apparent: the initial analysis, based on average hydrophobic strip content coupled with overall hydrophobic content, lacks detailed positional information. For example, an average hydrophobic content of "1" in strip 1 is not indicative of the relative positions of the one or more WYPH residues that would be present. This, of course, is remedied by visual inspection of the groupings assembled by strip scores and hydrophobic content alone. However, as the number of models increases, the amount of effort to analyze them visually becomes increasingly laborious. This suggests that further steps might be employed to good effect in the initial characterization prior to visual inspection. An additional step to consider is the use of another three 5-Å-wide strips overlaid horizontally atop the initial three vertical strips. This "six strip analysis" provides a Cartesian two-coordinate system, which better defines the positional information of the hydrophobic residues.

In addition to the initial "six strip analysis," another useful aid that might be considered in future work is use of a cladogram-style diagram to more easily visualize the similarity of the various proteins rather than simply a static class number.

References

Altschul, S. F., Madden, T. L., Schaffer, A. A., Zhang, J., Zhang, Z., Miller, W., and Lipman, D. J. (1997). Gapped BLAST and PSI-BLAST: A new generation of protein database search programs. *Nucleic Acids Res.* **25**, 3389–3402.

Bairoch, A., and Apweiler, R. (2000). The SWISS-PROT protein sequence database and its supplement TrEMBL in 2000. *Nucleic Acids Res.* **28**, 45–48.

Benson, D. A., Karsch-Mizrachi, I., Lipman, D. J., Ostell, J., and Wheeler, D. L. (2005). GenBank. *Nucleic Acids Res.* **33**, D34–D38.

Berman, H. M., Westbrook, J., Feng, Z., Gilliland, G., Bhat, T. N., Weissig, H., Shindyalov, I. N., and Bourne, P. E. (2000). The Protein Data Bank. *Nucleic Acids Res.* **28**, 235–242.

Birck, C., Chen, Y., Hulett, F. M., and Samama, J. P. (2003). The crystal structure of the phosphorylation domain in PhoP reveals a functional tandem association mediated by an asymmetric interface. *J. Bacteriol.* **185**, 254–261.

Chapman, B., and Chang, J. (2000). Biopython: Python tools for computational biology. *ACM SIGBIO Newslett.* **20**, 15–19.

Colovos, C., and Yeates, T. O. (1993). Verification of protein structures: Patterns of nonbonded atomic interactions. *Protein Sci.* **2**, 1511–1519.

DeLano, W. L. (2002). *The PyMOL molecular graphics system.* DeLano Scientific, Palo Alto, CA.

Eisenberg, D., Luthy, R., and Bowie, J. U. (1997). VERIFY3D: Assessment of protein models with three-dimensional profiles. *Methods Enzymol.* **277**, 396–404.

Fabret, C., Feher, V. A., and Hoch, J. A. (1999). Two-component signal transduction in *Bacillus subtilis*: How one organism sees its world. *J. Bacteriol.* **181**, 1975–1983.

Fiser, A., Feig, M., Brooks, C. L., 3rd, and Sali, A. (2002). Evolution and physics in comparative protein structure modeling. *Acc. Chem. Res.* **35**, 413–421.

Gough, J., Karplus, K., Hughey, R., and Chothia, C. (2001). Assignment of homology to genome sequences using a library of hidden Markov models that represent all proteins of known structure. *J. Mol. Biol.* **313**, 903–919.

Grebe, T. W., and Stock, J. B. (1999). The histidine protein kinase superfamily. *Adv. Microb. Physiol.* **41**, 139–227.

Hoch, J. A., and Varughese, K. I. (2001). Keeping signals straight in phosphorelay signal transduction. *J. Bacteriol.* **183**, 4941–4949.

Jones, D. T., and Swindells, M. B. (2002). Getting the most from PSI-BLAST. *Trends Biochem. Sci.* **27**, 161–164.

Kanehisa, M., Goto, S., Kawashima, S., and Nakaya, A. (2002). The KEGG databases at GenomeNet. *Nucleic Acids Res.* **30**, 42–46.

Kirton, S. B., Baxter, C. A., and Sutcliffe, M. J. (2002). Comparative modelling of cytochromes P450. *Adv. Drug Deliv. Rev.* **54**, 385–406.

Kojetin, D. J., Thompson, R. J., and Cavanagh, J. (2003). Sub-classification of response regulators using the surface characteristics of their receiver domains. *FEBS Lett.* **554**, 231–236.

Kyte, J., and Doolittle, R. F. (1982). A simple method for displaying the hydropathic character of a protein. *J. Mol. Biol.* **157**, 105–132.

Laskowski, R. A., Rullmannn, J. A., MacArthur, M. W., Kaptein, R., and Thornton, J. M. (1996). AQUA and PROCHECK-NMR: Programs for checking the quality of protein structures solved by NMR. *J. Biomol. NMR* **8,** 477–486.

Lewis, R. J., Muchova, K., Brannigan, J. A., Barak, I., Leonard, G., and Wilkinson, A. J. (2000). Domain swapping in the sporulation response regulator Spo0A. *J. Mol. Biol.* **297,** 757–770.

Marti-Renom, M. A., Stuart, A. C., Fiser, A., Sanchez, R., Melo, F., and Sali, A. (2000). Comparative protein structure modeling of genes and genomes. *Annu. Rev. Biophys. Biomol. Struct.* **29,** 291–325.

Ohta, N., and Newton, A. (2003). The core dimerization domains of histidine kinases contain recognition specificity for the cognate response regulator. *J. Bacteriol.* **185,** 4424–4431.

Ouzounis, C. A., Coulson, R. M., Enright, A. J., Kunin, V., and Pereira-Leal, J. B. (2003). Classification schemes for protein structure and function. *Nat. Rev. Genet.* **4,** 508–519.

Robinson, V. L., Wu, T., and Stock, A. M. (2003). Structural analysis of the domain interface in DrrB, a response regulator of the OmpR/PhoB subfamily. *J. Bacteriol.* **185,** 4186–4194.

Sola, M., Gomis-Ruth, F. X., Serrano, L., Gonzalez, A., and Coll, M. (1999). Three-dimensional crystal structure of the transcription factor PhoB receiver domain. *J. Mol. Biol.* **285,** 675–687.

Tzeng, Y. L., and Hoch, J. A. (1997). Molecular recognition in signal transduction: The interaction surfaces of the Spo0F response regulator with its cognate phosphorelay proteins revealed by alanine scanning mutagenesis. *J. Mol. Biol.* **272,** 200–212.

Wang, G., and Dunbrack, R. L., Jr. (2003). PISCES: A protein sequence culling server. *Bioinformatics* **19,** 1589–1591.

Wolfenden, R., Andersson, L., Cullis, P. M., and Southgate, C. C. (1981). Affinities of amino acid side chains for solvent water. *Biochemistry* **20,** 849–855.

Wu, C. H., Huang, H., Arminski, L., Castro-Alvear, J., Chen, Y., Hu, Z. Z., Ledley, R. S., Lewis, K. C., Mewes, H. W., Orcutt, B. C., Suzek, B. E., Tsugita, A., *et al.* (2002). The Protein Information Resource: An integrated public resource of functional annotation of proteins. *Nucleic Acids Res.* **30,** 35–37.

Xu, Q., Porter, S. W., and West, A. H. (2003). The yeast YPD1/SLN1 complex: Insights into molecular recognition in two-component signaling systems. *Structure* **11,** 1569–1581.

Zapf, J., Sen, U., Madhusudan, Hoch, J. A., and Varughese, K. I. (2000). A transient interaction between two phosphorelay proteins trapped in a crystal lattice reveals the mechanism of molecular recognition and phosphotransfer in signal transduction. *Structure* **8,** 851–862.

Zhao, R., Collins, E. J., Bourret, R. B., and Silversmith, R. E. (2002). Structure and catalytic mechanism of the *E. coli* chemotaxis phosphatase CheZ. *Nat. Struct. Biol.* **9,** 570–575.

Section II

Biochemical and Genetic Assays of Individual Components of Signaling Systems

[8]   Purification and Assays of *Rhodobacter capsulatus* RegB–RegA Two-Component Signal Transduction System

By Lee R. Swem, Danielle L. Swem, Jiang Wu, and Carl E. Bauer

Abstract

Two-component signal-transduction systems, composed of a histidine-sensor kinase and a DNA-binding response regulator, allow bacteria to detect environmental changes and adjust cellular physiology to live more efficiently in a broad distribution of niches. Although many two-component signal-transduction systems are known, a limited number of signals that stimulate these systems have been discovered. This chapter describes the purification and characterization of the predominant two-component signal-transduction system utilized by *Rhodobacter capsulatus,* a nonsulfur purple photosynthetic bacterium. Specifically, we explain the overexpression, detergent solubilization, and purification of the full-length membrane-spanning histidine-sensor kinase RegB. We also provide a method to measure autophosphorylation of RegB and discern the effect of its signal molecule, ubiquinone, on autophosphorylation levels. In addition we describe the overexpression and purification of the cognate response regulator RegA and a technique used to visualize the phosphotransfer reaction from RegB to RegA.

Introduction

The RegB–RegA two-component signal-transduction system is highly conserved among 28 species of proteobacteria (Elsen *et al.*, 2004). Extensive genetic analysis of RegB and RegA deletion mutants in *Rhodobacter capsulatus* has demonstrated that many energy-generating and energy-utilizing processes are controlled by the RegB–RegA system (Bird *et al.*, 1999; Du *et al.*, 1999; Dubbs and Tabita, 2003; Elsen *et al.*, 2000, 2004; Swem and Bauer, 2002; Swem *et al.*, 2001). Also, biochemical analysis has provided evidence that phosphorylated RegA binds DNA and directly controls the transcription of genes involved in these processes (Bird *et al.*, 1999; Du *et al.*, 1998). Operons within the RegB–RegA regulon include components required for photosynthesis, hydrogen utilization, cytochrome biogenesis, tetrapyrole biosynthesis, carbon fixation, nitrogen fixation, and dehydrogenase synthesis (Fig. 1) (Elsen *et al.*, 2004). The common theme among all members of this regulon is that they either generate reducing

METHODS IN ENZYMOLOGY, VOL. 422   0076-6879/07 $35.00

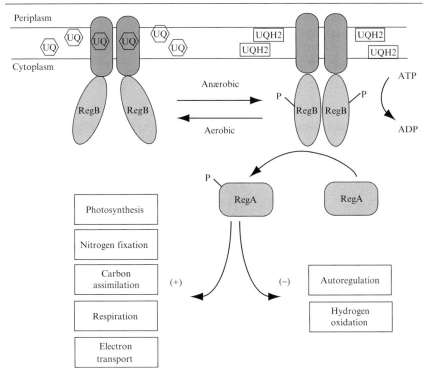

FIG. 1. Schematic of the RegB–RegA regulon. RegB is depicted as a membrane-bound histidine sensor kinase, and autophosphorylation is inhibited by ubiquinone (UQ) under aerobic conditions. Under anaerobic conditions the ubiquinone pool becomes reduced to the ubiquinol state (UQH2) and RegB is able to phosphorylate. The phosphate is transferred to RegA, which regulates multiple cellular processes.

power in the form of reduced ubiquinol or utilize reducing power either directly or indirectly from the ubiquinone pool (Fig. 1). Not coincidentally it was discovered that ubiquinone serves as the signal that regulates the RegB–RegA system (Swem *et al.*, 2006).

The sensor kinase, RegB, contains six membrane-spanning helices at the amino terminus followed by a cytosolic domain containing a conserved H-box site of phosphorylation, a redox box containing a redox-active cyteine residue, followed by an adenine triphosphate (ATP) kinase domain at the carboxyl terminus (Potter *et al.*, 2002; Swem *et al.*, 2003). The primary amino acid sequence of the membrane-spanning domain of RegB is not highly conserved with the exception of the second periplasmic loop that contains a universally conserved GGXXNPF motif (Swem *et al.*, 2006).

Mutational and biochemical analyses have demonstrated that this region interacts with the ubiquinone pool and that the presence of oxidized ubiquinone inhibits the autophosphorylation activity of isolated full-length RegB *in vitro* (Swem *et al.*, 2006). Interestingly, the reduced form of ubiquinone (ubiquinol) does not inhibit RegB autophosphorylation. The ubiquinone specificity of RegB allows this system to sense the cellular redox state of *R. capsulatus* and elicit physiological changes in response to changes in environmental oxygen tension. The autophosphorylation level of RegB directly controls the phosphorylation state of RegA, which affects its ability to bind DNA and regulate diverse oxygen-dependent cellular processes (Bird *et al.*, 1999; Du *et al.*, 1998).

This chapter describes the methods used to overexpress and isolate full-length RegB and its cognate response regulator, RegA. We also offer an assay to measure the signaling properties that ubiquinone exerts on RegB. Finally, we provide an assay for measuring RegB autophosphorylation and phosphate transfer to RegA.

Expression and Purification of RegB

Culture Media

> Luria-Bertani (LB) broth: 10 g tryptone, 5 g yeast extract, and 10 g NaCl in 1 liter dH_2O
> Terrific broth: 12 g tryptone, 24 g yeast extract, and 4 ml glycerol in 900 ml dH_2O. After autoclaving, add 100 ml of sterile 0.17 M KH_2PO_4 and 0.72 M K_2HPO_4 to bring the final volume to 1 liter.

Buffers

> RegB lysis buffer: 10 mM Tris, pH 8, and 100 mM NaCl
> Solubilization buffer: 20 mM Tris, pH 8.0, 20 mM imidazole, 300 mM NaCl, 20% glycerol, and 1% *n*-dodecyl-β-D-maltoside
> Wash buffer: 20 mM Tris, pH 8.0, 20 mM imidazole, 150 mM NaCl, and 10% glycerol
> Elution buffer: 20 mM Tris, pH 8.0, 200 mM imidazole, 150 mM NaCl, and 10% glycerol
> Storage buffer: 10 mM Tris, pH 8.0, 150 mM NaCl, and 50% glycerol
> 4× SDS-PAGE loading buffer: 200 mM Tris, pH 6.8, 8% SDS, 0.4% bromophenol blue, 40% glycerol, and 4% β-mercaptoethanol
> Western transfer buffer 1×: 25 mM Tris-base, 192 mM glycine, and 20% (v/v) methanol

Selection of RegB Overexpression Strains

Full-length *regB* is polymerase chain reaction (PCR) amplified from *R. capsulatus* genomic DNA using primers RegBFull*Nco*I (5'-TACCA-TGGTGAGGGCTGTCGACC) and RegBFull*Xho*I (5'-ATCTCGAGG-GCGGTGATCGGAACATTC). The PCR product is cloned into the *Nco*I and *Xho*I sites of pET28 (Novagen), which places RegB in-frame with a carboxyl terminus 6 His tag. The resulting plasmid, pET28RegBfull, is then transformed into the T7-based isopropyl-β-D-thiogalactopyranoside (IPTG)-inducible *Escherichia coli* expression strain, BL21(DE3). The transformants are selected for on LB agar plates containing 50 μg/ml kanamycin and grown at 37° overnight. Six to eight independent colonies are streaked for isolation on fresh LB-agar medium and grown overnight at 37°. Five-milliliter liquid Terrific broth cultures are started from each independent isolate and grown overnight with shaking (250 rpm) at 37°. The next day the cultures are diluted 1:50 in fresh Terrific broth and grown to an OD_{600} of 0.4 to 0.6, at which point IPTG is added to a final concentration of 500 μM. The cultures are induced for an additional 4 h of shaking (250 rpm) at 37°, and then 16 μl of each cell culture is mixed with 4 μl of 4× SDS-PAGE loading buffer, heated to 95° for 5 min, and separated on a 7.5% SDS-polyacrylamide gel. The SDS-polyacrylamide gel-separated protein is then transferred to a nitrocellulose membrane in 1× Western transfer buffer overnight at 30 V at 4°. The nitrocellulose membrane is probed by Western blot analysis with a horseradish peroxidase-conjugated His-tag antibody according to the manufacturer's instructions (Santa Cruz Biotech). Overexpressed RegB protein is visualized using the WestDura Super Signal substrate (Pierce), and a 50-kDa protein band is identified that is absent from a control strain containing the pET28 vector without the RegB open reading frame. Isolates judged to contain maximal expression of RegB are grown overnight in LB broth supplemented with 50 μg/ml kanamycin in the absence of IPTG and stored at −80° in 20% glycerol.

Optimization of RegB Induction Conditions

Overproduction of large amounts of full-length RegB in *E. coli* is determined to be lethal, leading to a very low yield of recombinant protein. To overcome lethality, a minimal amount of IPTG is utilized to overexpress RegB from the pET28 expression system. Optimum expression is determined by performing an IPTG killing experiment involving the plating of BL21(DE3)/pET28RegBfull onto LB plates containing IPTG at concentrations ranging from 10 μM to 1 mM. All IPTG concentrations above 200 μM resulted in nearly 100% lethality as judged by low plating efficiency. There is no observable inhibition of growth at IPTG concentrations of 20 μM or below. However, LB plates containing 75 μM IPTG give rise to very thin colonies that are translucent in appearance, presumably because of over-production of RegB. A concentration of 75 μM IPTG is thus used for overexpressing full-length RegB in liquid cultures.

Purification of Full-Length RegB

An overnight culture of BL21(DE3) containing plasmid pET28/ RegBfull is grown for 16 h with shaking (250 rpm) at 37° in LB supple-mented with 50 μg/ml kanamycin. The culture is diluted 1:50 into Terrific broth with 50 μg/ml kanamycin (1-liter batches in 2-liter flasks) and grown at 37° shaking (250 rpm) until an OD_{600} of 0.4–0.6 is reached, at which time IPTG is added to a final concentration of 75 μM. Protein expression is allowed to proceed for 4 h, shaking at 37° before the cells are harvested by centrifugation (7600g, 4°). The cell pellet is then stored at $-80°$ until protein purification is initiated.

For purification (Fig. 2), cell pellets obtained from 8 liters RegB over-expressed culture are gently thawed on ice in the presence of 100 ml lysis buffer until a homogeneous solution is obtained. Cells are lysed by two passages through the M-110L Micro Fluidizer Processor at 20,000 psi (Microfluidics). The cell lysate is clarified by centrifugation at 11,000g for 30 min at 4° to remove the outer membranes and any nonruptured *E. coli*. The supernatant, which contains RegB-loaded inner membranes, are pel-leted by centrifugation at 150,000g for 1 h at 4°. The inner membrane pellet is solubilized by the addition of 2 ml solubilization buffer per liter of lysed *E. coli* cells (typically 16 ml solubilization buffer). A Pyrex tissue grinder is used to homogenize the inner membranes in the presence of solubilization buffer. The solubilized membranes are shaken in a 50-ml conical tube attached to an orbital shaker, rotating at 100 rpm for 1 h at 24°. Any insoluble debris is then separated from the solubilized RegB protein by centrifugation at 150,000g for 1 h at 4°. The supernatant containing soluble full-length RegB protein is diluted 20-fold in wash buffer and allowed to incubate with 1 ml of settled charged nickel resin (Novagen) for 1 h at room temperature shaking at 40 rpm on an orbital shaker. The dilution of solu-bilized RegB with wash buffer is critical for the nickel affinity purification, as the high n-dodecyl-β-D-maltoside concentrations used for RegB solu-blization will inhibit His-tag/nickel affinity chromotagraphy. The RegB-loaded nickel resin is then pelleted by centrifugation at 1000g for 5 min at 4°. The supernatant is retained for SDS-PAGE analysis, and the resin is applied to a disposable gravity flow column and washed with 50 column volumes of wash buffer at 24°. Full-length RegB is eluted with 10 ml of

RegB purification scheme

Over-express full-length RegB in 8 liters of Terrific Broth with 4 h of IPTG
induction at 37°

↓

Harvest by centrifugation for 10 min at 7600×g at 4°

↓

Resuspend cell pellet in 100 ml of lysis buffer and disrupt cells with two passages
through a Micro Fluidizer processor at 20,000 psi

↓

Clarify cell lysate by low speed centrifugation at 11,000×g for 30 min at 4°

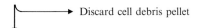 → Discard cell debris pellet

Concentrate inner membranes from low speed supernatant by centrifugation at 150,000×g
for 1 h at 4°

 → Discard supernatant

Extract RegB from inner-membrane high speed centrifugation pellet by addition of 16 ml
of solubilization buffer and shaking at 100 rpm for 1 h at 24°

↓

Remove insoluble debris by centrifugation at 150,000×g for 1 h at 4°

 → Discard pellet

Dilute supernatant containing solublized full-length RegB protein 20-fold in wash buffer
and incubate with 1 ml of charged nickel resin for 1 h at 24° while shaking at
40 rpm

↓

Collect RegB-loaded nickel resin by centrifugation at 1,000×g for 5 min at 4° and
wash resin in a gravity flow column with 50 column volumes of wash buffer at 24°

↓

Elute full-length RegB with 10 ml of elution buffer at 24°

↓

Fractions containing full-length RegB protein are pooled and dialyzed against
RegB storage buffer at 4° for 8–10 h and then stored at –20°

Fig. 2. RegB purification flowchart. A simple flowchart illustrating inner membrane
isolation and RegB solubilization and purification.

elution buffer at 24°, collected in 1-ml fractions, and analyzed for protein
content using a Bio-Rad protein assay (Bio-Rad). The RegB protein concentration in milligrams per milliliter is calculated using a bovine serum
albumin standard curve generated in the presence of the Bio-Rad protein

assay solution. The molecular mass of RegB (50 kDa) is then used to determine the molar concentration of protein. Fractions containing protein are separated on a 7.5% SDS-polyacrylamide gel to confirm the presence of RegB. Fractions containing RegB protein are pooled and dialyzed against storage buffer at 4° for 8–10 h and then stored at −20° until kinase assays are completed.

Expression and Purification of RegA

Buffers

Lysis/wash buffer: 20 mM Tris-HCl (pH 8.0), 500 mM NaCl, 0.1 mM EDTA, and 0.1% Triton X-100

Cleavage buffer: 20 mM Tris-HCl (pH 8.0), 500 mM NaCl, 0.1 mM EDTA, and 30 mM dithiothreitol

Elution buffer: 20 mM Tris-HCl (pH 8.0), 500 mM NaCl, and 0.1 mM EDTA

Dialysis buffer: 50 mM Tris-HCl (pH 8.0), 200 mM KCl, 10 mM MgCl$_2$, and 50% glycerol

Overexpression of RegA

The expression vector used to fuse RegA to the chitin-binding domain was constructed as described by Du *et al.* (1998). The RegA expression plasmid, pET29CBD::*regA,* is transformed into the T7-based IPTG-inducible expression strain, BL21(DE3). The RegA overexpression strain is grown for 14 to 16 h with shaking (250 rpm) at 37° in LB containing 50 μg/ml kanamycin. The culture is then diluted 1:50 in Terrific broth containing 50 μg/ml kanamycin (1-liter batches in 2-liter flasks). The cells are grown to an OD$_{600}$ of 0.4 to 0.6, at which time RegA expression is induced by the addition of 500 μM IPTG. RegA expression is allowed to proceed for 4 h, shaking (250 rpm) at 37°, and is then harvested by centrifugation at 7600g at 4° for 10 min. The cell pellet is stored at −80° until purification is initiated.

Purification of RegA

The cell pellet obtained from 4 liters of overexpressed RegA culture is resuspended in 50 ml lysis buffer. Cells are lysed by two passages through the M-110L Micro Fluidizer Processor at 20,000 psi (Microfluidics), and the cell lysate is clarified by centrifugation at 11,000g for 30 min at 4°. A 10-ml chitin affinity column is prepared by applying 15 ml of chitin bead slurry

(New England Biolabs) to a disposable column and allowed to pack by gravity flow. The column is equilibrated with 5 column volumes of lysis/wash buffer. The clarified RegA cell lysate is then applied to the column at a flow rate of 1 ml/min and subsequently washed with 10 column volumes of lysis/wash buffer. The final 1-ml volume of wash buffer is collected and analyzed for protein content using the Bio-Rad protein assay (Bio-Rad). If the wash buffer still contains protein contamination, the column is washed with an additional 10 column volumes or until no more contaminating protein is detected in the column flow through. The chitin column is then quickly purged with 3 column volumes of cleavage buffer and immediately capped off and stored for 10 h at 4°. The RegA protein is then eluted by flushing the column with 20 ml elution buffer and collected in 1-ml fractions. Each fraction is assayed for protein content using the Bio-Rad protein assay (Bio-Rad). Fractions containing protein (typically fractions 3–15) are separated by SDS-PAGE and stained with Coomassie blue to confirm the presence of a 20-kDa protein band corresponding to RegA. Fractions containing RegA protein are pooled and dialyzed overnight at 4° against dialysis buffer and stored at −20°. The RegA protein concentration in milligrams per milliliter is calculated using a bovine serum albumin standard curve generated in the presence of the Bio-Rad protein assay solution. The molecular mass of RegA (20 kDa) is then used to determine the molar concentration of protein.

RegB Kinase and Phosphotransfer Assays

Buffers

> 10× protein kinase buffer: 200 mM Tris, pH 8.0, 500 mM NaCl, 60 mM
> MgCl$_2$, 1 mM CaCl$_2$, and 1 M KCl
> 10× ATP cocktail: 10 mM unlabeled ATP and 0.5 μCi γ-^{32}P-labeled
> ATP (ICN Biomedical, 5000 Ci/mmol)

RegB Kinase Assays

RegB is diluted to a final concentration of 5–10 μM in 1× protein kinase buffer and then incubated at 37° for 20 min. The kinase reaction is initiated by the addition of a 1/10 volume of a 10× ATP cocktail that contains trace amounts of γ-^{32}P ATP. Aliquots are removed at times ranging between 0.5 and 16 min and quenched by the addition of 4× SDS-PAGE loading buffer. The samples are then separated by SDS-PAGE, analyzed by autoradiographic film, and quantified using a phosphor-imaging system (Typhoon

9200, Amersham Biosciences). It is extremely important to avoid boiling the quenched reaction prior to separation on SDS-PAGE, as the phosphorylated protein is heat labile.

Ubiquinone Inhibition of RegB

For ubiquinone inhibition experiments, benzoquinone (Sigma) or coenzyme Q_1 (Sigma) is solubilized in 95 % ethanol at a concentration of 167 mM and then diluted in water to a final concentration of 16.7 mM. Either benzoquinone or coenzyme Q_1 is then added to the RegB kinase reaction at a 20- to 200-fold molar excess over that of RegB. Maximal inhibition is observed at a concentration of 50:1 coenzyme Q_1 to RegB. An appropriate concentration (<1%) of ethanol is added to a RegB kinase control reaction that does not contain ubiquinone. Both reactions are incubated at 37° for 20 min prior to the addition of 10× ATP cocktail to initiate the phosphorylation reaction (Fig. 3). The kinase reactions are completed as described

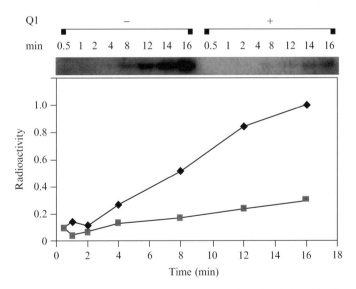

FIG. 3. Inhibitory effect of ubiquinone on RegB autophosphorylation. RegB autophosphorylation is assayed in the presence and absence of ubiquinone (Q_1). The concentration of ubiquinone (Q_1) used in the reaction is 200-fold molar excess to RegB. Reaction aliquots are removed at 0.5, 1, 2, 4, 8, 12, 14, and 16 min, quenched with 4× SDS-loading buffer, and separated by SDS-PAGE. RegB autophosphorylation is visualized and quantified using the Typhoon phosphor-imaging system (Amersham Biosciences). ■ represents the radioactivity of the reaction aliquots in the presence of ubiquinone. ◆ represents the radioactivity of the reaction aliquots in the absence of ubiquinone.

earlier, and autophosphorylation inhibition is quantified using a phosphor-imaging system (Typhoon 9200, Amersham Biosciences).

RegB Phosphorylation of RegA

To measure RegB phosphotransfer, a RegB kinase reaction is assembled as described earlier and the reaction is started by the addition of $10 \times$ ATP cocktail. The RegB protein is allowed to autophosphorylate for 15 min, before the addition of a 1 M equivalent of RegA is added to the reaction. Aliquots are removed at 0.5, 1, 1.5, 2, 3, and 5 min following RegA addition and are quenched by the addition of $4\times$ SDS-PAGE loading buffer. The reactions are placed on ice until being separated on a 10% SDS-polyacrylamide gel. The gel is then visualized using the Typhoon phosphor-imaging system (Amersham Biosciences).

Acknowledgment

This study is supported by National Institutes of Health Grant GM040941 awarded to C. E. Bauer.

References

Bird, T. H., Du, S., and Bauer, C. E. (1999). Autophosphorylation, phosphotransfer, and DNA-binding properties of the RegB/RegA two-component regulatory system in *Rhodobacter capsulatus*. *J. Biol. Chem.* **274**, 16343–16348.

Du, S., Bird, T. H., and Bauer, C. E. (1998). DNA binding characteristics of RegA: A constitutively active anaerobic activator of photosynthesis gene expression in *Rhodobacter capsulatus*. *J. Biol. Chem.* **273**, 18509–18513.

Du, S., Kouadio, J. L., and Bauer, C. E. (1999). Regulated expression of a highly conserved regulatory gene cluster is necessary for controlling photosynthesis gene expression in response to anaerobiosis in *Rhodobacter capsulatus*. *J. Bacteriol.* **181**, 4334–4341.

Dubbs, J. M., and Tabita, F. R. (2003). Interactions of the cbbII promoter-operator region with CbbR and RegA (PrrA) regulators indicate distinct mechanisms to control expression of the two cbb operons of *Rhodobacter sphaeroides*. *J. Biol. Chem.* **278**, 16443–16450.

Elsen, S., Dischert, W., Colbeau, A., and Bauer, C. E. (2000). Expression of uptake hydrogenase and molybdenum nitrogenase in *Rhodobacter capsulatus* is coregulated by the RegB-RegA two-component regulatory system. *J. Bacteriol.* **182**, 2831–2837.

Elsen, S., Swem, L. R., Swem, D. L., and Bauer, C. E. (2004). RegB/RegA, a highly conserved redox-responding global two-component regulatory system. *Microbiol. Mol. Biol. Rev.* **68**, 263–279.

Potter, C. A., Ward, A., Laguri, C., Williamson, M. P., Henderson, P. J., and Phillips-Jones, M. K. (2002). Expression, purification and characterisation of full-length histidine protein kinase RegB from *Rhodobacter sphaeroides*. *J. Mol. Biol.* **320**, 201–213.

Swem, D. L., and Bauer, C. E. (2002). Coordination of ubiquinol oxidase and cytochrome cbb (3) oxidase expression by multiple regulators in *Rhodobacter capsulatus. J. Bacteriol.* **184,** 2815–2820.

Swem, L. R., Elsen, S., Bird, T. H., Swem, D. L., Koch, H. G., Myllykallio, H., Daldal, F., and Bauer, C. E. (2001). The RegB/RegA two-component regulatory system controls synthesis of photosynthesis and respiratory electron transfer components in *Rhodobacter capsulatus. J. Mol. Biol.* **309,** 121–138.

Swem, L. R., Gong, X., Yu, C. A., and Bauer, C. E. (2006). Identification of a ubiquinone-binding site that affects autophosphorylation of the sensor kinase RegB. *J. Biol. Chem.* **281,** 6768–6775.

Swem, L. R., Kraft, B. J., Swem, D. L., Setterdahl, A. T., Masuda, S., Knaff, D. B., Zaleski, J. M., and Bauer, C. E. (2003). Signal transduction by the global regulator RegB is mediated by a redox-active cysteine. *EMBO J.* **22,** 4699–4708.

[9] Purification and Reconstitution of PYP-Phytochrome with Biliverdin and 4-Hydroxycinnamic Acid

By Young-Ho Chung, Shinji Masuda, and Carl E. Bauer

Abstract

PYP-phytochrome (Ppr) is a unique photoreceptor that contains a blue light-absorbing photoactive yellow protein (PYP) domain, a red light-absorbing phytochrome domain, and a histidine kinase domain. This chapter describes overexpression of Ppr in a strain of *Escherichia coli* that allows covalent attachment of substoichiometric amounts of biliverdin *in vivo*. Ppr is then fully reconstituted with biliverdin, followed by attachment of 4-hydroxycinnamic acid (*p*-coumaric acid), *in vitro*. Holo-Ppr with both chromophores is then isolated via an affinity tag and quantified for chromophore attachment by analysis of the absorption spectrum for biliverdin and 4-hydroxycinnamic acid. We also provide conditions for measuring autophosphorylation of Ppr.

Introduction

The purple photosynthetic bacterium *Rhodospirillum centenum* has a novel photoreceptor called PYP-phytochrome (Ppr) (Jiang *et al.*, 1999). Ppr has been shown to control the regulation of chalcone synthase expression in response to changes in light intensity (Jiang *et al.*, 1999). The amino terminus of Ppr is composed of a photoactive yellow protein (PYP) domain (Cusanovich and Meyer, 2003) that covalently attaches the blue light-absorbing chromophore, 4-hydroxycinnamic acid (*p*-coumaric acid). The PYP domain is followed by a bilin-binding domain similar to that observed in many prokaryotic phytochromes known to covalently attach the red light-absorbing chromophore, biliverdin (Montgomery and Lagarias, 2002). These blue light- and red light-absorbing chromophore attachment domains are followed by a carboxyl-terminal histidine kinase domain (Montgomery and Lagarias, 2002).

Because Ppr is ancestral to cyanobacterial and plant phytochromes (Montgomery and Lagarias, 2002), photochemical analysis of Ppr provides an ancestral view to the function of cyanobacterial and plant phytochromes. This chapter describes the overexpression, isolation, and *in vitro* reconstitution of apo-Ppr with biliverdin and with 4-hydroxycinnamic acid.

METHODS IN ENZYMOLOGY, VOL. 422   0076-6879/07 $35.00
DOI: 10.1016/S0076-6879(06)22009-3

We also provide spectral features of reconstituted Ppr and assays for autophosphorylation.

Vector Construct

The *R. centenum ppr* gene is amplified by polymerase chain reaction (PCR) and cloned into the *Nde*I–*Eco*RI site of pET28a(+) (Novagen, Madison, WI) to form the Ppr expression plasmid pET28ppr. To make biliverdin in the host cell as a chromophore, the *R. palustris hmuO* gene is PCR amplified with forward (5′-GG**CATATG**GTGGTGGAAGCAGCG-AAAC-3′) and reverse primers (5′-G**GGATCC**GTAGCGCTCTAGG-CGTCGAG-3′) that contain *Nde*I and *Bam*HI sites, respectively, and then cloned into *Nde*I–*Bam*HI sites of the expression plasmid pET11a (Novagen), resulting in the recombinant heme oxygenase expressing plasmid pEThmuO. Coexpression of both proteins is accomplished by cotrans-forming pET28ppr that is kanamycin resistant and pEThmuO that is ampicillin resistant into BL21(DE3) (Novagen). Cotransformed strains are selected on LB agar growth medium containing 50 μg/ml kanamycin and 100 μg/ml ampicillin and stored in 20% glycerol at $-80°$.

Preparation of 4-Hydroxycinnamic Acid Anhydride and Biliverdin

We use 0.25 M 4-hydroxycinnamic acid anhydride to reconstitute apo-Ppr with its blue light-absorbing chromophore. 4-Hydroxycinnamic acid anhydride is made according to Imamoto *et al.* (1995) and Kroon *et al.* (1996). To make 4-hydroxycinnamic acid anhydride, 0.5 ml of a 3 M dicy-clohexylcarbodiimide (DCC) solution (6.25 g DCC dissolved in 2.5 ml of N,N'-dimethylformamide [DMF]) is added to 4-hydroxycinnamic acid solu-tion, which is made by dissolving 0.164 g 4-hydroxycinnamic acid (Sigma) in 3.5 ml of DMF on ice and shaken at 4° overnight. Dicyclohexyl urea precipitate is removed by centrifugation for 2 min at 15,000g in an Eppen-dorf centrifuge, and the clarified supernatant containing 4-hydroxycinnamic acid anhydride is then used for *in vitro* reconstitution with Ppr (see later).

Biliverdin hydrochloride (Frontier Scientific, UT) is dissolved in dimethyl sulfoxide to 20 mM for use in reconstitution of apo-Ppr with biliverdin *in vitro*.

Overexpression and Reconstitution of apo-Ppr with Chromophores

Cells are streaked onto an LB agar plate from a $-80°$ frozen stock and incubated overnight at 37°. Five to 10 colonies are inoculated in 50 ml of LB liquid media containing 50 μg/ml kanamycin and 100 μg/ml ampicillin and incubated overnight at 37°. A volume of 20 ml of overnight culture is added to 1 liter of Terrific Broth media containing 50 μg/ml kanamycin and 100 μg/ml ampicillin. Cells are shaken at 37° until the OD_{600} reaches 0.4 to 0.6 at which point 0.5 ml of filter-sterilized 1 M isopropyl-β-D-thiogalactopyrano-side (Sigma) and 0.5 ml of 1 mM of δ-aminolevulinic acid hydrochloride (Sigma) are added to the culture. The culture is then shaken at 18° for 20 to 24 h. δ-Aminolevulinic acid is used to make hemes in the host cells as a precursor (Beale, 1990). Also, it has been reported that introduction of exogenous δ-aminolevulinic acid results in a rapid increase of spectropho-tometrically detectable phytochrome in seedlings where the phytochrome level was inhibited by gabaculine (Elich and Lagarias, 1987), as well as effective chromophore attachment in E. coli expressed phytochrome (Gambetta and Lagarias, 2001).

Cells are harvested by centrifugation at 3000g at 4° for 10 min and are resuspended in 30 ml of 4° 1× His-binding buffer (5 mM imidazole, 500 mM NaCl, 20 mM Tris-HCl, pH 7.9). Resuspended cells are disrupted three times with a cell cracker (Schmidt and Huttner, 1998) and then clarified by centrifugation at 27,000g at 4° for 30 min. About 40 ml of the supernatant is placed in a 200-ml beaker wrapped with aluminum foil to guard against light exposure and slowly mixed with a stirring bar at 4°. Ten-microliter aliquots of activated 0.25 M 4-hydroxycinnamic anhydride acid are slowly added at 5-min intervals until a total of 60 μl has been added. Next, 7 μl of biliverdin is added every 5 min for a total of 35 μl of biliverdin. Fully reconstituted Ppr is then affinity purified from other cellular proteins as described later.

Purification of Ppr Reconstituted with Chromophores

An FPLC system (Amersham Biosciences) at 4° is used for purification of reconstituted Ppr. The supernatant reconstituted with chromophores is applied to a 5-ml HiTrap chelating column (Amersham Biosciences). The column is washed with 20 column volume of binding buffer (5 mM imidaz-ole, 500 mM NaCl, 20 mM Tris-HCl, pH 7.9) and 70 column volume of 5.5% of 1× elute buffer (1 M imidazole, 500 mM NaCl, 20 mM Tris-HCl, pH 7.9). Reconstituted Ppr is then eluted with a gradient of 5.5 to 100% of elute buffer at a flow rate of 2 ml/min. Fractions with the green color of Ppr are pooled and checked by SDS-PAGE. Salt in the pooled fractions is removed by overnight dialysis against 20 mM Tris-HCl, pH 8.0, at 4°. Reconstituted Ppr is eluted at approximately 150 mM imidazole.

An Amicon Centriprep YM10 filter (10,000 MWCO, Millipore Co., Bedford, MA) is then used to concentrate purified Ppr typically to a volume of about 2 ml. The protein concentration is adjusted to 1 mg/ml with Tris-HCl, pH 8.0, to measure absorption spectra. Dialysis against storage buffer (50% glycerol, 20 mM Tris-HCl, pH 7.6, 50 mM KCl) is also performed for long-term storage at $-20°$. From this method, more than a 10-mg amount of homogeneous, spectrally active Ppr with both chromophores is readily produced from 1 liter of starting cell culture. Purified reconstituted Ppr is also typically more than 95% pure as measured by SDS-PAGE analysis.

Spectroscopic Measurements of holo-Ppr, Ppr-BV, and Ppr-pCA

Absorption spectra are recorded using an Agilent 8453E ultraviolet-visible spectroscopy system (Agilent Technologies, Germany). The absorption spectrum of Ppr fully reconstituted with only 4-hydroxycinnamic acid has a single peak at 431 nm (Fig. 1). This is contrasted by Ppr reconstituted with only biliverdin, which has absorbance bands at 700 and 394 nm, as is typical of other bacteriophytochromes that are in the Pr form in the ground state. The spectrum of Ppr fully reconstituted with both chromophores

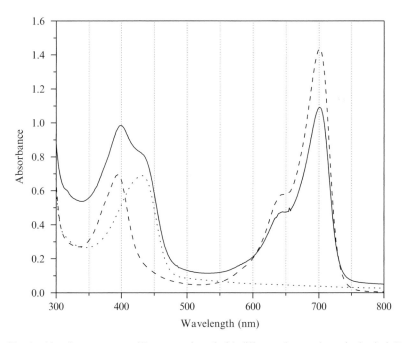

FIG. 1. Absorbance spectra of Ppr reconstituted with different chromophores in the dark. Ppr with both chromophores, including activated 4-hydroxycinnamic acid and biliverdin (solid line). Ppr with biliverdin (dashed line). Ppr with the activated 4-hydroxycinnamic acid (dotted line). (See color insert.)

(solid line in Fig. 1) displays absorbance maximum at 700 and 400 nm, plus a shoulder at 430 and at 640 nm.

In Vitro Autophosphorylation of Ppr

The *in vitro* phosphorylation experiment of reconstituted Ppr is performed under green safety light (Kodak 7B, Cat. No. 8070112) at room temperature. Protein in storage buffer is diluted with 10× kinase buffer (200 mM Tris-HCl, pH 7.8, 60 mM MgCl$_2$, 1 M NaCl) to yield a final concentration of 10 μM of Ppr in individual reaction mixtures. Half of the reaction mixtures is incubated in the dark and the other half is irradiated with specific light condition, such as red or blue light with intensity of 1 or 10 μmol/m^2/s. Phosphorylation of Ppr is initiated by adding 1/10 volume of 10× ATP mix composed of 36 μl of 10 mM ATP and 3 μl (0.024 mM) of [γ-^{32}P]ATP (>7000 Ci/mmol, MP Biochemicals). Immediately upon illuminating, 10-μl aliquots (2 μg of protein) are removed at time intervals, mixed with SDS-loading buffer, and placed on ice until analyzed by SDS-PAGE. The samples are subjected to SDS-PAGE, and the ^{32}P-labeled Ppr band is quantified by the Typhoon 9200 variable mode imager (Amersham Biosciences).

Acknowledgments

This study was supported by National Institutes of Health Grant GM040941 awarded to C. E. Bauer, by the Postdoctoral Fellowship Program of Korea Science & Engineering Foundation (KOSEF) to Y.-H. Chung and by Yamada Science Foundation to S. Masuda.

References

Beale, S. I. (1990). Biosynthesis of the tetrapyrrole pigment precursor, delta-aminolevulinic acid, from glutamate. *Plant Physiol.* **93**, 1273–1279.

Cusanovich, M. A., and Meyer, T. E. (2003). Photoactive yellow protein: A prototypic PAS domain sensory protein and development of a common signaling mechanism. *Biochemistry* **42**, 4759–4770.

Elich, T. D., and Lagarias, J. C. (1987). Phytochrome chromophore biosynthesis: Both 5-aminolevulinic acid and biliverdin overcome inhibition by gabaculine in etiolated *Avena sativa* L. seedlings. *Plant Physiol.* **84**, 304–310.

Gambetta, G. A., and Lagarias, J. C. (2001). Genetic engineering of phytochrome biosynthesis in bacteria. *Proc. Natl. Acad. Sci. USA* **98**, 10566–10571.

Imamoto, Y., Ito, T., Kataoka, M., and Tokunaga, F. (1995). Reconstitution photo-active yellow protein from apoprotein and p-coumaric acid derivatives. *FEBS Lett.* **374**, 157–160.

Jiang, Z., Swem, L. R., Rushing, B. G., Devanathan, S., Tollin, G., and Bauer, C. E. (1999). Bacterial photoreceptor with similarity to photoactive yellow protein and plant phytochromes. *Science* **285,** 406–409.

Kroon, A. R., Hoff, W. D., Fennema, H. P. M., Gijzen, J., Koomen, G. -J., Verhoeven, J. W., Crielaard, W., and Hellingwerf, K. (1996). Spectral tuning, fluorescence, and photoactivity in hybrids of photoactive yellow protein, reconstituted with native or modified chromophores. *J. Biol. Chem.* **271,** 31949–31956.

Montgomery, B. L., and Lagarias, J. C. (2002). Phytochrome ancestry: Sensors of bilins and light. *Trends Plant Sci.* **7,** 357–366.

Schmidt, A., and Huttner, W. B. (1998). Biogenesis of synaptic-like microvesicles in perforated PC12 cells. *Methods* **16,** 160–169.

[10] Oxygen and Redox Sensing by Two-Component
Systems That Regulate Behavioral Responses:
Behavioral Assays and Structural Studies of
Aer Using *In Vivo* Disulfide Cross-Linking

By BARRY L. TAYLOR, KYLIE J. WATTS, and MARK S. JOHNSON

Abstract

    A remarkable increase in the number of annotated aerotaxis (oxygen-seeking) and redox taxis sensors can be attributed to recent advances in bacterial genomics. However, *in silico* predictions should be supported by behavioral assays and genetic analyses that confirm an aerotaxis or redox taxis function. This chapter presents a collection of procedures that have been highly successful in characterizing aerotaxis and redox taxis in *Escherichia coli.* The methods are described in enough detail to enable investigators of other species to adapt the procedures for their use. A gas flow cell is used to quantitate the temporal responses of bacteria to a step increase or decrease in oxygen partial pressure or redox potential. Bacterial behavior in spatial gradients is analyzed using optically flat capillaries and soft agar plates (succinate agar or tryptone agar). We describe two approaches to estimate the preferred partial pressure of oxygen that attracts a bacterial species; this concentration is important for understanding microbial ecology. At the molecular level, we describe procedures used to determine the structure and topology of Aer, a membrane receptor for aerotaxis. Cysteine-scanning mutagenesis and *in vivo* disulfide cross-linking procedures utilize the oxidant $Cu(II)$-$(1,10$-phenanthroline$)_3$ and bifunctional sulfhydryl-reactive probes. Finally, we describe methods used to determine the boundaries of transmembrane segments of receptors such as Aer. These include 5-iodoacetamidofluorescein, 4-acetamido-4-disulfonic acid, disodium salt (AMS), and methoxy polyethylene glycol maleimide, a 5-kDa molecular mass probe that alters the mobility of Aer on SDS-PAGE.

Introduction

Sensing Oxygen and Redox Potential

    Behavioral responses to an oxygen gradient (aerotaxis) were first reported by van Leeuwenhoek in 1676 after he observed the accumulation of bacteria at the surface of pepper-water (Berg, 1975; Taylor, 1983b).

METHODS IN ENZYMOLOGY, VOL. 422   0076-6879/07 $35.00
Copyright 2007, Elsevier Inc. All rights reserved.   DOI: 10.1016/S0076-6879(06)22010-X

However, there was no systematic investigation of oxygen sensing until the 1880s when Engelmann reported that bacteria accumulated at the edges of a cover glass and around trapped air bubbles (Engelmann, 1881). Moreover, bacteria clustered near the chloroplasts of photosynthetic cells if the cells were illuminated with light, but not if they remained in the dark (Engelmann, 1883). Early investigators recognized not only that aerobic bacteria migrated to an air–liquid interface, but that anaerobes retreated from oxygen and microaerophilic bacteria accumulated at an intermediate position in an oxygen gradient. This position within the gradient was optimal for each bacterial species' metabolic lifestyle (Baracchini and Sherris, 1959). Today, it is known that aerotaxis is a key determinant in the ecology of many bacteria. The water column in a marine or estuarine environment contains horizontal veils of bacteria, with each species accumulated at a preferred oxygen concentration (Canfield and Des Marais, 1991; Donaghay, 1992; Jorgensen, 1982). The discovery of redox-sensing behavior in bacteria is more recent. In a spatial redox gradient, *Escherichia coli* and the microaerophile *Azospirillum brasilense* migrate to form a band at a preferred reduction potential (Bespalov *et al.*, 1996; Grishanin *et al.*, 1991; Taylor and Zhulin, 1998).

In *E. coli*, the same signal transduction mechanism is employed for aerotaxis and redox taxis. Both behaviors are a response to changes in the electron transport system (Bespalov *et al.*, 1996; Laszlo and Taylor, 1981; Taylor, 1983a,b; Taylor and Koshland, 1975; Taylor *et al.*, 1979). All redox molecules that elicit redox taxis, such as substituted quinones, divert electrons from the electron transport system and decrease proton motive force (Bespalov *et al.*, 1996). In *E. coli*, two receptors sense changes in the electron transport system: Aer (Bibikov *et al.*, 1997; Rebbapragada *et al.*, 1997) and Tsr (Rebbapragada *et al.*, 1997). The Aer receptor does not bind oxygen per se, but senses changes in the redox state of the electron transport system via an FAD cofactor (Bibikov *et al.*, 1997, 2000; Edwards *et al.*, 2006; Repik *et al.*, 2000). The Tsr (serine) chemoreceptor senses changes in proton motive force that result from perturbation of the electron transport system (Edwards *et al.*, 2006; Rebbapragada *et al.*, 1997). The common mechanism for aerotaxis and redox taxis has been designated energy taxis because the internal energy level of aerobic bacteria is dependent on proton motive force and the respiratory activity of the electron transport system. Other bacteria, such as *Bacillus subtilis,* use a different strategy for transducing an aerotaxis signal. The *B. subtilis* HemAT receptor binds oxygen directly so that aerotaxis is not dependent on changes in electron transport (Freitas *et al.*, 2003, 2005; Hou *et al.*, 2001).

With the rapid expansion of genomic studies, many open reading frames (ORFs) that are similar to the *aer* or *hemAT* genes have now been identified in bacteria as possible oxygen or redox sensors (Taylor and Zhulin, 1999). In addition, ORFs with sequences similar to the highly conserved signaling domain (annotated as MA or MCPsignal) of aerotaxis and chemotaxis receptors, but without a recognizable sensory input domain, have also been identified (http://genomics.ornl.gov/mist/ [Ulrich and Zhulin, 2007]). As a result, researchers who have not previously investigated aerotaxis and chemotaxis have an interest in screening motile bacteria for aerotaxis and redox taxis behavior. The goal of this chapter is to provide a step-by-step guide that a novice investigator can follow to measure aerotaxis and redox taxis. We also describe pitfalls and important controls that are well known in the chemotaxis community but are not readily accessible to new investigators.

The Two-Component System for Oxygen and Redox Sensing

The two-component histidine kinase system for *E. coli* chemotaxis is described elsewhere in this volume (Galperin and Nikolskaya, 2007; Wuichet *et al.*, 2007). Briefly, the Tsr, Tar, Trg, and Tap chemoreceptors modulate the autophosphorylation of the histidine kinase CheA when it is coupled to the chemoreceptor signaling domain via the CheW protein. Phospho-CheA phosphorylates the CheY protein, forming phospho-CheY, which temporarily reverses the rotational direction of the flagellar motors (Hess *et al.*, 1987, 1988). Adaptation to chemoeffectors is affected by methylation/demethylation of the transducing chemoreceptor [(Kehry *et al.*, 1983; Springer *et al.*, 1979); for a review see Bourret and Stock (2002)].

The *E. coli* Aer and Tsr receptors for aerotaxis and redox taxis use the same two-component phosphorylation cascade as chemotaxis (Fig. 1). However, unlike chemotaxis, adaptation of the Aer receptor to oxygen is methylation independent (Bibikov *et al.*, 2004; Niwano and Taylor, 1982). The mechanism of adaptation is not known, but Aer functions normally without the glutamyl residues that are methylated in chemotaxis (Bibikov *et al.*, 2004). In contrast, adaptation to oxygen in *B. subtilis* and *Halobacterium salinarium* is dependent on methylation (Lindbeck *et al.*, 1995; Wong *et al.*, 1995).

Structure of the Aer Oxygen and Redox Receptor

Escherichia coli chemoreceptors are Type I membrane receptors. Each chemoreceptor has two transmembrane segments that separate a periplasmic sensing domain from a cytosolic HAMP and signaling domain (see Tsr receptor in Fig. 1). The Aer receptor is different in that both its sensing and signaling domains are cytosolic. Each monomer of Aer consists of an N-terminal PAS sensing domain that is connected (via F1) to a membrane

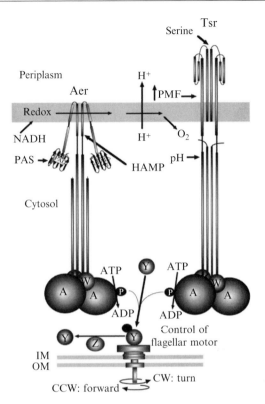

FIG. 1. Aerotaxis and chemotaxis signaling pathways in *E. coli*. Both Aer and Tsr are aerotaxis sensors, but Tsr is also a chemoreceptor sensing serine, temperature and pH. Aer has an FAD-binding PAS sensor that signals through a HAMP domain to the C-terminal signaling domain. Both receptors control the direction of flagellar rotation by altering the activity of the CheA histidine autokinase, which transphorylates the CheY tumble regulator (see text for details). A, CheA; W, CheW; Y, CheY; Z, CheZ; IM, inner membrane; OM, outer membrane; CW, clockwise; CCW, counter clockwise.

anchor (Fig. 1). The membrane anchor consists of two transmembrane segments divided by a short periplasmic loop (Amin *et al.*, 2006). The second of these transmembrane segments is linked to a cytosolic HAMP and signaling domain, both of which are homologous to the HAMP and signaling domains found in chemoreceptors.

Because the *E. coli* Aer protein is unstable when extracted from the membrane, most of what we know about the topography of Aer and its structure and function has been determined in whole cells, or in membrane vesicles, using a battery of *in silico*, genetic and biochemical techniques (Amin *et al.*, 2006; Bibikov *et al.*, 1997, 2000, 2004; Buron-Barral *et al.*, 2006;

Edwards *et al.*, 2006; Gosink *et al.*, 2006; Herrmann *et al.*, 2004; Ma *et al.*, 2004, 2005; Rebbapragada *et al.*, 1997; Repik *et al.*, 2000; Watts *et al.*, 2004, 2006a,b; Yu *et al.*, 2002). From these studies, we have reasonable structural models for the PAS domain (Repik *et al.*, 2000), the membrane anchor (Amin, 2006; Amin *et al.*, 2006), and the HAMP domain (K. J. Watts, unpublished data), as well as clues about the important regions for structural stability (Buron-Barral *et al.*, 2006; Herrmann *et al.*, 2004), interactions with other chemoreceptors (Gosink *et al.*, 2006), and signal transduction (Ma *et al.*, 2005; Repik *et al.*, 2000; Watts *et al.*, 2004, 2006a,b). Disulfide cross-linking, one of the more useful tools for determining structural topology, is reviewed elsewhere (Lai and Hazelbauer, 2007). We extended the use of this technique to map dimerization faces of the Aer HAMP domain in intact cells (K. J. Watts, unpublished findings). This chapter describes this technique and other disulfide-related methods used to determine the orientation and boundaries of the Aer receptor in the membrane.

Assays of Oxygen and Redox Sensing: General Considerations

The methods used to characterize behavioral responses in bacteria are classified according to the type of gradient that is used.

Spatial-Gradient Assays

These assays depend on the metabolic activity of the bacteria to consume an attractant in their immediate environment. The bacteria then migrate up the resulting gradient of attractant, seeking higher concentrations. The continuous consumption of attractant perpetuates the gradient and migration continues as an expanding band of bacteria. Typically, the bacteria are so concentrated that the band is visible without the use of a microscope. Examples of commonly used spatial-gradient assays for oxygen sensing include the swarm plate assay, where bacteria migrate through soft agar, and the capillary assay, where bacteria migrate through liquid medium. A redox spatial gradient can be formed by inserting an agarose plug containing quinone and ferricyanide (to keep the quinone oxidized) into a suspension of bacteria and hydroquinone (reduced quinone); bacteria respond to the resulting quinone–hydroquinone gradient and form a band around the agarose plug at the preferred reduction potential (Bespalov *et al.*, 1996).

Temporal-Gradient Assays

These assays are tedious but have the advantages that they are quantitative and do not depend on metabolism to form a gradient. They measure the length of time that cells take to adapt to a stimulus; stronger stimuli elicit a longer response. This temporal tuning is a requirement for bacterial chemotaxis, as bacteria are too small to navigate by comparing attractant concentrations at the front and back of the cell. They improve their accuracy of navigation by comparing changes in attractant concentration over time. If the concentration of attractant is increasing, they suppress directional changes and continue swimming in the "favorable" direction. If the attractant is decreasing, the bacteria tumble and randomly orient in a new direction. This type of navigation requires a short-term memory, and it is this memory that is measured in a temporal gradient assay. To perform a temporal assay, bacteria are placed in a microchamber and perfused with nitrogen gas for several minutes, followed by air or oxygen. The behavior of the bacteria in the temporal oxygen gradient is observed under a microscope. As the bacteria adapt, the time required for 50% of them to return to the prestimulus tumbling frequency is recorded. For a temporal redox assay, a redox agent such as a quinone is added directly to the cells, and the response to the change in reduction potential is measured.

Obtaining a Motile Strain

For studies of behavioral responses, we aim to have greater than 85% of the bacteria motile. We begin by screening for a motile strain. Bacteria are grown to midlog phase in an enriched medium such as LB broth (Lennox, 1955), or preferably, in tryptone broth. Aliquots (30–60 μl) of freshly grown cells are diluted in fresh medium (4 ml) and incubated at room temperature for 15–30 min. The acclimated bacteria (10^7 ml^{-1}; some investigators prefer 5×10^7 ml^{-1}) are then viewed under a dark-field microscope (800× magnification) and the percentage of cells with normal motility is calculated. If the cells are not highly motile, we either work with another strain or enrich the motility of the poorly motile strain. A culture may consist of motile cells and cells with defective motility. To identify motile clones, plate the cells and pick from individual colonies to inoculate tryptone soft agar (0.28%) (see recipe later). Incubate the soft agar at 30° in a humidified incubator for 14–20 h. The motile colonies form larger circles than colonies with defective motility. Pick cells from the edge of the larger colonies, inoculate 5 ml LB broth for growth, and then screen for motility under the microscope. For a more powerful selection procedure, inoculate the center of a tryptone soft agar plate with 2 μl of liquid culture. The motile cells will swim out during incubation. Allow the colony to expand until the outermost band nears the edge of the plate. The process may be repeated a second time using cells picked from the outer band to inoculate a fresh plate. An investigator should be aware that this selection procedure could isolate a motile contaminant or cells that have gained a gratuitous mutation.

After obtaining a motile strain, prepare multiple vials of frozen permanent cultures. Mix 700 μl of an early stationary-phase culture with 300 μl of sterile 50% glycerol (15% final glycerol concentration) in a sterile cryovial.

Store cryovials at −80° or in liquid nitrogen. For each new experiment, inoculate a fresh culture by scraping cells from the top of a frozen permanent (without allowing the surface of the frozen cells in the vial to thaw). Starting a new culture minimizes the risk of losing a plasmid from the cells or inadvertently selecting a mutant.

Growth of Bacteria for Behavioral Studies

The behavioral properties of *E. coli* strains vary with the phase of growth, media composition, aeration, and temperature. This necessitates a rigorous standardization of the growth conditions and preparation of the cells for behavioral measurements. Flagella synthesis in *E. coli* is temperature sensitive: cells are more motile at 30° than at 37°, and by 42°, flagella synthesis stops (Adler and Templeton, 1967; Silverman and Simon, 1977). Therefore, *E. coli* is routinely grown at 30° for most behavioral studies. The synthesis of flagella also varies with the phase of growth. We graph the percentage of motile cells versus density (or cell number) and use this information to standardize the density at which the cells will be harvested. For the growth conditions used in our laboratory, we harvest *E. coli* at OD = 0.35–0.45 when characterizing behavior. To improve aeration, we do not add more than 5 ml of growth medium per test tube (18 mm diameter) and shake the tubes vigorously (250 rpm) in a rotary water bath shaker. To prepare larger batches of cells, we add 30 ml of growth medium to a 250-ml flask or 150 ml to a 1-liter flask. Motility is deficient when cells grown in a rich medium such as LB broth are subcultured into minimal medium, such as H1 salts (Adler, 1973) with an added carbon source and auxotrophic requirements. This can be overcome by serially subculturing the cells in the minimal medium two or three times. A midexponential phase culture can then be kept at 4° for 1 to 2 weeks and used as an inoculum to start overnight cultures. An LB broth inoculum may be stored at 4° for 1 month.

For behavioral studies, harvested cells are usually washed and suspended in chemotaxis buffer, which supports optimal motility for *E. coli*.

Chemotaxis Buffer (J. S. Parkinson, Personal Communication)

Stock solution	Add	Final concentration
0.1 M EDTA (K$^+$)	0.1 ml	0.1 mM
0.1 M KPO$_4$ (pH 7.4)	10.0 ml	10 mM
1.0 M Na-lactate	1.0 ml	10 mM
Deionized H$_2$O	88 ml	

Autoclave, cool, and then add each of the following from sterile solutions.

Stock solution	Add (to above)	Final concentration
0.1 M MgSO$_4$	1.0 ml	1 mM
0.1 M (NH$_4$)$_2$SO$_4$	1.0 ml	1 mM

During the washing process, the flagella are easily damaged. Centrifuge the cells to form a firm, but not hard, pellet (e.g., 4800g for 3 min). Decant the supernatant and resuspend the pellet in chemotaxis buffer by gentle swirling. Do not vortex the cells. Repeat this step twice, resuspending the cells to the desired volume after the second wash.

Temporal Assay for Aerotaxis

Apparatus

Gas Flow Cell. A suitable perfusion chamber can be constructed from an aluminum block by a skilled machinist. The aluminum frame rests on the stage of an upright microscope (Fig. 2). A microscope slide (25 × 75 × 1 mm) is inserted to form the floor of the chamber and is replaced after one or two assays. A cover glass (24 × 60 mm, No. 1) forms the top of the chamber and is sealed to the aluminum frame using Lubriseal stopcock grease (Thomas Scientific, Swedesboro, NJ), which is oxygen impermeable. A gas line (Teflon, type FEP, Cole-Parmer Instrument, Vernon Hills, IL) enters a manifold that is separated from the main chamber by a polyurethane

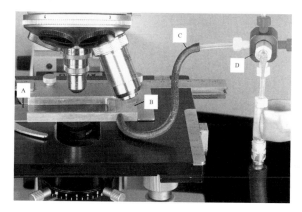

Fig. 2. Apparatus for aerotaxis temporal assays. (A) A glass microscope slide, onto which cells are placed, forms a modular base for the gas flow cell. (B) Aluminum flow cell. (C) Oxygen impermeable, flexible, butyl rubber tubing. (D) The four-port Hamilton valve used to toggle between aerobic and anaerobic gas mixtures. For a gas flow diagram see Fig. 4.

foam insert (approximately $22 \times 10 \times 4$ mm). The insert diffuses the gas flow to produce a laminar flow. Without the foam insert Bernoulli currents form in the gas stream and draw air into the chamber. The complete chamber is held in place on the microscope stage by the slide retainer clips, enabling the chamber to be positioned by the stage controls. Diagrams for construction of the chamber are shown in Fig. 3.

Microscope. Specialized dark-field optics are required to observe bacteria in the chamber. We use a Dialux Trinocular microscope (Leitz, Wetzlar, Germany), but similar optics may be available from other manufacturers. Most important is a long working distance (6.5 mm) dark-field objective (L32/ 0.4 [32×], UMK50, or N PLAN L50/0.5). We use a dry dark-field condenser. The microscope is fitted with a COHU solid state surveillance camera, Model 4815 (Cohu Inc., San Diego, CA) and the image is recorded using a commercial quality video recorder with a jog-wheel and single frame advance. A time-date generator (Model WJ-810; Panasonic, Secaucus, NJ) supplies a time stamp to the video recorder. Although we use an upright microscope, it may be possible to adapt an inverted microscope for use with the flow cell.

Gas Proportioner. For the temporal assay, the flow chamber is connected to a four-port Hamilton valve (#86729; Hamilton Co., Reno, NV), which is used to toggle between aerobic and anaerobic gases.

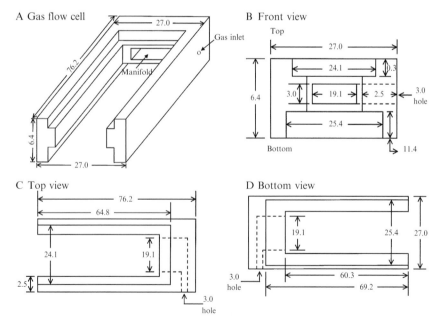

FIG. 3. Plans for constructing the gas flow cell shown in Fig. 2. (A) Perspective view as shown in Fig. 2. (B) Front, (C) top, and (D) bottom views. All measurements are given in millimeters.

One port of the valve is connected to a single gas flowmeter (Model FM-1050, tube #602; Matheson, Tri-Gas, Montgomeryville, PA) for nitrogen, and the other port is connected to a gas proportioner (Model 7300, tubes #602, Matheson, Tri-Gas) with a chamber for mixing nitrogen and oxygen (Fig. 4). Both the flowmeter and the gas proportioner are fitted with stainless steel inlet tubes that supply gases from cylinders of prepurified compressed gases (Matheson, Tri-Gas). Gas entering the flowmeter is humidified via inline bubblers composed of gas dispersion tubes in gas washing bottles (500 ml, Kimax) containing deionized water (Fig. 4). The humidified gas prevents the sample from drying out on the microscope stage during the temporal assay. Residual oxygen is removed from the anaerobic gas by directing it through an Oxy-Trap (Alltech Associates, Deerfield, IL) and an indicating Oxy-Trap (Alltech Associates) before it reaches the gas washing bottle. It is critical that connections do not permit

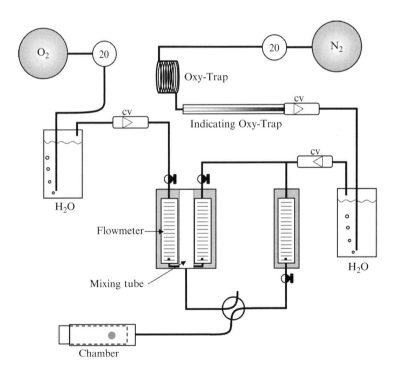

Fig. 4. Gas proportioner used to supply gas mixtures to the flow cell for aerotaxis temporal assays. Traces of oxygen are removed from prepurified nitrogen with in-series Oxy-Traps. There is a color change in the indicating Oxy-Trap when the proximal Oxy-Trap is exhausted (see text for details). Gas regulators are shown as small circles with the selected pressure (20 pounds per square inch; equivalent to 1.409 kg per square centimeter) indicated within the circle. cv, one-way check valve.

atmospheric oxygen to diffuse into the system. Where practical, lines are stainless steel with gas-tight fittings or, if more flexibility is needed, butyl rubber (Fisher Scientific) and FEP-type Teflon (Cole-Parmer Instruments) tubing, which have low permeabilities to oxygen.

The flow rate through the gas proportioner can be calibrated by directing the gas flow from the proportioner into an inverted graduated cylinder filled with water. Fill the cylinder by submerging it in a tub of water, and invert it, while keeping the open end underneath the water surface. Determine the volume of water displaced by the gas during a convenient time interval. Repeat at a range of flowmeter settings and plot the flow rate as a function of the meter settings. To complete the calibration process, the oxygen:nitrogen mixture flowing from the proportioner should be measured directly at different settings of the proportioner. Connect the proportioner output to a 60-ml syringe and insert a calibrated oxygen monitor (Hudson RCI monitor 5800 equipped with an oxygen sensor 5803; Hudson RCI, Temecula, CA) into the syringe barrel. Of note, water vapor exerts its own pressure on the gas so that water-saturated air is 19.6% oxygen rather than 20.9%. The gas mixture can be accurately varied over a 20-fold range. Use medical grade compressed air and purified nitrogen for oxygen concentrations down to 1%. Oxygen mixtures greater than 21% are obtained by mixing pure oxygen and nitrogen, and lower oxygen mixtures are obtained by combining premixed oxygen (e.g., 5%, 0.1%, oxygen in nitrogen, Matheson, Tri-Gas) with nitrogen.

Procedure

1. Cells should be freshly grown at 30° to midexponential phase (OD = 0.35–0.45), washed twice, and resuspended in chemotaxis buffer, as described earlier. To characterize aerotaxis responses, grow *E. coli* in a defined medium such as H1 salts (Adler, 1973) supplemented with a carbon source (e.g., 1% glycerol) and auxotrophic requirements. For preliminary screening, we may culture the strains in LB medium (Lennox, 1955) or tryptone broth, and dilute the cells in the same medium. The density of the diluted cells should result in 50–100 bacteria in the field of view when viewed through the microscope (800× magnification). Before using a higher density of cells, each investigator should compare the results at lower and higher density. At higher densities the observations may be skewed toward those cells that are most active.

2. For reliable results, the cells must be allowed to adapt to the fresh medium and the temperature used for the assay. During growth, bacteria preferentially consume serine and aspartate in a rich culture medium. If the harvested cells are resuspended in fresh LB medium, there is a prolonged

smooth-swimming response to reintroduced serine and aspartate. A 20-min adaptation time is usually sufficient, but adaptation should be confirmed by examining a 10-μl sample under the microscope. Adapted cells should show the random motility pattern of unstimulated cells. If there is a brief smooth-swimming response when the sample is placed on a microscope slide, it may indicate that the cells had depleted all available oxygen during the incubation period. In this case, aerate the cells during the incubation period with mild agitation.

3. For the aerotaxis assay, add 5–10 μl of the preadapted culture to a clean microscope slide and spread evenly over a 6-mm circle. With a smaller sample volume (5 μl) the smear is thinner and the culture equilibrates more rapidly after a change in the composition of the gas phase. However, thinner smears will evaporate more rapidly and adding 10-μl of the culture is more convenient for some experiments.

4. Insert the microscope slide into the floor of the gas flow cell and position the flow cell so that the microscope objective is focused on cells near the surface of the slide (bottom of the smear). Confirm that the cells are visible in the video monitor and that the plane of focus is the same in the microscope and video system.

5. Observe the motility of the bacteria. If there has been a significant increase in nonmotile cells, the bacteria may be adhering to the slide surface. Try a fresh slide or, if necessary, clean the slide with 95% alcohol or acetone. Always handle slides by the edges; we use four layers of Kimwipes (Kimberly-Clark Professional, Roswell, GA) when cleaning the surface. Amino acids from finger prints cause artifactual behavior by the bacteria. An investigator should have a thorough knowledge of the steady-state motility pattern of the strain. In an isotropic environment, *E. coli* cells that are wild type for aerotaxis swim for about 1 s, tumble briefly, and then resume swimming in a new direction, forming a three-dimensional random walk (Berg and Brown, 1972). Other species use different strategies for random motility. During prescreening, we evaluate the motility of *E. coli* by estimating whether the bacterial speed and tumbling frequency are within a normal range. If motility is normal, video record the prestimulus motility as a reference.

6. Initiate the temporal aerotaxis assay by perfusing the flow chamber with humidified and oxygen-free nitrogen gas from a gas proportioner (Figs. 2 and 4). Record the behavior of the cells with the time-date generator on. The swimming speed decreases and the tumbling frequency increases briefly in response to the declining oxygen concentration (Fig. 5). The tumbling frequency may then fall to near zero after the cells become anaerobic because the proton motive force is insufficient to support tumbling (Khan and Macnab, 1980). After 2 to 3 min, the cells acclimate to these conditions,

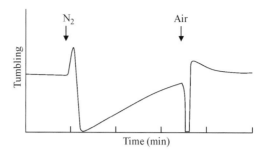

FIG. 5. Graph representing the behavioral response of *E. coli* to changes in the concentration of oxygen. Changes in the perfusing gas are indicated by arrows. Modified from Taylor (1983a) (see text for details).

adapt to anaerobiosis, and resume random motility. When the cells are fully adapted to anaerobiosis (3 min), switch the gas to 21% oxygen or stop the flow of nitrogen and allow air to diffuse into the chamber. Continue to record the behavior of the bacteria. The increase in oxygen suppresses tumbling for 60–75 s in wild-type *E. coli*. This is followed by a tumbling overshoot and restoration of random motility (Fig. 5). The assay can be repeated on the same cells by reperfusing the flow cell with nitrogen for 3 min. For precise measurements, record the aerotaxis responses in triplicate on each of three days (*n* = 9). Responses to air that are less than 40 s may indicate that the bacteria were not fully anaerobic. Check the gas supply line for a leak or exhausted Oxy-Trap oxygen scrubbers.

7. Replay the video record to quantitate the response, using the time stamp on the recording or a stop watch to time the response. Calculate the tumbling response from the start of tumbling until 50% of the bacteria return to the prestimulus tumbling frequency. For the smooth swimming response, there is a clearly demarcated acceleration of swimming speed when oxygen reaches the bacteria. Time the response from the acceleration to the 50% recovery point. The resulting measurements are a reliable estimate of behavioral responses (Spudich and Koshland, 1975) when made by an experienced investigator. It is preferable to observe the whole field of bacteria in the microscope and learn to concentrate on noting tumbles (abrupt direction changes) wherever they occur in the field rather than following individual bacteria. Novice investigators may wish to first use a stop watch to make replicate timings of a recorded response until the measurements are reliable. We routinely verify response times in a double-blind assay by an independent investigator. For more quantitative measurements, it is also possible to use computerized motion analysis (Lindbeck *et al.*, 1995; Wright *et al.*, 2006; Yu *et al.*, 2002).

Notes. The temporal aerotaxis assay can be adapted to species other than *E. coli.* For microaerophiles, it is important to be aware that while oxygen at low partial pressures may be an attractant, 21% oxygen in air may be a repellent (Johnson *et al.*, 1997; Zhulin *et al.*, 1996). As a result, if the bacteria are cycled between nitrogen and air, there may not be a net response. To avoid false negatives, determine the preferred partial pressure for a species using a capillary aerotaxis assay (see later). Then measure the behavioral response when the bacteria are cycled between nitrogen and the preferred concentration of oxygen for the species.

It is more convenient to quantitate the smooth swimming response than the tumbling response because the smooth response is longer. For precise measurements of the tumbling response, the bacteria should be preadapted (3 min) to a known concentration of oxygen (21% oxygen for *E. coli*) before perfusing with nitrogen (Rebbapragada *et al.*, 1997).

Spatial-Gradient Capillary Assay for Aerotaxis

Procedure

1. Examine under a dark-field microscope a fresh, midlog phase culture (OD = 0.3–0.45) grown in minimal medium. If the culture is highly motile, harvest the cells, wash twice, and resuspend to the same volume in chemotaxis buffer. Verify that more than 85% of the bacteria are motile after resuspension. Cells grown in LB medium can also be used in the capillary assay without harvesting. However, there may be a loss of resolution.

Depending on the individual strain, it may be necessary to modify the experimental conditions. Some *E. coli* strains stick to the surface, flocculate, or aggregate in capillary tubes. This is generally more problematic in minimal media, but rich media may have other interfering effects from nutrients and amino acids that compete with the aerotactic signal. For example, the *E. coli* Tsr receptor senses both serine and oxygen. While this receptor mediates an aerotactic band in rich LB medium, the band is more defined in chemotaxis buffer, which does not contain serine. If serine sensing is eliminated in Tsr, for example, by a Tsr-R64C missense mutation (Lee *et al.*, 1988), the defined band can be reproduced in LB medium (K. Pierre and M. S. Johnson, unpublished observation).

In addition to good motility, which is required for all behavioral capillary assays, aerotaxis capillary assays depend on adequate respiration to generate a steep oxygen gradient. It is therefore important that the culture is well aerated during growth and harvested in midexponential phase.

2. Resuspend bacteria to OD = 0.4–0.8 and transfer 0.5–1 ml to a 1.5-ml microcentrifuge tube. Insert one end of an optically flat, open-ended, rectangular, glass capillary (0.1 × 1.0 × 50-mm VitroTube, P/N 5010; VitroCom, Inc., Mountain Lakes, NJ) into the microcentrifuge tube and allow the suspension to rise by capillary action. Wipe the outside of the capillary. Immediately position the capillary (or multiple capillaries) on a microscope slide and place the slide on the stage of a dark-field microscope. Use a low-power objective (PL2.5/0.08, Leitz) and position the capillary so that the upper meniscus at the time of filling is at the edge of the field of view. We find that debris is more likely to accumulate at the lower meniscus under the influence of gravity. Video record the response (magnification 62×). Wild-type cells typically form a sharp aerotaxis band within 10 min. An aerotaxis band is characterized by a dense accumulation of cells at a distance from the air–liquid interface with clear zones flanking the band (Fig. 6A; also see Movie 1 at the *Methods in Enzymology* companion web site [http://books.elsevier.com/companions/9780123738516]).

3. To confirm that the band is formed by aerotaxis, observe the behavior of individual bacteria by increasing the magnification using a long working distance objective (total magnification 800×). True aerotactic bands have highly motile cells within the band that occasionally swim out of the band, but immediately tumble and return to the band. Pseudo-aerotactic bands may also be observed (Fig. 6B). They are typically aggregates of nonmotile cells that form on the anaerobic side of the oxic/anoxic

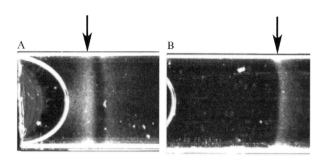

Fig. 6. Aerotactic band in a spatial oxygen gradient in a capillary. (A) Tsr-mediated aerotaxis band in *E. coli* BT3388 (*aer tsr tar trg tap*) containing pKP2 (expressing Tsr-R64C). Note a clear zone to the right of the band (arrow) that is often present when cells are in chemotaxis buffer. There are relatively few bacteria in this zone when viewed under high magnification (a video of this is available as Movie 1 at the *Methods in Enzymology* companion web site [http://books. elsevier.com/companions/9780123738516]). (B) An example of a pseudo-aerotaxis band (arrow) formed by *E. coli* BT3388 containing vector alone (pTrc99A). Under high magnification, the cells in a pseudo-aerotactic band are nonmotile (see Movie 2 at the *Methods in Enzymology* companion web site [http://books.elsevier.com/companions/9780123738516]).

boundary. Because cells at this boundary do not have the energy to swim, or the aerotactic sensing machinery to escape, the number of cells within this band increases over time. While it is not possible to unequivocally identify a pseudo-band without observing the behavior of individual bacteria under higher magnification, most pseudo-bands do not have the clear "avoidance" zone on the anoxic side of the band. Instead, they have an undefined border that tapers off toward the anoxic side of the band (Fig. 6B; also see Movie 2 at the *Methods in Enzymology* companion web site [http://books.elsevier. com/companions/9780123738516]).

4. The capillary assay is a qualitative, or at best semiquantitative, measure of aerotaxis. It is possible to distinguish among mutant strains that have a strong aerotaxis response, a weak response, or no response. Also, this assay can provide specific information about the preferred concentration of oxygen to which a strain is attracted. Obligate aerobes migrate to the highest oxygen concentrations at, or near, the air–liquid interface, anaerobes move to the interior of the capillary, and species with other metabolic lifestyles aggregate at points in between (Johnson *et al.*, 1997; Wong *et al.*, 1995; Zhulin *et al.*, 1996). *E. coli* strains that are wild type for aerotaxis (aer^+ tsr^+) band at 5% oxygen in the capillary assay (M. S. Johnson, unpublished observation). Strains with only the Aer receptor for aerotaxis seek a higher concentration of oxygen and band closer to the meniscus. Strains that have only the Tsr receptor band further from the meniscus than wild-type cells (Rebbapragada *et al.*, 1997). The following section describes use of the capillary assay to precisely determine the preferred concentration of oxygen for a species.

Using a Capillary to Determine the Preferred Partial Pressure of Oxygen for Bacteria

The preferred oxygen partial pressure for a bacterial strain can be measured directly using an oxygen electrode to measure the oxygen concentration in an aerotactic band. However, this method has limitations. First, the surface tension around the electrode partially distorts the aerotactic band. Second, an oxygen needle electrode lacks the sensitivity and stability to give accurate readings at oxygen partial pressures below 3.5 torr (0.5% oxygen). Alternatively, if a bacterial strain bands at a consistent position in the oxygen gradient, it is possible to estimate the oxygen concentration within the band by lowering the oxygen partial pressure at the gas–meniscus boundary and determining the oxygen partial pressure where the center of the aerotactic band reaches the meniscus. We describe both methods here.

Direct Measurement of Preferred Partial Pressure with an Oxygen Electrode

1. Grow cells in an appropriate minimal medium to midexponential phase, wash twice, and resuspend in chemotaxis buffer to OD = 0.6–0.8, as described previously.

2. Insert one end of an optically flat glass capillary (0.5 × 5.0 × 100 mm VitroTube, VitroCom) into a 1.5-ml microcentrifuge tube containing 1 ml of bacterial suspension. Allow the capillary to fill. Clean the capillary surface with a Kimwipe, tape the capillary onto a microscope slide, and tape the slide onto a microscope stage with the stage clamps removed. The capillary must be elevated above the stage so that the stage does not obstruct movement of the needle electrode. Position the capillary under a low-power objective of the dark-field microscope so that the meniscus is at the edge of the field of view. An aerotaxis band is expected to form in 10–15 min. Confirm under high magnification (800×) that the band is formed by aerotaxis (see earlier discussion).

3. Clamp a blunt-ended 29-gauge (~0.35 mm wide) CUSTOM needle Clarke electrode (S/N 050895–8; Diamond General Corp, Ann Arbor, MI) onto a micromanipulator (Märzhäuser; Fine Science Tools, Inc., Foster City, CA) capable of advancing the electrode in 10-μm increments along the long axis of the capillary. Align the needle electrode with the capillary so that it is at the center of, and perpendicular to, the meniscus (Fig. 7A). Carefully perforate the meniscus; this causes distortion of the band due to surface tension around the needle, but the band is not disrupted (Fig. 7B;

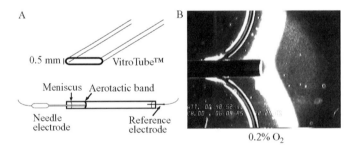

FIG. 7. Determining the preferred partial pressure of oxygen for *A. brasiliense* using an oxygen needle electrode. (A) A flat capillary tube (VitroTube), and the placement of needle and reference electrodes as viewed from above the capillary. (B) A needle oxygen electrode inserted into the aerotactic band, and the partial pressure of oxygen at this location. A micromanipulator was used to advance the oxygen electrode in 10-μm intervals. The inserted needle electrode distorts, but does not disrupt, the band (a video showing the meniscus being perforated is available as Movie 3 at the *Methods in Enzymology* companion web site [http://books.elsevier.com/companions/9780123738516]).

see also Movie 3 at the *Methods in Enzymology* companion web site [http://books.elsevier.com/companions/9780123738516]). After the electrode stabilizes, record the oxygen concentration and advance the manipulator by 10 μm. Repeat the process until the anaerobic side of the band is reached. Withdraw the electrode in 10-μm steps and compare these readings with those of the advancing electrode. The profile for meniscus entry and exit will be different, as the departing electrode pulls the meniscus along with it.

4. The partial pressure of oxygen at the center of the aerotactic band is defined as the preferred oxygen partial pressure for that particular strain. Although this value is different from the partial pressure of oxygen at the aerobic and anaerobic boundaries of the band, cells in the center of the band generally have the best motility.

Estimating Preferred Oxygen Partial Pressure by Varying the External Oxygen Partial Pressure

In this method, the capillary is placed in a gas perfusion chamber, and, as the external oxygen partial pressure is lowered in steps, the band moves predictably toward the meniscus (Fig. 8B). The external partial pressure of oxygen when the center of the aerotactic band reaches the meniscus is an estimate for the preferred oxygen partial pressure. The accuracy of these

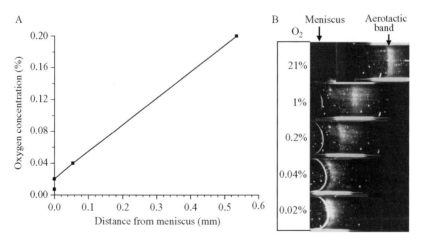

FIG. 8. Estimating the preferred oxygen partial pressure by varying the external oxygen partial pressure. The obligate anaerobe, *Desulfovibrio vulgaris*, was placed in a flat capillary tube (see Fig. 7A), and the distance between the meniscus and the aerotactic band was measured at decreasing external concentrations of oxygen (see text for details). (A) Plot (modified from Johnson *et al.*, 1997) of the four lowest oxygen concentrations perfused through the flow cell (0.2, 0.04, 0.02, and 0.007%) versus the banding distance from the meniscus. (B) Band movement toward the meniscus as the external oxygen partial pressure is lowered. Frames were freeze-captured from a video taken by S. Patel and M. S. Johnson. See text for details and Johnson *et al.* (1997) for a discussion of these data.

estimates has been verified empirically for the microaerophile *A. brasiliense* and the obligate anaerobe *Desulfovibrio vulgaris* (Johnson *et al.*, 1997; Zhulin *et al.*, 1996). Subsequent experiments revealed that, at the preferred oxygen partial pressure, *A. brasiliense* generated a maximum proton motive force and that *D. vulgaris* exhibited oxygen-dependent growth.

Procedure

1. Prepare the capillary ($0.1 \times 1.0 \times 50$ mm, VitroCom) as for an aerotaxis capillary assay (see earlier discussion), but place the capillary and microscope slide in the gas flow cell described for the temporal aerotaxis assay.

2. In this method, the distance of the aerotaxis band from the air–liquid interface will be measured on the display monitor of the microscope video recording system. To calibrate the monitor, focus the microscope on a stage micrometer, measure the monitor image with a flexible plastic ruler, and calculate the magnification factor. A photographic image can be substituted for the video image.

3. After the capillary is positioned so that the meniscus is at the edge of the field of view, perfuse the flow cell with medical-grade compressed air or 21% oxygen in nitrogen. Higher initial oxygen partial pressures may be used for obligate aerobes such as *B. subtilis* that form an aerotaxis band at the meniscus in air (Wong *et al.*, 1995). Wait for a band to form, and measure the distance between the center of this band and the meniscus.

4. Lower the oxygen partial pressure of the perfusing gas by adjusting the rates of air and nitrogen flow using the gas proportioner described earlier. Wait for the band to stabilize at a new position and remeasure the distance to the meniscus. Repeat the procedure, each time decreasing the oxygen concentration by an equal step (e.g., 21%, 19%, 17% ...) until the band forms at the air–liquid interface (Fig. 8B). The band will reach the meniscus near the preferred oxygen partial pressure; at concentrations below this, the band will shrink and eventually disappear (see Zhulin *et al.*, 1996).

5. Plot the oxygen partial pressure versus band distance from the meniscus. Because artifacts may occur when the aerotactic band reaches the meniscus, extrapolate the plot to determine the preferred oxygen partial pressure at the intersection of the plot with the *y* axis (Fig. 8A) (Johnson *et al.*, 1997; Zhulin *et al.*, 1996).

Spatial-Gradient Soft Agar Plate Assays for Aerotaxis

General Considerations

In soft agar assays, bacteria at the inoculation site consume nutrients, generating a spatial gradient. In response to the gradient, bacteria swim out from the inoculation site, moving through the network of agar strands as they seek a more favorable microenvironment. The gradient is maintained by continued consumption, and the bacteria move out in an expanding ring that is visible to the unaided eye. More than one concentric ring may be visible. Each ring results from chemotaxis to a specific molecule. Succinate and tryptone soft agar are used to screen for Aer-mediated responses. Succinate plates are made with minimal medium and succinate as the carbon source (Bibikov *et al.*, 1997). In *E. coli*, the Aer receptor uses both energy sensing and oxygen sensing to sense succinate. Respiration by the growing cells creates both a succinate gradient and an oxygen gradient, with the lowest oxygen partial pressure at the bottom of the plate. Aerotactic colonies form a dome-shaped colony in the soft agar, with a focused ring at the outer edge (wild-type Aer in Fig. 9, right panel). The center of the dome is at the surface of the agar, and the outer edge is located at the bottom of the plate, where the oxygen is depleted fastest. Nonaerotactic colonies, however, do not move out as far from the inoculation site and lack an aerotactic ring at the edge (Fig. 9). It should be noted that, in the presence of Tar, succinate plates do not detect Tsr-mediated aerotaxis (Bibikov *et al.*, 2000).

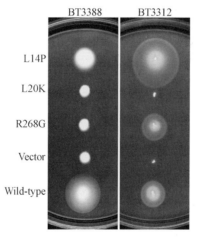

FIG. 9. Functional rescue of aerotaxis defects and distinctive Aer-related colony morphologies on soft agar plates. IPTG-inducible plasmids carrying the indicated Aer missense mutations were tested for function in *E. coli* strains BT3388 (*aer tsr tar trg tap*) and BT3312 (*aer tsr*). BT3388 colonies were inoculated into tryptone soft agar containing 20 μM IPTG and 50 μg/ml ampicillin, while BT3312 colonies were inoculated into succinate minimal soft agar without IPTG. Both plates were incubated at 30° for 14 h. The Aer-L14P mutant produced a large nonaerotactic swarm in tryptone agar due to constant tumbling (a clockwise bias). Many clockwise-biased mutants (in BT3388) are rescued by the presence of other chemoreceptors (in BT3312), forming super-swarmer colonies relative to the size of the wild-type Aer control. The null aerotaxis defect of Aer-R268G, but not Aer-L20K, is rescued by the presence of chemoreceptors in BT3312.

Tryptone soft agar is also used for aerotaxis assays and for observing phenotypic differences associated with mutants that have an increased or decreased frequency of tumbling (clockwise flagellar rotation) (Bibikov *et al.*, 2004; Buron-Barral *et al.*, 2006; Gosink *et al.*, 2006; Watts *et al.*, 2006a,b) (Fig. 9, left panel; Fig. 10). For wild-type *E. coli* on tryptone agar, an outer ring is formed by bacteria sensing serine through the Tsr chemoreceptor, and a second, inner ring by bacteria sensing aspartate through the Tar chemoreceptor. Aerotaxis results in wild-type *E. coli* forming dome-shaped colonies.

Aerotaxis in *E. coli* can be observed using Aer or Tsr as the sole receptor, or using Aer or Tsr in the presence of other chemoreceptors. Typically, we use *E. coli* BT3312 (*aer tsr*) or *E. coli* BT3388 (*aer tsr tar trg tap*) as the host strain and express the Aer or Tsr protein from a plasmid (pGH1, Aer; pKP1 or pKP2, Tsr [Watts *et al.*, 2004] and K. Pierre [see Fig. 6A]). Similar strains have also been used in aerotaxis studies (Buron-Barral *et al.*, 2006). We screen plasmid-expressed Aer in an *aer tsr* host on succinate agar (Fig. 9, right panel) and plasmid-expressed Aer in an *aer tsr tar trg tap* host on tryptone agar (Fig. 9, left panel; Fig. 10). The soft agar

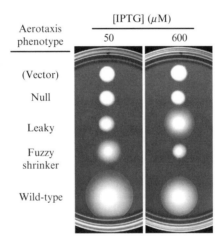

FIG. 10. Phenotypes of aerotaxis mutants on tryptone soft agar plates. Mutant plasmids were tested in strain UU1250, which lacks Aer and the four conventional MCP family chemoreceptors (Tap,Tar, Trg, Tsr). Colonies were transferred to tryptone soft agar plates containing 50 μg/ml ampicillin and either 50 or 600 μM IPTG to induce Aer expression. Plates were incubated at 30° for 17 h. Plasmids were pCJ30 (vector), pMB1-L220P (null), pMB1-V246A (leaky), pMB1-N228S (fuzzy), and pSB20 (wild type). Reproduced from Buron-Barral *et al.* (2006). Copyright © 2006 American Society for Microbiology. All rights reserved. Used with permission.

plate assays are not suitable for screening Tsr-mediated aerotaxis and we use capillary or temporal assays for Tsr-mediated aerotaxis. Deleting an abundant chemoreceptor (Tar or Tsr) lowers the tumbling frequency and impairs the Aer-mediated aerotaxis response, yielding a smaller colony size on tryptone or succinate soft agar. This is particularly evident when both Tar and Tsr are absent. A normal tumbling frequency can be restored by overexpressing the Aer protein (Bibikov *et al.*, 1997; Rebbapragada *et al.*, 1997). The concentration of Aer expressed should be the minimum necessary to restore aerotaxis because, like many membrane proteins, its over-expression is harmful to the cell. To determine an appropriate induction level, prepare a series of tryptone soft agar plates with increasing concentrations of the inducer for *aer* expression (e.g., 0 to 1000 μM for isopropyl-β-D-thiogalactoside [IPTG]). In subsequent experiments, include the inducer at the minimum concentration that gave an optimal colony size on soft agar plates. The same concentration of inducer may be used in succinate or tryptone soft agar assays. We do not usually add IPTG to swarm plates when *E. coli* BT3312/pGH1 cells are used for the assay. These cells express Aer at 10-fold the chromosomal expression level of Aer (D. Salcedo and M. S. Johnson, unpublished data). We routinely add 20 μM IPTG when no chemoreceptors are expressed in the host (*E. coli* BT3388/pGH1).

Succinate Soft Agar

Solution A	Add	Final concentration
Agar	2.8 g	0.28%[a]
NaCl	5 g	85.5 mM
Deionized H$_2$O to 500 ml		

[a] Calibrate individual agar lots (0.2–0.3%) to find the concentration that gives a preferred consistency.

Solution B	Add	Final concentration
K$_2$HPO$_4$	1.1 g	10 mM
KH$_2$PO$_4$	0.48 g	10 mM
(NH$_4$)$_2$SO$_4$	0.13 g	1 mM
Deioinized H$_2$O to 450 ml		

1. Dissolve solutions A and B, autoclave separately, then cool to 50°.
2. Add supplements (from presterilized stock solutions) into solution B: final pH 7.0.

Stock supplement	Volume	Final concentration
1.25 M succinate, pH 7.6	24 ml	30 mM
0.1% thiamine HCl	1 ml	1 μg/ml
0.1 m MgSO$_4$	10 ml	1 mM
Required amino acids		0.1 mM
Deionized H$_2$O to 50 ml		

3. Combine solutions A and B. If required, add antibiotics at half the concentration used for liquid growth media or solid agar plates. Mix thoroughly.
4. Pour 20–40 ml per Petri dish (10 cm diameter).
5. Allow plates to harden for at least 4 h at room temperature. Swarm plates may be placed in a plastic bag and stored at room temperature for a maximum of 2–3 days.
6. The pH and concentrations of phosphate, EDTA, and NaCl were selected to optimize motility of *E. coli* (Adler and Templeton, 1967; J. S. Parkinson, personal communication).

Tryptone Soft Agar

Ingredient	Add	Final concentration
BactoTryptone	10 g	1%
NaCl	5 g	85.5 mM
Agar	2.8 g[a]	0.28%
Deionized H$_2$O to 1 liter		

[a]Calibrate individual agar lots (0.2–0.3%) to find the concentration that gives a preferred consistency.

1. Autoclave and cool to 50°. If required, antibiotics can be added at half the concentration used for liquid growth media or solid agar plates.
2. If required, add 500 μl of 0.1% thiamine stock solution.
3. Pour up to 40 ml per Petri dish (10 cm diameter).
4. Allow plates to harden for at least 4 h at room temperature. Swarm plates may be placed in a plastic bag and stored at room temperature for a maximum of 2–3 days.

Procedure

1. Inoculate soft agar plates from fresh colonies or overnight cultures. Use a toothpick to inoculate from fresh colonies: stab to the bottom of the

soft agar for best reproducibility. Alternatively, use a pipettor with a sterile tip. Inoculate 2 μl of liquid culture into the soft agar with a vertical line of bacteria, starting about one-third of the distance from the bottom of the plate. Do not touch the bottom or leave residual liquid on the top surface. Bacteria preferentially spread in liquid at either surface, instead of swimming through the agar. Inoculate up to eight colonies per plate.

2. Incubate at 30–35° overnight in a humidified incubator. Standard conditions are 30° for 16–24 h for succinate plates and 12–20 h for tryptone plates. Place a flat dish of water in the incubator for humidity. Use care in transporting plates to and from the incubator. Sudden movements can disturb the agar, leading to artifacts in the shape of the colony.

3. At the end of incubation, view the colonies on a light box that provides indirect back lighting of the colonies. The fine structure of the colonies is not visible without well-adjusted lighting. Our preferred light box is described elsewhere. Photograph the colonies on the light box (Parkinson, 2007).

Notes. In interpreting the results of soft agar assays, use the wild-type Aer colony as a reference. The wild-type colony has a dome shape and a distinct ring of cells at the outer edge (Fig. 9). Colonies in which the cells are not aerotactic will typically have a cylindrical shape and lack the outer ring of cells. Cells in the smallest colonies may be nonmotile or smooth swimming (Wolfe and Berg, 1989) or have an inverted response to oxygen (Watts *et al.*, 2006b). Cells with a higher than normal tumbling frequency form fuzzy colonies of intermediate size (Fig. 10) (Buron-Barral *et al.*, 2006; Watts *et al.*, 2006b). Some of the mutant Aer proteins are associated with larger than normal colony size when expressed in an *aer tsr* host ("super swarms," Fig. 9) (Ma *et al.*, 2005). Further characterization is possible on soft agar plates by varying the induction of *aer* (Fig. 10) and by comparing the behavior of an Aer mutant in a receptorless host with the behavior of the mutant in an *aer tsr* (Watts *et al.*, 2006b) (Fig. 9) or an *aer* (Buron-Barral *et al.*, 2006) host. [For a discussion, see Buron-Barral *et al.* (2006).] Ultimately, the Aer mutants must be characterized by analyzing their behavior using temporal assays and flagellar rotation profiles, in addition to using genetic tests for dominance, suppression, and rescue (Buron-Barral *et al.*, 2006; Gosink *et al.*, 2006; Rebbapragada *et al.*, 1997; Watts *et al.*, 2004, 2006a,b).

Although soft agar plates may be the preferred method for initial screening of mutants for aerotaxis phenotype, it may be difficult to interpret gene-inactivation experiments in the presence of more than one aerotaxis transducer. In *E. coli* temporal assays, aerotaxis was not inactive in *aer* mutants unless both the *aer* and the *tsr* genes were mutated (Rebbapragada *et al.*, 1997). Initial studies indicate other species may also have more than one aerotaxis receptor, making it difficult to identify a specific receptor gene. For some species, it may be possible to clone a candidate aerotaxis receptor gene in an *E. coli* Δ*aer* Δ*tsr* strain and look for restoration of aerotaxis.

Spatial Assays for Redox Taxis

Background

Escherichia coli cells migrate to form a sharply defined band at a preferred reduction potential (Bespalov *et al.*, 1996). Bacteria swimming out of either side of the band tumble and return to the preferred conditions at the site of the band. This behavioral response was named redox taxis (Bespalov *et al.*, 1996). Redox molecules, such as substituted quinones, elicit redox taxis by affecting the electron transport system and proton motive force. In *E. coli*, redox taxis is mediated by the Aer and Tsr receptors, which also mediate aerotaxis (Rebbapragada *et al.*, 1997). As with aerotaxis, spatial-gradient and temporal-gradient assays may be used to measure redox taxis. These experiments are performed most easily under the microscope in a gas flow cell to maintain a strict anaerobic environment, free from the interfering response to oxygen. The spatial gradient can be generated using the method described previously (Bespalov *et al.*, 1996; Grishanin *et al.*, 1991), incorporating simplified modifications described for the agarose-in-plug bridge assay (Yu and Alam, 1997).

Stock Solutions

Prepare fresh 100 m*M* solutions of 1,4-benzoquinone (or another substituted quinone) and 1,4-hydroquinone (or another substituted hydroquinone) in ethanol. Keep the quinone solution at $-20°$.

Procedure

1. Grow cells aerobically in LB medium at $30°$ to OD = 0.35 to 0.45. Wash the cells twice and resuspend to 6×10^8 cells ml^{-1} in chemotaxis buffer supplemented with 20 m*M* sodium lactate.
2. Transfer 1 μl of the stock hydroquinone solution to 1 ml of the washed cells to yield 100 μ*M* hydroquinone and 6×10^8 cells ml^{-1}.
3. Break a microscope slide coverslip into pieces by squeezing the coverslip inside a folded paper towel. Place four of these pieces onto a microscope slide, and lay another coverslip over these pieces; position the pieces at the four corners underneath the coverslip.

4. Prepare a 2%, low-melting temperature agarose mixture containing 100 μM oxidized quinone and 2 mM ferricyanide (to maintain the quinone in the oxidized state). The mixture can be made in a 1.5-ml microcentrifuge tube by adding 0.01 g agarose to 0.5 ml of the quinone/ferricyanide mixture dissolved in chemotaxis buffer. Heat the microcentrifuge tube at 70° in a hot water bath with periodic shaking until the agarose is melted (Yu and Alam, 1997).
5. Remove the coverslip (step 3), transfer 10 μl of the agarose mixture onto the slide, and cover the agarose with the coverslip.
6. Place the slide in the gas flow cell described earlier and allow the plug to equilibrate with nitrogen gas for 5 min.
7. Without removing the slide from the flow cell, pipette the bacteria/hydroquinone mixture underneath the coverslip so that the liquid surrounds the plug with no air bubbles.
8. Using a low-power objective of a dark-field microscope (overall magnification, 62×), monitor the area circumscribing the agarose plug. A band of cells should be evident within 10 min as a gradient of oxidized and reduced quinone develops (Bespalov *et al.*, 1996) (Fig. 11).

Notes. The method of generating a redox gradient is applicable for other bacterial species as well. Moreover, the protocol may be modified to include different artificial electron acceptors, such as N,N,N',N'-tetramethyl-*p*-phenylenediamine (TMPD) ([Grishanin *et al.*, 1991]; see Fig. 11).

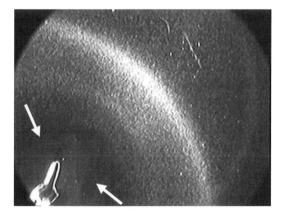

Fig. 11. Band formation by *Azospirillum brasilense* in a redox gradient. An agar plug (between arrows) containing 200 μM N,N,N',N'-tetramethyl-p-phenylenediamine (TMPD) and 10 mM ferricyanide was placed in medium containing *A. brasilense*, 200 μM TMPD, and 5 mM sodium malate as described in Grishanin *et al.* (1991). The photo was freeze-captured from a video record by V. Bespalov and I. B. Zhulin.

Temporal Assay for Redox Taxis

This assay is similar to the temporal assay for chemotaxis (Spudich and Koshland, 1975). A redox chemical is added to a drop of bacterial culture on a microscope slide and the response of bacteria to the resulting change in reduction potential of the preparation is video recorded (Bespalov *et al.*, 1996). The magnitude of the behavioral response is dependent on the reduction potential of the chemoeffector. Under aerobic conditions, oxidized quinones are repellents and reduced quinones do not elicit a response (Bespalov *et al.*, 1996). The reductants, 2-mercaptoethanol, dithiothreitol, and sodium borohydride, elicit a brief smooth-swimming response. Table I summarizes the behavioral responses of *E. coli* to selected redox compounds (Bespalov *et al.*, 1996). To determine whether a newly derived strain exhibits redox taxis, we routinely add 1,4-benzoquinone to the bacteria.

Stock Solution

Prepare a fresh 100 m*M* solution of 1,4-benzoquinone (or another substituted quinone) in ethanol and store at −20°.

Working Solution

Every 20 min, dilute an aliquot of the stock solution 1:1000 in double deionized water for a 100 μ*M* working solution.

Procedure

1. Grow cells in an appropriate minimal medium to midexponential phase, wash twice, and resuspend in chemotaxis buffer as described earlier. Dilute the cells in chemotaxis buffer to $1-2 \times 10^7$ ml^{-1} and stand at room temperature 10 min.
2. Dispense 9 μl of the preadapted cell suspension onto a microscope slide and transfer to the stage of a dark-field microscope (L32/0.4, long working distance [6.5 mm] objective, Leitz). Allow 2 min for temperature equilibration. Using a micropipettor, add 1 μl working solution to the drop of cell suspension on the microscope slide and mix rapidly with a disposable pipette tip.
3. Video record the behavioral response of the bacteria. Calculate the tumbling response from the end of mixing to the time at which 50% of the bacteria return to the prestimulus tumbling frequency (see earlier discussion).
4. For a negative control, add 1 μl of a water:ethanol solution corresponding to the concentration in the quinone solution. If an

E. coli strain has a smooth bias (low tumbling frequency), assays of the behavioral response to an attractant may be impaired. It is possible to add 1 μl of 2 mM CoCl$_2$ to the cell suspension to increase tumbling frequency, before addition of the quinone. If the preparation does not respond to CoCl$_2$, discard the preparation and use a fresh cell preparation.

TABLE I

BEHAVIORAL RESPONSES OF *E. COLI* TO REDOX COMPOUNDS[a]

Compound	Reduction potential (mV)[b]	Behavioral response[c]	
		Threshold (μM)	Adaptation time (s)
1,4-Benzoquinone (BQ)			
2,6-Dichloro-BQ	ND[d]	1	>250
BQ	+99	2	>250
2-Methyl-BQ	+23	2	150
2,6-Dimethyl-BQ	−80	4	60
2,3-Dimethoxy-5-methyl-BQ	ND	6	50
Tetramethyl-BQ	−240	40	NR[d]
1,4-Naphthoquinone (NQ)			
5-Hydroxy-NQ	−93	0.2	150
NQ	−140	1	110
5-Hydroxy-2-methyl-NQ	−156	2	70
2-Methyl-NQ	−203	5	60
2-Hydroxy-NQ	−415	400	NR
Reductant			
2-Mercaptoethanol	ND	1	15
Dithiothreitol	ND	1	15
Sodium borohydride	ND	0.1	15
Other redox compounds			
Methyl viologen	−446	NR	NR
Methylene blue	+11	NR	NR
Potassium ferricyanide	+420	NR	NR

[a] Reproduced from Bespalov *et al.* (1996). Copyright © 1996 National Academy of Sciences (USA). All rights reserved.

[b] Data from Wardman (1989).

[c] The behavioral responses of *E. coli* MM335 were measured in a temporal gradient assay. The oxidized quinones tested were repellents and the reductants were attractants. The time interval was determined for 50% of the cell population to return to prestimulus behavior after the addition of the compound to be tested (final concentration, 20 μM). This is defined as the adaptation time. Quinones were dissolved in dimethyl sulfoxide or in ethanol. The final concentration of dimethyl sulfoxide and ethanol in water solutions in the assay was <0.01% and did not cause a behavioral response in *E. coli*.

[d] ND, not determined; NR, no response.

Disulfide Cross-Linking *In Vivo* to Elucidate the Structure of Aer

Background

There are notable differences between the structure of Aer and Type I chemoreceptors (Fig. 1), which have periplasmic ligand-binding domains, and cytosolic signaling domains. Aer has a PAS sensory input domain and a signaling domain that are both in the cytosol. In order to understand structure–function relationships in Aer, we investigated the topology and boundaries of the Aer membrane anchor and the structure of the HAMP domain, which links the membrane anchor to the signaling domain.

The Aer receptor is unstable after it is solubilized in detergent (Bibikov *et al.*, 2000; Repik *et al.*, 2000) and is unsuitable for structural studies by nuclear magnetic resonance (NMR) or X-ray diffraction. Cysteine and disulfide scanning presents an alternative, albeit indirect, strategy that has been used to infer the three-dimensional organization of membrane proteins (Bass *et al.*, 1999; Falke *et al.*, 1988; Lee *et al.*, 1994; Lynch and Koshland, 1991). In these techniques, residues are serially replaced by cysteine, which serves as a marker for the original residue. In the presence of an oxidant, an introduced cysteine can form a disulfide bond with another cysteine (approximately 1 in 10^5 collisions result in disulfide formation [Careaga and Falke, 1992]). The rate and extent of cross-linking can reveal the proximity between two introduced cysteines. If this proximity of serial cysteine replacements has a regular periodicity, it may be used to map the secondary structure of the region (Lee *et al.*, 1994). Bi- and trifunctional sulfhydryl reactive probes can be used as molecular rulers to estimate the nearest distance between two cysteines (Loo and Clarke, 2001) and to infer the tertiary and quaternary organization of a protein (Studdert and Parkinson, 2004, 2005). Hydrophilic monofunctional sulfhydryl reactive probes can also be used to determine the solvent accessibility of an introduced cysteine (Bass *et al.*, 1999; Boldog and Hazelbauer, 2004; Mehan *et al.*, 2003).

Originally, the oxidant Cu(II)-(1,10-phenanthroline)$_3$ was used to oxidize and crosslink chemoreceptor thiols in membrane vesicles (Falke *et al.*, 1988; Falke and Koshland, 1987). More recently, oxidants such as diamide (Kosower and Kosower, 1995; Studdert and Parkinson, 2004), iodine (Lee *et al.*, 1995; Pakula and Simon, 1992), and Cu(II)-(1,10-phenanthroline)$_3$ (Lee *et al.*, 1995; Lai and Hazelbauer, 2007) have been used successfully to oxidize and cross-link protein thiols *in vivo*. However, systematic, *in vivo*, disulfide scanning studies of topology have been reported only for periplasmic and membrane regions, not for cytosolic regions. One might expect disulfide scanning to be more difficult in the cytosol for a number of reasons. The *E. coli* cytosol is a reducing environment, regulated by the thioredoxin and the glutaredoxin systems to maintain the intracellular redox potential between −260 and −280 mV (Gilbert, 1990; Ritz and Beckwith, 2001). This is lower than the midpoint potential for a free cysteine/cystine (CS/CSSC) redox couple (−220 mV [Jocelyn, 1967]). In addition, the relatively low concentration of protein sulfhydryls is unfavorable for disulfide formation within a cytosolic protein. However, in *E. coli*, the midpoint potential for disulfide bond formation within a protein varies from −125 to −270 mV, depending on the adjacent amino acids (Jacob *et al.*, 2003). Steric constraints can also affect the accessibility of a cysteine to an oxidant or a disulfide to a reductant (Careaga and Falke, 1992).

We used Cu(II)-(1,10-phenanthroline)$_3$ to map dimerization faces of the cytosolic Aer HAMP domain in intact cells (K. J. Watts, unpublished data). In practice, the helical faces were resolved easily by the cross-linking periodicity, which indicated that the Aer-HAMP AS2 region has a coiled-coil structure. *In vivo* cross-linking data are consistent with the NMR structure of the *Archaeoglobus fulgidus* HAMP domain (Hulko *et al.*, 2006).

Before initiating these studies, all native cysteines must be replaced with a silent replacement (usually alanine) to yield a "Cys-less" protein, followed by the serial substitution of cysteine at each residue in the protein segment under investigation.

Site-Directed Mutagenesis for Cysteine Replacement

Procedures

1. Site-directed mutagenesis is performed using a QuikChange site-directed mutagenesis kit (Stratagene, La Jolla, CA), which contains proofreading *PfuTurbo* DNA polymerase. The newer generation kit, QuikChangeII, offers a higher fidelity DNA polymerase, *PfuUltra*. In our experience, *PfuTurbo* and *PfuUltra* both work well for the amplification of double-stranded plasmids; however, other high-fidelity DNA polymerases (with $3' \rightarrow 5'$ exonuclease proofreading activity) may also be used.

2. Design primers for site-directed mutagenesis according to instructions for the QuikChange kit. As a general rule, design primers based on *E. coli* codon usage when expressing a foreign gene in *E. coli* (Goetz and Fuglsang, 2005; Medigue *et al.*, 1991). This is because rare codons, especially in clusters, can result in low levels of protein expression (Kim and Lee, 2006; Peng *et al.*, 2004). There is no problem associated with the choice of a cysteine codon for protein expression in *E. coli*, since, in highly expressed *E. coli* genes, tgc and tgt encode for 61.15 and 38.85% of the cysteines, respectively (Medigue *et al.*, 1991). In our experience, primer purification or

5′ phosphorylation is not necessary for efficient site-directed mutagenesis. The reaction size can also be reduced to 25 μl to conserve reagents.

3. After site-directed mutagenesis, electrophorese 2 μl of the amplified DNA on a 1% agarose gel. If a product of the correct size is observed, digest the DNA with Dpn 1 for 3 h at 37° and then transform competent bacteria with the product.

4. Check eight colonies for function on soft agar plates, with the appropriate aerotaxis controls. If a mutant created by site-directed mutagenesis is nonaerotactic on swarm plates, check the protein expression by Western blot before verifying the DNA sequence.

Disulfide Cross-Linking in the Cytosol Using Copper Phenanthroline

The procedure used for *in vivo* cross-linking of the Aer protein is based on a published method (Hughson and Hazelbauer, 1996). Reagents and their preparation are described elsewhere (Hughson and Hazelbauer, 1996; Lai and Hazelbauer, 2007).

Procedure

1. Grow BT3312 (*aer tsr*) cells expressing a plasmid-borne Aer protein with a single cysteine replacement in H1 minimal medium supplemented with 30 mM succinate, 0.1% casamino acids, and 100 μg ml^{-1} ampicillin. At OD = 0.4–0.6, induce Aer expression with 50 μM IPTG for 3 h.

2. Remove 80 μl cells, mix with 80 μl of 600 μM Cu(II)-(1,10-phenanthroline)$_3$ [hereafter, CuPhe], incubate the reaction at 23° for various time intervals (1, 2, 5, 10, 15 min), and quench with 40 μl stop solution containing 100 mM Tris, 40 mM NaH$_2$PO$_4$ (pH 7.8), 50 mM EDTA, 50 mM N-ethylmaleimide (NEM; Sigma, St. Louis, MO), 5% SDS, 20% sucrose, and 125 μg/ml bromophenol blue. Boil the sample for 5 min and then place on ice. For the zero time control, add water in place of CuPhe and immediately quench with stop solution.

3. Run samples on SDS-PAGE (8% acrylamide [Schagger and von Jagow, 1987]) under nonreducing conditions, electroblot onto nitrocellulose, and probe with a 1:130,000 dilution of primary Aer$_{2-166}$ antisera overnight (Repik *et al.*, 2000), followed by a 1:10,000 dilution of secondary antibody (goat antirabbit HRP conjugate; Bio-Rad) for 30 min. Visualize the Aer band by adding 4 ml of SuperSignal West Pico Chemiluminescent Substrate (Pierce Chemical, Rockford, IL) for 4 min, and expose to Kodak BioMax Light Film (Kodak, Rochester, NY) from 5 s to 5 min (as needed).

4. Digitize the film image (IS-1000, Alpha Innotech Corp, San Leandro, CA), quantitate the bands, and estimate the percentage of cross-linking by comparing the intensity of the cross-linked dimer band to the sum of the

intensities of monomer and dimer bands. BT3312/pGH1 (wild-type Aer) and BT3312/pMB1 (Aer-C193S/C203A/C253A, Aer-[Cys-less]) are used as positive and negative cross-linking controls, respectively (Ma *et al.*, 2004).

5. Plot the time course of cross-linking for each introduced cysteine in Aer to determine the linear range of the reaction for that residue. It is important to use the initial velocity of cross-linking reactions to compare the relative proximity between cognate cysteines at different sites. For cytosolic cross-linking of Aer single cysteine replacements, we use the 10-min time point to compare the percentage of cross-linked residues because, with few exceptions, this point is within the linear range of the reaction.

6. At 23°, disulfide bonds can form between distant thiols as a result of random collisions between proteins diffusing laterally throughout the lipid bilayer. These random collisions can sometimes be distinguished from specific, *in situ* interactions by lowering the reaction temperature below the lipid phase transition temperature. For many membrane proteins, this phase transition lowers the rate of lateral diffusion (Guan *et al.*, 2002; Overath *et al.*, 1975). To distinguish disulfide bonds between adjacent thiols, we conduct duplicate experiments at 23 and 4°. For low temperature experiments, precool the cells and CuPhe separately at 4° for 10 min, initiate the reaction by transferring 80 μl of chilled CuPhe to 80 μl of chilled cells, and incubate the mixture for another 20 min at 4°. Quench the reaction with stop solution containing NEM (as described earlier), and boil for 5 min. Compare the extent of cross-linking at 23° for 10 min with that at 4° for 20 min.

Notes. Because cysteines can cross-link during the denaturation and boiling steps, if the concentration of NEM is inadequate to block free sulfhydryl groups, a parallel control can be included to test for artifactual cross-linking during the denaturation step. Incubate cells with 10 mM NEM for 10 min prior to adding the CuPhe and inspect the resulting Western blots to verify that these samples do not form cross-linked products.

Differentiating Intra- from Interdimeric Disulfide Bonds

For proteins such as chemoreceptors, in which dimers are organized into higher order structures (Kim *et al.*, 2002; Studdert and Parkinson, 2004), interdimeric disulfide bonds may be falsely interpreted as intradimeric bonds. This is particularly likely if the oligomeric structure is stable. In this case, interdimeric collisions may not decrease significantly when the experiments are repeated at lower receptor concentrations or lower temperatures, and intra- and interdimeric faces might not be distinguished. We used a different tactic to differentiate inter- from intradimeric faces in

Aer (Amin, 2006). We introduced into the receptor two different cysteine replacements, each known to readily form disulfide bonds, and determined whether higher order complexes were trapped when both replacements were oxidized. This strategy required that (i) the intradimeric specificity of one of the disulfide bonds was known and that (ii) the two replacements were in different regions of the protein so they could not cross-link with each other. In Aer, V260C (in the proximal signaling domain) is positioned at the dimer interface (K. Watts, unpublished observation), and we used this residue as a reference for intradimeric cross-links (Amin, 2006). Because V260C resides in the cytosol, it could be paired with introduced cysteines in the periplasmic and membrane regions without incidental cross-linking between the two.

Procedure

1. Choose a reference Aer protein that has a substituted cysteine that (i) cross-links readily, (ii) is located in a region where it will not cross-link with additional cysteines that will be investigated for interdimeric cross-linking, and (iii) is likely to be on the intradimeric interface. This decision should be supported by previous cross-linking data. For example, residues with the highest probability of cross-linking and that follow a 3 + 4 cross-linking periodicity would be expected to be on the dimer interface of a coiled-coil region.

2. Make binary cysteine substitutions that pair the cysteine with an unknown pattern of cross-linking with the reference cysteine that forms intradimeric cross-links.

3. Use SDS-PAGE to separate the complexes formed by these constructs in response to the oxidant CuPhe. If both cysteine replacements form a disulfide bond with the same monomer, complexes larger than dimers should not form. If the cysteines cross-link with different monomers, tetramers and higher order complexes will form.

4. Once intra- and interdimeric cross-linking has been determined, verify these data by pairing different combinations of all intradimeric cross-linkers, or all interdimeric cross-linkers. All of these combinations should yield dimers exclusively. If any of these combinations form larger complexes, one of the two cysteines must collide both within and between dimers. This residue can be identified by assaying further binary combinations of introduced cysteines.

Notes. Once compiled, these data can be useful in differentiating intradimeric surfaces from those that face outward. At best, one might infer intra- and interdimeric collisional faces, as well as regions that reside between these faces, or are highly dynamic.

In Vivo Cross-Linking Using Bifunctional Sulfhydryl-Reactive Linkers

The close proximity required for two cysteines to form a disulfide bond can be a limiting factor in structural studies. If necessary, the length of a cross-linking arm can be extended by using homobifunctional thiol linkers. These linkers have some flexibility, they come in assorted sizes, and they have been used as molecular rulers to estimate the minimum distance between two cysteines (e.g., Loo and Clarke, 2001). They may also be useful to confirm putative spatial relationships in a protein that are predicted by modeling programs or X-ray diffraction. A novel extension of the bifunctional linker was reported by Studdert and Parkinson (2004, 2005), who used a trifunctional linker [tris-(2-maleimidoethyl)amine] to target cysteines in intact *E. coli*; the reagent cross-linked monomers from three different chemoreceptor dimers within a trimer of dimers complex.

One caveat should be heeded when using these linkers as molecular rulers: the distance values given by the manufacturer are determined for a fully extended molecule, which may be rare in solution. Of 32 different bifunctional linkers studied by stochastic dynamics simulations, only 2 had an average distance within 0.5 Å of the cited distance. For example, Pierce Chemical lists bis-maleimideohexane (BMH) as having a span of 16.1 Å, but dynamic simulations indicate that its average length is 10.16 ± 2.41 Å (Green *et al.*, 2001). However, bis-maleimidoethane (BMOE) is listed as 8 Å (Pierce Chemical) and its dynamic simulated average length is 8.18 ± 0.75 Å. We have used both of these homobifunctional thiol reagents to cross-link cysteines between dimers *in vivo* to demonstrate a hexamer formation, consistent with a trimer of dimers structure for the Aer receptor (Amin, 2006). Both linkers contain a flexible alkyl chain flanked by maleimido groups, they can transverse both outer and inner membranes of *E. coli*, and they do not require an oxidant to cross-link cysteines. The protocol differs slightly from that used for disulfide cross-linking.

Procedure

1. Grow cells expressing Aer with introduced cysteines in H1 minimal medium and induce Aer expression with 50 μM IPTG as described previously.
2. Combine 74 μl of cells (OD = 0.8) and 6 μl (270 μM final concentration) of BMH or BMOE (1 mg ml^{-1} stock in dimethyl sulfoxide). Incubate for 10 to 30 min at 23°, before quenching with 20 μl of the stop solution (as described earlier, without NEM) containing a final concentration of 80 mM β-mercaptoethanol.
3. Boil samples for 5 min, separate proteins on SDS-PAGE, and visualize Aer by Western blot.

4. Incubate a parallel control sample with 80 mM β-mercaptoethanol for 10 min before the addition of BMH or BMOE.

Notes. If the intent is to cross-link subunits to form higher order oligomers, two or more cysteines can be strategically introduced into one monomer. Also, the probability of trapping can be increased by incubating the cells at 30° for 30 min, with the caveat that a greater number of random collisions will occur.

Determining the Boundaries of Transmembrane Segments in Receptors

In *E. coli*, Type I chemoreceptors have two transmembrane segments flanking a large periplasmic (ligand-binding) domain. Aer, however, has a single 38 residue hydrophobic segment that is barely long enough to transverse the membrane twice (Bibikov *et al.*, 1997, 2000; Rebbapragada *et al.*, 1997; Repik *et al.*, 2000). As a result, the topology of the membrane anchor in Aer must be quite different from the transmembrane segments of other chemoreceptors. To determine the overall topology of the membrane module in Aer, we first used membrane vesicles to estimate the solvent accessibility of introduced cysteines. Using two hydrophilic, sulfhydryl-reactive probes, we found that four residues in the central region of the membrane module were solvent accessible (Amin *et al.*, 2006). We next used an *in vivo* approach to prove that these central, solvent-accessible residues were exposed to the periplasm. Data are consistent with an Aer membrane module that has two transmembrane segments (TM1, residues 164 to 183; TM2, residues 188 to 205) that flank a short periplasmic loop (residues 184 to 187) (Amin *et al.*, 2006).

Accessibility Studies in Membrane Vesicles

Preparation of Membrane Vesicles

Bacterial membranes containing wild-type or mutated Aer receptors are prepared as published (Butler and Falke, 1998) with several modifications.

1. Inoculate 5 ml of overnight cultures into 250 ml H1 minimal medium. Grow with vigorous agitation at 30°, induce with 0.6 mM IPTG at OD = 0.4, and grow for an additional 3 h.
2. Harvest cells by centrifugation at 10,000g for 10 min at 4° and resuspend in 4 ml of low-salt buffer (100 mM sodium phosphate, pH 7.0, 10% glycerol, 10 mM EDTA, 1 mM 1,10-phenanthroline containing freshly added 50 mM dithiothreitol [DTT], and 2 mM PEFAbloc; Centerchem., Norwalk, CT).

3. Disrupt cells with three freeze–thaw cycles, and lyse by sonication (Branson Sonifier cell disrupter 200; Danbury, CT) at 60% power (three 15-s bursts with 45-s pauses) in an ice/salt bath.
4. Centrifuge at 12,000g for 20 min at 4°. Save the supernatant, and centrifuge at 485,000g for 20 min to pellet the membranes. Wash the pellet three times as follows. Resuspend the membrane fraction in high-salt buffer 1 (20 mM sodium phosphate, pH 7.0, 2 M KCl, 10% glycerol, 10 mM EDTA, 1 mM 1,10-phenanthroline, containing freshly added 5 mM DTT and 2 mM PEFAbloc) and centrifuge. Resuspend the pellet in high-salt buffer 2 (same as high-salt buffer 1 except without DTT and 1,10-phenanthroline). Recentrifuge, resuspend in the final buffer (20 mM sodium phosphate, pH 7.0, 10% glycerol, 0.1 mM EDTA, 2 mM PEFAbloc), centrifuge, and resuspend in 200 μl of final buffer.
5. Transfer 10-μl aliquots into 20 tubes, freeze in a dry ice/ethanol bath, and store at −80°. Measure the concentration of protein in the membrane with the BCA assay (Pierce Chemical) using bovine serum albumin as the standard. Dilute frozen membrane preparations to the required concentration with final buffer prior to use.

Accessibility Studies with 5-Iodoacetamidofluorescein

This protocol is a modification of previous accessibility studies with 5-Iodoacetamidofluorescein (5-IAF) (Boldog and Hazelbauer, 2004; Butler and Falker, 1998; Danielson et al., 1997).

Procedure

1. For each sample, add 1 μl 6.25 mM 5-IAF (made up in dimethyl formamide) to 11.5 μl membrane vesicles containing 5 to 15 μg protein in final buffer. This yields a final concentration of 0.5 mM 5-IAF. To determine cysteine accessibility, prepare parallel reactions under native and denaturing conditions.
2. Incubate the native sample at 23° for 5 min, and stop the reaction by adding 12.5 μl of 2× sample buffer with 5% β-mercaptoethanol. Boil this sample for 5 min at 100°.
3. While incubating the native sample (step 2), determine the accessibility under denaturing conditions by boiling a parallel reaction mixture (step 1) in the presence of 1% SDS (1.4 μl of 10% SDS) at 100° for 5 min. Stop the reaction by adding 11.1 μl 2× sample buffer with 5% β-mercaptoethanol and boil for an additional 5 min at 100°.
4. Separate proteins on SDS PAGE and immediately place the wet gel on a UV light box and analyze the fluorescein fluorescence with

a gel documentation system (Alpha-Innotech Corp). Estimate the total protein in each Aer-[5-IAF] band by subsequently staining the gel with Coomassie blue. Calculate the percentage accessibility for each residue (ratio of the native to denatured Aer fluorescence divided by the ratio of native to denatured Aer protein × 100).

Accessibility Studies with mPEG-MAL

SDS-treated membrane vesicles contain a sulfhydryl-reactive protein that comigrates with Aer on SDS-PAGE. To eliminate the competing signal from this unknown protein, we also used the hydrophilic sulfhydryl reactive reagent, methoxy polyethylene glycol maleimide (mPEG-MAL; mPEG-MAL 5000, Nektar Therapeutics, Huntsville, AL) (Lu and Deutsch, 2001) to estimate the membrane boundaries of Aer. This reagent has an actual molecular mass of 5 kDa, but Aer-mPEG-MAL complexes exhibit an apparent molecular mass of ~10 kDa higher than unreacted Aer so the two forms can be separated on SDS-PAGE and visualized on Western blots. Western blotting eliminates background signals from other sulfhydryl-containing proteins that may have similar mobilities on SDS-PAGE.

Procedure

1. For each sample, add 1.3 μl 50 mM mPEG-MAL (in water) to 11.7 μl membrane vesicles containing 4 to 6 μg protein in final buffer. This yields a final concentration of 5 mM mPEG-MAL. To determine the cysteine accessibility, prepare two parallel reactions under native and denaturing conditions.
2. For native conditions, incubate one sample for 1 h at 23° and the other at 4° to limit diffusion into the membrane; at 4°, mPEG-MAL does not cross membranes, even when incubated for 24 h (Lu and Deutsch, 2001). Incubation with mPEG-MAL for longer than 1 h does not increase the amount of mPEG-MAL bound.
3. To determine the reactivity of the denatured sample, boil a third reaction mixture from step 1 in the presence of 1% SDS at 100° for 5 min.
4. Stop the reactions by the addition of 6.75 μl SDS sample buffer containing 5% β-mercaptoethanol. The reducing agent quenches unreacted mPEG-MAL, but does not cleave Aer–mPEG-MAL adducts. Boil the samples for 5 min, electrophorese them on SDS-PAGE, and analyze them by Western blot.
5. Estimate the accessibility to mPEG-MAL by comparing the accessibilities under native and denaturing conditions as described for the 5-IAF studies.

Note. As of September 29, 2006, Necktar Therapeutics announced that it will no longer sell mPEG-MAL; it can now be obtained from CreativeBiochem (Winston-Salem, NC).

Determining Periplasmic Accessibility in Intact Cells

Although mPEG-MAL is too large to penetrate outer porins, mPEG-MAL can be used to determine whether an accessible residue is periplasmic or cytosolic by (1) preblocking accessible periplasmic cysteines with a small porin-permeable probe, (2) denaturing the cells in SDS, and (3) reacting the denatured protein with mPEG-MAL. For the blocking step, we used 4-acetamido-4-disulfonic acid, disodium salt (AMS; Molecular Probes, Inc., Eugene, OR). It penetrates the outer but not the inner membrane of intact cells and reacts with accessible periplasmic cysteines.

Procedure

1. Incubate intact cells expressing Aer with a single cysteine replacement in the region of interest with or without AMS (8 mM final concentration) in final buffer for 45 min at 23°.
2. Remove unreacted AMS by washing the cells three times in final buffer.
3. Add SDS and mPEG-MAL to a final concentration of 1% and 10 mM, respectively, and immediately disrupt the cells by heating the sample at 100° for 5 min. Incubate the lysate for another 15 min at 23° before stopping the reaction with 10 μl SDS sample buffer containing 5% β-mercaptoethanol.
4. Boil the samples for 5 min, separate on SDS-PAGE, Western blot, and analyze the bands to determine whether the introduced cysteine is blocked by AMS. Blocked cysteines will not show the mobility shift (from Aer-mPEG-MAL) evident in the unblocked sample.

Acknowledgments

We thank J. S. Parkinson for many helpful discussions and for sharing his laboratory procedures for soft agar plate assays. We thank D. N. Amin, G. L. Hazelbauer, I. B. Zhulin, V. Bespalov, and J. J. Falke for helpful discussions. We acknowledge support from the National Institute of General Medical Sciences (GM29481) to B. L. Taylor and a National Medical Test Bed award to M. S. Johnson.

The following video files can be viewed at the Methods in Enzymology web site, http://books.elsevier.com/companions/9780123738516.

MOVIE 1. Tsr-mediated aerotaxis band in a flat capillary tube containing chemotaxis buffer. *E. coli* BT3388 (*aer tsr tar trg tap*) cells expressed the mutant Tsr-R64C protein from the plasmid, pKP2 (Tsr-R64C does not sense serine). At low power (62×), the cells form an aerotactic band with a clearing zone on the (right) anaerobic side of the band. At higher magnification (800×), motile cells can be seen near the meniscus, but both cell density and motility increase in the center of the aerotactic band. Cells leaving the band tumble and return to the band. To the right of the clearing zone, cells are anaerobic, immotile, and therefore trapped.

MOVIE 2. Pseudo band formed by nonmotile, anaerobic *E. coli* cells in chemotaxis buffer. When viewed under low magification (62×), BT3388 (*aer tsr tar trg tap*) cells containing vector alone appear to form an aerotactic band. However, under higher magnification (800×), one can see that although cells are motile between the meniscus and the band, those cells that reach the band become immotile. Therefore, the pseudo band is formed by trapped, not aerotactic cells.

MOVIE 3. Perforation of the meniscus in a capillary by a needle oxygen electrode. A well-defined aerotactic band formed by *Azospirillum brasilense* becomes distorted as the electrode perforates the meniscus. Although the event disturbs some cells at the anaerobic (right) edge of the band, they quickly return to reform a well-defined boundary.

References

Adler, J. (1973). A method for measuring chemotaxis and use of the method to determine optimum conditions for chemotaxis by *Escherichia coli. J. Gen. Microbiol.* **74,** 77–91.

Adler, J., and Templeton, B. (1967). The effect of environmental conditions on the motility of *Escherichia coli. J. Gen. Microbiol.* **46,** 175–184.

Amin, D. N. (2006). Membrane organization and multimeric interactions of the Aer receptor in *E. coli*. Ph.D. Dissertation. p. 167. Loma Linda University, Loma Linda, CA.

Amin, D. N., Taylor, B. L., and Johnson, M. S. (2006). Topology and boundaries of the aerotaxis receptor Aer in the membrane of *Escherichia coli. J. Bacteriol.* **186,** 894–901.

Baracchini, O., and Sherris, J. C. (1959). The chemotactic effect of oxygen on bacteria. *J. Pathol. Bacteriol.* **77,** 565–574.

Bass, R. B., Coleman, M. D., and Falke, J. J. (1999). Signaling domain of the aspartate receptor is a helical hairpin with a localized kinase docking surface: Cysteine and disulfide scanning studies. *Biochemistry* **38,** 9317–9327.

Berg, H. C. (1975). Chemotaxis in bacteria. *Annu. Rev. Biophys. Bioeng.* **4,** 119–136.

Berg, H. C., and Brown, D. A. (1972). Chemotaxis in *Escherichia coli* analysed by three-dimensional tracking. *Nature* **239,** 500–504.

Bespalov, V. A., Zhulin, I. B., and Taylor, B. L. (1996). Behavioral responses of *Escherichia coli* to changes in redox potential. *Proc. Natl. Acad. Sci. USA* **93,** 10084–10089.

Bibikov, S. I., Barnes, L. A., Gitin, Y., and Parkinson, J. S. (2000). Domain organization and flavin adenine dinucleotide-binding determinants in the aerotaxis signal transducer Aer of *Escherichia coli. Proc. Natl. Acad. Sci. USA* **97,** 5830–5835.

Bibikov, S. I., Biran, R., Rudd, K. E., and Parkinson, J. S. (1997). A signal transducer for aerotaxis in *Escherichia coli. J. Bacteriol.* **179,** 4075–4079.

Bibikov, S. I., Miller, A. C., Gosink, K. K., and Parkinson, J. S. (2004). Methylation-independent aerotaxis mediated by the *Escherichia coli* Aer protein. *J. Bacteriol.* **186,** 3730–3737.

Boldog, T., and Hazelbauer, G. L. (2004). Accessibility of introduced cysteines in chemoreceptor transmembrane helices reveals boundaries interior to bracketing charged residues. *Protein Sci.* **13,** 1466–1475.

Bourret, R. B., and Stock, A. M. (2002). Molecular information processing: Lessons from bacterial chemotaxis. *J. Biol. Chem.* **277,** 9625–9628.

Buron-Barral, M., Gosink, K. K., and Parkinson, J. S. (2006). Loss- and gain-of-function mutations in the F1-HAMP region of the *Escherichia coli* aerotaxis transducer Aer. *J. Bacteriol.* **188**, 3477–3486.

Butler, S. L., and Falke, J. J. (1998). Cysteine and disulfide scanning reveals two amphiphilic helices in the linker region of the aspartate chemoreceptor. *Biochemistry* **37**, 10746–10756.

Canfield, D. E., and Des Marais, D. J. (1991). Aerobic sulfate reduction in microbial mats. *Science* **251**, 1471–1473.

Careaga, C. L., and Falke, J. J. (1992). Thermal motions of surface alpha-helices in the D-galactose chemosensory receptor: Detection by disulfide trapping. *J. Mol. Biol.* **226**, 1219–1235.

Danielson, M. A., Bass, R. B., and Falke, J. J. (1997). Cysteine and disulfide scanning reveals a regulatory alpha-helix in the cytoplasmic domain of the aspartate receptor. *J. Biol. Chem.* **272**, 32878–32888.

Donaghay, P. L., Rimes, H. M., and Sieburth, J. M. (1992). Simultaneous sampling of fine scale biological, chemical and physical structure in stratified waters. *Arch. Hydrobiol. Beih. Ergebn. Limnol.* **36**, 97–108.

Edwards, J. C., Johnson, M. S., and Taylor, B. L. (2006). Differentiation between electron transport sensing and proton motive force sensing by the Aer and Tsr receptors for aerotaxis. *Mol. Microbiol.* **62**, 823–837.

Engelmann, T. W. (1881). Neue methode zur untersuchung der sauerstoffausscheidung pflanzlicher und tierischer organismen. *Botanische Zeitung* **39**, 441–448.

Engelmann, T. W. (1883). Bacterium photometricum: Ein beitrag zur vergleichenden physiologie des licht und fabensinnes. *Pfluegers Arch Gesamte Physiol Menschen Tiere.* **42**, 183–186.

Falke, J. J., Dernburg, A. F., Sternberg, D. A., Zalkin, N., Milligan, D. L., and Koshland, D. E., Jr. (1988). Structure of a bacterial sensory receptor: A site-directed sulfhydryl study. *J. Biol. Chem.* **263**, 14850–14858.

Falke, J. J., and Koshland, D. E., Jr. (1987). Global flexibility in a sensory receptor: A site-directed cross-linking approach. *Science* **237**, 1596–1600.

Freitas, T. A., Hou, S., and Alam, M. (2003). The diversity of globin-coupled sensors. *FEBS Lett.* **552**, 99–104.

Freitas, T. A., Saito, J. A., Hou, S., and Alam, M. (2005). Globin-coupled sensors, protoglobins, and the last universal common ancestor. *J. Inorg. Biochem.* **99**, 23–33.

Galperin, M. Y., and Nikolskaya, A. N. (2007). Identification of sensory and signal-transducing domains in two-component signaling systems. *Methods Enzymol.* **422**(3), (this volume).

Gilbert, H. F. (1990). Molecular and cellular aspects of thiol-disulfide exchange. *Adv. Enzymol. Relat. Areas Mol. Biol.* **63**, 69–172.

Goetz, R. M., and Fuglsang, A. (2005). Correlation of codon bias measures with mRNA levels: Analysis of transcriptome data from *Escherichia coli. Biochem. Biophys. Res. Commun.* **327**, 4–7.

Gosink, K. K., del Carmen Buron-Barral, M., and Parkinson, J. S. (2006). Signaling interactions between the aerotaxis transducer Aer and heterologous chemoreceptors in *Escherichia coli. J. Bacteriol.* **188**, 3487–3493.

Green, N. S., Reisler, E., and Houk, K. N. (2001). Quantitative evaluation of the lengths of homobifunctional protein cross-linking reagents used as molecular rulers. *Protein Sci.* **10**, 1293–1304.

Grishanin, R. N., Chalmina, I. I., and Zhulin, I. B. (1991). Behaviour of *Azospirillum brasilense* in a spatial gradient of oxygen and in a "redox" gradient of an artificial electron acceptor. *J. Gen. Microbiol.* **137**, 2781–2785.

Guan, L., Murphy, F. D., and Kaback, H. R. (2002). Surface-exposed positions in the trans-membrane helices of the lactose permease of *Escherichia coli* determined by intermolecular thiol cross-linking. *Proc. Natl. Acad. Sci. USA* **99,** 3475–3480.

Herrmann, S., Ma, Q., Johnson, M. S., Repik, A. V., and Taylor, B. L. (2004). PAS domain of the Aer redox sensor requires C-terminal residues for native-fold formation and flavin adenine dinucleotide binding. *J. Bacteriol.* **186,** 6782–6791.

Hess, J. F., Oosawa, K., Kaplan, N., and Simon, M. I. (1988). Phosphorylation of three proteins in the signaling pathway of bacterial chemotaxis. *Cell* **53,** 79–87.

Hess, J. F., Oosawa, K., Matsumura, P., and Simon, M. I. (1987). Protein phosphorylation is involved in bacterial chemotaxis. *Proc. Natl. Acad. Sci. USA* **84,** 7609–7613.

Hou, S., Freitas, T., Larsen, R. W., Piatibratov, M., Sivozhelezov, V., Yamamoto, A., Meleshkevitch, E. A., Zimmer, M., Ordal, G. W., and Alam, M. (2001). Globin-coupled sensors: A class of heme-containing sensors in Archaea and Bacteria. *Proc. Natl. Acad. Sci. USA* **98,** 9353–9358.

Hughson, A. G., and Hazelbauer, G. L. (1996). Detecting the conformational change of transmembrane signaling in a bacterial chemoreceptor by measuring effects on disulfide cross-linking *in vivo. Proc. Natl. Acad. Sci. USA* **93,** 11546–11551.

Hulko, M., Berndt, F., Gruber, M., Linder, J. U., Truffault, V., Schultz, A., Martin, J., Schultz, J. E., Lupas, A. N., and Coles, M. (2006). The HAMP domain structure implies helix rotation in transmembrane signaling. *Cell* **126,** 929–940.

Jacob, C., Giles, G. I., Giles, N. M., and Sies, H. (2003). Sulfur and selenium: The role of oxidation state in protein structure and function. *Angew Chem. Int. Ed. Engl.* **42,** 4742–4758.

Jocelyn, P. C. (1967). The standard redox potential of cysteine-cystine from the thiol-disulphide exchange reaction with glutathione and lipoic acid. *Eur. J. Biochem.* **2,** 327–331.

Johnson, M. S., Zhulin, I. B., Gapuzan, M. E., and Taylor, B. L. (1997). Oxygen-dependent growth of the obligate anaerobe *Desulfovibrio vulgaris* Hildenborough. *J. Bacteriol.* **179,** 5598–5601.

Jorgensen, B. B. (1982). Ecology of the bacteria of the sulphur cycle with special reference to anoxic-oxic interface environments. *Philos. Trans. R. Soc. Lond. B Biol. Sci.* **298,** 543–561.

Kehry, M. R., Bond, M. W., Hunkapiller, M. W., and Dahlquist, F. W. (1983). Enzymatic deamidation of methyl-accepting chemotaxis proteins in *Escherichia coli* catalyzed by the *cheB* gene product. *Proc. Natl. Acad. Sci. USA* **80,** 3599–3603.

Khan, S., and Macnab, R. M. (1980). The steady-state counterclockwise/clockwise ratio of bacterial flagellar motors is regulated by protonmotive force. *J. Mol. Biol.* **138,** 563–597.

Kim, S., and Lee, S. B. (2006). Rare codon clusters at 5′-end influence heterologous expression of archaeal gene in *Escherichia coli. Protein Expr. Purif.* **50,** 49–57.

Kim, S. H., Wang, W., and Kim, K. K. (2002). Dynamic and clustering model of bacterial chemotaxis receptors: Structural basis for signaling and high sensitivity. *Proc. Natl. Acad. Sci. USA* **99,** 11611–11615.

Kosower, N. S., and Kosower, E. M. (1995). Diamide: An oxidant probe for thiols. *Methods Enzymol.* **251,** 123–133.

Lai, W-C., and Hazelbauer, G. L. (2007). Analyzing transmembrane chemoreceptors using *in vivo* disulfide formation between introduced cysteines. *Methods Enzymol* **423,** in press.

Laszlo, D. J., and Taylor, B. L. (1981). Aerotaxis in *Salmonella typhimurium*: Role of electron transport. *J. Bacteriol.* **145,** 990–1001.

Lee, G. F., Burrows, G. G., Lebert, M. R., Dutton, D. P., and Hazelbauer, G. L. (1994). Deducing the organization of a transmembrane domain by disulfide cross-linking: The bacterial chemoreceptor Trg. *J. Biol. Chem.* **269,** 29920–29927.

Lee, G. F., Lebert, M. R., Lilly, A. A., and Hazelbauer, G. L. (1995). Transmembrane signaling characterized in bacterial chemoreceptors by using sulfhydryl cross-linking *in vivo. Proc. Natl. Acad. Sci. USA* **92,** 3391–3395.

Lee, L., Mizuno, T., and Imae, Y. (1988). Thermosensing properties of *Escherichia coli* tsr mutants defective in serine chemoreception. *J. Bacteriol.* **170**, 4769–4774.

Lennox, E. S. (1955). Transduction of linked genetic characters of the host by bacteriophage P1. *Virology* **1**, 190–206.

Lindbeck, J. C., Goulbourne, E. A., Jr., Johnson, M. S., and Taylor, B. L. (1995). Aerotaxis in *Halobacterium salinarium* is methylation-dependent. *Microbiology* **141**(Pt. 11), 2945–2953.

Loo, T. W., and Clarke, D. M. (2001). Determining the dimensions of the drug-binding domain of human P-glycoprotein using thiol cross-linking compounds as molecular rulers. *J. Biol. Chem.* **276**, 36877–36880.

Lu, J., and Deutsch, C. (2001). Pegylation: A method for assessing topological accessibilities in Kv1.3. *Biochemistry* **40**, 13288–13301.

Lynch, B. A., and Koshland, D. E., Jr. (1991). Disulfide cross-linking studies of the transmembrane regions of the aspartate sensory receptor of *Escherichia coli*. *Proc. Natl. Acad. Sci. USA* **88**, 10402–10406.

Ma, Q., Johnson, M. S., and Taylor, B. L. (2005). Genetic analysis of the HAMP domain of the Aer aerotaxis sensor localizes flavin adenine dinucleotide-binding determinants to the AS-2 helix. *J. Bacteriol.* **187**, 193–201.

Ma, Q., Roy, F., Herrmann, S., Taylor, B. L., and Johnson, M. S. (2004). The Aer protein of *Escherichia coli* forms a homodimer independent of the signaling domain and flavin adenine dinucleotide binding. *J. Bacteriol.* **186**, 7456–7459.

Medigue, C., Rouxel, T., Vigier, P., Henaut, A., and Danchin, A. (1991). Evidence for horizontal gene transfer in *Escherichia coli* speciation. *J. Mol. Biol.* **222**, 851–856.

Mehan, R. S., White, N. C., and Falke, J. J. (2003). Mapping out regions on the surface of the aspartate receptor that are essential for kinase activation. *Biochemistry* **42**, 2952–2959.

Niwano, M., and Taylor, B. L. (1982). Novel sensory adaptation mechanism in bacterial chemotaxis to oxygen and phosphotransferase substrates. *Proc. Natl. Acad. Sci. USA* **79**, 11–15.

Overath, P., Brenner, M., Gulik-Krzywicki, T., Shechter, E., and Letellier, L. (1975). Lipid phase transitions in cytoplasmic and outer membranes of *Escherichia coli*. *Biochim. Biophys. Acta* **389**, 358–369.

Pakula, A. A., and Simon, M. I. (1992). Determination of transmembrane protein structure by disulfide cross-linking: The *Escherichia coli* Tar receptor. *Proc. Natl. Acad. Sci. USA* **89**, 4144–4148.

Parkinson, J. S. (2007). A "bucket of light" for viewing bacterial colonies in soft agar. *Methods Enzymol.* **423**, in press.

Peng, L., Xu, Z., Fang, X., Wang, F., Yang, S., and Cen, P. (2004). Preferential codons enhancing the expression level of human beta-defensin-2 in recombinant *Escherichia coli*. *Protein Pept. Lett.* **11**, 339–344.

Rebbapragada, A., Johnson, M. S., Harding, G. P., Zuccarelli, A. J., Fletcher, H. M., Zhulin, I. B., and Taylor, B. L. (1997). The Aer protein and the serine chemoreceptor Tsr independently sense intracellular energy levels and transduce oxygen, redox, and energy signals for *Escherichia coli* behavior. *Proc. Natl. Acad. Sci. USA* **94**, 10541–10546.

Repik, A., Rebbapragada, A., Johnson, M. S., Haznedar, J. O., Zhulin, I. B., and Taylor, B. L. (2000). PAS domain residues involved in signal transduction by the Aer redox sensor of *Escherichia coli*. *Mol. Microbiol.* **36**, 806–816.

Ritz, D., and Beckwith, J. (2001). Roles of thiol-redox pathways in bacteria. *Annu. Rev. Microbiol.* **55**, 21–48.

Schagger, H., and von Jagow, G. (1987). Tricine-sodium dodecyl sulfate-polyacrylamide gel electrophoresis for the separation of proteins in the range from 1 to 100 kDa. *Anal. Biochem.* **166**, 368–379.

Silverman, M., and Simon, M. I. (1977). Bacterial flagella. *Annu. Rev. Microbiol.* **31**, 397–419.

Springer, M. S., Goy, M. F., and Adler, J. (1979). Protein methylation in behavioural control mechanisms and in signal transduction. *Nature* **280**, 279–284.

Spudich, J. L., and Koshland, D. E., Jr. (1975). Quantitation of the sensory response in bacterial chemotaxis. *Proc. Natl. Acad. Sci. USA* **72**, 710–713.

Studdert, C. A., and Parkinson, J. S. (2004). Crosslinking snapshots of bacterial chemoreceptor squads. *Proc. Natl. Acad. Sci. USA* **101**, 2117–2122.

Studdert, C. A., and Parkinson, J. S. (2005). Insights into the organization and dynamics of bacterial chemoreceptor clusters through *in vivo* crosslinking studies. *Proc. Natl. Acad. Sci. USA* **102**, 15623–15628.

Taylor, B. L. (1983a). How do bacteria find the optimal concentration of oxygen? *Trends Biochem. Sci.* **8**, 438–441.

Taylor, B. L. (1983b). Role of proton motive force in sensory transduction in bacteria. *Annu. Rev. Microbiol.* **37**, 551–573.

Taylor, B. L., and Koshland, D. E., Jr. (1975). Intrinsic and extrinsic light responses of *Salmonella typhimurium* and *Escherichia coli*. *J. Bacteriol.* **123**, 557–569.

Taylor, B. L., Miller, J. B., Warrick, H. M., and Koshland, D. E., Jr. (1979). Electron acceptor taxis and blue light effect on bacterial chemotaxis. *J. Bacteriol.* **140**, 567–573.

Taylor, B. L., and Zhulin, I. B. (1998). In search of higher energy: Metabolism-dependent behaviour in bacteria. *Mol. Microbiol.* **28**, 683–690.

Taylor, B. L., and Zhulin, I. B. (1999). PAS domains: Internal sensors of oxygen, redox potential, and light. *Microbiol. Mol. Biol. Rev.* **63**, 479–506.

Ulrich, L. E., and Zhulin, I. B. (2007). MiST: A microbial signal transduction database. *Nucleic Acids Res.* **35**, D386–D390.

Wardman, P. (1989). Reduction potentials of one-electron couples involving free radicals in aqueous solution. *J. Phys. Chem. Ref. Data* **18**, 1637–1755.

Watts, K. J., Johnson, M. S., and Taylor, B. L. (2006a). Minimal requirements for oxygen sensing by the aerotaxis receptor Aer. *Mol. Microbiol.* **59**, 1317–1326.

Watts, K. J., Ma, Q., Johnson, M. S., and Taylor, B. L. (2004). Interactions between the PAS and HAMP domains of the *Escherichia coli* aerotaxis receptor Aer. *J. Bacteriol.* **186**, 7440–7449.

Watts, K. J., Sommer, K., Fry, S. L., Johnson, M. S., and Taylor, B. L. (2006b). Function of the N-terminal cap of the PAS domain in signaling by the aerotaxis receptor Aer. *J. Bacteriol.* **188**, 2154–2162.

Wolfe, A. J., and Berg, H. C. (1989). Migration of bacteria in semisolid agar. *Proc. Natl. Acad. Sci. USA* **86**, 6973–6977.

Wong, L. S., Johnson, M. S., Zhulin, I. B., and Taylor, B. L. (1995). Role of methylation in aerotaxis in *Bacillus subtilis*. *J. Bacteriol.* **177**, 3985–3991.

Wright, S., Walia, B., Parkinson, J. S., and Khan, S. (2006). Differential activation of *Escherichia coli* chemoreceptors by blue-light stimuli. *J. Bacteriol.* **188**, 3962–3971.

Wuichet, K., Alexander, R. P., and Zhulin, I. B. (2007). Comparative genomic and protein sequence analyses of a complex system controlling bacterial chemotaxis. *Methods Enzymol.* **422**(1), (this volume).

Yu, H. S., and Alam, M. (1997). An agarose-in-plug bridge method to study chemotaxis in the Archaeon Halobacterium salinarum. *FEMS Microbiol. Lett.* **156**, 265–269.

Yu, H. S., Saw, J. H., Hou, S., Larsen, R. W., Watts, K. J., Johnson, M. S., Zimmer, M. A., Ordal, G. W., Taylor, B. L., and Alam, M. (2002). Aerotactic responses in bacteria to photoreleased oxygen. *FEMS Microbiol. Lett.* **217**, 237–242.

Zhulin, I. B., Bespalov, V. A., Johnson, M. S., and Taylor, B. L. (1996). Oxygen taxis and proton motive force in *Azospirillum brasilense*. *J. Bacteriol.* **178**, 5199–5204.

[11]   Two-Component Signaling in the Virulence of *Staphylococcus aureus*: A Silkworm Larvae-Pathogenic Agent Infection Model of Virulence

By KENJI KUROKAWA, CHIKARA KAITO, and KAZUHISA SEKIMIZU

Abstract

Staphylococcus aureus is a pathogenic bacterium that causes abscesses, pneumonia, endocarditis, and food poisoning. *S. aureus* is also one of the resident flora of the endotherm and colonizes the host by skillfully evading its defense mechanism. Identification of attenuated mutants of *S. aureus* in an animal infection model is useful for investigating its adaptability and pathogenesis. This chapter describes a staphylococcal two-component *SA0614–SA0615* system, which was identified using a silkworm larvae infection model.

Introduction

Staphylococcus aureus is a resident flora of the endotherm, which colonizes and grows in the nose and on the skin of adult humans. Although *S. aureus* rarely causes life-threatening diseases in healthy adults, a break in the skin or other injury may allow the bacteria to produce infections that cause abscesses not only on the skin, but also in internal organs, such as the heart (endocarditis) (Tenover and Gorwitz, 2006). Staphylococcal infections including abscesses around foreign materials and bacteremia are often found in individuals with a weakened immune system, or intravenous catheters or prosthetic heart valve, after surgery, or in steroid users. Some *S. aureus* strains produce a variety of compounds that are toxic to animal cells and cause toxinoses, such as TSS, which causes toxic shock syndrome, exfoliatin toxins, which cause scaled skin syndrome, and enterotoxins, which cause food poisoning. Furthermore, *S. aureus* has received increasing attention because it has rapidly gained resistance to various antibiotics, making it difficult to treat (Hiramatsu *et al.*, 2001). Further understanding of this bacterium is important because *S. aureus* strains often cause infections in hospitalized patients.

Many studies have investigated the genes and components that contribute to the pathogenicity and adaptability of staphylococci. Biochemical methods have identified many toxins produced by these bacteria (Bohach, 2006), and genetic methods have been developed to study mutants (Novick, 1991).

METHODS IN ENZYMOLOGY, VOL. 422   0076-6879/07 $35.00
                    DOI: 10.1016/S0076-6879(06)22011-1

Furthermore, the genomes of *S. aureus* strains N315 and Mu50 were published in 2001 (Kuroda *et al.*, 2001), and the sequences of at least nine strains are currently available in public databases (http://gib.genes.nig.ac.jp/). Understanding of the functions of the genes related to the pathogenicity and adaptability of *S. aureus* is also expected to be accelerated by genome analysis in closely related bacteria.

The two-component signaling system is involved in the environmental adaptability of bacteria. Most of the two-component signaling systems are composed of a transmembrane sensor kinase and a cognate transcriptional regulator that is phosphorylated by the sensor kinase, altering its ability to regulate transcription. Genome analysis of *S. aureus* revealed that it has 16 sets of homologs of the two-component signaling system. Of these, *agr*, *sae*, *arl*, and *srr* are involved in regulation of toxin production (Cheung *et al.*, 2004; Novick, 2003). Also, the *vraSR* genes are involved in vancomycin resistance (Kuroda *et al.*, 2003), and the *yycFG* genes are essential for staphylococcal cell growth (Dubrac and Msadek, 2004; Martin *et al.*, 1999).

Additional research is necessary for understanding the pathogenicity and adaptability of *S. aureus*. For this purpose, it is important to study the function of genes in animal models of infection. Hence, we established a model that uses the larvae of the silkworm, *Bombyx mori*.

Silkworm Larvae Infection Model

Animal models are necessary for studies on infection by pathogenic bacteria. Previous studies have used mammals such as mouse, rat, rabbit, dog, and monkey, but they are expensive and pose ethical problems. Therefore, we have developed a model of infection using silkworms.

Because of the long history of sericulture, methods of breeding and growing genetically homogeneous silkworms have already been developed. Thus, a steady supply of silkworms can be available all year. The large size of silkworm larvae allows quantitative evaluation of pathogens and drugs by injection (up to 100 μl) into the hemolymph. In addition, the midgut or liver (the fat body in the silkworm) can be removed for pharmacological experiments. The genome of the silkworm has been sequenced (Mita *et al.*, 2004; Xia *et al.*, 2004), specific genetic methods have been developed (Goldsmith *et al.*, 2005), and a variety of variant strains have been isolated, some of which are available from the SilkwormBase of the National BioResource Project of Japan (http://www.shigen.nig.ac.jp/silkwormbase/index.jsp). Therefore, it is expected that the silkworm will become a more common animal model.

We found that silkworm larvae were killed by injection into the hemolymph of pathogenic bacteria and fungi, such as *S. aureus*, *Streptococcus pyogenes*, *Pseudomonas aeruginosa*, *Stenotrophomonas maltophilia*,

Vibrio cholerae, *Candida albicans*, and *Candida tropicalis* (Hamamoto *et al.*, 2004; Kaito *et al.*, 2002, 2005). The killing by these microorganisms can be prevented by administration of antibiotics or antifungal agents (Hamamoto *et al.*, 2004), suggesting that death is due to the growth of the pathogenic agent, and, therefore, that it should be possible to use this system to identify novel antibiotics and antifungal agents. Most of the chemicals that inhibit the growth of pathogens *in vitro*, however, are also toxic to mammals or have poor pharmacokinetics. Use of the silkworm larvae infection system as a screen should help avoid these problems. Furthermore, the silkworm system allows for the identification of genes involved in the pathogenicity of *S. aureus*.

Pathogenicity-Related Genes That Can Be Identified in the Silkworm Infection Assay

We searched for mutants of *S. aureus* with attenuated pathogenicity in the silkworm larvae infection model. This new model system should allow identification of novel pathogenicity-related genes and clarification of their functions because the scale of screening can be increased. Two other invertebrates, *Drosophila* and nematode, have also been reported as model systems for infection (Bae *et al.*, 2004; Begun *et al.*, 2005; Needham *et al.*, 2004).

The silkworm, an insect, avoids infection via the innate immune system and lacks an acquired immune system. Previous studies have examined how the silkworm combats infection, including the prophenoloxidase cascade triggered by peptidoglycan or β-1,3-glucan and pattern-recognition proteins (Ochiai and Ashida, 2000; Tsuchiya *et al.*, 1996; Yoshida *et al.*, 1996), induction of antimicrobial peptides (Yamakawa and Tanaka, 1999), and nodule formation (Ohta *et al.*, 2006). The silkworm infection model allows the evaluation of specific aspects of staphylococcal pathogenicity in humans. Therefore, it should be possible to use this model along with preexisting assays for mammalian cells to determine the function of previously uncharacterized genes. In addition, mutants of *S. aureus* that lack or have reduced abilities to kill the silkworm larvae, are sensitized to an innate immune response, or have impaired cell growth *in vivo* can be isolated with this model. Similarly, mutants with high pathogenicity, can escape or do not induce the innate immune system, or increase the ability to be killed by the silkworm larvae can be identified. Therefore, studies in the silkworm model should help reveal how pathogenic bacteria and fungi cause infections, how the innate immune system functions, and mechanisms of survival and death in insects. These findings should also help in the development of novel agents for curing and preventing infectious diseases.

Identification of Genes Involved in the Killing of Silkworm
  Larvae by Bacteria

There are two methods for identifying genes involved in the pathogenicity and adaptability of bacteria in the silkworm larvae infection model.
The first of these uses reverse genetics, in which targeted deletion mutants
are constructed and injected into silkworm larvae to find the mutants with
pathogenicity (Kaito et al., 2005). We previously reported that a deletion
mutant of agr, one of the genes of the 16 two-component signaling systems
in the staphylococcal genome, reduced the ability of S. aureus to kill
silkworm larvae (Kaito et al., 2005).

The second method utilizes forward genetics; however, it is very difficult
to identify attenuated mutants from mutagen-treated cells using decreased
killing ability alone because approximately 100,000 larvae must be injected
to obtain 10 attenuated mutants. Furthermore, additional silkworm larvae
are needed for evaluating complementation by genome library plasmids,
which is necessary to identify the mutated genes. To increase the ability to
obtain attenuated mutants, we searched for genes that both decrease the
pathogenicity and the resistance to a drug that nonspecifically inhibits
bacterial cell growth. This is expected to increase the frequency of attenuated mutants because the selected drug-sensitive mutants are then used as
the source of mutated genes in the silkworm infection assay. Also, such
genes can be identified by the ability to complement the drug sensitivity
of the mutant from genome library plasmids, where it is possible to identify
the responsible plasmid within a few days from 30,000 attenuated mutant-
derived transformants by measuring bacterial cell growth on agar plates
in vitro. Thus, we can examine whether the identified mutation simultaneously causes both drug sensitivity and attenuated virulence using the
complemented strain and its deletion mutant.

We screened using chlorpromazine (CPZ) (unpublished observations),
a psychotropic drug. CPZ is lipophilic and forms micelles that act on cell
membranes (Tanji et al., 1992; Wajnberg et al., 1988). CPZ is also reported
to induce oxidative stress (Eghbal et al., 2004). Thus, the CPZ-sensitive
mutants are expected to include mutants of transmembrane or cell membrane-associated proteins as well as proteins involved in the oxidative stress
response. We obtained 16 attenuated mutants from 100 CPZ-sensitive
mutants. Complementation analysis revealed that 1 of the 16 mutants had
a mutation in the SA0614 gene, which encodes a two-component signaling
system regulator homolog (our unpublished observations). This chapter
describes the protocol used to identify this two-component signaling
gene. Use of other drugs for selecting for attenuated mutants should
allow identification of other pathogenicity and adaptability genes.

Identification of an *SA0614* Response Regulator Mutant by Monitoring
CPZ Sensitivity and Ability to Kill Silkworm Larvae

1. RN4220 cells are grown overnight at 30° in 0.2 ml LB medium (1%
bactotryptone, 0.5% yeast extract, and 1% NaCl) containing 0.4% ethyl-
methanesulfate (Sigma) in 96-well plates (Inoue *et al.*, 2001). Cells are
serially diluted, plated on an LB agar plate, and grown at 30°. Colonies are
transferred to sterile woolen cloth, stamped on LB agar plates with and
without 0.15 m*M* CPZ (Shionogi, Osaka, Japan), and grown overnight
at 37°. Cells that cannot grow on the CPZ-containing plate are selected.
Frequency to isolate a CPZ-sensitive mutant among mutagen-treated
cells is set about 1 per 1000. Also, one colony from one well of the
ethylmethanesulfate-treated cell culture is selected.

2. CPZ-sensitive mutants are collected, and their ability to kill silkworm
larvae is determined (see later).

3. After obtaining CPZ-sensitive mutants with attenuated virulence,
genes that complement CPZ sensitivity are identified by plasmid complementation analysis (Inoue *et al.*, 2001; Li *et al.*, 2004; Matsuo *et al.*, 2003).
To construct a plasmid genome library, *S. aureus* chromosomal DNA is
partially digested with DNase I and blunt ended with T4 DNA polymerase
and the large (Klenow) fragment of DNA polymerase I. The 2- to 4-kbp
DNA fragments of them are prepared by agarose gel electrophoresis,
inserted into the *Sma*I site of the pKE515 shuttle vector (Li *et al.*, 2004),
and used to transform *Escherichia coli* cells. The plasmid library is prepared from resultant *E. coli* transformants. Transformation of *S. aureus*
cells with the library plasmids was described previously (Inoue *et al.*, 2001).
The genomic DNA fragment cloned on the plasmid that complements the
CPZ sensitivity is sequenced, and genes in the fragment are determined
using the genome database on the Web (http://gib.genes.nig.ac.jp/). The
responsible gene for complementation is determined, where individual
genes on the genomic fragment can be amplified by polymerase chain
reaction methods and cloned into a plasmid.

4. The corresponding chromosomal gene locus on the parent and
mutant strains is sequenced and compared to examine for the presence of
mutations. During this process, the mutation in the *SA0614* response regulator gene is determined. To examine whether the *SA0614* gene mutation is
responsible for both attenuated virulence and CPZ sensitivity, phage transduction experiments using phage 80α or phage 11 can be performed (Murai
et al., 2006; Novick, 1991). In these experiments, a drug resistance marker
gene can be inserted near the *SA0614* gene and used for selection of
transductants. The resulting transductants are then examined to determine
whether CPZ sensitivity and attenuated virulence cannot be separated.

Also, a deletion mutant for the mutated gene can be constructed and examined for its ability to affect CPZ sensitivity and virulence in the silkworm. Using this latter method, we evaluated the *SA0614* as well as the *SA0615* gene, which encodes a two-component signaling system sensor protein. We found that mutants of these genes show similar phenotypes for CPZ sensitivity, virulence in silkworm larvae, and cell wall integrity (see later), suggesting that they function together as a two-component signaling system (unpublished observations).

Silkworm Larvae Infection Assay

1. Fertilized silkworm eggs (HuYo × TukubaNe; Ehime Sansyu, Yawatahama, Japan) are grown until the fourth molted larvae in a clean room at 27° on Silkmate 2S, a paste that includes antibiotics (Nosan Corporation, Yokohama, Japan). It takes 14 to 15 days from the hatched eggs. It is possible to delay the hatching of silkworm eggs and the growth of larvae by placing them at a low temperature. Generally, we use disposable plastic packs to grow the larvae. In each pack, 30 to 500 larvae are bred, depending on their size.

2. The fourth molted larvae are collected and starved for 1 day, during which time the larvae eliminate the antibiotics. Starving the larvae for 2 days is tolerant for the infection assays.

3. An artificial diet lacking antibiotics (Katakura Industries, Tokyo, Japan) is given to the shed fifth instar larvae the evening before the day of bacterial injection.

4. The next day, the fifth instar larvae are injected with bacteria. Larvae with obviously small body sizes or that have not completed ecdysis are not used. Larvae in which the body weight is even are used if required. Bacterial suspensions in 0.9% NaCl (1×10^8 to 5×10^9 colony-forming units [CFU]/ml *S. aureus* cells) are injected into the silkworm hemolymph. Typically, 50 μl of suspension is injected using a 1-ml disposal syringe with a 27-gauge needle (Terumo, Tokyo, Japan), and each group includes 5 to 10 larvae. During the injection, the silkworm larva is held in one hand, with its head to one side, and is positioned to open the space in the crest of its back (Fig. 1A). The syringe is held in the other hand with the needle tip window surface up. The needle is inserted between the crest of the back in the direction of the tail (Fig. 1B), and the sample solution is injected in quick motion. Because the silkworm has an open blood vascular system, administered bacterial cells spread through the whole body rapidly. Practice is necessary to ensure that samples are injected only into the hemolymph because the midgut occupies most of the body of the silkworm larva. Red ink can be used to practice the injection; when injected into the

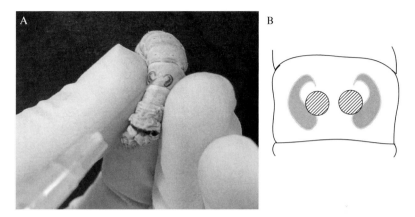

Fig. 1. Injection into silkworm larvae. (A) Handling of silkworm larva. (B) Injection points are marked with hatched circles.

hemolymph, the body surface becomes red, which does not happen if the injection is made into the midgut.

5. After the injection, the larvae for each group are placed in plastic packs with Kimwipes at the bottom. Feeding is not carried out after the injection. The plastic packs are placed in a safety cabinet (AirTech Japan, Tokyo, Japan) maintained at 27° and 50% humidity for 3 to 5 days. The temperature can be increased to 37°, although this may alter the pathogenicity. At a higher temperature, silkworm larvae can breed for 3 days.

6. The number of surviving silkworm larvae is determined at appropriate times. If it is difficult to determine whether the larvae are alive or dead, they can be stimulated with a pair of tweezers. Attention is paid to whether the larvae are paralyzed immediately following the injection, die after a period of time, or recover. Figure 2 shows representative results for the silkworm infection assay using the *SA0614* two-component signaling regulator mutant.

Measurement of the Number of Bacteria in Silkworm Hemolymph

1. The tip of a pipette tip for a 5-ml pipettor is cut with sterilized scissors and inserted into a 15-ml plastic centrifuge tube.
2. Two or three abdominal legs of silkworm larvae infected with bacteria are removed with scissors. The larvae are then placed into the pipette tip from the tail and pushed into the tip.
3. Several minutes later, 200 μl of hemolymph is collected from the bottom of the centrifuge tube.

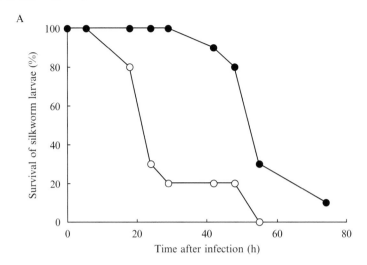

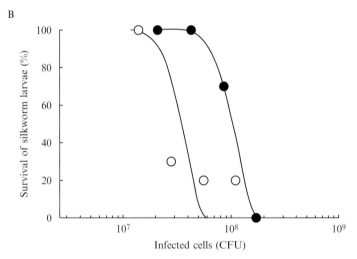

FIG. 2. Typical results for silkworm infection assays. (A) Time course analysis. Ten silkworm larvae were injected with eightfold diluted overnight cultures of the parent strain RN4220 (2.8×10^7 CFU, open circle) or ZS0270 mutant (4.3×10^7 CFU, closed circle) that had amino acid substitutions in the *SA0614* gene. Larval survival was monitored for 3 days. (B) Dose–response analysis. Ten silkworm larvae were injected with severalfold diluted overnight cultures of the parent strain RN4220 (open circle) or ZS0270 (closed circle) mutant as indicated. Larval survival was monitored 24 h after infection.

4. The hemolymph is serially diluted with saline and spread onto Bacto tryptic soy broth (TSB, Becton Dickinson) agar plates.
5. The plates are incubated for 1 day at 37° (or at another appropriate temperature), after which the number of colonies is counted. After 30 h, 10^7 S. aureus strain RN4220 cells proliferate into approximately 10^9 cells.

Defect in Cell Wall Integrity of the SA0614 Response Regulator Mutant

Chlorpromazine is a membrane-disrupting reagent that decreases membrane fluidity (Tanji et al., 1992) and kills bacteria at concentrations near its critical micelle concentration (Murai et al., 2006; Wajnberg et al., 1988). Studies have revealed that the sensitivity of the mutant and parental RN4220 strains do not differ in the presence of ionic detergents, including sodium dodecyl sulfate and deoxycholate (which kill bacteria at concentrations below the critical micelle concentration). In contrast, and similar to findings with CPZ, the mutant bacteria are much less sensitive to the nonionic detergents Triton-X 100 and Nonidet P-40 at concentrations near the critical micelle concentration.

We suspect that this differential sensitivity to nonionic detergents is due to a change in the cell wall composition or quantity. To explore this further, we examined the effect of lysozyme, a peptidoglycan hydrolase, on bacterial growth. Growth at a higher concentration of lysozyme was impaired in the mutant but not the parental RN4220 cells. We also compared the peptidoglycan-induced melanization of the parental and mutant cells using silkworm larvae plasma-high sensitive reagent (SLP-HS kit; Wako Pure Chemical Industries, Osaka, Japan) and a Wako Toxinometer ET-2000, which measures the intensity of melanization (blackening) caused by contact between the peptidoglycan and the larval peptidoglycan recognition protein. We found a decrease in the melanization-inducing activity of the mutants, suggesting that the dynamics of cell wall peptidoglycan, including biosynthesis or degradation, are altered in the mutants (unpublished observations).

Detergent and Lysozyme Sensitivity Test

Twofold dilution series of reagents (100 μl per well) are prepared in 96-well plates. Bacterial cultures are grown overnight and diluted 10,000-fold with TSB medium. A 100-μl aliquot of the diluted bacteria culture is added to the reagent wells and incubated for 18 h at 37°. As a control for bacterial growth, the bacterial cells are grown in TSB medium without drugs.

TSB medium supplemented with drugs and incubated without bacterial cells is used as a blank. The bacterial growth is measured at 600 nm with spectrophotometer.

Melanization-Inducing Activity of Bacterial Peptidoglycan

The amount of bacterial cell wall peptidoglycan can be calculated by measuring the melanization (blackening) reaction using the SLP-HS reagent kit (Wako) (Tsuchiya *et al.*, 1996) and an ET-2000 Toxinometer (Wako). Briefly, a bacterial cell culture (200 μl) at exponential or stationary phase is boiled for 20 min in 10% trichloroacetic acid, cooled on ice, and centrifuged at 20,000g for 30 min at 4°. The pellet is washed twice with 200 μl of deionized distilled water, suspended in 28 μg/ml trypsin dissolved in wash buffer (100 mM Tris-HCl buffer, pH 7.5, 20 mM MgCl$_2$, 1 mM CaCl$_2$), and incubated for 12 h at 37°. The digested cell wall suspension is centrifuged at 20,000g for 30 min at 4°, and the pellet is washed twice with deionized distilled water and suspended in 200 μl of deionized distilled water. Next, the digested cell wall is sonicated (duty cycle 90, output 5–10, 1 min, Branson Sonifier 450). The resulting suspension is diluted 100-fold with the SLP diluent (Wako), and a 200-μl aliquot is added to the SLP-HS reagent (Wako). Reaction between SLP-HS reagent and peptidoglycan results in blackening of the suspension, and the absorbance is measured with the ET-2000 Toxinometer. Peptidoglycan from *S. aureus* (Wako) is used as a standard.

References

Bae, T., Banger, A. K., Wallace, A., Glass, E. M., Aslund, F., Schneewind, O., and Missiakas, D. M. (2004). *Staphylococcus aureus* virulence genes identified by bursa aurealis mutagenesis and nematode killing. *Proc. Natl. Acad. Sci. USA* **101,** 12312–12317.

Begun, J., Sifri, C. D., Goldman, S., Calderwood, S. B., and Ausubel, F. M. (2005). *Staphylococcus aureus* virulence factors identified by using a high-throughput *Caenorhabditis elegans*-killing model. *Infect. Immun.* **73,** 872–877.

Bohach, G. A. (2006). *Staphylococcus aureus* exotoxins. *In* "Gram-Positive Pathogens" (V. A. Fischetti, R. P. Novick, J. J. Ferretti, D. A. Portnoy, and J. I. Rood, eds.), pp. 464–477. ASM Press, Washington, DC.

Cheung, A. L., Bayer, A. S., Zhang, G., Gresham, H., and Xiong, Y. Q. (2004). Regulation of virulence determinants *in vitro* and *in vivo* in *Staphylococcus aureus*. *FEMS Immunol. Med. Microbiol.* **40,** 1–9.

Dubrac, S., and Msadek, T. (2004). Identification of genes controlled by the essential YycG/YycF two-component system of *Staphylococcus aureus*. *J. Bacteriol.* **186,** 1175–1181.

Eghbal, M. A., Tafazoli, S., Pennefather, P., and O'Brien, P. J. (2004). Peroxidase catalysed formation of cytotoxic prooxidant phenothiazine free radicals at physiological pH. *Chem. Biol. Interact.* **151,** 43–51.

Goldsmith, M. R., Shimada, T., and Abe, H. (2005). The genetics and genomics of the silkworm, *Bombyx mori. Annu. Rev. Entomol.* **50,** 71–100.

Hamamoto, H., Kurokawa, K., Kaito, C., Kamura, K., Manitra Razanajatovo, I., Kusuhara, H., Santa, T., and Sekimizu, K. (2004). Quantitative evaluation of the therapeutic effects of antibiotics using silkworms infected with human pathogenic microorganisms. *Antimicrob. Agents Chemother.* **48,** 774–779.

Hiramatsu, K., Cui, L., Kuroda, M., and Ito, T. (2001). The emergence and evolution of methicillin-resistant *Staphylococcus aureus. Trends Microbiol.* **9,** 486–493.

Inoue, R., Kaito, C., Tanabe, M., Kamura, K., Akimitsu, N., and Sekimizu, K. (2001). Genetic identification of two distinct DNA polymerases, DnaE and PolC, that are essential for chromosomal DNA replication in *Staphylococcus aureus. Mol. Genet. Genomics* **266,** 564–571.

Kaito, C., Akimitsu, N., Watanabe, H., and Sekimizu, K. (2002). Silkworm larvae as an animal model of bacterial infection pathogenic to humans. *Microb. Pathog.* **32,** 183–190.

Kaito, C., Kurokawa, K., Matsumoto, Y., Terao, Y., Kawabata, S., Hamada, S., and Sekimizu, K. (2005). Silkworm pathogenic bacteria infection model for identification of novel virulence genes. *Mol. Microbiol.* **56,** 934–944.

Kuroda, M., Kuroda, H., Oshima, T., Takeuchi, F., Mori, H., and Hiramatsu, K. (2003). Two-component system VraSR positively modulates the regulation of cell-wall biosynthesis pathway in *Staphylococcus aureus. Mol. Microbiol.* **49,** 807–821.

Kuroda, M., Ohta, T., Uchiyama, I., Baba, T., Yuzawa, H., Kobayashi, I., Cui, L., Oguchi, A., Aoki, K., Nagai, Y., Lian, J., Ito, T., *et al.* (2001). Whole genome sequencing of meticillin-resistant *Staphylococcus aureus. Lancet* **357,** 1225–1240.

Li, Y., Kurokawa, K., Matsuo, M., Fukuhara, N., Murakami, K., and Sekimizu, K. (2004). Identification of temperature-sensitive *dnaD* mutants of *Staphylococcus aureus* that are defective in chromosomal DNA replication. *Mol. Genet. Genomics* **271,** 447–457.

Martin, P. K., Li, T., Sun, D., Biek, D. P., and Schmid, M. B. (1999). Role in cell permeability of an essential two-component system in *Staphylococcus aureus. J. Bacteriol.* **181,** 3666–3673.

Matsuo, M., Kurokawa, K., Nishida, S., Li, Y., Takimura, H., Kaito, C., Fukuhara, N., Maki, H., Miura, K., Murakami, K., and Sekimizu, K. (2003). Isolation and mutation site determination of the temperature-sensitive *murB* mutants of *Staphylococcus aureus. FEMS Microbiol. Lett.* **222,** 107–113.

Mita, K., Kasahara, M., Sasaki, S., Nagayasu, Y., Yamada, T., Kanamori, H., Namiki, N., Kitagawa, M., Yamashita, H., Yasukochi, Y., Kadono-Okuda, K., Yamamoto, K., *et al.* (2004). The genome sequence of silkworm, *Bombyx mori. DNA Res.* **11,** 27–35.

Murai, N., Kurokawa, K., Ichihashi, N., Matsuo, M., and Sekimizu, K. (2006). Isolation of a temperature-sensitive *dnaA* mutant of *Staphylococcus aureus. FEMS Microbiol Lett.* **254,** 19–26.

Needham, A. J., Kibart, M., Crossley, H., Ingham, P. W., and Foster, S. J. (2004). *Drosophila melanogaster* as a model host for *Staphylococcus aureus* infection. *Microbiology* **150,** 2347–2355.

Novick, R. P. (1991). Genetic systems in staphylococci. *Methods Enzymol.* **204,** 587–636.

Novick, R. P. (2003). Autoinduction and signal transduction in the regulation of staphylococcal virulence. *Mol. Microbiol.* **48,** 1429–1449.

Ochiai, M., and Ashida, M. (2000). A pattern-recognition protein for beta-1,3-glucan: The binding domain and the cDNA cloning of beta-1,3-glucan recognition protein from the silkworm, *Bombyx mori. J. Biol. Chem.* **275,** 4995–5002.

Ohta, M., Watanabe, A., Mikami, T., Nakajima, Y., Kitami, M., Tabunoki, H., Ueda, K., and Sato, R. (2006). Mechanism by which *Bombyx mori* hemocytes recognize microorganisms: Direct and indirect recognition systems for PAMPs. *Dev. Comp. Immunol.* **30,** 867–877.

Tanji, K., Ohta, Y., Kawato, S., Mizushima, T., Natori, S., and Sekimizu, K. (1992). Decrease by psychotropic drugs and local anaesthetics of membrane fluidity measured by fluorescence anisotropy in *Escherichia coli*. *J. Pharm. Pharmacol.* **44**, 1036–1037.

Tenover, F. C., and Gorwitz, R. J. (2006). The epidemiology of *Staphylococcus* infections. *In* "Gram-Positive Pathogens" (V. A. Fischetti, R. P. Novick, J. J. Ferretti, D. A. Portnoy, and J. I. Rood, eds.), pp. 526–534. ASM Press, Washington, DC.

Tsuchiya, M., Asahi, N., Suzuoki, F., Ashida, M., and Matsuura, S. (1996). Detection of peptidoglycan and beta-glucan with silkworm larvae plasma test. *FEMS Immunol. Med. Microbiol.* **15**, 129–134.

Wajnberg, E., Tabak, M., Nussenzveig, P. A., Lopes, C. M., and Louro, S. R. (1988). pH-dependent phase transition of chlorpromazine micellar solutions in the physiological range. *Biochim. Biophys. Acta* **944**, 185–190.

Xia, Q., Zhou, Z., Lu, C., Cheng, D., Dai, F., Li, B., Zhao, P., Zha, X., Cheng, T., Chai, C., Pan, G., Xu, J., *et al.* (2004). A draft sequence for the genome of the domesticated silkworm (*Bombyx mori*). *Science* **306**, 1937–1940.

Yamakawa, M., and Tanaka, H. (1999). Immune proteins and their gene expression in the silkworm, *Bombyx mori*. *Dev. Comp. Immunol.* **23**, 281–289.

Yoshida, H., Kinoshita, K., and Ashida, M. (1996). Purification of a peptidoglycan recognition protein from hemolymph of the silkworm, *Bombyx mori*. *J. Biol. Chem.* **271**, 13854–13860.

[12] TonB System, *In Vivo* Assays and Characterization

By Kathleen Postle

Abstract

The multiprotein TonB system of *Escherichia coli* involves proteins in both the cytoplasmic membrane and the outer membrane. By a still unclear mechanism, the proton-motive force of the cytoplasmic membrane is used to catalyze active transport through high-affinity transporters in the outer membrane. TonB, ExbB, and ExbD are required to transduce the cytoplasmic membrane energy to these transporters. For *E. coli*, transport ligands consist of iron-siderophore complexes, vitamin B_{12}, group B colicins, and bacteriophages T1 and ϕ80. Our experimental philosophy is that data gathered *in vivo*, where all known and unknown components are present at balanced chromosomal levels in the whole cell, can be interpreted with less ambiguity than when a subset of components is overexpressed or analysed *in vitro*. This chapter describes *in vivo* assays for the TonB system and their application.

Introduction

In Gram-negative bacteria, the TonB system is responsible for energizing transport events at the outer membrane using the proton-motive force of the cytoplasmic membrane (reviewed in Braun, 2003; Faraldo-Gomez and Sansom, 2003; Postle and Kadner, 2003; Wiener, 2005). In the *Escherichia coli* K-12 cytoplasmic membrane, TonB exists in a complex with two integral cytoplasmic membrane (CM) proteins, ExbB and ExbD. At the outer membrane (OM), TonB directly contacts and transduces energy to outer membrane active transporters (FepA, FecA, FhuA, FiuA, Cir, and FhuE for the transport of iron–siderophores; BtuB for the transport of vitamin B_{12}). TonB-gated transporters (TGTs) consist of 22-stranded β barrels with the lumen of each barrel occluded by a $\sim$150 residue amino-terminal domain, known as the cork. The function of TonB appears to involve release or rearrangement of the cork, and its associated bound ligand, from the barrel into the periplasmic space (Ma *et al.*, 2007; Devanathan and Postle, unpublished results). From there, siderophores or vitamin B_{12} are captured by binding proteins and delivered to ABC transporters in the cytoplasmic membrane. The TGTs also serve as outer membrane receptors for a variety of bacteriophages and protein toxins called colicins (reviewed in Braun *et al.*, 2002).

METHODS IN ENZYMOLOGY, VOL. 422   0076-6879/07 $35.00
DOI: 10.1016/S0076-6879(06)22012-3

In vitro study of the TonB system has been limited by an inability to fully reconstitute the system. The amino terminus of TonB is essential for its activity (Jaskula *et al.*, 1994; Larsen and Postle, 2001; Larsen *et al.*, 1999); however, it has not yet been possible to purify full-length TonB with the amino terminus intact. Nonetheless, the TonB carboxy terminus (~90 residues) has been crystallized and its nuclear magnetic resonance (NMR) structure determined, both alone (Chang *et al.*, 2001; Kodding *et al.*, 2005; Sean Peacock *et al.*, 2005) and in the presence of TGTs (Pawelek *et al.*, 2006; Shultis *et al.*, 2006). Although the structures have provided important food for thought, the correlation between *in vitro* and *in vivo* results is not good, suggesting that the most meaningful and unambiguously interpretable data will come from *in vivo* studies for the time being. For just one example out of several, while *in vivo* studies identify the five aromatic amino acids of the TonB carboxy terminus as virtually the only residues playing a role in recognition of TGTs, in all of the crystal/NMR structures solved to date, the side chains of the aromatic residues are buried (reviewed in Postle and Larsen, 2007; Wiener, 2005).

Selection For and Against the *tonB* Gene

The *tonB* gene is unusually versatile in that both mutant and wild-type *tonB* can be directly selected. Mutations are selected by resistance to bacteriophage and colicins or by resistance to TonB-dependent antibiotics (Hantke and Braun, 1978; Larsen *et al.*, 1993; Nikaido and Rosenberg, 1990). Selection against *tonB* mutants uses chromium (Larsen *et al.*, 1994; Wang and Newton, 1969a).

Selecting Against tonB

Strains lacking *tonB* are highly pleiotropic, with phenotypes that include a lack of high-affinity iron transport, lack of vitamin B_{12} transport, tolerance to group B colicins, and tolerance to bacteriophages T1 and ϕ80. It is worth remembering that, due to overall iron starvation, other aspects of *E. coli* physiology for *tonB* strains are abnormal as well. For example, enzymes involved in aromatic amino acid synthesis are derepressed because most aromatic amino acid pathway substrates are being diverted to the synthesis of enterochelin (McCray and Herrmann, 1976). Similarly, under iron starvation conditions, certain tRNAs are undermodified such that their impaired function is capable of deattenuating the tryptophan operon (Buck and Griffiths, 1982).

Selection with a single TonB-dependent bacteriophage or colicin will yield mutations that inactivate either the receptor for that phage or TonB. Mutations are not generally recovered in ExbB or ExbD, primarily because

TolQ and TolR proteins can replace their function sufficiently that the cells will be killed by the selective agents (Bradbeer, 1993; Braun, 1989; Skare *et al.*, 1993). Selection in a TolQ/R background is complicated by the compromised state of the OM and the compensatory secretion of capsular material that characterizes these strains. Secretion of capsular material can itself confer physical protection against phage. Selection with multiple TonB-dependent bacteriophage or colicins that recognize different TGTs (e.g., colicin B, which uses FepA as a receptor together with bacteriophage φ80, which uses FhuA as receptor) specifically targets mutations to *tonB*. Selections for *tonB* mutants have dropped into disuse largely because the occurrence of knockout missense mutants has been limited to the amino-terminal TonB transmembrane domain, with the majority of those recovered involving insertions, large deletions, or nonsense mutations. A note of caution about using bacteriophage T1: Although bacteriophage T1 is a TonB-dependent bacteriophage, its use is not advised except under special containment conditions. The extraordinary stability of this phage has led to the maxim: "if you are using T1 in your lab, pretty soon everyone in the building will be using T1"—not a good way to maintain collegial bridges.

Selecting for TonB$^+$

Even though iron is an essential nutrient, *tonB* strains can grow without problem because of the degree to which iron contaminates chemicals/growth media and the apparent existence of a low-affinity iron uptake system. However, because chromium competes with the low-affinity iron uptake system, *tonB* strains are unusually sensitive to chromium (Wang and Newton, 1969a). Thus resistance to chromium can serve as a selection for wild-type TonB system function. This selection has been used successfully to obtain second site *exbB* suppressors of *tonB* missense mutations (Larsen *et al.*, 1994).

To select for TonB$^+$ phenotypes, a culture of *tonB* cells is grown to saturation, washed twice in 1× M9 salts, and plated on chromium agar plates (1.5% Difco agar in M9 minimal salts medium supplemented with 0.3% vitamin-free casamino acids, 0.2% glucose, 4.0 μg ml^{-1} thiamine, 10 mM MgSO$_4$, and 100 μM CrCl$_3$). In the early days, growth of *tonB* strains in minimal media required the addition of citrate, even though *E. coli* does not use citrate as a carbon source and a *tonB* strain cannot transport ferric-citrate. This was most likely due to chromium contamination in the media and is no longer a problem (Wang and Newton, 1969b). It is important to use vitamin-free casamino acids and other chemicals with the lowest Fe contamination possible for this selection to work, as supplemental iron allows growth regardless of *tonB* status.

Precautions for Experiments Where TonB System Proteins Are Expressed from Plasmids

Mutations in genes are most easily generated and mobilized in plasmids. In a multicomponent system, it is often most straightforward to assay mutant proteins expressed from plasmids in strains where the cognate gene has been knocked out (Datsenko and Wanner, 2000). However, when this approach is taken to study the TonB system, it is important to ensure that only chromosomal levels of all components are expressed, from either the chromosome or a regulatable promoter.

TonB and ExbD Exhibit Dominant-Negative Gene Dosage Effects

In the initial efforts to clone *tonB* (early 1980s), three research groups (Higgins, Kadner, and Postle) all had made the same discovery, a fact that became apparent once notes were informally compared: it was impossible to select TonB plasmid clones by growth on chromium plates in a *tonB* background (the standard selection for wild-type TonB). This phenomenon was later described more fully where it was noted that overexpression of TonB from multicopy plasmids caused a dominant decrease in detectable TonB activity (Mann *et al.*, 1986). TonB was shown to be unstable in the absence of ExbB also thus supplying the rationale for dominant negativity (Fischer *et al.*, 1989; Skare and Postle, 1991). When TonB is overexpressed, there is insufficient ExbB to stabilize the excess TonB. The proteolyzed TonB peptides become competitive at sites where wild-type TonB would normally interact (Howard *et al.*, 2001; Jaskula *et al.*, 1994). It is therefore important to maintain balanced chromosomal levels of TonB, ExbB, and ExbD (Higgs *et al.*, 2002a) due to the requirement for ExbB to stabilize both ExbD and TonB. Indeed ExbD also exhibits dominant-negative gene dosage, while ExbB does not (Bulathsinghala and Postle, unpublished observations). Chromosomal levels of expression can be accomplished by the use of genes expressed from the pBAD promoter (Guzman *et al.*, 1995).

One or More Proteins in the TonB System Remain to be Identified

One might wonder whether overexpression of TonB, ExbB, and ExbD proteins simultaneously could lead to interpretable data about the function(s) of the complex. Such an approach is impeded by the apparent existence of at least one unidentified protein in the system. Even though TonB is a stable protein ($t_{1/2} = 90-120$ min), in the absence of continued protein synthesis TonB activity decays with a half-life of 15–20 min (Kadner and McElhaney, 1978; Skare and Postle, 1991). The other known proteins of the TonB system, ExbB, ExbD, and the outer membrane transporter FepA, are chemically stable for over 90 min when expressed at normal chromosomal levels (Skare and Postle, 1991; unpublished results). Proton-motive force, which energizes the TonB system, remains undisturbed when protein synthesis is inhibited (Skare and Postle, unpublished results). The conclusion is that there must be an unidentified protein of short half-life that participates in energizing TonB. The unidentified protein has not turned up in selections, leading to speculation that this unstable protein has either redundant function or is essential for cell growth. Either way, until the unidentified protein has been fully characterized, there is a possibility that simultaneous overexpression of ExbB, ExbD, and TonB might exceed the capacity of the unidentified protein. Where TonB system protein levels are not chromosomal, skewed assay results cannot be excluded.

Phenotypic Assays for the TonB System

Because the TonB system is involved in signal transduction (with the signal being energy), several of the assays for its activity are of necessity indirect and based on the role that TonB plays in the transport of ligands and colicins, and irreversible phage adsorption (Hancock and Braun, 1976). Different assays have different windows of sensitivity (Fig. 1). Spot titers of phage or colicin preparations on lawns of bacteria are used to estimate the level of TonB activity in a given strain. TonB activity is also used as an indirect measure of ExbB and ExbD activity (Held and Postle, 2002). As few as 1 TonB molecule per cell can be detected by vitamin B_{12}-dependent growth in *metE* strains or by sensitivity to bacteriophage $\phi80$. Less sensitive but more discriminating, relative rates of iron transport can distinguish between ~10 and 100% TonB activity. Disk assays for siderophore utilization are unreliable because less TonB can result in larger zones of growth due to diffusion of the siderophore in the plates (Larsen *et al.*, 2003a). Given their variations in sensitivity, it is recommended that an entire panel of assays be applied to each mutant. Because *tonB* strains are iron stressed, they hyperexcrete the siderophore enterochelin (also known as enterobactin). The excess enterochelin can be readily detected on chrome-azurol-S plates (Schwyn and Neilands, 1987), but is also apparent in turning both liquid and solid media slightly pink, the color of Fe-enterochelin. With each of these assays, TonB (or ExbB or ExbD) can be classed as fully active, partially active, or completely inactive; however, little is revealed about the reasons for the inactivity.

Colicin/Phage Sensitivity by Spot Titer Assays

If a plasmid-encoded TonB system protein is being assayed, the first step is to determine the inducer concentration required for chromosomal levels of expression by immunoblot (described further later). TonB system mutants

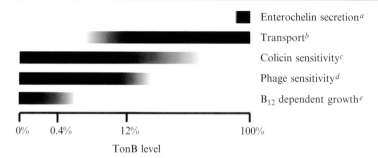

[a]Based on zones of clearing on CAS indicator plates
[b]Includes [^{55}Fe]-siderophore uptake and irreversible ϕ80 adsorption assays
[c]Based on spot titer assays with the TonB-dependent colicins B, D, la, and M
[d]Based on spot titer assays with bacteriophage ϕ80
[e]Based on ability of *metE* strains to grow in the presence of 5 pM vitamin B$_{12}$

FIG. 1. Comparison of TonB phenotypic assays. Cells expressing wild-type TonB protein at 100, 12, or 0.4% of chromosomally encoded levels were evaluated in phenotypic assays. The black region depicts the range of TonB over which a given assay can discriminate. For example, while transport assays can discriminate between ~10 and 100% TonB activity, they cannot tell the difference between something less than 10% activity and zero activity. The transition from black to gray identifies regions where the exact end point is not clear because more intermediate levels of TonB were not tested. The enterochelin hypersecretion assay lacks a gray zone because it is insensitive and unsuitable for use in all but the broadest phenotypic characterizations (e.g., is the TonB system active or not?). Note that the 0.4% level corresponds approximately to 1 TonB per cell under iron-replete conditions. Figure adapted from Postle and Larsen (2004) and reprinted with permission.

are often somewhat unstable and require increased concentrations of inducer relative to the wild type to achieve chromosomal levels of expression. The basic procedure is to subculture a saturated overnight LB culture of the strain to be tested at 1:100 in T-broth with appropriate antibiotics and inducer (arabinose in our experiments). When the culture reaches midexponential phase ($A_{550} = 0.4$ on a Spectronic 20 with a 1.5-cm path length), 3.0 ml tempered molten T-top agar (50–60°) is added to 200 μl cells in a sterile 13-mm glass tube. The tube is vortexed rapidly but gently and the contents quickly poured onto a T-plate equilibrated at room temperature. *Note*: antibiotics and inducer identical to those in the liquid culture have been previously added to both the T-plates and the T-top agar.

Just before the cells are combined with molten agar, an aliquot is precipitated with an equal volume of cold 20% (w/v) trichloroacetic acid (TCA) for later analysis by immunoblot to confirm that the proteins in question are expressed at chromosomal levels. Cells in TCA are incubated on ice for at least 15 min, centrifuged in a microfuge, and the supernatant aspirated.

Pellets from 0.5 ml of cells are suspended in 25 μl 1 M Tris-HCl at pH 7.8 and 25 μl 2× gel sample buffer. We have learned that without TCA precipitation, TonB and ExbD proteins are proteolytically degraded at some point during lysis and boiling in Laemmli gel sample buffer (Laemmli, 1970; Skare and Postle, 1991).

After the T-top agar has solidified undisturbed for at least 5 min, 10-μl aliquots of diluted phage or colicin are spotted onto the plates. If the highest dilution is spotted first, the need to change to a fresh tip after spotting a given dilution can be avoided. Plates are incubated upright at 37° for 18 h and results recorded as the highest dilution at which clearing is evident. These assays are performed in triplicate.

Notes on dilution: Each agent must be diluted beyond its capacity to clear a lawn such that at least one dilution that does not clear the lawn is tested. Bacteriophage ϕ80 is usually diluted 10-fold in sterile λ-Ca^{2+} buffer and, as an alternative, can be fully titered to determine relative plating efficiencies (Larsen *et al.*, 2003a). Colicins are usually diluted 5-fold in λ-Ca^{2+} buffer. It is generally a good idea to make fresh dilutions for each assay; in the case of colicin M, it is imperative due to its instability. Colicin M can be stabilized by the addition of 0.1% Triton X-100. If other diluents are used, such as 1× M9 minimal salts, it is important to remember that colicin M requires Ca^{2+} (Schaller *et al.*, 1981). *tonB* strains should show no evident clearing even with undiluted colicin.

λ-Ca^{2+} buffer: 1.0 ml 1 M Tris-HCl, pH 7.9 (10 mM), 2.0 ml 1 M MgSO$_4$ (20 mM), 1.0 ml 0.5 M CaCl$_2$ (5 mM), 96 ml distilled H$_2$O. *Note*: This can be made from sterile solutions added to sterile bottles or prepared nonsterilely and autoclaved.

Colicin Preparation

Colicins are protein toxins synthesized by some strains of *E. coli* to kill sensitive strains in what could be considered intraniche warfare. TonB-dependent group B colicins are generally large proteins consisting of three domains: a receptor-binding domain where the TGTs serve as receptors, a translocation domain that facilitates movement of the colicin across the OM via the transporter/receptor, and a toxic domain that determines how the unfortunate *E. coli* will be killed. The means of death are many and varied and have been reviewed in Braun *et al.* (2002) and Cao and Klebba (2002).

Colicins are encoded on plasmids and expressed from SOS promoters. Thus colicin expression is induced in late exponential phase ($A_{550} = 1.0$) by the addition of mitomycin C to a final concentration of 1 μg ml^{-1}.

Because mitomycin C is carcinogenic, gloves should be worn and care taken to avoid direct contact with media that contain it. Cells are incubated for 4 h with shaking (or less if lysis occurs), collected by centrifugation, suspended in 1/10 volume of $1\times$ M9 medium, and lysed by two passages through a French press at $\sim$20,000 psi. The lysate is centrifuged to remove unbroken cells and debris (20 min at $\sim$11, 000g) and the supernatant is filtered through 0.45-μm filters to sterilize (procedure adapted from Pugsley and Reeves, 1977). Aliquots can be frozen at $-20°$ until use. Special handling is needed for colicin M, which is unstable when diluted (Schaller $et\ al.$, 1981). Colicin Ia is expressed from pColIa-CA53 (Davies and Reeves, 1975), colicin D is expressed from pColD-CA23 (Davies and Reeves, 1975), colicin M is expressed from pTO4 (Olschlager $et\ al.$, 1984), and colicin B is expressed from pES3 (Pressler $et\ al.$, 1986).

Vitamin B_{12} Nutritional Assay

Unlike iron, vitamin B_{12} (cyanocobalamin) is not a required nutrient. Nonetheless, it fits the profile of TonB-dependent transport ligands by being sufficiently large, scarce, and important that it is actively transported by TGT.

There are two methionine synthases in $E.\ coli$, MetE and MetH. It is advantageous for $E.\ coli$ to use MetH (if it can obtain vitamin B_{12}) because the MetH turnover rate is 100 times faster than that of MetE (Whitfield $et\ al.$, 1970). Because MetH is vitamin B_{12} dependent, a nutritional requirement for vitamin B_{12} becomes apparent in $metE$ strains grown in the absence of methionine. These growth assays, like sensitivity to bacteriophage ϕ80, are capable of detecting very low level of TonB activity (Larsen $et\ al.$, 2003a).

Vitamin B_{12}-dependent $metE$ strains are grown to midexponential phase in M9 minimal medium, supplemented as in the iron transport assays described later, with the exception that the final growth medium contains $\sim$90 μM FeCl$_3$ • 6H$_2$O. The cells are harvested in midexponential phase and washed twice in unsupplemented $1\times$ M9 salts. Two hundred microliters of cells is added to 5 ml melted methionine-free top agar (0.75%) in an M9 base supplemented with defined amino acids (Gerhardt $et\ al.$, 1994). The top agar also contains 0.01% tetrazolium to allow enhanced detection of growth zones. The top agar is allowed to solidify on methionine-free supplemented plates (1.5% agar), and sterile disks containing 5 μl of 100 or 500 nM freshly prepared vitamin B_{12} or, as a control, 27 mM methionine are placed in triplicate on the solidified top agar. Growth zones are measured 18 h postinoculation. The larger the zone around the vitamin B_{12}-containing disks, the more TonB activity. $tonB$ strains cannot grow except around the disk containing methionine.

[55]Fe-Ferrichrome Transport Assay

Overview. Colicin and phage sensitivity assays develop over a 12- to 18-h time span and are thus very sensitive to low levels of TonB system activity. In contrast, assays to determine relative rates of iron-siderophore or vitamin B_{12} transport take place over a short time frame and are thus able to discriminate in the range between $\sim$10 and 100% TonB activity where colicin and phage sensitivity assays are less useful.

Ferrichrome crosses the OM through the TonB-gated transporter known as FhuA. This siderophore produced by the fungus *Ustilago sphaerogena* is, unlike vitamin B_{12}, available commercially (Sigma) and, unlike enterochelin—the siderophore that is native to *E. coli*—is very stable. The ferrichrome transport protocol described here has been adapted (Koster and Braun, 1990) and is also described in Larsen and Postle (2001). The best strains to use for Fe-ferrichrome transport assays might be those mutated in the *aroB* locus to prevent synthesis of enterochelin precursors that could serve as alternative siderophores (Hantke, 1990). If necessary, the *aroB* locus can be transferred into strains by P1 transduction from strain CAG18450 [MG1655 *zhf*-50::Tn*10*] (Higgs *et al.*, 2002b; Singer *et al.*, 1989). The *aroB* strain is grown overnight in 5 ml LB at 37° and subcultured 1:100 into 10 ml 1× M9 medium (see later) with higher levels of iron ($\sim$100 μM). However, *aroB* strains generally grow poorly in minimal medium no matter how much iron is added, and if *aroB*[+] strains are used instead, it does not appear to affect the results significantly. Both enterochelin and ferrichrome require the TonB system for transport into cells. Furthermore, it seems reasonable to assume that the degree of exchange of [55]Fe between enterochelin and ferrichrome taking place during the short time course of the transport experiment is limited. For specific investigations on the ferrichrome transporter FhuA, *aroB* strains should probably be employed.

Another assay provides discrimination among TonB system levels similar to the Fe-ferrichrome transport assay, but is not nearly as straightforward. Bacteriophage ϕ80 adsorbs to *E. coli* in two steps: an initial reversible adsorption that is energy independent, followed by irreversible adsorption that is TonB system dependent (Hancock and Braun, 1976). The basic experiment is to add ϕ80 to a sensitive strain and, at various times afterward, withdraw samples that are vortexed vigorously to remove unadsorbed phage and then centrifuged. Supernatants representing the unadsorbed phage population are then titered and the rate of adsorption determined (Skare and Postle, 1991). Because this assay is significantly more work than transport assays and more difficult to standardize, it has largely fallen into disuse.

^{55}Fe-Ferrichrome Transport Assay Protocol

1. After growth of a 10-ml culture to $A_{550} = 0.5$ (midexponential phase) at 37° in side-arm flasks with vigorous aeration, cells are centrifuged at room temperature and suspended in an equal volume of $1\times$ M9 salts, 0.4% glucose, and 0.1 mM nitrilotriacetate to block the nonspecific low-affinity iron uptake system. The suspended cells are incubated in disposable 50-ml tubes with shaking at 30° for 5 min to adapt to the new lower temperature. At 37° transport rapidly becomes nonlinear.

2. Immediately prior to initiation of transport, a 0.5-ml sample is removed and precipitated with TCA preparative to immunoblot analysis to confirm chromosomal levels of expression.

3. While shaking continues at 30°, transport is initiated by the addition of 150 pmol ^{55}Fe-ferrichrome per milliliter of cells. Each tube is vortexed briefly and triplicate 0.5-ml samples are harvested by filtration onto Whatman G/FC glass microfiber filters in a vacuum manifold.

4. Each sample is rapidly washed three times with 5 ml 0.1 M LiCl, and the filters are placed on blotting paper to dry. This is the 1-min sample. Additional triplicate samples are taken at 4, 7, and 10 (sometimes also 13 and 16) min. After drying, filters are placed in scintillation vials with 3 ml scintillant. Filters are counted in a liquid scintillation counter, and relative rates of transport are determined from the raw counts. Data can be reported as cpm per 1×10^8 cells or per A_{550} ml. Because ^{55}Fe is a low-energy X-ray emitter, it can be counted in a 0-400 window using a Beckman LS6500 multipurpose liquid scintillation counter. In our experiments, there is less variation between triplicate samples if the filters are incubated overnight in scintillant prior to counting.

5. TCA-precipitated samples are resolved on 11% SDS-PAGE and immunoblotted for the protein of interest to confirm that the assayed cultures were all expressing near-chromosomal levels of proteins in question (e.g., TonB mutants). Because ferrichrome transport is an indirect assay of TonB system functioning, mutations in TonB, ExbB, and ExbD can all be assayed.

Growth Medium Recipe. For 100 ml $1\times$ M9 medium (all solutions are sterile):

10 ml $10\times$ M9 salts (Shedlovsky and Brenner, 1963)
1 ml 50% glycerol (if using arabinose to induce gene expression from arabinose promoter on a plasmid) or 2 ml 20% glucose
2 ml 20% casamino acids, preferably lowest iron contamination available
1 ml 4 mg/ml L-tryptophan (missing from casamino acids)

200 μl vitamin B$_1$ at 2 mg/ml
100 μl 1 M MgSO$_4$
100 μl 0.5 M CaCl$_2$
50 μl FeCl$_3$ • 6H$_2$O at 1 mg/ml (1.85 μM)
85 ml sterile dH$_2$O
Antibiotics as needed

Preparation of ^{55}Fe-Ferrichrome

1. Prepare 1 mM iron-free ferrichrome (FW = 687.7) stock using 10 mM HCl as solvent. The 1 mM ferrichrome solution can be frozen at $-20°$ and used repeatedly if stored in a container that protects it from light. (Dissolving the entire contents of the bottle of ferrichrome [Sigma F8014] allows for easy storage and eliminates the need to weigh out tiny amounts.)

2. Dilute ^{55}FeCl$_3$ to 150 μM with 0.5 M HCl. Because the nuclide is provided in small volumes, the expanded volume also reduces losses due to evaporation. Specific activity of the ^{55}FeCl$_3$ as it comes from the supplier can range from 20 to 120 mCi/mg. The half-life of ^{55}Fe is 2.7 years.

3. Add 5 μl 1 mM ferrichrome and 5 μl 150 μM ^{55}FeCl$_3$ to 240 μl 10 mM HCl. The final concentration is 20 μM ferrichrome loaded with ^{55}FeCl$_3$ with a 6.7:1 excess of ferrichrome.

4. Incubate the ^{55}Fe-ferrichrome solution at 37° for 15 min and place on ice for the remainder of the experiment.

Typical results in this assay are that the rate of transport remains linear over the time course, with 10,000 to 20,000 cpm incorporated by 10 min. Strains lacking TonB bind only a few hundred counts per minute, an amount that remains unchanged over the course of the assay.

Mechanistically Informative Assays

Mechanistically informative TonB system assays generally require the use of antisera directed against the component proteins to determine steady-state levels of mutant proteins, localize proteins, analyze assembly, and detect conformational changes. Even the phenotypic assays described earlier require knowledge of steady levels of mutant proteins in order to interpret the results unambiguously. To generate specific antisera, we obtained purified TonB as inclusion bodies of a hybrid TrpC-TonB-G26D protein that carries a signal sequence mutation (Skare *et al.*, 1989) and, as technologies improved, were also able to purify His-tagged versions of ExbB and ExbD. Overexpressed OM transporters were generous gifts from colleagues (Clive Bradbeer and Dick van der Helm). Colicin B was purified during a collaboration to determine its crystal structure (Hilsenbeck *et al.*, 2004). We generated monoclonal antibodies directed against TonB as part of a collaboration (Larsen *et al.*, 1996) and made the remaining antisera ourselves.

To obtain high specificity antisera with low cross-reactivity, we prescreen several rabbits for the lowest possible level of cross-reactivity in their preimmune sera. Selected rabbits are each surgically implanted subcutaneously by staff veterinarians with a golf ball-sized whiffle ball, which serves as a chamber through which antigen can be injected and red blood cell-free immune serum can be retrieved (Hillam *et al.*, 1974; Ried *et al.*, 1990). The proteins used as antigens do not need to be especially pure—the only important criterion is that they be free of contaminants at the same molecular mass as the antigen on the SDS gel from which they will be excised. After confirming that a purified protein band on a one-dimensional gel has no extraneous proteins (using two-dimensional electrophoresis), the band containing the antigen is excised, emulsified, and injected without any adjuvant other than the polyacrylamide gel itself. The band of interest is identified by Coomassie stain, which does not need to be removed. Freund's adjuvant should not be used due to the deleterious effect it has on the animals and the generation of many cross-reactive immunoglobulin species in the resulting antiserum.

In Vivo *Formaldehyde Cross-Linking*

Formaldehyde can cross both OM and CM (it efficiently cross-links GroEL protein in our experiments). The degree of cross-linking can be controlled by time, temperature, and pH. Cross-links are broken by boiling for 5 to 10 min (Prossnitz *et al.*, 1988; Skare *et al.*, 1993). Monomeric formaldehyde inserts a CH_2 group between two reactive residues (K, Y, H, C $\gg$ R, W) and is thus a probe for close protein contacts through diverse side chains (Means and Feeney, 1971). Failure to observe a cross-linked complex is not evidence that protein–protein interaction is not occurring elsewhere between noncross-linkable residues. In addition, detectable cross-linked complexes might not represent the entire set of interactions with other proteins. Protein–protein interactions could occur in which the correct amino acids are not positioned correctly to allow cross-linking or the interactions may be too transient to trap efficiently.

In vivo, TonB can cross-link to OM proteins Lpp, OmpA, and FepA and to CM proteins ExbB and ExbD (Higgs *et al.*, 1998, 2002b; Skare *et al.*, 1993). The OM TonB complexes appear to represent interactions with nonenergized TonB (Ghosh and Postle, 2005). Cross-links between ExbB and TonB assay accurately whether correct transmembrane domain relationships have been maintained (Larsen *et al.*, 1994). In addition, *in vivo* formaldehyde cross-linking detects oligomeric forms of ExbB and ExbD (Higgs *et al.*, 1998, 2002b) and can be used to rule out assembly defects as the cause of mutant phenotypes. Cross-linked complexes are detected by immunoblot and manifest as specific bands with molecular masses that are larger than the monomer.

ExbB/ExbD and especially TonB have a low abundance in the cell. Under conditions of extreme induction (knockout of the Fur transcriptional repressor) the highest levels of TonB achieved are 1350 ± 400 copies per cell (Higgs *et al.*, 2002a). Because formaldehyde cross-linking is inefficient (Skare *et al.*, 1993), it is not currently feasible to purify cross-linked complexes of proteins expressed at chromosomal levels and subject them to mass spectrometric analysis to identify sites of interaction.

For *in vivo* cross-linking, 1 ml of midexponential phase cells ($A_{550} = 0.5$) is harvested by centrifugation for 5 min in a microcentrifuge at room temperature. The cell pellet is suspended in 938 μl 100 mM phosphate buffer, pH 6.8. Tris or any buffer with free amino groups should not be used. These experiments should be carried out in a hood and wearing protective gloves to avoid contact with formaldehyde, a probable carcinogen that is very toxic by inhalation, ingestion, or skin contact. To initiate cross-linking, 62 μl of 16% monomeric *p*-formaldehyde (final concentration = 1.0%) is added to the suspended cell pellet with vortexing and the incubation is continued at room temperature for 15 min in the hood. Monomeric *p*-formaldehyde is obtained in sealed glass ampoules from Electron Microscopy Services (215-646-1566), web site www.emsdiasum.com. It can be stored at room temperature for 1 month in a small brown vial after opening. Treated cells are collected by centrifugation for 5 min at room temperature in a microcentrifuge. Pellets are suspended in 50 μl 1× Laemmli gel sample buffer (Laemmli, 1970) and heated to 60° for 5 min. Control samples may be boiled for 5 to 10 min to break cross-links. Samples may be frozen at $-20°$ but best results seem to arise from immediate analysis by SDS-polyacrylamide gel electrophoresis.

Sucrose Equilibrium Density Gradient Fractionation

There are many protocols available for fractionation of *E. coli* membranes by sucrose density gradient fractionation. The method we use (Osborn *et al.*, 1972) is briefly summarized here with the TonB-specific features noted. Cells are grown to $A_{550} = 0.5$, chilled to 4°, and 14 A_{550} ml are harvested, suspended in 2 ml 10 mM HEPES, pH 7.8, and lysed by two passes through a French press at 20,000 psi. Unbroken cells are removed by low-speed centrifugation (600g for 5–10 min). It is especially important to keep the

g force well below the 13,000g for 15 min that will pellet OM efficiently (Nikaido, 1994). Eight hundred fifty microliters of the lysate is loaded onto the top of a 30–56% sucrose gradient for centrifugation in a MLS-50 swinging bucket rotor in a Beckman Optima Max Ultracentrifuge at 50,000 rpm (238,000g) for 6 h. Although this protocol can be scaled up for larger swinging bucket rotors, such as the SW40 (Letain and Postle, 1997), the tiny MLS-50 swinging bucket rotor is now used because it has a more advantageous clearing factor (k). The decreased centrifugation times help limit proteolysis of TonB system proteins, especially the mutant variants that can be unstable even in whole cells. The gradient is prepared by layering sucrose solutions in 10 mM HEPES, pH 7.8, beginning with a small volume of 56% (w/w) sucrose (0.2 ml) and adding successive layers of 50, 45, 40, 35, and 30% (w/w) sucrose (0.8 ml for each layer). Flotation experiments are performed by adjusting the lysate or a fraction from a previous gradient to 56% sucrose and loading it on a small cushion of 61% sucrose, with layers of 56%, 50%, 45%, and so on sucrose layered on top (Letain and Postle, 1997). Fractions are collected (0.25 ml), and 100 μl is immediately precipitated with an equal volume of 4° 20% TCA for immunoblot analysis with the remainder frozen for subsequent determination of specific gravity and NADH oxidase activity. The TCA-precipitated pellet is suspended in 50 μl gel sample buffer (Laemmli, 1970), and 15 μl is resolved on a 1.0-mm-thick slab gel with 20 wells. TonB is detected in both the CM and the OM (Larsen *et al.*, 2003b; Letain and Postle, 1997). Established markers are used to identify fractions containing the CM and the OM. NADH oxidase activity (Osborn *et al.*, 1972) measured in each fraction identifies the CM fractions typically found at a sucrose density of 1.16–1.18. The distinctive pattern of highly expressed OM proteins OmpF and OmpA in the stained immunoblot between 35 and 40 kDa identifies the OM fractions typically found at a sucrose density of 1.22–1.25 (Letain and Postle, 1997; Osborn *et al.*, 1972). Polyvinylidene fluoride (PVDF) membranes used for the immunoblot analysis are stained following chemiluminescent detection of proteins of interest in 0.1% (w/v) Coomassie brilliant blue R, 50% methanol, 7% glacial acetic acid and destained in the same solution but without the Coomassie blue.

In addition to the location of the OM fractions, the characteristic pattern seen on the stained PVDF membranes also provides information about how well the sucrose density gradient formed. Even though protein is denser than lipid or any biological membrane, most of the soluble proteins will be located in the top fractions of the gradient because of their small size and the viscosity of the sucrose. Large protein complexes such as ribosomes will be found at increasingly higher sucrose densities as the run times increase. Given enough time, they will pellet at the bottom of the gradient. For this reason, initial characterizations of a protein as being membrane associated are best confirmed by flotation in a sucrose density gradient.

Pmf-Responsive Conformational Changes

The amino terminal ~140 amino acids of TonB can be protected from proteinase K added exogenously to intact spheroplasts if two conditions are met: (1) ExbB, ExbD, TonB, and the pmf were all intact at the time the spheroplasts were produced and (2) pmf is subsequently collapsed by protonophores DNP or CCCP (Larsen *et al.*, 1994, 1999). The result is a ~25-kDa proteinase K-resistant fragment that derives from the TonB amino terminus (Fig. 2). In the absence of ExbB/D or if the TonB transmembrane domain carries a mutation that inactivates TonB, such as TonB-ΔVal17, TonB-His20Tyr, or TonB-Ser16Leu, the addition of proteinase K following treatment with the protonophore CCCP results in complete degradation of the TonB. The TonB carboxy terminus is not required for the pmf-responsive conformational change that leads to proteinase K resistance. The proteinase K-resistant fragment is formed by as few as TonB amino acids 1−168 (full-length TonB is 239 amino acids). (Smaller amino-terminal TonB fragments have not been tested.) Formation of the proteinase K-resistant fragment serves as an indicator of overall potential for TonB to be energized or otherwise managed by ExbB/D.

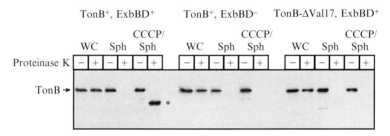

FIG. 2. TonB conformations differ *in vivo* depending on whether pmf is present or absent. The position of full-length TonB is indicated by an arrow. The asterisk indicates the position of the proteinase K-resistant fragment indicative of prior TonB energization. WC, whole cells; Sph, spheroplasts; CCCP/Sph, spheroplasts pretreated with protonophore CCCP to collapse the pmf prior to proteinase K treatment. Strain backgrounds in which experiments were performed are indicated above each set. The proteinase K-resistant fragment does not form when TonB carries a mutant transmembrane domain (ΔVal17) or if ExbB/D is absent. Figure adapted from Larsen *et al.* (1999) and reprinted with permission.

TABLE I
SOLUTIONS FOR SPHEROPLAST FORMATION[a]

Buffer[a]	H_2O	1 M TrisAc, pH 8.2	2 M sucrose	1 M MgSO$_4$	500 mM EDTA
#1	8.0	0.5	1.25	0.2	—
#2	5.5	2.0	2.5	—	0.01
#3	8.0	2.0	—	—	—
#4	6.6	2.0	1.25	0.2	—
#5	9.5	0.5	—	—	0.05

[a] Buffers are made in 10-ml aliquots using distilled deionized H_2O. Volumes are in milliliters.

Escherichia coli can be converted to spheroplasts when the OM is loosened, the peptidoglycan largely degraded exogenously by added lysozyme, and the intact cytoplasmic membrane stabilized by osmoprotectants according to several protocols with small variations (Randall and Hardy, 1986; Witholt *et al.*, 1976). We follow the Randall and Hardy (1986) procedure with minor modifications. Recipes for spheroplasting buffers are shown in Table I.

Bacteria are grown in M9 minimal salts with casamino acids medium described earlier to $A_{550} = 0.5$ and six 1-ml aliquots are pelleted in microcentrifuge tubes. If strains are lacking ExbB/D or are *aroB* it may be necessary to increase the amount of Fe added to $\sim$90 μM. While cells are growing, the lysozyme solution (2 mg/ml in distilled H_2O) and proteinase K solution can be made. (For proteinase K, start with 2 mg/ml—the precise amount may need to be titrated in the assay as too much will also degrade the "proteinase K-resistant fragment.") All solutions should be well chilled before beginning the experiment and all cell samples should remain on ice throughout.

1. Tubes 1 + 2: suspend in 500 μl buffer 1, place on ice. These are *whole cells*.
2. Tubes 3−6: suspend in 250 μl buffer 2 by gently pipetting the buffer over the pellet repeatedly. This is key.
3. Tubes 3−6: add 20 μl freshly prepared lysozyme at 2 mg/ml in dH_2O.
4. Tubes 3−6: immediately add 250 μl buffer 3.
5. Tubes 3−6: incubate on ice 5 min with tops off for subsequent addition of Mg^{2+}.
6. Tubes 3−6: add 10 μl 1.0 M MgSO$_4$ to stabilize the spheroplasts. Mix gently.

7. Tubes 3–6: spin 5 min in microcentrifuge.

8. Tubes 3–6: aspirate supernatants but leave just a bit in the tube.

9. Tubes 3 + 4: suspend in 0.5 ml buffer 4 by gently pipetting just above the pellet. This is key. Tubes 3 + 4 are *intact spheroplasts*.

10. Tubes 5 + 6: suspend in 500 μl buffer 5 and vortex very well. Add 10 μl 1 M MgSO$_4$. These are *lysed spheroplasts*.

11. Tubes 2, 4, and 6: add 6.3 μl of proteinase K solution at 2.0 mg/ml (final concentration 25 μg/ml). Mix by pipetting up and down; when all three are done, mix gently and tap to get *all* liquids to the bottom of the tube. Incubate on ice for 15 min.

12. All tubes: add 5 μl freshly prepared 100 mM phenylmethanesulfonyl fluoride to inactivate the proteinase K.

13. All tubes: add 500 μl chilled 20% TCA and analyze samples by immunoblot.

As an alternative to the lysed spheroplasts, at step 10, tubes 5 and 6 are suspended in 500 μl buffer 4 and carbonyl cyanide *m*-chlorophenylhydrazone (CCCP; 50 μM) or dinitrophenol (DNP; 10 mM) is added. Control experiments require two additional tubes of unlysed spheroplasts with solvent for each protonophore added (ethanol or dimethyl sulfoxide, respectively). Because lysis of spheroplasts can also collapse the CM proton gradient (Larsen *et al.*, 1994), it is essential to monitor the integrity of spheroplasts treated with protonophores. For this control, cells are grown to early exponential phase and treated with 1 mg/ml isopropyl-thio-β-D-galactoside to induce expression of the cytoplasmic enzyme β-galactosidase prior to conversion to spheroplasts at $A_{550} = 0.5$. Enzyme levels (Miller, 1972) are determined for pelleted spheroplasts that were treated with protonophores and the corresponding supernatants to confirm that 85% or more of the β-galactosidase remains in the pellet (Larsen *et al.*, 1999).

Dividing the Energy Transduction Cycle into Two Halves

Of the several known siderophores that *E. coli* K12 can use to obtain iron, it synthesizes only enterochelin and its intermediates. TonB does not transduce energy to nonligand-bound transporters (Larsen *et al.*, 1994, 1999). Therefore, in an *aroB* strain, the progression of TonB from an energy-transducing conformation to a discharged conformation is prevented (Larsen *et al.*, 1999). Because of this, the energy transduction cycle can be functionally divided into two stages by use of *aroB* strains. For example, if a conformational change is prevented in an *aroB* strain, then it likely occurs following TonB transduction of energy to OM transporters.

Ability of TonB Cys Substitutions to Form Dimers

The use of Cys substitutions to monitor *in vivo* conformational dynamics and protein–protein relationships has been very powerful in *E. coli* (Falke and Hazelbauer, 2001; Kaback and Wu, 1997). Because most of the TonB system interactions take place in the oxidizing environment of the periplasm, the spontaneous formation of disulfide-linked dimers is a powerful tool for dissection of *in vivo* interactions. Cys substitutions at TonB carboxy-terminal aromatic amino acids F202C, W213C, Y215C, or F230C spontaneously form disulfide-linked dimers in the CM *in vivo* (Ghosh and Postle, 2005). These dimers do not form if ExbB/D are absent or if the TonB carries a transmembrane domain mutation in addition to the Cys substitution. Lack of dimer formation could result from failure of an energy-transducing conformation to form or from folding of the TonB carboxy terminus so rapidly that disulfide-linked dimers cannot be trapped. Current data suggest that the latter is most likely the case (Larsen *et al.*, 2007).

In this protocol, the TonB F202C dimers (for example) are expressed from a plasmid-encoded arabinose promoter at chromosomal levels as determined previously. Midexponential-phase cultures are harvested by centrifugation at 4° and lysed by boiling for 5–10 min in the presence of 50 mM iodoacetamide to ensure that disulfide cross-links detected occurred prior to lysis (Cadieux and Kadner, 1999; Ghosh and Postle, 2005). Samples are electrophoresed on nonreducing SDS-polyacrylamide gels, and TonB dimers are detected by immunoblotting. The dimers appear to exist in three different conformations that appear as bands with three different apparent molecular masses (Fig. 3).

It is unfortunately not possible to directly assay the effect of protonophore DNP or CCCP on this process because the (1) disulfide cross-linked dimers preexist in cells and (2) newly synthesized TonB requires pmf for export across the cytoplasmic membrane (Skare *et al.*, 1989). This precludes experiments where TonB expression is induced in the presence of protonophores.

Potentially Mechanistically Informative Assays

The protocols described previously have been adapted and developed using chromosomal levels of all the TonB system components. Some interesting and potentially very useful *in vivo* assays have been developed where the meaning of results obtained is less clear either because some components are overexpressed or because additional experimental results have raised questions about older experiments. These assays are described briefly so that future readers are made alert to the possibility of eventual refinements that will be more definitively interpretable.

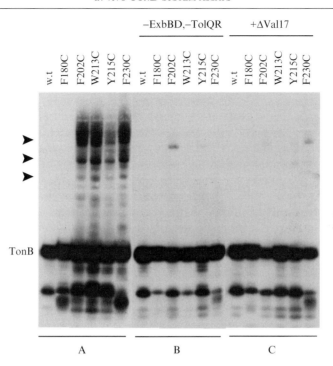

Fig. 3. *In vivo* formation of disulfide-linked dimers requires pmf mediators ExbB and ExbD and the TonB transmembrane domain. Plasmids encoding TonB and TonB-Cys substitutions were expressed at chromosomal levels in A [KP1344 (Δ*tonB::blaM*)] or B [KP1440 (Δ*tonB::blaM*, *exbB*::Tn*10*, *tolQ*(am))]. (C) The plasmid-encoded TonB parent and TonB-Cys substitutions also carry a deletion of Val17 in the transmembrane domain that prevents TonB activity. They are expressed at chromosomal levels in KP1344. An immunoblot of strains resolved on a nonreducing 11% SDS-polyacrylamide gel is shown. Arrows point to positions of three disulfide-linked dimers formed by TonB with Cys substitutions at F202, W213, Y215, and F230. Figure adapted from Ghosh and Postle (2005) and reprinted with permission.

Interaction between TonB and OM Transporter BtuB In Vivo

TonB-gated transporters share a reasonably high degree of similarity. Most notably, at the amino terminus each has a pentapeptide with the consensus D/E T X X V, known as the TonB box. Certain proline substitutions in this region inactivate the transporters completely (Barnard *et al.*, 2001; Heller *et al.*, 1988; Schoffler and Braun, 1989). Second site suppressors of TonB box mutations occur at TonB-Gln160. Direct contact between TonB-Q160 and the TonB box of transporters has been demonstrated *in vivo* (Cadieux and Kadner, 1999; Ogierman and Braun, 2003). However, in those experiments, both the OM transporter and the three known cytoplasmic proteins (TonB/ExbB/ExbD) were expressed from plasmids where the ratios of the proteins to one another were unknown and where they were each almost certainly overexpressed relative to any unknown proteins in the system. Because TonB does not transduce energy to nonligand-bound transporters *in vivo* (Larsen *et al.*, 1999), results showing that the presence or absence of transport ligand has little effect on disulfide formation are consistent with the idea that overexpression may not accurately reflect normal interactions (Ogierman and Braun, 2003). In another study, TonB interactions with mutant TonB boxes were increased in the presence of ligand, but contrary to expectation, contact with the mutant inactive TonB box still occurred at high levels, although the residues contacted were different (Cadieux and Kadner, 1999). Although not the expected result, it may well correctly reflect what occurs *in vivo*. The problem is that since the system is unbalanced, it is not possible to know with certainty.

Substrate-Induced Signaling by OM Transporters

Because TonB does not transduce energy to nonligand-bound OM transporters, there must be a signal to indicate ligand occupancy. Ligand-enhanced disulfide cross-link formation between TonB and the TonB box suggested that, in the case of the BtuB transporter, the TonB box is involved (Cadieux and Kadner, 1999). It was subsequently demonstrated *in vivo* that Cys substitutions at TonB box residues showed up to a ~6-fold increase in labeling by the thiol-specific reagent 1-biotinamido-4-[4′(maleimidomethyl)cyclohexane-carboxamido]butane in the presence of the ligand compared to its absence. The increase in labeling reflects an increase in periplasmic accessibility for the TonB box; however, the presence and absence of TonB had no effect on extent of labeling (Cadieux *et al.*, 2003). These experiments were carried out in the presence of what was almost certainly a *tonB* phenocopy due to highly overexpressed BtuB (on high copy number pUC8 backbone). Thus, while there may be no difference in TonB box accessibility whether or not TonB is present, this question could not be answered with this experimental system.

In Vivo Shuttling between CM and OM

As noted earlier, TonB fractionates with both the CM and the OM. This behavior suggested that TonB might shuttle to the OM to transduce stored potential energy to OM transporters (Letain and Postle, 1997). An experimental system was set up where the levels of all TonB system components were at chromosomal levels and the accessibility of the TonB amino terminus to the periplasm could be determined *in vivo* by whether and under what circumstances it could be labeled with Oregon Green Maleimide (OGM), which is marketed and has been used successfully as a membrane-impermeant reagent. TonB that was capable of associating with the OM was labeled by OGM, whereas TonB that could not associate (due to deletion of the carboxy terminus) remained unlabeled. Both full-length and truncated TonB were situated apparently identically in the CM complex of ExbB/D, as each could cross-link to ExbB and each was conformationally responsive to pmf (assays described earlier). It was concluded that TonB shuttles between CM and OM during energy transduction.

There is potentially a problem with this interpretation. Unpublished results indicate that both cytoplasmically localized TonB lacking its signal sequence and the cytoplasmic protein GroEL can be labeled with OGM, and therefore OGM is not absolutely membrane impermeant in this experimental system. Thus, even *in vivo* experiments where all the TonB system components are balanced can be subject to potential difficulties in interpretation. The explanation for the original results in light of these new data appears to be that the ExbB/D complex protects the extreme amino terminus of truncated TonB from labeling via the cytoplasmic route (Savenkova and Postle, in preparation).

Acknowledgments

I am very grateful to colleagues Volkmar Braun, Clive Bradbeer, Robert Kadner, Joe Neilands, Tony Pugsley, and Peter Reeves for their pioneering work on the TonB system phenotypic assays, which we did not invent, only adapt. Past members of the laboratory, Penelope Higgs, Joydeep Ghosh, Ray Larsen, Tracy Letain, and Jonathan Skare, deserve special recognition for their development of the more mechanistically informative *in vivo* TonB system assays. I am also grateful to the National Institutes of General Medical Sciences and the National Science Foundation for their support over many years. Finally, I thank Ray Larsen, Marta Manning, Anne Ollis, Kyle Kastead, and Charles Bulathsinghala for critical reading of the manuscript.

References

Barnard, T. J., Watson, M. E., Jr., and McIntosh, M. A. (2001). Mutations in the *Escherichia coli* receptor FepA reveal residues involved in ligand binding and transport. *Mol. Microbiol.* **41,** 527–536.

Bradbeer, C. (1993). The proton motive force drives the outer membrane transport of cobalamin in *Escherichia coli. J. Bacteriol.* **175,** 3146–3150.

Braun, V. (1989). The structurally related *exbB* and *tolQ* genes are interchangeable in conferring *tonB*-dependent colicin, bacteriophage, and albomycin sensitivity. *J. Bacteriol.* **171,** 6387–6390.

Braun, V. (2003). Iron uptake by *Escherichia coli. Front. Biosci.* **8,** s1409–s1421.

Braun, V., Patzer, S. I., and Hantke, K. (2002). Ton-dependent colicins and microcins: Modular design and evolution. *Biochimie* **84,** 365–380.

Buck, M., and Griffiths, E. (1982). Iron mediated methylthiolation of tRNA as a regulator of operon expression in *Escherichia coli. Nucleic Acids Res.* **10**, 2609–2624.

Cadieux, N., and Kadner, R. J. (1999). Site-directed disulfide bonding reveals an interaction site between energy-coupling protein TonB and BtuB, the outer membrane cobalamin transporter. *Proc. Natl. Acad. Sci. USA* **96**, 10673–10678.

Cadieux, N., Phan, P. G., Cafiso, D. S., and Kadner, R. J. (2003). Differential substrate-induced signaling through the TonB-dependent transporter BtuB. *Proc. Natl. Acad. Sci. USA* **100**, 10688–10693.

Cao, Z., and Klebba, P. E. (2002). Mechanisms of colicin binding and transport through outer membrane porins. *Biochimie* **84**, 399–412.

Chang, C., Mooser, A., Pluckthun, A., and Wlodawer, A. (2001). Crystal structure of the dimeric C-terminal domain of TonB reveals a novel fold. *J. Biol. Chem.* **276**, 27535–27540.

Datsenko, K., and Wanner, B. (2000). One-step inactivation of chromosomal genes in *Escherichia coli* K-12 using PCR products. *Proc. Natl. Acad. Sci. USA* **97**, 6640–6645.

Davies, J. K., and Reeves, P. (1975). Genetics of resistance to colicins in *Escherichia coli* K-12: Cross-resistance among colicins of group A. *J. Bacteriol.* **123**, 102–117.

Falke, J. J., and Hazelbauer, G. L. (2001). Transmembrane signaling in bacterial chemoreceptors. *Trends Biochem. Sci.* **26**, 257–265.

Faraldo-Gomez, J. D., and Sansom, M. S. (2003). Acquisition of siderophores in gram-negative bacteria. *Nat. Rev. Mol. Cell. Biol.* **4**, 105–116.

Fischer, E., Günter, K., and Braun, V. (1989). Involvement of ExbB and TonB in transport across the outer membrane of *Escherichia coli*: Phenotypic complementation of *exb* mutants by overexpressed *tonB* and physical stabilization of TonB by ExbB. *J. Bacteriol.* **171**, 5127–5134.

Gerhardt, P., Murray, R. G. E., Wood, W. A., and Krieg, N. R. (1994). "Methods For General and Molecular Bacteriology." American Society for Microbiology Press, Washington, DC.

Ghosh, J., and Postle, K. (2005). Disulphide trapping of an *in vivo* energy-dependent conformation of *Escherichia coli* TonB protein. *Mol. Microbiol.* **55**, 276–288.

Guzman, L. M., Belin, D., Carson, M. J., and Beckwith, J. (1995). Tight regulation, modulation, and high-level expression by vectors containing the arabinose P-BAD promoter. *J. Bacteriol.* **177**, 4121–4130.

Hancock, R. W., and Braun, V. (1976). Nature of the energy requirement for the irreversible adsorption of bacteriophages T1 and ⌀80 to *Escherichia coli. J. Bacteriol.* **125**, 409–415.

Hantke, K. (1990). Dihydroxybenzoylserine: A siderophore for *E. coli. FEMS Microbiol. Lett.* **55**, 5–8.

Hantke, K., and Braun, V. (1978). Functional interaction of the *tonA/tonB* receptor system in *Escherichia coli. J. Bacteriol.* **135**, 190–197.

Held, K. G., and Postle, K. (2002). ExbB and ExbD do not function independently in TonB-dependent energy transduction. *J. Bacteriol.* **184**, 5170–5173.

Heller, K. J., Kadner, R. J., and Günter, K. (1988). Suppression of the *btuB*451 mutation by mutations in the *tonB* gene suggests a direct interaction between TonB and TonB-dependent receptor proteins in the outer membrane of *Escherichia coli. Gene* **64**, 147–153.

Higgs, P. I., Larsen, R. A., and Postle, K. (2002a). Quantitation of known components of the *Escherichia coli* TonB-dependent energy transduction system: TonB, ExbB, ExbD, and FepA. *Mol. Microbiol.* **44**, 271–281.

Higgs, P. I., Letain, T. E., Merriam, K. K., Burke, N. S., Park, H., Kang, C., and Postle, K. (2002b). TonB interacts with nonreceptor proteins in the outer membrane of *Escherichia coli. J. Bacteriol.* **184**, 1640–1648.

Higgs, P. I., Myers, P. S., and Postle, K. (1998). Interactions in the TonB-dependent energy transduction complex: ExbB and ExbD form homomultimers. *J. Bacteriol.* **180**, 6031–6038.

Hillam, R. P., Tengerdy, R. P., and Brown, G. L. (1974). Local antibody production against the murine toxin of *Yersinia pestis* in a golf ball-induced granuloma. *Infect. Immun.* **10,** 458–463.

Hilsenbeck, J. L., Park, H., Chen, G., Youn, B., Postle, K., and Kang, C. (2004). Crystal structure of the cytotoxic bacterial protein colicin B at 2.5 A resolution. *Mol. Microbiol.* **51,** 711–720.

Howard, S. P., Herrmann, C., Stratilo, C. W., and Braun, V. (2001). *In vivo* synthesis of the periplasmic domain of TonB inhibits transport through the FecA and FhuA iron side-rophore transporters of *Escherichia coli. J. Bacteriol.* **183,** 5885–5895.

Jaskula, J. C., Letain, T. E., Roof, S. K., Skare, J. T., and Postle, K. (1994). Role of the TonB amino terminus in energy transduction between membranes. *J. Bacteriol.* **176,** 2326–2338.

Kaback, H. R., and Wu, J. (1997). From membrane to molecule to the third amino acid from the left with a membrane transport protein. *Quart. Rev. Biophys.* **30,** 333–364.

Kadner, R. J., and McElhaney, G. (1978). Outer membrane-dependent transport systems in *Escherichia coli*: Turnover of TonB function. *J. Bacteriol.* **134,** 1020–1029.

Kodding, J., Killig, F., Polzer, P., Howard, S. P., Diederichs, K., and Welte, W. (2005). Crystal structure of a 92-residue C-terminal fragment of TonB from *Escherichia coli* reveals significant conformational changes compared to structures of smaller TonB fragments. *J. Biol. Chem.* **280,** 3022–3028.

Koster, W., and Braun, V. (1990). Iron(III) hydroxamate transport of *Escherichia coli*: Restoration of iron supply by coexpression of the N- and C-terminal halves of the cytoplasmic membrane protein FhuB cloned on separate plasmids. *Mol. Gen. Genet.* **223,** 379–384.

Laemmli, U. K. (1970). Cleavage of structural proteins during the assembly of the head of bacteriophage T4. *Nature* **227,** 680–685.

Larsen, R. A., Chen, G. J., and Postle, K. (2003a). Performance of standard phenotypic assays for TonB activity, as evaluated by varying the level of functional, wild-type TonB. *J. Bacteriol.* **185,** 4699–4706.

Larsen, R. A., Deckert, G. E., Kastread, K. A., Devanathan, S., Keller, K. L., and Postle, K. (2007). His_{20} provides the sole functionally-significant side chain in the essential TonB transmembrane domain. *J. Bacteriol.* In press.

Larsen, R. A., Letain, T. E., and Postle, K. (2003b). *In vivo* evidence of TonB shuttling between the cytoplasmic and outer membrane in *Escherichia coli. Mol. Microbiol.* **49,** 211–218.

Larsen, R. A., Myers, P. S., Skare, J. T., Seachord, C. L., Darveau, R. P., and Postle, K. (1996). Identification of TonB homologs in the family *Enterobacteriaceae* and evidence for conservation of TonB-dependent energy transduction complexes. *J. Bacteriol.* **178,** 1363–1373.

Larsen, R. A., and Postle, K. (2001). Conserved residues Ser(16) and His(20) and their relative positioning are essential for TonB activity, cross-linking of TonB with ExbB, and the ability of TonB to respond to proton motive force. *J. Biol. Chem.* **276,** 8111–8117.

Larsen, R. A., Thomas, M. G., and Postle, K. (1999). Protonmotive force, ExbB and ligand-bound FepA drive conformational changes in TonB. *Mol. Microbiol.* **31,** 1809–1824.

Larsen, R. A., Thomas, M. T., Wood, G. E., and Postle, K. (1994). Partial suppression of an *Escherichia coli* TonB transmembrane domain mutation (ΔV17) by a missense mutation in ExbB. *Mol. Microbiol.* **13,** 627–640.

Larsen, R. A., Wood, G. E., and Postle, K. (1993). The conserved proline-rich motif is not essential for energy transduction by *Escherichia coli* TonB protein. *Mol. Microbiol.* **10,** 943–953.

Letain, T. E., and Postle, K. (1997). TonB protein appears to transduce energy by shuttling between the cytoplasmic membrane and the outer membrane in Gram-negative bacteria. *Mol. Microbiol.* **24,** 271–283.

Ma, L., Kaserer, W. A., Annamalai, R., Scott, D. C., Jin, B., Jiang, X., Xiao, Q., Maymani, H., Massia, L. M., Ferreira, L. C., Newton, S. M., and Klebba, P. E. (2007). Evidence of ball-and-chain transport of ferric enterobactin through FepA. *J. Biol. Chem.* **282,** 397–406.

Mann, B. J., Holroyd, C. D., Bradbeer, C., and Kadner, R. J. (1986). Reduced activity of TonB-dependent functions in strains of *Escherichia coli. FEMS Lett.* **33,** 255–260.

McCray, J. W., Jr., and Herrmann, K. M. (1976). Derepression of certain aromatic amino acid biosynthetic enzymes of *Escherichia coli* K-12 by growth in Fe3+-deficient medium. *J. Bacteriol.* **125,** 608–615.

Means, G. E., and Feeney, R. E. (1971). "Chemical Modification of Proteins." Holden-Day, San Francisco, CA.

Miller, J. H. (1972). "Experiments in Molecular Genetics." Cold Spring Harbor, NY: Cold Spring Harbor Laboratory Press, Cold Spring Harbor, NY.

Nikaido, H. (1994). Isolation of outer membranes. *Methods Enzymol.* **235,** 225–234.

Nikaido, H., and Rosenberg, E. Y. (1990). Cir and Fiu proteins in the outer membrane of *Escherichia coli* catalyze transport on monomeric catechols: Study with β-lactam antibiotics containing catechol and analogous groups. *J. Bacteriol.* **172,** 1361–1367.

Ogierman, M., and Braun, V. (2003). Interactions between the outer membrane ferric citrate transporter FecA and TonB: Studies of the FecA TonB box. *J. Bacteriol.* **185,** 1870–1885.

Olschlager, T., Schramm, E., and Braun, V. (1984). Cloning and expression of the activity and immunity genes of colicins B and M on ColBM plasmids. *Mol. Gen. Genet.* **196,** 482–487.

Osborn, M. J., Gander, J. E., Parisi, E., and Carson, J. (1972). Mechanism of assembly of the outer membrane of *Salmonella typhimurium. J. Biol. Chem.* **247,** 3962–3972.

Pawelek, P. D., Croteau, N., Ng-Thow-Hing, C., Khursigara, C. M., Moiseeva, N., Allaire, M., and Coulton, J. W. (2006). Structure of TonB in complex with FhuA, *E. coli* outer membrane receptor. *Science* **312,** 1399–1402.

Postle, K., and Kadner, R. J. (2003). Touch and go: Tying TonB to transport. *Mol. Microbiol.* **49,** 869–882.

Postle, K, and Larsen, R (2004). The TonB, ExbB, and ExbD Proteins. *In* "Iron Transport in Bacteria" (J. H. Crosa, A. R. Mey, and S. M. Payne, eds.), pp. 96–112. ASM Press, Washington, DC.

Postle, K., and Larsen, R. A. (2007). TonB-dependent energy transduction between outer and cytoplasmic membranes. *Biometals.*

Pressler, U., Braun, V., Wittmann-Liebold, B., and Benz, R. (1986). Structural and functional properties of colicin B. *J. Biol. Chem.* **261,** 2654–2659.

Prossnitz, E., Nikaido, K., Ulbrich, S. J., and Ames, G. F.-L. (1988). Formaldehyde and photoactivatable cross-linking of the periplasmic binding protein to a membrane component of the histidine transport system of *Salmonella typhimurium. J. Biol. Chem.* **263,** 17917–17920.

Pugsley, A. P., and Reeves, P. (1977). Comparison of colicins B-K260 and D-CA23: Purification and characterization of the colicins and examination of colicin immunity in producing strains. *Antimicrob. Agents Chemother.* **11,** 345–358.

Randall, L. L., and Hardy, S. J. (1986). Correlation of competence for export with lack of tertiary structure of the mature species: A study *in vivo* of maltose-binding protein in *E. coli. Cell* **46,** 921–928.

Ried, J. L., Walker-Simmons, M. K., Everard, J. D., and Diani, J. (1990). Production of polyclonal antibodies in rabbits is simplified using perforated plastic golf balls. *Biotechniques* **12,** 661–666.

Schaller, K., Dreher, R., and Braun, V. (1981). Structural and functional properties of colicin M. *J. Bacteriol.* **146,** 54–63.

Schoffler, H., and Braun, V. (1989). Transport across the outer membrane of *Escherichia coli* K12 via the FhuA receptor is regulated by the TonB protein of the cytoplasmic membrane. *Mol. Gen. Genet.* **217,** 378–383.

Schwyn, B., and Neilands, J. B. (1987). Universal chemical assay for the detection and determination of siderophores. *Anal. Biochem.* **160,** 47–56.

Sean Peacock, R., Weljie, A. M., Peter Howard, S., Price, F. D., and Vogel, H. J. (2005). The solution structure of the C-terminal domain of TonB and interaction studies with TonB box peptides. *J. Mol. Biol.* **345,** 1185–1197.

Shedlovsky, A., and Brenner, S. (1963). A chemical basis for the host-induced modification of T-even bacteriophages. *Proc. Natl. Acad. Sci. USA* **50,** 300–305.

Shultis, D. D., Purdy, M. D., Banchs, C. N., and Wiener, M. C. (2006). Outer membrane active transport: Structure of the BtuB: TonB complex. *Science* **312,** 1396–1399.

Singer, M., Baker, T. A., Schnitzler, G., Deischel, S. M., Goel, M., Dove, W., Jaacks, K. J., Grossman, A. D., Erickson, J. W., and Gross, C. A. (1989). A collection of strains containing genetically linked alternating antibiotic resistance elements for genetic mapping of *Escherichia coli. Microbiol. Rev.* **53,** 1–24.

Skare, J. T., Ahmer, B. M. M., Seachord, C. L., Darveau, R. P., and Postle, K. (1993). Energy transduction between membranes: TonB, a cytoplasmic membrane protein, can be chemically cross-linked *in vivo* to the outer membrane receptor FepA. *J. Biol. Chem.* **268,** 16302–16308.

Skare, J. T., and Postle, K. (1991). Evidence for a TonB-dependent energy transduction complex in *Escherichia coli. Mol. Microbiol.* **5,** 2883–2890.

Skare, J. T., Roof, S. K., and Postle, K. (1989). A mutation in the amino terminus of a hybrid TrpC-TonB protein relieves overproduction lethality and results in cytoplasmic accumulation. *J. Bacteriol.* **171,** 4442–4447.

Wang, C. C., and Newton, A. (1969a). Iron transport in *Escherichia coli*: Relationship between chromium sensitivity and high iron requirement in mutants of *Escherichia coli. J. Bacteriol.* **98,** 1135–1141.

Wang, C. C., and Newton, A. (1969b). Iron transport in *Escherichia coli*: Roles of energy-dependent uptake and 2,3-dihydroxybenzoylserine. *J. Bacteriol.* **98,** 1142–1150.

Whitfield, C. D., Steers, E. J., Jr., and Weisbach, H. (1970). Purification and properties of 5-methyltetrahydropteroyltriglutamate-homocysteine transmethylase. *J. Biol. Chem.* **245,** 390–401.

Wiener, M. C. (2005). TonB-dependent outer membrane transport: Going for Baroque? *Curr. Opin. Struct. Biol.* **15,** 394–400.

Witholt, B., Boekhout, M., Brock, M., Kingma, J., Heerikhuizen, H. V., and Leij, L. D. (1976). An efficient and reproducible procedure for the formation of spheroplasts from variously grown *Escherichia coli. Anal. Biochem.* **74,** 160–170.

[13] Biochemical Characterization of Plant Ethylene Receptors Following Transgenic Expression in Yeast

By G. ERIC SCHALLER and BRAD M. BINDER

Abstract

The ethylene receptors of plants are related to and originated from bacterial histidine kinases. As such they represent a system by which one can study not only how the ethylene signal is perceived and its signal transduced, but also how bacterial two-component systems have been adapted for signal transduction in a eukaryote. Much of the biochemical characterization of the ethylene receptors, including the demonstration of kinase activity, ethylene binding, and interaction with other signaling components, has relied on the ability of the receptors to be functionally expressed in transgenic yeast. This chapter describes some of the key approaches used for such work, with a special emphasis on techniques employed to analyze ethylene binding. In many cases the approaches used in transgenic yeast may also be used for studies of the receptors in the native plant.

Introduction

The simple gas ethylene serves as a diffusible hormone in plants (Abeles *et al.*, 1992; Mattoo and Suttle, 1991). Ethylene regulates seed germination, seedling growth, leaf and petal abscission, organ senescence, fruit ripening, and responses to stress and pathogens. Ethylene is perceived by a receptor family (Fig. 1A), which in the model plant Arabidopsis consists of five members, ETR1, ERS1, ETR2, ERS2, and EIN4 (Bleecker, 1999; O'Malley *et al.*, 2005; Schaller and Kieber, 2002). The ethylene receptors have similar overall structures with transmembrane domains near their N termini and histidine kinase-like domains in their C-terminal halves, but can be divided into two subfamilies based on phylogenetic analysis and some shared structural features (Bleecker, 1999; Chang and Stadler, 2001; Schaller and Kieber, 2002). All five receptor members contain three highly conserved transmembrane domains that incorporate the ethylene-binding site and a GAF domain of unknown function in their N-terminal halves (Hall *et al.*, 1999; O'Malley *et al.*, 2005; Rodriguez *et al.*, 1999; Schaller and Bleecker, 1995). The subfamily 1 receptors ETR1 and ERS1 have a highly conserved histidine kinase domain containing all the required

METHODS IN ENZYMOLOGY, VOL. 422   0076-6879/07 $35.00
     DOI: 10.1016/S0076-6879(06)22013-5

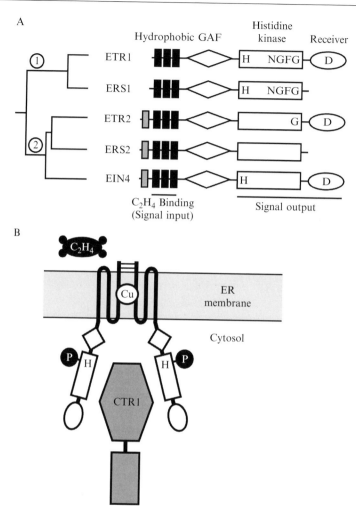

FIG. 1. Characteristics of Arabidopsis ethylene receptors. (A) Primary structure of the five-member ethylene-receptor family. Black bars represent transmembrane segments. Gray bars represent putative signal sequences. Diamonds indicate GAF domains. Rectangles indicate histidine kinase domains. Ovals indicate receiver domains. The conserved phosphorylation sites on histidine (H) and aspartate (D) are indicated if present. Conserved motifs (N, G1, F, G2) within the histidine kinase domain are indicated if present. There are two subfamilies of ethylene receptors (subfamily 1 and 2) based on sequence and phylogenetic analysis. Positions of the ethylene binding (input domain) and output domain are based on biochemical and genetic analysis. (B) Biochemical model for the ethylene receptor ETR1. The receptor is a homodimer with disulfide linkages near the N termini, contains a copper cofactor for ethylene binding within the transmembrane domains, and has been shown to autophosphorylate on the conserved histidine within the kinase domain. The histidine kinase and receiver domains of ETR1 interact with the Raf-like kinase CTR1 (shown in gray).

motifs essential for kinase functionality, with histidine kinase activity for both having been demonstrated *in vitro* (Gamble *et al.*, 1998; Moussatche and Klee, 2004). The subfamily 2 receptors ETR2, ERS2, and EIN4 lack residues considered essential for histidine kinase activity and have instead been proposed to act as serine/threonine kinases (Moussatche and Klee, 2004). ERS1 may act as a bifunctional kinase with both histidine and serine/threonine kinase activity (Moussatche and Klee, 2004). Some of the ethylene receptors (ETR1, ETR2, and EIN4) possess a receiver domain in addition to a histidine kinase-like domain.

Although the ethylene receptors share a similar modular structure to many bacterial histidine kinases, incorporating an input domain for ligand binding and an output domain with kinase activity, the extent to which histidine kinase activity (and a His-Asp phosphorelay) plays a role in signaling by the receptors is still a matter of contention (Mason and Schaller, 2005). To date histidine kinase activity has only been implicated in modulating the signaling pathway, rather than being instrumental in the pathway (Binder *et al.*, 2004; Qu and Schaller, 2004; Wang *et al.*, 2003). Instead the key signaling element acting immediately downstream of the receptors appears to be the Raf-like kinase CTR1 (Kieber *et al.*, 1993), which physically interacts with the receptors to form a signaling complex (Clark *et al.*, 1998; Gao *et al.*, 2003; Huang *et al.*, 2003) (Fig. 1B). Additional research, however, is needed to fully resolve the precise role that histidine kinase activity plays in signal transduction by the receptors.

Much of our knowledge on the biochemical characteristics of the receptors has come from the analysis of proteins transgenically expressed in yeast. Transgenic expression was a key strategy leading to the discoveries that the receptors formed disulfide-linked homodimers (Hall *et al.*, 2000; Schaller *et al.*, 1995), that they bound ethylene (Hall *et al.*, 2000; O'Malley *et al.*, 2005; Schaller and Bleecker, 1995), and that the ethylene-binding site contained a copper cofactor (Rodriguez *et al.*, 1999), none of which could be predicted from the primary sequence (Fig. 1B). Transgenic expression in yeast has also been used to define and characterize the kinase activities of the receptors (Gamble *et al.*, 1998; Moussatche and Klee, 2004; Zhang *et al.*, 2004). Furthermore, transgenic expression in yeast has been used as a means to characterize interactions of the receptors with other proteins, most notably the Raf-like kinase CTR1, which is an essential component of the primary ethylene signaling pathway (Cancel and Larsen, 2002; Clark *et al.*, 1998; Gao *et al.*, 2003; Huang *et al.*, 2003). This chapter focuses on the use of yeast as a tool to facilitate the biochemical analysis of the receptors, with an emphasis on the methods used to study ethylene binding to the receptors.

Transgenic Expression of Ethylene Receptors in Yeast

Transgenic expression in yeast has proved useful for the analysis of many plant proteins and has allowed for their expression in a functional form where transgenic expression in bacteria has failed. In addition to allowing for expression to higher levels than is typically possible in plants and thereby facilitating biochemical analysis, transgenic expression also allows for the analysis of the protein in isolation from other plant proteins. This last point is proving to be of increasing importance because plant genome projects have revealed that many plant proteins belong to multimember families arising in part from genome duplications (Bowers *et al.*, 2003; Paterson *et al.*, 2005). Thus it is often not possible to obtain biochemical information about an individual isoform in the native plant because enzymatic and binding data will represent the average of multiple isoforms. The reader is referred to general references on the use of yeast for basic methods (e.g., Amberg *et al.*, 2005; Fink and Guthrie, 2002). Here we mention a few of the approaches that have facilitated analysis of the ethylene receptors.

For ethylene receptors, both full-length and individual domains have been expressed in yeast, and expression has been from constitutive as well as inducible promoters (Gamble *et al.*, 1998; Hall *et al.*, 2000; Moussatche and Klee, 2004; O'Malley *et al.*, 2005; Rodriguez *et al.*, 1999; Schaller and Bleecker, 1995; Schaller *et al.*, 1995; Zhang *et al.*, 2004). Use of the constitutive ADH promoter drives expression of full-length ETR1 to approximately 100-fold higher levels than the native expression level found in Arabidopsis, based on immunological analysis (Schaller and Bleecker, 1995). Use of a constitutive promoter simplifies growth of the transgenic yeast, which is accomplished by first growing an overnight starter culture in the appropriate dropout medium, diluting this the following day into a larger volume of growth medium, and letting this grow overnight to an OD_{600} of 0.7–1.0.

For galactose induction of the transgene, an overnight starter culture is also used, but the yeast is then transferred to growth medium containing 2% galactose/2% raffinose for induction. Glucose (dextrose), the usual carbon source in growth media, is not used during induction because of glucose repression of the GAL promoter. To remove the glucose, the starter culture is pelleted by centrifugation, washed twice with the gal/raf medium by resuspending and repelleting, and then diluted into the final growth medium for induction. Growth in the gal/raf medium is slower than that in glucose-containing growth medium and, if desired, a large amount of starter culture can be used followed by 2–4 h induction in the gal/raf medium (e.g., a 50-ml starter culture pelleted and then added to 100 ml gal/raf medium).

A convenient means to circumvent glucose repression of the GAL promoter is to use a yeast strain such as sc295 (*MATa GAL4 GAL80 ura3-52 leu2-3,112 reg1-501 gal1 pep4-3*). This strain is *gal1* (cannot metabolize the galactose inducer), *reg1* (GAL promoters not glucose repressed), and *pep4-3* (protease deficiency). The additional use of the pMTL4c plasmid (LEU) can enhance expression from the GAL1 promoter by overproducing GAL4 protein, but many do not see any stimulatory effect from this plasmid. With the yeast strain sc295, one can use a standard glucose-containing growth medium and simply induce expression of your gene by the addition of 0.5% galactose. This approach was used in characterization of histidine kinase activity after expression of the receptor as a GST fusion using the vector pEG(KT) (Gamble *et al.*, 1998, 2002; Mitchell *et al.*, 1993).

Because expression of different constructs can affect the yeast growth rate, the growth rate needs to be determined for each construct and the amount of inoculant from the starter culture adjusted accordingly.

Histidine Kinase Activity

Overview

Initial experiments using fusion proteins to the histidine kinase domain of the ethylene receptor ETR1 did not yield kinase activity when expressed and purified from *Escherichia coli* (Gamble *et al.*, 1998). However, when GST fusions to the soluble domain of ETR1 were expressed in yeast (*Saccharomyces cerevisiae*), affinity purified, and assayed, autophosphorylation was observed (Gamble *et al.*, 1998). The phosphorylated residue was shown to be histidine based on several independent criteria. First, the incorporated phosphate was resistant to treatment with 3 M NaOH, but was sensitive to 1 M HCl; this sensitivity to acid but not alkali is indicative of phosphohistidine (Duclos *et al.*, 1991). Second, autophosphorylation was abolished by site-directed mutation of the predicted site of phosphorylation (His353). The kinase activity was shown to derive from ETR1, rather than a contaminant, because mutagenesis of catalytic residues within the kinase domain abolished autophosphorylation.

Additional members of the ethylene receptor family have also now been examined by taking a similar approach, but using *Schizosaccharomyces pombe* rather than *S. cerevisiae* for transgenic expression (Moussatche and Klee, 2004; Xie *et al.*, 2003; Zhang *et al.*, 2004). The use of different yeast expression systems may prove useful as we found that the receptor ERS1 was susceptible to proteolysis in *S. cerevisiae* (unpublished results), but this was apparently not a problem when expressed in *S. pombe*.

One advantage with using yeast for the analysis of histidine kinase activity is that contamination from other histidine kinases is not a major concern due to the limited number of two-component signaling elements in yeast; for example, *S. cerevisiae* only contains one His-kinase, SLN1, that participates in a His-Asp phosphorelay (Posas *et al.*, 1996). However, contamination may pose a more serious problem when examining potential ser/thr kinase activity found in some diverged eukaryotic histidine kinases (Moussatche and Klee, 2004; Popov *et al.*, 1992, 1993; Zhang *et al.*, 2004) due to the large number of ser/thr kinases found in yeast. We detected a low level of ser/thr phosphorylation (based on acid/base stability of the phosphoamino acid) on histidine kinases expressed in and purified from yeast, following an *in vitro* kinase assay. This was at lower levels than the histidine phosphorylation and, unlike the histidine phosphorylation, was not affected by site-directed mutations in the kinase, consistent with this being a contaminating activity. The basal level of ser/thr phosphorylation was also found when we expressed a bacterial histidine kinase in yeast, again consistent with this being a contaminating activity (unpublished results). Another potential source of contamination is HSP70, which copurified with receptor fusions expressed in *S. pombe* (Moussatche and Klee, 2004). HSP70 has ATPase activity and also forms a phosphorylated intermediate, with evidence suggesting that this is on a histidine residue (Hiromura *et al.*, 1998; Lu *et al.*, 2006). Whether HSP70 activity can result in phosphorylation of the ethylene receptors is unknown. Thus, there are several potential sources of contaminating activities in yeast, and elimination of phosphorylation by site-directed mutation is essential to showing that the activity originates from the kinase of interest.

Several different approaches have been used to assay for the kinase activity of ethylene receptors following their affinity purification from yeast. Kinase activity has been analyzed with samples still attached to glutathione agarose beads (Gamble *et al.*, 1998, 2002) as well as in solution after elution from the beads (Moussatche and Klee, 2004; Xie *et al.*, 2003; Zhang *et al.*, 2004). We provide a protocol for assaying kinase activity of the protein while attached to the beads (Gamble *et al.*, 1998, 2002). This procedure simplifies removal of unincorporated radionuclide so that minimal radioactivity is present during subsequent gel electrophoresis.

Histidine Kinase Assay Protocol

1. Grow a 5-ml overnight culture in the appropriate dropout medium (e.g., for yeast strain sc295 containing the pMTL4C plasmid and an ETR1 construct in pEG-KT, use yeast minimal media with ura-, leu-).

2. Inoculate 100 ml of dropout medium containing 0.5% galactose with 0.5–2 ml (depending on growth rate) of overnight culture and allow to grow 12–15 h until OD_{600} is 0.7–1.0.

3. Harvest yeast by centrifugation at 3000g for 10 min. Wash the pellet by resuspending in 50 ml cold water and pellet again by centrifugation.

4. Weigh yeast and resuspend in 1 ml extraction buffer (50 mM Tris, pH 7.6, 1 mM EDTA, 100 mM NaCl, 10% glycerol; make the extraction buffer 1 mM phenylmethylsulfonyl fluoride [PMSF] from a 100 mM stock in isopropanol just before use). Weighing yeast allows one to equilibrate yeast concentrations if multiple samples are used.

5. Break open the yeast by beating with cold glass beads. Bead beating may be accomplished using microfuge tubes placed into a vortexer with a Turbo Mix (Scientific Industries) attachment. Alternatively, one can use a vortexer with a test tube holder attached to its head, in which case the yeast sample is placed in a Pyrex culture tube; this method is especially useful when multiple samples need to be processed. Add an equal volume of cold glass beads to the yeast suspension (mark the level of the yeast solution and pour in beads to that level), and vortex for 8 min at 4°.

6. Transfer supernatant to a microfuge tube using a 1-ml tip pushed close to the bottom of the tube to avoid aspirating beads. Rinse beads with 0.5 ml extraction buffer, vortex, and add to collection tube.

7. Centrifuge sample at 3000g for 5 min, 4°, and save supernatant.

8. Centrifuge at 100,000g for 30 min, 4°, and save supernatant. This step removes contaminating membranes that increase the background of nonspecific kinase activity.

9. Prepare glutathione agarose beads (Sigma) for affinity purification of the fusion protein. The quantity of beads depends in part on the number of subsequent kinase reaction assays; we use 60 μl of hydrated beads per reaction. When hydrating the beads, 35–40 mg glutathione agarose will result in 450 μl hydrated beads. The hydrated beads are washed three times, with the last wash being performed with extraction buffer. For the washes we pellet the beads for 30 s at 1500g in a microfuge and remove the supernatant with a Hamilton syringe or a gel-loading pipette tip.

10. Transfer yeast supernatant (approximately 1 ml) to a microfuge tube with the washed glutathione agarose beads. Bind the GST fusion to the beads by incubating 30 min at room temperature using a rocker platform.

11. Remove supernatant and wash beads by mixing with buffer and then pelleting beads for 30 s at 1500g in a microfuge. Wash three times with 1 ml Tris-buffered saline (20 mM Tris, pH 7.4, at 25°, 137 mM NaCl; TBS), 0.1% Tween 20, letting the last wash shake for 3 min before removal. Wash once in 1 ml TBS, letting this wash sit for 5 min on ice. The beads can

be divided into aliquots at this stage if multiple kinase reactions are planned; do this by adding buffer, mixing, and, while beads are in suspension, pipetting a set volume into microfuge tubes.

12. A single kinase reaction is performed directly with the beads in a solution volume of 30 μl. The typical kinase assay buffer is 50 mM Tris (pH 7.6), 50 mM KCl, 5 mM MnCl$_2$, 2 mM dithiothreitol (DTT), 10% glycerol, 0.5 mM ATP (containing 20–30 μCi γ^{32}P-ATP). The histidine kinase activity of the ethylene receptors showed greater activity with manganese than with magnesium as the divalent cation. A mixture of unlabeled and labeled ATP is used so that the assay is performed close to the K_m for the enzyme and to avoid artifactual chemical labeling that can occur when only labeled ATP is used. The reaction is started by the addition of the ATP.

13. Incubate the reaction for the desired time (e.g., 45 min) at room temperature.

14. Stop the reaction by the addition of 0.5 ml TBS, 10 mM EDTA. EDTA is used in the buffer to chelate the divalent cation. Pellet the resin briefly in a microfuge by low-speed centrifugation. Remove the supernatant with a loading tip and dispose to radioactive waste.

15. Wash the beads with 0.5 ml TBS. This will remove the majority of the unincorporated radionuclide. (For washing, vortex on low to resuspend beads and centrifuge at low speed in a microfuge and remove wash to radioactive waste.)

16. Resuspend resin in 25 μl SDS-PAGE loading buffer. The SDS in the loading buffer will elute the GST fusion from the beads. The sample is loaded onto the SDS-PAGE gel with a gel-loading tip so as not to take up the glutathione beads. Following SDS-PAGE (Laemmli, 1970), the gel is typically blotted to an Immobilon nylon membrane (Millipore).

Isolation of Receptors for Use in Ethylene-Binding Assays

Overview

The ready diffusion of ethylene across membranes facilitated initial ethylene-binding experiments because one did not need to perform a biochemical purification of the receptor in order to test its binding. Binding assays could be performed with intact living organisms. This approach was used in the initial work to characterize ethylene binding in plants (Jerie et al., 1979; Sisler, 1979) and subsequently to characterize the ethylene receptor ETR1 following its transgenic expression in yeast (Schaller and Bleecker, 1995). More recently, it has been used as a basis by which the extant of ethylene binding was examined across many species (Wang *et al.*, 2006). We provide a protocol here for the growth and isolation of intact yeast cells to be used in binding experiments.

It is, however, often useful to either partially purify the receptor or to completely purify the receptor for more in-depth biochemical analysis. We therefore provide two protocols, each resulting in an increased level of purity for the ethylene receptor following its transgenic expression in yeast. It is important to note that during purification the copper cofactor required for ethylene binding is lost from the receptor. Thus to maintain and/or regenerate ethylene-binding ability it is necessary to include the metal cofactor during purification. This ability to reconstitute the binding ability allowed for a detailed analysis as to which metals could function in ethylene binding. Notably, it was found that in addition to copper, silver could also be incorporated into the receptor and function in ethylene binding (Rodriguez *et al.*, 1999). This discovery provided a potential explanation for the long-known effect of silver as an inhibitor of ethylene responses (Beyer, 1976): silver can apparently replace the normal copper cofactor and even bind ethylene, but because of its larger size disrupts the normal conformation of the receptor such that it cannot transduce the ethylene signal after binding.

We give a protocol for isolating yeast membranes. This procedure was useful in determining that the receptors ETR1 and ERS1 formed disulfide-linked dimers (Hall *et al.*, 2000; Schaller *et al.*, 1995) and in determining which metals served as cofactors for binding ethylene (Rodriguez *et al.*, 1999). In addition, we give a protocol for solubilization of receptors from the yeast membranes. The solubilization protocol works equally well for full-length, truncated, and GST-labeled receptors. Solubilization is a necessary step prior to affinity purification of tagged receptors and allowed for the determination that there was one copper molecule and thus one ethylene-binding site per receptor homodimer (Rodriguez *et al.*, 1999). The detergents SB-16 or lysophosphatidylcholine are used for solubilization; other detergents such as Triton X-100, CHAPS, octylglucopyranoside, and SDS were found to eliminate ethylene-binding activity (Binder, unpublished results). Lysophosphatidylcholine also allows for solubilization of ETR1 from Arabidopsis (Gao *et al.*, 2003). Lysophosphatidylcholine was previously found to allow for solubilization with activity of another multipass membrane protein from plants, the proton-pumping ATPase (H^+-ATPase) (Serrano, 1984), and thus appears to be one of the better detergents for solubilizing these hydrophobic proteins, potentially because of its similarity to the phospholipids of the membrane bilayer.

Protocol for Isolating Intact Yeast Cells

1. Start a 5.0-ml yeast culture using a colony of interest in the corresponding liquid dropout media (e.g., Trp⁻ for pYcDE vector) and grow overnight (or until confluent) in shaker at 30°.
2. Use 5.0 ml of this culture to start a 500-ml culture (a 100-fold dilution of the O/N culture) in dropout media and grow for approximately 17 h or until OD_{600} gets close to 1.0.
3. Harvest yeast cells by centrifugation at 1500*g* for 5 min at 4°.
4. Wash pellet with 50 ml of sterile water (or at least 5.0 ml per gram of yeast). To do this resuspend the yeast with the water, transfer to a 50-ml centrifuge tube, and pellet at 1500*g* for 5 min. Discard the supernatant and weigh the yeast pellet. Multiply the weight of the yeast by 5 and then bring the total weight to this value with sterile water. This will result in the resuspension of the yeast to a concentration of 1.0 g of cell tissue per 5.0 ml water.
5. Collect cells (1.0 g of cells/filter; i.e., use 5 ml of the yeast suspension) by vacuum filtration onto glass fiber filters (47 mm GF/C, Whatman). Put each filter with cells into a 35-mm Petri dish.

Protocol for Isolating Yeast Membranes

1. Resuspend yeast cells isolated as described earlier to a final concentration of 1 gram per 5 ml of extraction buffer consisting of 50 m*M* Tris-HCl (pH 8.0 at 4°), 100 m*M* NaCl, 1% dimethyl sulfoxide (DMSO), 10% glycerol, 1 m*M* PMSF (added fresh from a 100 m*M* stock in isopropanol). For maximal ethylene binding, copper sulfate needs to be added. This is a convenient step to do this. A concentration of 600 μM seems to work best for stabilizing ETR1 ethylene-binding activity if added prior to disrupting the cells.

2. Disrupt yeast cells using a bead-beating mill (Biospec Products, Bartlesville, OK). A 45-ml yeast suspension (1 g per 5 ml buffer) is placed in a 90-ml bead chamber, and an equal volume of small glass beads (0.5 mm diameter) are added to fill the chamber. The beads should be precooled in a freezer. The chamber is placed in an inner jacket filled with 50% (v/v) glycerol precooled to −20°. This is placed in an outer jacket also filled with −20° 50% (v/v) glycerol. The suspension is disrupted in the mill for 1 min, followed by a 30-s cool-down period. Repeat disruption procedure two more times. The extract is removed and the beads rinsed thoroughly with the extraction buffer.

3. Centrifuge the extract at 10,000*g* for 10 min at 4° and save the supernatant.

4. Centrifuge the supernatant at 100,000g for 30 min and resuspend the membrane pellet in a buffer of choice. A variety of buffers can be used. A buffer combination that has yielded good binding activity is 10 mM MES, pH 6.0, 10% (w/v) sucrose, 1% (v/v) DMSO. However, if solubilization is carried out, a solubilization buffer (50 mM NaPO$_4$, pH 7.2, 150 mM NaCl, 2 mM DTT, 5% [w/v] sucrose) should be used. If copper has not been added during isolation of the membranes, it should be added to a final concentration of 300 μM copper sulfate for maximal ethylene binding. If silver salts are to be employed, it is essential to eliminate halides since they form insoluble salts with silver. To ensure a homogeneous suspension, use three passes in a Wheaton homogenizer.

5. Rapidly freeze the membranes in liquid nitrogen. This step has been found to increase ethylene-binding activity. The frozen pellets can be stored at $-80°$ or thawed and used.

Protocol for Solubilization of Membranes

1. Thaw the frozen yeast membrane pellet on ice. For maximal ethylene binding, copper must be added prior to solubilization. Therefore, if copper has not been added up to this point, then it should be added now (300 μM copper sulfate). Before solubilizing, be sure to take a portion of the membrane preparation for binding activity assays and other analyses as needed.

2. Bring the membrane suspension to room temperature and add either SB-16 or lysophosphatidylcholine to a final concentration of 5 mg ml^{-1} while stirring the mixture. Stir for 15 min.

3. Centrifuge at 100,000g for 30 min. Remove the supernatant, resuspend the pellet in the original volume as described earlier, and treat with detergent a second time. The supernatants can be pooled.

4. Following solubilization, affinity purification of tagged proteins may take place by standard methods (Rodriguez *et al.*, 1999; Smith and Johnson, 1988), providing that detergent is maintained in the solutions to prevent aggregation of the receptor.

Ethylene-Binding Activity

The method for ethylene binding is based on that used by Sisler (1979) for assaying binding of ethylene in plant tissues. This method is readily applicable to intact organisms such as plants and transgenic yeast as well as to extracts from these organisms. Radiolabeled ethylene is available from American Radiolabeled Chemicals (ARC) labeled with either ^{3}H or [14]C. The [^{14}C]ethylene is sold as a gas in a break vial at a specific activity of 50–100 mCi/mmol. After purchase, it needs to be bound to 0.25 M mercuric perchlorate (see next section on preparation and considerations when using mercuric perchlorate). To bind to the mercuric perchlorate, the ampoule is placed into an airtight sealed chamber with a magnetic hammer (e.g., iron weight suspended by use of a magnet) along with the mercuric perchlorate. To accomplish this task, we have used sealed Ball jelly jars with a glass scintillation vial glued to the inside bottom to hold the mercuric perchlorate. The ampoule is broken and the gaseous ethylene is allowed to bind for 24 h in the sealed container; the mercuric perchlorate is stirred throughout this time with a magnetic stir bar. Binding takes place in a hood so that when finished any residual unbound ethylene is vented for air disposal. The mercuric perchlorate with bound ethylene is then aliquoted into screw cap tubes and stored at $-20°$. This may be diluted for subsequent use with mercuric perchlorate to a convenient concentration.

[^{3}H]Ethylene is available commercially from ARC at a higher specific activity than [^{14}C]ethylene and is sold already bound to mercuric perchlorate. However, in practice these conveniences are offset by the readiness with which the ^{3}H is lost from the radiolabeled ethylene, potentially due to its exchange with water and other compounds. This results in decreasing specific activity of the [^{3}H]ethylene and greater likelihood of contamination.

Because ethylene is a gas, concentrations in the literature are usually reported in μl liter^{-1} (or parts per million) and thus it is necessary to convert among units of radioactivity, molarity, and gaseous volume. In one case, we used sufficient mercuric perchlorate so that after the initial binding 1 μl of the stock solution contained 0.248 μCi of [^{14}C]ethylene, or 4.36 nmol, because the specific activity of the [^{14}C]ethylene purchased was 56.9 mCi/mmol. The gaseous volume can be determined based on $PV = nRT$ (the most convenient form of the gas constant here being $R = 82.0575$ cm^3 atm K^{-1} mol^{-1}). For gaseous ethylene at 25° and 1 atmosphere pressure, 1.0 μl of gas equals 41 nmol of ethylene. For the stock solution described earlier, 1.0 μl of [^{14}C]C_2H_4 stock solution = 0.248 μCi = 4.36 nmol = 0.107 μl of [^{14}C]C_2H_4.

The ethylene-binding assay is performed in a sealed Ball jelly jar, which serves as the ethylene-binding chamber (Fig. 2). One can use large or small jars depending on the number of samples that will be put in the jar. A glass scintillation vial is glued to the inside bottom of the jar and serves to hold the mercuric perchlorate with bound ethylene. A rubber sleeve stopper (e.g., Fisher; stopper size No. 13) is inserted into the metal lid of the jar to allow for the injection of materials by use of a syringe with needle.

Although intact organisms can be placed directly into the jar, aqueous solutions need to be applied in a manner that gives maximal surface area and access to the ethylene gas. For example, yeast membranes or

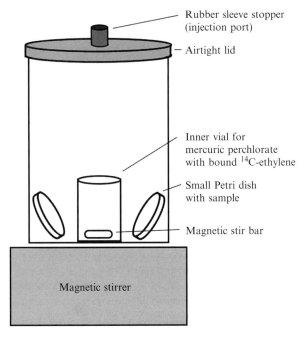

FIG. 2. Experimental setup for ethylene-binding assay.

solubilized receptors can be mixed in a tube with cellulose (0.75 ml solution with 0.6 g cellulose works well). Alternatively, 0.75 ml of the sample can be applied to a 2.5×20-cm piece of Whatman No. 1 filter paper that has been rolled into a tube and inserted into a 1.5-ml microfuge tube. Care must be taken to have the suspension all soak into the paper rather than accumulate at the bottom of the centrifuge tube.

Ethylene-Binding Assay Protocol

1. Perform the ethylene-binding assay in a fume hood. Add an appropriate amount of [^{14}C]ethylene trapped in mercuric perchlorate to the vial inside the binding chamber. Use a dilution such that the final volume added is 100 μl mercuric perchlorate solution or less. There should be a stirring bar in the vial inside the binding chamber.

2. Place samples around the vial in the binding chamber. For a 230-ml binding chamber, use a maximum of four samples. Close the chamber and place it on top of a stir plate.

3. Inject 1.0 ml of saturated LiCl into the vial inside the chamber to release bound ethylene and incubate samples for 4 h (stirring on the lowest setting). If controls are being run with nonradioactive ethylene, use a second chamber to which is added 1 ml liter^{-1} of [^{12}C]ethylene in addition to [^{14}C]ethylene. Add the nonradioactive ethylene first and then the LiCl.

4. Open the binding chamber and remove samples. Let samples air for 5 min from the time the chamber was opened to dissipate unbound ethylene. Lightly shaking the samples helps in this dissipation.

5. Put each sample into a separate 230-ml jelly jar along with a scintillation vial containing 300 μl of mercuric perchlorate. Close tightly and incubate at 65° for 90 min. This incubation releases the ethylene bound to the sample.

6. Let jars sit at room temperature for 17–24 h to let the mercuric perchlorate in the vial trap the ethylene released by the sample.

7. Open the jars in the hood. Remove scintillation vials and add 10 ml of scintillation fluid to each. Shake thoroughly to mix the mercuric perchlorate and scintillation fluid. Incubate for 15 min in the dark prior to measuring counts on the scintillation counter.

Considerations When Working with Mercuric Perchlorate

Mercuric perchlorate (0.25 M in 12% perchloric acid) is synthesized from red mercuric oxide and 70% perchloric acid and is used for reversible binding of ethylene (Young *et al.*, 1952). Because it contains mercury as well as perchloric acid, great care must be taken in its use. The mercury salts are considered highly toxic. Perchloric acid and mercuric oxide are corrosive and are damaging to skin, eyes, and mucous membranes. Perchloric acid can be explosive and shock sensitive when heated and is also a powerful oxidizer capable of reacting with or igniting organic materials. For these reasons, researchers should check with their safety department before use.

Containers containing perchloric acid solutions should be kept tightly sealed and uncontaminated because dehydration and/or contamination can cause explosion or fire. Containers should be clearly labeled with warnings of the hazardous nature (e.g., "Highly Toxic, Oxidizer, Corrosive–authorized personnel only"). Mercuric perchlorate and perchloric acid should be stored separately from organics and organic acids. Note, however, that at concentrations of less than 72.5% and when used at room temperature that perchloric acid is normally stable and precautions for work with perchloric acid are similar to those of other mineral acids.

Disposal is also a consideration. The binding of [^{14}C]ethylene to mercuric perchlorate generates a mixed waste (radioactivity and mercury) and these must be separated before disposal can take place. In most cases, the radioactive ethylene (typically less than 2000 dpm or 0.0009 μCi in the sample) can be released from the sample. This is done in the hood by adding saturated LiCl to the solution for air disposal. Following release, the radioactivity level remaining should be measured in the sample to determine if it is below the acceptable background level (check current limits with safety department). If radioactivity remains in the solution, then the mercury must be separated out. This is accomplished by adding dissolved sodium sulfide (calculate the amount of mercury present in the solution and then add an excess molar quantity of dissolved sodium sulfide). Mercury sulfide will precipitate as a black residue. This can be removed by filtration. If necessary the perchlorate can be removed by adding KOH, which gives insoluble potassium perchlorate and which can be filtered or centrifuged out. The pH of the solution for disposal can be adjusted if necessary by adding 1 M Tris, pH 8 (measure pH with pH paper and dispose of paper with mercuric perchlorate waste).

Protocol for Preparation of Mercuric Perchlorate Solution (Adapted from Young et al., 1952)

1. Perform all work in a designated chemical fume hood lined with absorbent pads. Check with local Environmental Health and Safety for appropriate pads and spill kit to assure compatibility.
2. Wear chemically resistant gloves (heavy-duty nitrile or neoprene; do not use PVA gloves), chemical splash goggles, and a splash resistant apron when making up the solution. Do not get in eyes, on skin, or on clothing. Avoid prolonged or repeated exposure. Immediately wipe up any spills or splashes.
3. When preparing a stock solution of mercuric perchlorate (0.25 M in 12% perchloric acid), all operations should be performed in the hood, including weighing of the red mercuric oxide.
4. Carefully dilute 43 ml of 70% redistilled perchloric acid to approximately 63 ml by adding the acid to 20 ml water. Place the diluted perchloric acid in a glass mortar.
5. Carefully weigh out 13.55 g of red mercuric oxide and slowly add it to the diluted perchloric acid in small portions. Carefully grind each portion in the mortar to dissolve without caking before adding the next portion.
6. Filter with suction using a glass fiber filter in a Buchner funnel. Make up to 250 ml with water.
7. Rinse contaminated glassware and tools with water and collect the rinse for disposal. Wipe down with pads (moisten with water if need be) and pick up spill pads from hood; dispose of as hazardous waste if contaminated.

Acknowledgments

We thank Edward Sisler who pioneered the development of procedures for measuring ethylene binding and who has made his expertise in these matters freely available to us over the years. We also acknowledge our debt to our scientific father Tony Bleecker and to our scientific grandfather Hans Kende, two exemplary scientists who recently passed away. We thank the Department of Energy, the National Science Foundation, and the USDA-NRICGP for their support of our research on ethylene signaling.

References

Abeles, F. B., Morgan, P. W., and Saltveit, M. E., Jr. (1992). "Ethylene in Plant Biology." Academic Press, San Diego.

Amberg, D. C., Burke, D., and Strathern, J. N. (2005). "Methods in Yeast Genetics: A Cold Spring Harbor Laboratory Course Manual." Cold Spring Harbor Laboratory Press, Cold Spring Harbor, NY.

Beyer, Jr., E. M. (1976). Silver ion, a potent inhibitor of ethylene action in plants. *Plant Physiol.* **58**, 268–271.

Binder, B. M., O'Malley, R. C., Wang, W., Moore, J. M., Parks, B. M., Spalding, E. P., and Bleecker, A. B. (2004). Arabidopsis seedling growth response and recovery to ethylene: A kinetic analysis. *Plant Physiol.* **136**, 2913–2920.

Bleecker, A. B. (1999). Ethylene perception and signaling: An evolutionary perspective. *Trends Plant Sci.* **4**, 269–274.

Bowers, J. E., Chapman, B. A., Rong, J., and Paterson, A. H. (2003). Unravelling angiosperm genome evolution by phylogenetic analysis of chromosomal duplication events. *Nature* **422**, 433–438.

Cancel, J. D., and Larsen, P. B. (2002). Loss-of-function mutations in the ethylene receptor ETR1 cause enhanced sensitivity and exaggerated response to ethylene in Arabidopsis. *Plant Physiol.* **129**, 1557–1567.

Chang, C., and Stadler, R. (2001). Ethylene hormone receptor action in Arabidopsis. *Bioessays* **23**, 619–627.

Clark, K. L., Larsen, P. B., Wang, X., and Chang, C. (1998). Association of the Arabidopsis CTR1 Raf-like kinase with the ETR1 and ERS1 ethylene receptors. *Proc. Natl. Acad. Sci. USA* **95**, 5401–5406.

Duclos, B., Marcandier, S., and Cozzone, A. J. (1991). Chemical properties and separation of phosphoamino acids by thin-layer chromatography and/or electrophoresis. *Methods Enzymol.* **201**, 10–21.

Fink, G. R., and Guthrie, C. (2002). Guide to yeast genetics and molecular cell biology, part B. *Methods Enzymol.* **350**.

Gamble, R. L., Coonfield, M. L., and Schaller, G. E. (1998). Histidine kinase activity of the ETR1 ethylene receptor from Arabidopsis. *Proc. Natl. Acad. Sci. USA* **95**, 7825–7829.

Gamble, R. L., Qu, X., and Schaller, G. E. (2002). Mutational analysis of the ethylene receptor ETR1: Role of the histidine kinase domain in dominant ethylene insensitivity. *Plant Physiol.* **128**, 1428–1438.

Gao, Z., Chen, Y. F., Randlett, M. D., Zhao, X. C., Findell, J. L., Kieber, J. J., and Schaller, G. E. (2003). Localization of the Raf-like kinase CTR1 to the endoplasmic reticulum of Arabidopsis through participation in ethylene receptor signaling complexes. *J. Biol. Chem.* **278**, 34725–34732.

Hall, A. E., Chen, Q. G., Findell, J. L., Schaller, G. E., and Bleecker, A. B. (1999). The relationship between ethylene binding and dominant insensitivity conferred by mutant forms of the ETR1 ethylene receptor. *Plant Physiol.* **121,** 291–299.

Hall, A. E., Findell, J. L., Schaller, G. E., Sisler, E. C., and Bleecker, A. B. (2000). Ethylene perception by the ERS1 protein in Arabidopsis. *Plant Physiol.* **123,** 1449–1458.

Hiromura, M., Yano, M., Mori, H., Inoue, M., and Kido, H. (1998). Intrinsic ADP-ATP exchange activity is a novel function of the molecular chaperone, Hsp70. *J. Biol. Chem.* **273,** 5435–5438.

Huang, Y., Li, H., Hutchison, C. E., Laskey, J., and Kieber, J. J. (2003). Biochemical and functional analysis of CTR1, a protein kinase that negatively regulates ethylene signaling in Arabidopsis. *Plant J.* **33,** 221–233.

Jerie, P. H., Shaari, A. R., and Hall, M. A. (1979). The compartmentalization of ethylene in developing cotyledons of *Phaseolus vulgaris* L. *Planta* **144,** 503–507.

Kieber, J. J., Rothenberg, M., Roman, G., Feldman, K. A., and Ecker, J. R. (1993). *CTR1*, a negative regulator of the ethylene response pathway in Arabidopsis, encodes a member of the Raf family of protein kinases. *Cell* **72,** 427–441.

Laemmli, U. K. (1970). Cleavage of structural proteins during the assembly of the head of bacteriophage T4. *Nature* **227,** 680–685.

Lu, Y., Hu, Q., Yang, C., and Gao, F. (2006). Histidine 89 is an essential residue for Hsp70 in the phosphate transfer reaction. *Cell Stress Chaperones* **11,** 148–153.

Mason, M. G., and Schaller, G. E. (2005). Histidine kinase activity and the regulation of ethylene signal transduction. *Can. J. Bot.* **83,** 563–570.

Mattoo, A. K., and Suttle, J. C. (1991). "The Plant Hormone Ethylene." CRC Press, Boca Raton, FL.

Mitchell, D. A., Marshall, T. K., and Deschenes, R. J. (1993). Vectors for inducible over-expression of glutathione S-transferase fusion proteins in yeast. *Yeast* **9,** 715–723.

Moussatche, P., and Klee, H. J. (2004). Autophosphorylation activity of the Arabidopsis ethylene receptor multigene family. *J. Biol. Chem.* **279,** 48734–48741.

O'Malley, R. C., Rodriguez, F. I., Esch, J. J., Binder, B. M., O'Donnell, P., Klee, H. J., and Bleecker, A. B. (2005). Ethylene-binding activity, gene expression levels, and receptor system output for ethylene receptor family members from Arabidopsis and tomato. *Plant J.* **41,** 651–659.

Paterson, A. H., Bowers, J. E., Van de Peer, Y., and Vandepoele, K. (2005). Ancient duplication of cereal genomes. *New Phytol.* **165,** 658–661.

Popov, K. M., Kedishvili, N. Y., Zhao, Y., Shimomura, Y., Crabb, D. W., and Harris, R. A. (1993). Primary structure of pyruvate dehydrogenase kinase establishes a new family of eukaryotic protein kinases. *J. Biol. Chem.* **268,** 26602–26606.

Popov, K. M., Zhao, Y., Shimomura, Y., Kuntz, M. J., and Harris, R. A. (1992). Branched-chain alpha-ketoacid dehydrogenase kinase: Molecular cloning, expression, and sequence similarity with histidine protein kinases. *J. Biol. Chem.* **267,** 13127–13130.

Posas, F., Wurgler-Murphy, S. M., Maeda, T., Witten, E. A., Thai, T. C., and Saito, H. (1996). Yeast HOG1 MAP kinase cascade is regulated by a multistep phosphorelay mechanism in the SLN1-YPD1-SSK1 "two component" osmosensor. *Cell* **86,** 865–875.

Qu, X., and Schaller, G. E. (2004). Requirement of the histidine kinase domain for signal transduction by the ethylene receptor ETR1. *Plant Physiol.* **136,** 2961–2970.

Rodriguez, F. I., Esch, J. J., Hall, A. E., Binder, B. M., Schaller, G. E., and Bleecker, A. B. (1999). A copper cofactor for the ethylene receptor ETR1 from Arabidopsis. *Science* **283,** 996–998.

Schaller, G. E., and Bleecker, A. B. (1995). Ethylene-binding sites generated in yeast expressing the Arabidopsis *ETR1* gene. *Science* **270,** 1809–1811.

Schaller, G. E., and Kieber, J. J. (2002). Ethylene. *In* "The Arabidopsis Book" (C. Somerville and E. Meyerowitz, eds.), Vol. DOI/10.1199/tab.0071. American Society of Plant Biologists, Rockville, MD.

Schaller, G. E., Ladd, A. N., Lanahan, M. B., Spanbauer, J. M., and Bleecker, A. B. (1995). The ethylene response mediator ETR1 from Arabidopsis forms a disulfide-linked dimer. *J. Biol. Chem.* **270,** 12526–12530.

Serrano, R. (1984). Purification of the proton pumping ATPase from plant plasma membranes. *Biochem. Biophys. Res. Commun.* **121,** 735–740.

Sisler, E. C. (1979). Measurement of ethylene binding in plant tissue. *Plant Physiol.* **64,** 538–542.

Smith, D. B., and Johnson, K. S. (1988). Single-step purification of polypeptides expressed in *E. coli* as fusions with glutathione S-transferase. *Gene* **67,** 31–40.

Wang, W., Esch, J. J., Shiu, S.-H., Agula, H., Binder, B. M., Chang, C., Patterson, S. E., and Bleecker, A. B. (2006). Identification of important regions for ethylene binding and signaling in the transmembrane domain of the ETR1 ethylene receptor of Arabidopsis. *Plant Cell.* **18,** 3429–3442.

Wang, W., Hall, A. E., O'Malley, R., and Bleecker, A. B. (2003). Canonical histidine kinase activity of the transmitter domain of the ETR1 ethylene receptor from Arabidopsis is not required for signal transmission. *Proc. Natl. Acad. Sci. USA* **100,** 352–357.

Xie, C., Zhang, J. S., Zhou, H. L., Li, J., Zhang, Z. G., Wang, D. W., and Chen, S. Y. (2003). Serine/threonine kinase activity in the putative histidine kinase-like ethylene receptor NTHK1 from tobacco. *Plant J.* **33,** 385–393.

Young, R. E., Pratt, H. K., and Biale, J. B. (1952). Manometric determination of low concentrations of ethylene with particular reference to plant material. *Anal. Chem.* **24,** 551–555.

Zhang, Z. G., Zhou, H. L., Chen, T., Gong, Y., Cao, W. H., Wang, Y. J., Zhang, J. S., and Chen, S. Y. (2004). Evidence for serine/threonine and histidine kinase activity in the tobacco ethylene receptor protein NTHK2. *Plant Physiol.* **136,** 2971–2981.

[14] Structure of SixA, a Histidine Protein Phosphatase of the ArcB Histidine-Containing Phosphotransfer Domain in *Escherichia coli*

By Toshio Hakoshima and Hisako Ichihara

Abstract

Escherichia coli protein SixA was the first identified histidine protein phosphatase that dephosphorylates the histidine-containing phosphotransfer (HPt) domain of histidine kinase ArcB. The crystal structures of the free and tungstate-bound forms of SixA revealed an α/β architecture with a fold unlike those previously described in eukaryotic protein phosphatases, but related to a family of phosphatases containing the arginine-histidine-glycine (RHG) motif at their active sites. Compared with these RHG phosphatases, SixA lacks an extra α-helical subdomain that forms a lid over the active site, thereby forming a relatively shallow groove important for accommodating the kidney-shaped four-helix bundle of the HPt domain. Sequence database searches revealed that a single SixA homolog was found in a variety of bacteria, where two homologs were found in some bacteria while no homolog was found in others. No SixA homologs were found in the majority of firmicutes and euryarchaea. Structure-based examination and multiple alignment of sequences revealed SixA active residues from loop $\beta 1$-H2, which might assist in the identification of SixA homologs among RHG phosphatases even with poor amino acid identity.

Introduction

Two-component signaling systems play a crucial role in cellular adaptation to the environment in microorganisms such as bacteria, yeasts, and fungi, as well as in plants. The signaling systems are made up of three protein domains possessing histidine or aspartate as their active residue; the histidine protein kinase (HPK) and the histidine-containing phosphotransfer (HPt) domains contain a histidine residue and the receiver domain contains an aspartate residue (Robinson *et al.*, 2000). In a substantial subset of two-component signaling systems, these domains are combined to form a multistep His-Asp phosphorelay in which phosphates are transferred from histidine to aspartate and aspartate to histidine residues (Burbulys *et al.*, 1991; Mizuno, 1998; Wurgler-Murphy and Saito, 1997).

METHODS IN ENZYMOLOGY, VOL. 422   0076-6879/07 $35.00
    DOI: 10.1016/S0076-6879(06)22014-7

SixA was found to exhibit protein histidine phosphatase activity involving the dephosphorylation of phosphohistidine at the HPt domain C terminus of sensor kinase ArcB engaged in anaerobic responses in *Escherichia coli* cells (Fig. 1) (Matsubara and Mizuno, 2000; Ogino *et al.*, 1998). Interestingly, SixA possesses the arginine-histidine-glycine (RHG) motif at its active site. This RHG signature is found in phosphatases possessing a common catalytic mechanism involving a covalent phospho-histidine intermediate. This new protein phosphatase is distinct from previously reported protein phosphatases of two-component signaling systems such as RapA, RapB (Perego *et al.*, 1994), Spo0E (Ohlsen *et al.*, 1994), and CheZ (Hess *et al.*, 1988) in that these act as protein aspartate phosphatases involved in the dephosphorylation of phosphoaspartate residues of the response regulators.

ArcB is a hybrid sensor kinase involved in the anaerobic respiratory control (Arc) response. The cytoplasmic region of ArcB includes HPK, receiver, and HPt domains (Fig. 1). The response regulator for ArcB is the transcription factor ArcA, which has a receiver domain and, in the phosphorylated form, functions as a repressor of the *sdh*CDAB operon. SixA phosphatase activity toward ArcB has been shown to be specific to phosphorylated His717 in the C-terminal HPt domain (Matsubara and Mizuno, 2000). In the absence of oxygen, *E. coli* cells can utilize the TCA cycle and the electron transport system for ATP synthesis in the presence of an electron acceptor such as nitrate. Under anaerobic conditions, SixA dephosphorylates ArcB so that phosphate transfer from ArcB to ArcA is interrupted. This event results in derepression of the *sdh*CDAB operon encoding the succinate dehydrogenase complex involved in the TCA cycle.

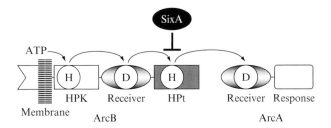

FIG. 1. The anaerobic response pathway of *E. coli*. ArcB is a hybrid sensor kinase that consists of an extracellular domain, a transmembrane helix, and intracellular HPK, receiver and HPt domains. The response regulator, ArcA and OmpR consist of receiver and response domains. HPK and HPt domains possess an active histidine residue (H) and receiver domains an active aspartate residue (D). The response domain of the response regulators comprises a HTH domain involved in transcriptional regulation through specific DNA binding. SixA acts on the HPt domain.

In yeast and plants, HPt domains were found in the form of isolated small proteins such as Ypd1 in yeast (Posas *et al.*, 1996) and AHPs in plants (Suzuki *et al.*, 1998).

It is clear that an understanding of protein phosphatase activity is important in delineating the regulation of multistep signaling systems, as they can provide additional regulatory checkpoints in various pathways. This chapter describes the structure and function of *E. coli* SixA and discusses the potential means by which SixA homologs can be identified in other organisms. Comparison of SixA with other related phosphatase structures is also addressed.

Overall Structure

The three-dimensional structures of the free and tungstate-bound forms of *E. coli* SixA were determined by X-ray crystallography at 2.06 and 1.90 Å resolution, respectively (Hamada *et al.*, 2005). The tungstate ion is a good mimic of the substrate phosphate (Heo *et al.*, 2002). SixA consists of 161 amino acid residues that form an $\alpha\beta$ architecture and comprises a six-stranded β sheet ($\beta 1-\beta 6$), five α helices (H2 and H3 and H5–H7), and two short 3_{10} helices (H1 and H4) (Fig. 2). The β sheet, formed by four-stranded ($\beta 1$-$\beta 4$-$\beta 2$-$\beta 3$) parallel and two-stranded ($\beta 5,\beta 6$) antiparallel

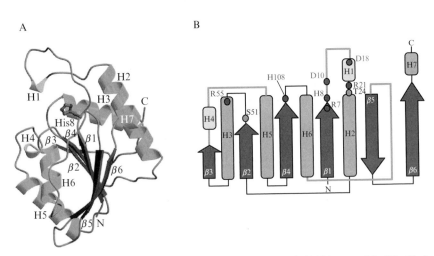

FIG. 2. Structure of the protein histidine phosphatase SixA. (A) Ribbon model of the SixA structure. The side chain of active residue His8 is shown as a stick model, and the α helices (green), 3_{10} helices (blue) and β strands (red) are labeled. (B) Folding topology of SixA. Color codes are the same as in A. Loops H4-H5, $\beta 1$-H2, and H6-$\beta 5$ that frequently contain insertions in other RHG phosphatases are colored orange (see text). (See color insert.)

associations, is sandwiched between a pair of long α helices (H2, H3) on one side and a pair of long α helices (H5, H6) on the other side that form extensive hydrophobic cores. The carboxyl end of strand $\beta1$ contains the active residue His8 of the RHG motif that is located at the shallow groove.

Like many α/β proteins that act as phosphatases, phosphorylases, or negatively charged nucleotide-binding enzymes, the SixA α/β arrangement directs all N termini of four α helices toward the carboxyl end of the parallel β strands. In particular, helices H3 and H6 have their N termini pointing toward His8. The dipole moments aligned along these α helices likely contribute to attracting the negatively charged phosphate group of the protein substrate toward the active site that is located at the carboxyl end of the parallel β sheet and is highly positively charged due to localized basic residues at the site (Fig. 3A). Interestingly, SixA is an acidic protein with an isoelectric point of 4.2. Indeed, the overall molecular surface is mostly negatively charged (Fig. 3B). This indicates that the electric field around SixA pushes the negatively charged phosphohistidine residue toward the active site.

Phosphoglycerate mutase (PGM) is representative of enzymes possessing a similar α/β fold containing the RHG motif at the active site (Jedrzejas, 2000). The cofactor-dependent phosphoglycerate mutase (dPGM) superfamily includes more distant relatives, such as ubiquitous PGMase,

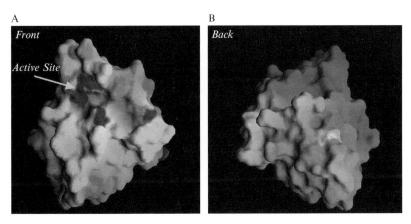

Fɪɢ. 3. Electrostatic molecular surface of SixA. (A) Front view of the electrostatic molecular surface of SixA. Regions of the surface that possess positive electrostatic potentials are colored blue, while those possessing negative potentials are colored red (blue = 10 $k_B T$, red = −10 $k_B T$, where k_B is Boltzmann's constant and T is the absolute temperature). The tungstate ion is shown as a stick model. Acidic and basic residues are labeled. (B) Back view of the electrostatic molecular surface of SixA. (See color insert.)

eukaryotic fructose-2,6-bisphosphatase (F26BPase), prostatic acid phosphatase (PAPase), *E. coli* bacterial periplasmic phosphatase, glucose 1-phosphate phosphatase, alminium-inducible proteins (Ais), and *Bacillus stearothermophilus,* the YhfR gene product (PhoE). SixA shares limited sequence homology (~14%) with the RHG phosphatase domains of PGMase superfamily members, while these enzymes share a similar fold of the phosphatase core domains.

Compared with SixA, the RHG phosphatases possess larger RHG phosphatase domains that have insertions that form an extra α-helical subdomain. The extra α-helical subdomain of yeast PGMase (Campbell *et al.*, 1974) is formed primarily by a long inserted segment between strand β2 and helix H5 (Figs. 2B and 4A). Parts of two loops, β1-H2 and H6-β5,

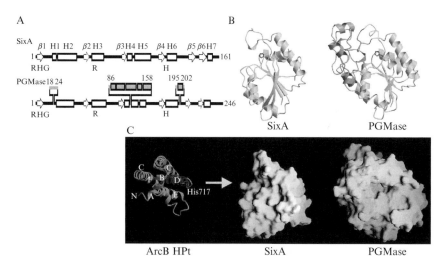

FIG. 4. Comparison of SixA and PGMase. (A) Comparison of the linear arrangement of structural elements and insertion. Helices and β strands are represented with arrows and rectangles, respectively. Inserted segments forming the helical domain are colored orange as in Fig. 2B. (B) Comparison of the RHG phosphatase folds found in SixA (PDB code 1A0B) and PGMase (PDB code 5PGM), the prototype of the RHG enzymes. Ribbon representation of the corresponding RHG phosphatase folds colored pale green. Each active histidine residue is shown as a ball-and-stick model. The extra helical domains forming the active sites of the RHG phosphatase domains are colored orange. (C) The convexity of the ArcB HPt domain (with ribbon representation) is matched with the grooved molecular surface of SixA (with surface representation), whereas the α-helical subdomain of yeast PGMase generates an active site that forms a deep pocket for binding to its small substrate 3-phosphoglycerate and prevents docking to the HPt domain. The active histidine residue (His717) of the HPt domain is shown as a ball-and-stick model. The extra helical domain is colored yellow. The active histidine residues of SixA and PGMase are colored blue-green. (See color insert.)

which may have additional inserted segments, contribute to form the extra subdomain. These insertion sites are also found in both rat F26BPase and rat PAPase and seem to be common among RHG phosphatases that act against small substrates. Detailed examination of these insertions is discussed later. Excluding the insertion and deletion sites, the folds of the RHG phosphatase domains are essentially identical (Fig. 4B): the superposition of SixA on the phosphatase core domains of yeast PGMase, rat F26BPase, and rat PAPase results in best fit with root mean square deviations ranging from 2.7 to 3.5 Å for corresponding C_α carbon atoms.

Conservation of Catalytic Machinery and Active Site

SixA possesses a relatively open active site located alongside a prominent shallow groove and dephosphorylates the phosphate group bound to the active histidine residue, His717, of the HPt domain of ArcB (Ogino *et al.*, 1998). The crystal structure of the ArcB HPt domain revealed that the HPt domain adopts a somewhat kidney-shaped structure formed by a four-helix bundle core fold, with the active histidine residue located at the middle of helix D (Kato *et al.*, 1997). There is a general shape match between the grooved molecular surface of SixA and the convexity of the ArcB HPt domain (Fig. 4C). In contrast to the SixA open active site, each extra α-helical subdomain of other RHG phosphatases covers the active site so that the RHG phosphatase domains form a deep pocket at the active site to trap requisite small substrates.

The RHG (residues 7–9) motif at the C-terminal edge of the β sheet is surrounded by a long loop between strand β1 and helix H2 (loop β1-H2; residues 8–17) and two short loops comprising β2-H3 (residues 51–53) and β4-H6 (residues 108–110). In addition to Arg7 and His8 of the RHG motif, the active site contains conserved residues Asp18, Arg21, Ser51, Arg55, and His108 (Fig. 5A). Among these candidates for the catalytic residues, two histidines (His8 and His108) and two arginines (Arg7 and Arg55) contact with the tungstate ion directly and therefore are likely to be essential for the catalysis, suggesting an in-line nucleophilic attack by His8 (Fig. 5B). Superposition of the active site residues of SixA and each RHG catalytic domain reveals that the relative positions and side chain conformations of two histidines and two arginines (His8, His108, Arg7, and Arg55) are well preserved and suggests that the catalytic mechanism of these enzymes is identical (Hamada *et al.*, 2005). The essential role of these conserved residues in catalytic activity has been investigated by mutation studies of RHG phosphatases (Lin *et al.*, 1992; Tauler *et al.*, 1990).

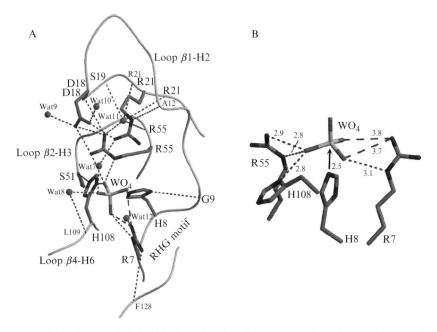

FIG. 5. Active site of SixA. (A) The active site of the tungstate-bound form of SixA. Side chains are shown as stick models with large labels. Main-chain tracings are shown with tubes colored light green. Hydrogen bonds are indicated by dotted lines. Water molecules are shown with small labels (Wats). Residues that participate in hydrogen-bonding interactions through their main chains are indicated with small labels at the main-chain tracings. (B) Close-up stereo view of residues interacting with the tungstate ion at the active site. Hydrogen-bonding and ion-pairing interactions are indicated by dotted lines and broken lines, respectively. Bonding distances are also shown. The contact between the tungsten atom and the N_ε nitrogen atom of His8 is indicated by an arrow. (See color insert.)

These four residues are involved in a hydrogen-bonding/ion-pairing network that adjusts their positions and conformations suitable for catalytic activity. The network involves many main chain amide groups of the loops forming the active site and the Ser51 side chain that is directly hydrogen bonded to His108. Asp21 fixes the side chain conformations of Arg55 and Arg21 by hydrogen-bonding interactions. These three arginines are responsible for forming the positively charged active site and possibly play a role in delocalizing the charges of the transition state expected during phosphate transfer to His8 and also in recruiting and binding the negatively charged substrate phosphate group.

Given the lack of structural evidence of the complex between SixA and its substrate protein, the mechanism by which SixA exerts its specific activity remains unknown. Preliminary model building suggests that the active site pocket of SixA located alongside a relatively shallow groove exhibits a general shape suitable for accommodating the substrate ArcB HPt domain folded into a four-helix bundle. Two contact sites were proposed for interaction between SixA and the AtcB HPt domain in a docking model with a modeled phosphate bound to ArcB active residue His717 on helix D and SixA active residue His8 in an in-line mechanism (Hamada et al., 2005). Site 1 is located at SixA loop β1-H2 contacting with helices D and C of the HPt domain, and site 2 is located at SixA helices H2 and H7 with HPt helix B. Interestingly, the former site involves a SixA-characteristic loop containing the SixA determinant residues as mentioned previously. Further experimental studies are required to clarify the specificity of this novel phosphatase.

Sequence Analysis of SixA Homologs

The dPGM superfamily contains a variety of phosphatases exhibiting both broad and narrow substrate specificity. Moreover, these phosphatases seem to be abundant in microbial genomes. A preliminary challenge for functional annotation based on sequence comparison suggests that SixA forms a distant evolutionary relationship and bears no particularly close relationship to any well-characterized family (Rigden, 2003). Utilizing data derived from the various genome sequence projects, we identified SixA homologs following database searches with BLAST (Altschul et al., 1990) and PSI-BLAST (Altschul et al., 1997) at the National Center for Biotechnology Information (NCBI) site. We searched putative homologs of E. coli SixA from bacteria, archaeal, fungi, and yeast. SixA homologs were found within bacteria, and a total of 98 putative homologs of E. coli SixA were identified from bacteria and archaea for the following analyses (Fig. 6). Interestingly, two copies of SixA homologs were found in three species of α-proteobacteria Mesorhizobium loti, Rhizobium etli, and Sinorhizobium meliloti and in one species of γ-proteobacteria Hahella chejuensis. In contrast to proteobacteria, we failed to find any candidate within most of genera of Firmicutes and other bacteria families (Fusobacteria, Planctomicesm, Chlamydiae, Thermotogae, and Chloroflexi). Moreover, SixA seems to be rare in archaea. This trend in SixA distribution may be correlated with the distribution of HPt proteins. Indeed, the SMART database (Letunic et al., 2006; Schultz et al., 1998) contains a small number (20) of archaeal HPt proteins out of a total of 1444 HPt proteins that contain 1380 bacterial proteins. These archaeal HPt proteins are chemotaxis sensor kinases rather than ArcB homologs.

The collected amino acid sequences of SixA homologs were aligned by MAFFT (Katoh et al., 2002, 2005) with the NW-NS-i option (Fig. 7). Members of the RHG phosphatase family Ais, PhoE, and dPGM were included

\<Bacteria\>

alpha-proteobacteria
- Rickettsia −
- Wolbachia −
- Pelagibacter −
- Anaplasma −
- Ehrlichia −
- Neorickettsia −
- Mesorhizobium loti + (2)
- Mesorhizobium sp. +
- Sinorhizobium + (2)
- Agrobacterium +
- Rhizobium + (2)
- Brucella +
- Bradyrhizobium +
- Rhodopseudomonas +
- Nitrobacter +
- Bartonella −
- Caulobacter +
- Silicibacter +
- Rhodobacter +
- Jannaschia +
- Roseobacter +
- Zymomonas −
- Novosphingobium +
- Sphingopyxis +
- Erythrobacter +
- Gluconobacter +
- Rhodospirillum +
- Magnetospirillum −

beta-proteobacteria
- Neisseria −
- Chromobacterium +
- Ralstonia +
- Burkholderia +
- Bordetella −
- Rhodoferax +
- Polaromonas +
- Nitrosomonas +
- Nitrosospira +
- Azoarcus +
- Dechloromonas +
- Thiobacillus +
- Methylobacillus +

gamma-proteobacteria
- Escherichia +
- Salmonella +
- Yersinia +
- Erwinia +
- Photorhabdus +
- Buchnera −
- Wigglesworthia −
- Blochmannia −
- Sodalis +
- Haemophilus +
- Pasteurella +
- Mannheimia +
- Xylella +
- Xanthomonas +
- Vibrio +
- Photobacterium +
- Pseudomonas +
- Psychrobacter +
- Acinetobacter +
- Shewanella +
- Idiomarina +
- Colwellia +
- Pseudoalteromonas +
- Saccharophagus +
- Coxiella +

- Legionella +
- Methylococcus −
- Francisella −
- Nitrosococcus −
- Hahella + (2)
- Chromohalobacter +
- Alcanivorax +
- Baumannia −

delta/epsiron-proteobacteria
- Geobacter +
- Pelobacter −
- Desulfovibrio −
- Lawsonia −
- Bdellovibrio +
- Desulfotalea +
- Anaeromyxobacter +
- Myxococcus +
- Syntrophus −
- Helicobacter +
- Wolinella +
- Thiomicrospira +
- Campylobacter +

Actinobacteria
- Mycobacterium +
- Corynebacterium +
- Nocardia +
- Rhodococcus +
- Streptomyces +
- Tropheryma −
- Leifsonia +
- Propionibacterium +
- Thermobifida +
- Frankia +
- Bifidobacterium +
- Symbiobacterium −
- Rubrobacter +
- marine actinobacterim clade −

Cyanobacteria
- Synechocystis +
- Synechococcus +
- Thermosynechococcus +
- Gloeobacter +
- Anabaena +
- Nostoc +
- Prochlorococcus −
- Trichodesmium +

Bacteroidetes/Chlorobium
- cytophaga +
- Porphyromonas −
- Salinibacter −
- Chlorobium +
- Pelodictyon +

Spirochaetales
- Borrelia −
- Treponema −
- Leptospira +

Deinococcus-Thermus group
- Deinococcus −
- Thermus +

Aquificae
- Aquifex +

Fibrobacters/Acidobacteria
- Acidobacteria +

Fusobacteria
- Fusobacterium −

Planctomices
- Rhodopirellula −

Chlamydiae
- Chlamydia −
- Chlamydophila −
- Parachlamydia −

Thermotogae
- Thermotoga −

Chloroflexi
- Dehalococcoides −

Firmicutes
- Bacillus −
- Oceanobacillus −
- Geobacillus −
- Staphylococcus −
- Listeria −
- Lactococcus −
- Streptococcus −
- Lactobacillus −
- Enterococcus −
- Clostridium −
- Carboxydothermus −
- Desulfitobacterium +
- Thermoanaerobacter −
- Moorella −
- Mycoplasma −
- Ureaplasma −
- Phytoplasma −
- Mesoplasma −

\<Archaea\>

Euryarchaeota
- Methanococcus −
- Methanosarcina −
- Methanococcoides −
- Methanospirillum −
- Methanobacterium −
- Methanosphaera −
- Methanopyrus −
- Archaeoglobus +
- Halobacterium −
- Haloarcula −
- Haloquadratum −
- Natronomonas −
- Thermoplasma −
- Picrophilus −
- Pyrococcus −
- Thermococcus −

Crenarchaeota
- Aeropyrum +
- Sulfolobus +
- Pyrobaculum +

Nanoarchaeota
- Nanoarchaeum −

FIG. 6. A list of bacteria and archaea in which SixA homologs were found. Each genus name is assigned + or − to indicate the presence or absence, respectively, of a SixA homolog. Species that contain two copies of SixA homologs are marked (2) and include α-proteobacteria *Mesorhizobium loti MAFF 303099, Rhizobium etli CFN 42,* and *Sinorhizobium meliloti 1021* and γ-proteobacteria *Hahella chejuensis KCTC 2396.* Classification of bacteria and archaea is according to NCBI taxonomy (http://www.ncbi.nlm.nih.gov/). (See color insert.)

in the alignment to define characteristic residues of SixA phosphatases. The alignment reveals that Asp18 and Arg21 are well conserved in SixA homologs but not within others. Thus, these two residues are determinant residues for SixA homologs. As mentioned earlier, these residues contribute to form the active site: Arg21 is a member of three arginines forming the positively charged active site and stabilizing the transition state of the

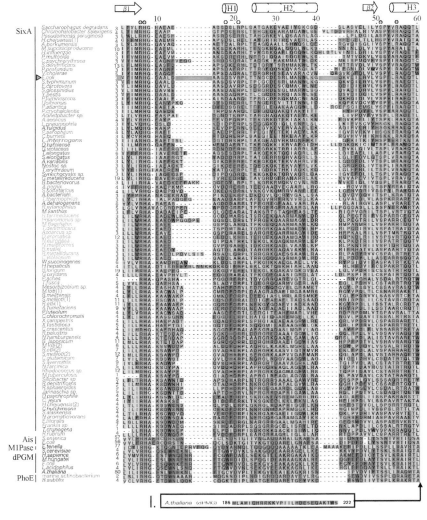

FIG. 7. (*continued*)

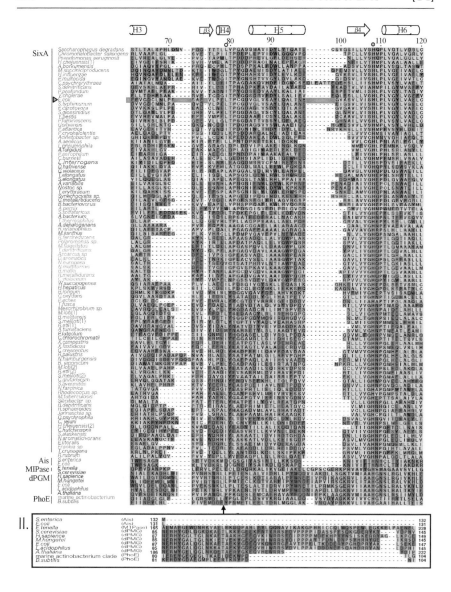

phosphate transfer reaction and Asp18 plays a role of fixing two of these arginines. Both Asp18 and Arg21 are located at the C-terminal region of loop β1-H2 containing helix H1 (Fig. 2B). This loop contains the RHG motif at the N-terminal region and is important for forming the SixA active

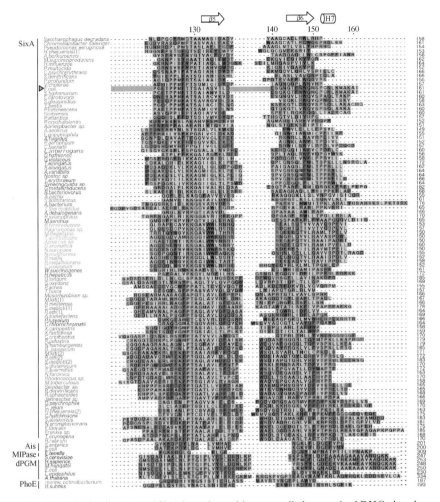

FIG. 7. Multiple alignment of SixA homologs with some well-characterized RHG phosphatases. Sequence data with significant similarity to *E. coli* SixA were collected following database searches using BLAST with E-value <0.001 and PSI-BLAST employing five iterations with E-value <0.005 at the NCBI site (http://www.ncbi.nlm.nih.gov/blast/psiblast.cgi). Following alignment of the collected sequences by MAFFT (Katoh *et al.*, 2002, 2005), alignment was slightly modified by visual inspection. Bacterial proteins are colored magenta (α-proteobacteria), orange (β-proteobacteria), red (γ-proteobacteria), blue (δ/ε-proteobacteria), cyan (actinobacteria), green (cyanobacteria), ruby (bacteroidetes/ chlorobium), and other colors (others). Archaeal and eukaryotic proteins are in gray and black, respectively. Numbers at the beginning and end of each sequence indicate amino acid position of the protein. Aromatic (W, F, Y) residues are colored red, aliphatic (V, L, I, M) pink, acidic and amide (D, E, N, Q) blue, basic (K, R, H) green, cysteine (C) yellow, and others (S, T, P, G, A) gray. (See color insert.)

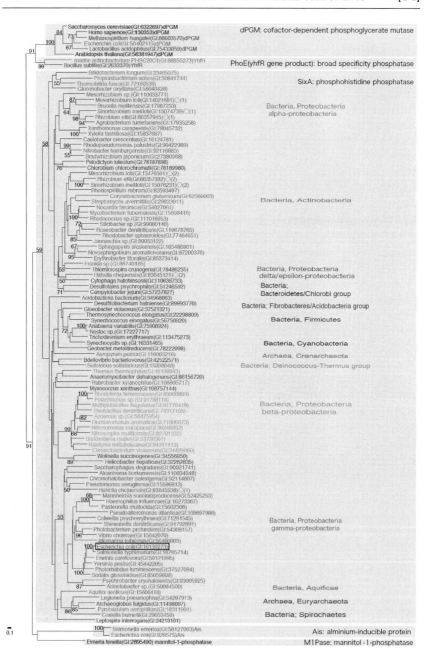

site pocket. The appropriate positioning and orientation of these two residues seem to be reliant on formation of short 3_{10}-helix H1 within the relatively long β1-H2 loop.

The alignment also shows that active residues Arg7, His8, Arg55, His108, and Ser51 represent common determinant residues for RHG phosphatases containing more distant relatives in addition to SixA homologs. Moreover, the presence of the last glycine of the RHG motif is not an absolute prerequisite for these enzymes and is frequently replaced with alanine. In the *E. coli* SixA structure, Gly9 contacts with Leu23, which is strictly limited to leucine or isoleucine in the RHG phosphatases.

The presence or absence of the insertion at loop β3-H5 represents one determinant for detecting SixA homolog candidates. One exception is Ais, which lacks this insertion. *E. coli* SixA possesses a conserved aliphatic residue, Leu79, which should be essential in forming 3_{10}-helix H4 within this short β3-H5 loop. This short helix contributes to active site formation by contacting His108 and Ser51. This aliphatic residue may be characteristic of SixA and is not conserved in Ais, as is the case with Asp18 and Arg21.

To investigate the evolutionary relationships between SixA and other RHG phosphatases, a molecular phylogenetic tree was constructed by the neighbor-joining (NJ) method (Saitou and Nei, 1987) using unambiguously aligned sites of a multiple alignment (Fig. 8). As reported previously (Rigden, 2003), SixA forms a monophyletic cluster distant from other families, including the Ais family. Within the SixA family, most members of each of the three proteobacteria are clustered. Two copies of SixA homologs of three species in α-proteobacteria are separated into two groups that form roughly two clusters, while two copies of SixA homologs of *Hahella chejuensis* in γ-proteobacteria are distantly separated from each other and from any copy of α-proteobacteria. Archaeal SixA homologs are included in the bacterial SixA families.

FIG. 8. An unrooted NJ tree of SixA and some well-characterized RHG phosphatases. Each sequence is indicated by the source name and the GI number in NCBI. For clarity, each genus contains one representative species. From the multiple alignments, 114 unambiguously aligned sites were used for the calculation of evolutionary distances. The evolutionary distance between every pair of aligned sequences was calculated as the maximum likelihood (ML) estimate (Felsenstein, 1996) using the JTT model (Jones *et al.*, 1992) for the amino acid substitutions. Based on these distances, an NJ tree was constructed for all of the sequences included in the alignment. The statistical significance of the NJ tree topology was evaluated by bootstrap analysis (Felsenstein, 1985) with 1000 iterative tree constructions. Bootstrap probability of a cluster is only shown at the root node of the cluster when the value is equal to or greater than 50%. The bar under the tree corresponds to 0.1 amino acid substitutions/site. Phylogenetic analysis and tree drawing were carried out using the XCED program package (http://www.biophys.kyoto-u.ac.jp/~katoh/programs/align/xced/). (See color insert.)

Eukaryotic Histidine Phosphatases

At present, we failed to identify SixA homologs in yeast, fungi, and plants with significant probability. However, identification of HPt proteins and domains with high probability is also problematic given the short amino acid length. We believe that SixA homologs in a variety of bacteria and archaea may possess distinct target proteins containing a HPt domain in addition to or in lieu of ArcB homologs. It is likely that currently accumulated genome data are insufficient to fully characterize SixA evolutionary relationships and that the identification of further SixA homologs should follow with the acquisition of more data. Further structural and functional studies of these candidates are essential in increasing our understanding of diverged two-component signaling pathways.

Acknowledgments

We thank Dr. Kei-ichi Kuma (Kyoto University, ICR) for his helpful advice in sequence analyses. This research was supported in part by a Protein 3000 project on Signal Transduction to TH from the MESSC of Japan.

References

Altschul, S. F., Gish, W., Miller, W., Meyers, E. W., and Lipman, D. J. (1990). Basic local alignment search tool. *J. Mol. Biol.* **215**, 403–410.
Altschul, S. F., Madden, T. L., Schäffer, A. A., Zhang, J., Zhang, Z., Miller, W., and Lipman, D. J. (1997). Gapped BLAST and PSI-BLAST: A new generation of protein database search programs. *Nucleic Acids Res.* **25**, 3389–3402.
Burbulys, D., Trach, K. A., and Hoch, J. A. (1991). Initiation of sporulation in *B. subtilis* is controlled by a multicomponent phosphorelay. *Cell* **64**, 545–552.
Campbell, J. W., Watson, H. C., and Hodgson, G. I. (1974). Structure of yeast phosphoglycerate mutase. *Nature* **250**, 301–303.
Felsenstein, J. (1985). Confidence-limits on phylogenies: An approach using the bootstrap. *Evolution* **39**, 783–791.
Felsenstein, J. (1996). Inferring phylogenies from protein sequences by parsimony, distance, and likelihood methods. *Methods Enzymol.* **266**, 418–427.
Hamada, K., Kato, M., Shimizu, T., Ihara, K., Mizuno, T., and Hakoshima, T. (2005). Crystal structure of the protein histidine phosphatase SixA in the multistep His-Asp phosphorelay. *Genes Cells* **10**, 1–11.
Heo, Y. S., Ryu, J. M., Park, S. M., Park, J. H., Lee, H. C., Hwang, K. Y., and Kim, J. (2002). Structural basis for inhibition of protein tyrosine phosphatases by Keggin compounds phosphomolybdate and phosphotungstate. *Exp. Mol. Med.* **34**, 211–223.

Hess, J. F., Oosawa, K., Kaplan, N., and Simon, M. I. (1988). Phosphorylation of three proteins in the signaling pathway of bacterial chemotaxis. *Cell* **53,** 79–87.

Jedrzejas, M. J. (2000). Structure, function, and evolution of phosphoglycerate mutases: Comparison with fructose-2,6-bisphosphatase, acid phosphatase, and alkaline phosphatase. *Prog. Biophys. Mol. Biol.* **73,** 263–287.

Jones, D. T., Taylor, W. R., and Thornton, J. M. (1992). The rapid generation of mutation data matrices from protein sequences. *Comput. Appl. Biosci.* **8,** 275–282.

Katoh, K., Kuma, K., Toh, H., and Miyata, T. (2005). MAFFT version 5: Improvement in accuracy of multiple sequence alignment. *Nucleic Acids Res.* **33,** 511–518.

Katoh, K., Misawa, K., Kuma, K., and Miyata, T. (2002). MAFFT: A novel method for rapid multiple sequence alignment based on fast Fourier transform. *Nucleic Acids Res.* **30,** 3059–3066.

Kato, M., Mizuno, T., Shimizu, T., and Hakoshima, T. (1997). Insights into multistep phosphorelay from the crystal structure of the C-terminal HPt domain of ArcB. *Cell* **88,** 717–723.

Matsubara, M., and Mizuno, T. (2000). The SixA phospho-histidine phosphatase modulates the ArcB phosphorelay signal transduction in *Escherichia coli. FEBS Lett.* **470,** 118–124.

Mizuno, T. (1998). His-Asp phosphotransfer signal transduction. *J. Biochem. (Tokyo)* **123,** 555–563.

Letunic, I., Copley, R. R., Pils, B., Pinkert, S., Schultz, J., and Bork, P. (2006). SMART 5: Domains in the context of genomes and networks. *Nucleic Acids Res.* **34,** D257–D260.

Lin, K., Li, L., Correia, J. J., and Pilkis, S. J. (1992). Arg-257 and Arg-307 of 6-phospho fructo-2-kinase/fructose-2,6-bisphosphatase bind the C-2 phospho group of fructose-2, 6-bisphosphate in the fructose-2,6-bisphosphatase domain. *J. Biol. Chem.* **267,** 19163–19171.

Ogino, T., Matsubara, M., Kato, N., Nakamura, Y., and Mizuno, T. (1998). An *Escherichia coli* protein that exhibits phosphohistidine phosphatase activity towards the HPt domain of the ArcB sensor involved in the multistep His-Asp phosphorelay. *Mol. Microbiol.* **27,** 573–585.

Ohlsen, K. L., Grimsley, J. K., and Hoch, J. A. (1994). Deactivation of the sporulation transcription factor Spo0A by the Spo0E protein phosphatase. *Proc. Natl. Acad. Sci. USA* **91,** 1756–1760.

Perego, M., Hanstein, C., Welsh, K. M., Djavakhishvili, T., Glaser, P., and Hoch, J. A. (1994). Multiple protein-aspartate phosphatases provide a mechanism for the integration of diverse signals in the control of development in *B. subtilis. Cell* **79,** 1047–1055.

Posas, F., Wurgler-Murphy, S. M., Maeda, T., Witten, E. A., Thai, T. C., and Saito, H. (1996). Yeast HOG1 MAP kinase cascade is regulated by a multistep phosphorelay mechanism in the SLN1-YPD1-SSK1 "two-component" osmosensor. *Cell* **86,** 865–875.

Rigden, D. J. (2003). Unexpected catalytic site variation in phosphoprotein phosphatase homologues of cofactor-dependent phosphoglycerate mutase. *FEBS Lett.* **536,** 77–84.

Robinson, V. L., Buckler, D. R., and Stock, A. M. (2000). A tale of two components: A novel kinase and a regulatory switch. *Nat. Struct. Biol.* **7,** 626–633.

Saitou, N., and Nei, M. (1987). The neighbor-joining method: A new method for reconstructing phylogenetic tress. *Mol. Biol. Evol.* **4,** 406–425.

Schultz, J., Milpetz, F., Bork, P., and Ponting, C. P. (1998). SMART, a simple modular architecture research tool: Identification of signaling domains. *Proc. Natl. Acad. Sci. USA* **95,** 5857–5864.

Suzuki, T., Imamura, A., Ueguchi, C., and Mizuno, T. (1998). Histidine-containing phospho-transfer (HPt) signal transducers implicated in His-to-Asp phosphorelay in Arabidopsis. *Plant Cell Physiol.* **39,** 1258–1268.

Tauler, A., Lin, K., and Pilkis, S. J. (1990). Hepatic 6-phosphofructo-2-kinase/fructose-2,6-bisphosphatase: Use of site-directed mutagenesis to evaluate the roles of His-258 and His-392 in catalysis. *J. Biol. Chem.* **265,** 15617–15622.

Wurgler-Murphy, S. M., and Saito, H. (1997). Two-component signal transducers and MAPK cascades. *Trends Biochem. Sci.* **22,** 172–176.

[15] Triggering and Monitoring Light-Sensing Reactions in Protein Crystals

By PIERRE-DAMIEN COUREUX and ULRICH K. GENICK

Abstract

Many bacterial photoreceptors signal via histidine kinases. The light-activated nature of these proteins provides unique experimental opportunities to study their molecular mechanisms of signal transduction. One of these opportunities is the combined application of X-ray crystallography and optical spectroscopy in protein crystals. By combining these two methods it is possible to correlate protein structure to protein function in a way that is exceedingly difficult or impossible to achieve in most other experimental systems. This chapter is divided into two parts. The first part provides a brief overview of light-regulated histidine kinases and the most important techniques for studying the structure of photocycle intermediates by crystallography. The second part of the chapter is dedicated to practical advice on how to select, mount, activate, and monitor the structural and spectroscopic responses of photoreceptor crystals. This chapter is intended for readers who want to start using these experimental tools themselves or who wish to understand enough about the techniques to critically evaluate the work of others.

Light-Regulated Histidine Kinases: A Brief Introduction

Recent genome sequencing efforts have led to an explosion in the number of bacterial photoreceptors. Many of these photoreceptors signal via histidine kinases. Light signaling via histidine kinases involves, at a minimum, two domains: a light-sensing domain and the kinase domain. In some cases these two domains are part of a single polypeptide chain (Jiang *et al.*, 1999; Vierstra and Davis, 2000), whereas in others the two domains reside in two separate proteins. In some of the latter cases the interaction between the polypeptide chain containing the light-sensing and kinase domains is mediated by a transducer protein so that the light-sensing and kinase domains signal to one another without coming into direct physical contact (Spudich, 2006). In any of these cases both the light-sensing and the kinase domains are functionally and structurally independent and can be studied separately. As a matter of fact, at the time this chapter was written there have been no successful structure determinations of light-regulated histidine kinase domains in complex with their cognate light-sensing domains.

METHODS IN ENZYMOLOGY, VOL. 422   0076-6879/07 $35.00
        DOI: 10.1016/S0076-6879(06)22015-9

The high degree of conservation in the fold and function of the histidine kinase domains (Bilwes *et al.*, 1999, 2001) is juxtaposed by considerable diversity in the light-sensing domains. This chapter discusses the two main classes of light-sensing domains that signal through bacterial histidine kinases. Given the large number of light-regulated histidine kinases our list is necessarily incomplete and given the rapid rate with which new classes of light-sensing domains are discovered, even a list that is complete now would likely become incomplete in the near future. Therefore, instead of aiming to be comprehensive we use a small number of examples to highlight general principles. Also, excellent review articles discussing recent advances in our understanding of the function and biology of these families of light-regulated histidine kinases are available and are referenced.

Microbial Rhodopsins

Microbial rhodopsins (Spudich, 2006) form a large family of light-activated retinal-binding proteins. While their biological functions are diverse (Spudich, 2006), the overall molecular architecture of most sensory rhodopsins is highly similar (Fig. 1). Microbial rhodopsins consist of seven

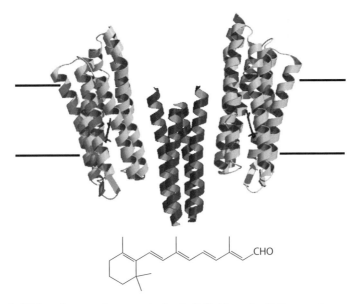

FIG. 1. Ribbon diagram of sensory rhodopsin II (light gray) with its seven transmembrane helix fold and the transducer molecule HtrII (dark gray) in their natural 2:2 complex. The figure is based on the deposited coordinates (PDB 2F95) reported by Moukhametzianov *et al.* (2006). The retinal chromophore can be seen edge on buried in the protein interior and is also shown separately as a chemical line drawing.

transmembrane helices. The retinal chromophore is attached to the protein via a lysine Schiff base and buried inside an internal pocket formed by the seven helices. While the structure of microbial rhodopsins is highly conserved, their cellular functions and molecular modes of action are quite diverse. Microbial rhodopsins can be divided into two general classes: those that use light to pump protons (Lanyi and Schobert, 2004) and ions (Essen, 2002) and those that use light to collect information about the environment. The latter of the two, generally called sensory rhodopsins, are further distinguished by their mechanism of signal transduction. Some regulate histidine kinase activity via transmembrane transducers homologous to bacterial chemotaxis receptors, whereas others transmit signals by regulating ion channel activity (Nagel *et al.*, 2005) or by activating a soluble transducer protein (Jung *et al.*, 2003). Guided by the theme of this volume, this chapter focuses on those rhodopsins that signal via histidine kinases, namely sensory rhodopsin I and II (SR I and SR II) (Bogomolni and Spudich, 1987). SR I and SR II were the first sensory rhopdopsins found in bacteria, and SR II has now been studied very extensively by both structural (Gordeliy *et al.*, 2002; Luecke *et al.*, 2001; Moukhametzianov *et al.*, 2006) and spectroscopic techniques (Bergo *et al.*, 2000; Losi *et al.*, 1999; Swartz *et al.*, 2000), providing insights into the molecular mechanism of signal transduction at a level of detail available for only a handful of other sensory proteins. An excellent review of the recent advances in the study of SR I and SR II has been prepared by Spudich (2006). Briefly, both SR I and SR II exist as 2:2 dimers with their cognate transducer molecules HtrI and HtrII, respectively. These transducer molecules have pronounced structural and functional homology to chemotaxis receptors. Light-driven isomerization of the retinal chromophore results in a protein conformational change involving the coordinated tilting and twisting of helices in the sensory rhodopsin and its cognate tranducers. This structural change (Moukhametzianov *et al.*, 2006) propagates through the long intracellular α helices of the transducer to regulate the activity of CheA. CheA in turn regulates bacterial swimming through the well-known CheA−CheY pathway that stimulates or suppresses the stochastic switching between clockwise and counterclockwise rotations of the bacterial flagellum.

PAS/GAF/LOV Domains

Since the early 1990s there has been explosive growth in the number of newly discovered bacterial photoreceptors. A very large fraction of these new photoreceptors employs light-sensing domains from the PAS/GAF/ LOV family of folds.

Using sequence information alone, PAS (Taylor and Zhulin, 1999), GAF (Hurley, 2003), and LOV (Crosson *et al.*, 2003) domains were initially identified as separate domain families, but once structures for members of each of these families had been determined, it became clear that all three belong to one large fold family.

PAS/GAF/LOV domains (Fig. 2) employ chromophores as chemically diverse as *p*-coumaric acid (Baca *et al.*, 1994), flavin mononucleotide (Christie *et al.*, 1999), and a range of linear tetrapyrrole derivatives (Lamparter *et al.*, 2004; Vierstra and Davis, 2000). In the case of bacterial phytochromes and their tetrapyrrole chromophores, diversity is further enhanced by variations in the chromophore attachment site (Lamparter *et al.*, 2004). Despite this

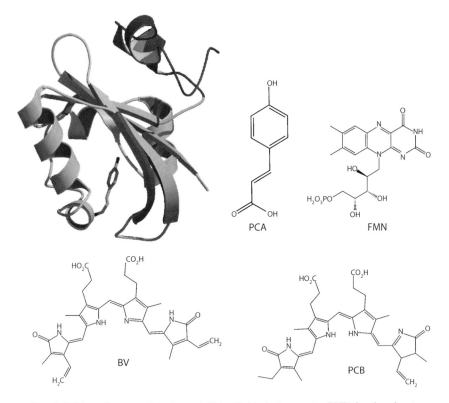

FIG. 2. Ribbon diagram of the bacterial blue light photoreceptor PYP showing the structural separation between the chromophore-binding core conserved across all PAS/GAF/LOV domains (light gray) and the N-terminal subdomain implicated in signal transmission (dark gray). PYP's *p*-coumaric acid chromophore is shown as a stick figure in the environment of the protein fold and as a chemical line drawing (PCA). Other chromophores employed by PAS/GAF/LOV domains, including biliverdin (BV), flavin mononucleotide (FMN), and phycocyanobilin, are shown as chemical line drawings.

tremendous chemical diversity and variability in the chemical mechanism and location of the chromophore attachment site, the location of the chromophore-binding pocket and the secondary structure elements that form this pocket are conserved across the entire family of PAS/GAF/LOV domain photoreceptors.

The PAS/GAF/LOV domain fold is characterized by a central anti-parallel β sheet with 2-1-5-4-3 topology. The structural core of PAS/GAF/LOV domains containing the chromophore-binding pocket is formed by one face of this β sheet and by a series of secondary structure elements that connect strands 2 and 3. In most PAS/GAF/LOV domains the "back side" of the central β sheet is covered by a small 20–50 amino acid subdomain. This subdomain shows little structural conservation, can be attached to either the N or the C terminus of the domain's core, and usually is not required for proper folding or the primary photochemistry of the domain (Harigai et al., 2001; Harper et al., 2003, 2004; Vreede et al., 2003). However, this subdomain appears to play a key role in signaling. In the two PAS/GAF/LOV light-sensing domains that have been studied most extensively, namely the blue light photoreceptor PYP (Kyndt et al., 2004) and the phototropin LOV domain (Harper et al., 2003), light activation of the chromophore leads to the unfolding and dissociation of this subdomain from the domain core (Harper et al., 2003; van der Horst et al., 2001). In the case of the LOV domain it was shown that disruption of the interface between this subdomain and the core of the domain was sufficient to affect regulation of the kinase partner of this light-sensing domain (Harper et al., 2004).

In addition to their ability to regulate the activity of histidine kinases, PAS/GAF/LOV domains can also regulate a broad range of other regulatory domains, including chemotaxis receptors (Watts et al., 2006), diguanylate cyclases (Kyndt et al., 2004; Losi, 2004), serine/threonine kinases (Fankhauser, 2000; Harper et al., 2004), and transcription factors (Semenza, 1999).

This ability to communicate via so many different signaling partners, combined with the ability to employ a very diverse set of sensory cofactors, likely contributed to the remarkable evolutionary success of the PAS/GAF/LOV domain family (see Wuichet et al., 2007).

How PAS/GAF/LOV domains are able to achieve this flexibility is still an open question. We suspect that the ability of PAS/GAF/LOV domains to transmit signals between virtually any combination of sensory cofactor and signaling partner is unlikely to be based on signal transmission through intricate networks of specific intramolecular interaction. Instead we suspect that the intra- and interdomain communication of PAS/GAF/LOV domains operates on a generic mechanism based on a simple geometric or energetic principle. The structural separation of the sensory function and the signaling function into the structurally independent core and signaling subdomain may be the key to such a generic signaling mechanism. Transmission of the signal between these two subdomains appears to be achieved by the very general and nonspecific mechanism of destabilizing the interface between the domain core and the small subdomain. Association of the signaling subdomain and the sensory domain appears to be already tentative even in the absence of a signal (Harper *et al.*, 2003), and any additional disturbance, regardless of its molecular nature, will lead to dissociation of the subdomain and generation of the signal. How the release of this subdomain regulates the activity of a targeted kinase or any of the other myriad of output domains regulated by PAS/GAF/LOV domains is another highly interesting but unresolved question.

Photoreceptors and Kinetic Crystallography: A Near Perfect Match

Kinetic crystallography (for a review, see Bourgeois and Royant, 2005) is the attempt to extend protein crystallography to the study of structural intermediates that occur during the functional cycle of a protein. It turns out that the molecular properties of many of the light-sensing domains employed by bacterial histidine-kinase photoreceptors are ideally suited for kinetic crystallography experiments. First, protein photoreceptors are activated by light and, unlike the small molecules or protein partners that trigger the majority of biological signaling cascades, light can be delivered to the photoreceptor molecules in a crystal with relative ease. The technology of light generation has advanced to the point that experiments on photoreceptors are now constrained largely by the quality of the samples and by the fundamental laws of physics and photochemistry and not by the performance of experimental equipment.

A second key advantage of photoreceptors is that the chromophore of a photoreceptor also provides a convenient and efficient spectroscopic probe to monitor the photochemical and protein conformational responses to light activation.

A third advantage is the optical transparency of protein crystals, which allows light to be delivered to molecules throughout the crystal and to record both ultraviolet (UV)-visible absorption and fluorescence spectra in the crystal under precisely the same experimental conditions under which the structure of the protein is observed. This allows the direct correlation of protein structure and sensory properties in a way that is difficult or impossible to replicate in most other experimental systems.

Finally, unlike mammalian rhodopsin photoreceptors, where light activation results in the release of the retinal chromophore and requires a complex set of biochemical steps to regenerate the dark-adapted state, light-sensing domains signaling through bacterial histidine kinases generally display self-contained photocycles. That is, the chromophores of these photoreceptors remain attached to the protein throughout the entire functional cycle and, after performing their signaling function, the receptors return to their original state without the aid of auxiliary factors. Such a self-contained photocycle is highly desirable for both structural and spectroscopic studies. Samples do not have to be maintained in complete darkness during purification and crystallization and the light-response reaction can be triggered repeatedly during measurements so that overall data quality can be enhanced by averaging spectral and structural data over multiple activation cycles.

Kinetic Crystallography: Two Alternative Strategies

When protein photoreceptors are activated by light the absorbed photon triggers a series of photochemical and protein structural changes jointly referred to as the photocycle of the protein. During this photocycle, various intermediates will accumulate and decay. If the rates of buildup and decay for these intermediates are sufficiently different—as they often are—it will be possible to identify time windows in which a single intermediate will dominate the ensemble of conformational states in the sample. The goal is to identify those time windows, to collect crystallographic data in them, and to thereby determine the structure of the intermediate.

In photoreceptors the lifetimes of intermediates cover the femtosecond (e.g., double-bond isomerization) to second (return to dark-adapted state) range, and pulsed laser sources that can deliver light fast enough to accumulate even the shortest lived of those intermediates are now available commercially. However, collection of a crystallographic data set typically requires minutes to hours. To still determine the structures of protein photocycle intermediates two basic strategies are possible: One can either speed up data collection or one can stabilize the intermediate so that the lifetime of this intermediate exceeds the time required for data collection. The first of the two strategies is generally referred to as true time-resolved crystallography. However, speeding up the standard oscillation method (also known as rotation methods) of data collection (Arndt and Wonacott, 1977) beyond its standard timescale of several seconds to minutes is not trivial. The oscillation method involves the continuous rotation of the crystal in a monochromatic X-ray beam during data collection. To avoid the superposition of the recorded reflections on the detector this rotation is interrupted after a sweep of approximately one degree of rotation and the detector is read out to produce a diffraction image. Depending on the internal symmetry of the protein crystal, between 30 and 180 images obtained from consecutive one-degree sweeps make up a complete data set. Extending this method to the millisecond or microsecond timescale, where many of the interesting protein structural changes in photoreceptors take place, seems out of the question for now. To still determine structures for these short-lived intermediates, true time-resolved crystallography employs the Laue technique (Moffat, 2003) of data collection. For a review, see Bourgeois and Royant (2005). Briefly, in the Laue method, polychromatic X-ray beams are used to expose stationary crystals. The polychromatic X-ray beam eliminates the need for crystal rotation during data collection, and a single exposure of a static crystal can be used to collect a substantial portion of a complete data set. The current generation of synchrotron X-ray sources can be operated in so-called single-bunch mode, in which they generate polychromatic X-ray pulses as short as 100 ps and deliver sufficient photon flux to obtain diffraction data from a single X-ray pulse. By combining and synchronizing these X-ray sources with femtosecond laser sources, crystallographic data collection on photocycle intermediates can be performed in a manner analogous to standard pump-probe laser spectroscopy with subnanosecond time resolution. The laser pulse activates the crystal, the photoreceptor molecules in the crystal then progress through their photocycle, and when the intermediate of interest accumulates, crystallographic data are collected by the X-ray pulse. In the case of a protein with a self-contained photocycle the sample then relaxes to its starting state so that this cycle can be repeated. While the data collection principle is rather simple, its implementation is extremely challenging. Also, the quality of diffraction data obtained from a single 100-ps pulse is too low to be useful. Therefore, diffraction images are typically generated by accumulating the X-ray intensities of 10–20 pump-probe-relax cycles on a detector before the image is read out (Ren et al., 2001). To increase data redundancy such diffraction images are collected for many different crystal orientations. For the collection of a complete data set a crystal may then need to be exposed several hundred or even several thousand times to the activating laser beam and X-ray pulse. Not many crystals are able to tolerate these X-ray and laser doses while maintaining the exquisite crystal order required for Laue data collection. Therefore, while picosecond time-resolved Laue crystallography has produced spectacular results for the bacterial photoreceptor protein PYP (Ren et al., 2001) and the oxygen carrier myoglobin (Schotte *et al.*, 2003), it will remain to be seen if true time-resolved crystallography experiments can be extended to other experimental systems with less ideal sample properties.

Trapping Intermediates for X-Ray Crystallography

As described at the beginning of the previous section the structure of photocycle intermediates can also be determined by slowing down or bringing to a halt the functional cycle of a protein when the intermediate of interest has accumulated.

The two most popular techniques for achieving this arrest in the functional cycle are chemical and physical trapping. Chemical trapping is of greater importance in the study of enzymes where substrate analogs can be used to generate stable enzyme–substrate adducts for structural studies. In the study of light-activated proteins, chemical trapping is confined largely to the generation of site-directed mutants that arrest at a particular point in the photocycle. The stabilization and structure determination of the M-state photocycle intermediate of bacteriorhodopsin, with the help of the D96N mutant (Luecke *et al.*, 1999), provide one example of chemical trapping applied to a light-activated protein.

Physical trapping, specifically trapping of intermediates at cryogenic temperatures, is by far the most popular technique for photoreceptor proteins. The conceptually simplest form of cryotrapping is the accumulation of an intermediate at room temperature followed by rapid freezing to cryogenic temperature. These cryogenic temperatures freeze out the cooperative motions of networks of neighboring residues required for protein conformational changes and protein function (Rasmussen *et al.*, 1992) so that, once frozen, photoreceptors are unable to progress through their photocycle. While examples of this activate-then-freeze approach to light-activated proteins exist (Schobert *et al.*, 2003), the most common cryotrapping method employed to photoreceptors is the freeze-then-activate method (Genick *et al.*, 1998). This method may appear counterintuitive, as it "traps" before it activates, so a small excursion into the physical basis of this technique may be helpful. Occurring within a few hundred femtoseconds to a few picoseconds, chromophore isomerizations in photoreceptor proteins are among the fastest of all chemical reactions. For comparison even the simplest protein conformational changes, such as side chain rotations, take place on timescales thousands to millions of times slower. Then, by physical necessity, the initial photochemical reactions of photoreceptors cannot be accompanied by large concerted side chain motions, and protein conformational changes during the primary photochemical reaction are generally limited to displacements of less than 0.2–0.3 Å in the first shell of residues surrounding the chromophore. This 0.2- to 0.3-Å distance scale corresponds to the atomic mean free path inside a protein, which is the average distance a molecular group can travel before it collides with its nearest neighbor. In other words, on the timescale of the photochemical reaction, the structure of the photoreceptor's protein component remains essentially "frozen" and the only molecular group undergoing substantial structural changes is the chromophore itself. Following this rationale, freezing of photoreceptors to cryogenic temperatures should not hinder the primary photochemical reaction. Indeed, the efficiency of light-driven chromophore isomerization does not appear to diminish at liquid nitrogen temperature (Birge *et al.*, 1988). However, the cryogenic temperature does block subsequent conformational changes in the protein. The freeze-then-activate method therefore traps photoreceptors in a state in which the primary photochemical reaction is complete, but the surrounding protein has not been able to adjust to the new chromophore conformation.

It is then possible to raise the sample temperature to allow successively larger protein conformational changes. The idea is that to undergo these ever-larger structural changes, the protein has to overcome ever-higher energy barriers. By adjusting the temperature, one can select which height of energy barrier the protein is able to surmount. As a result it is then possible to step through the functional cycle of a photoreceptor by illuminating the sample at liquid nitrogen temperature, by using light to drive the initial photochemical reaction, and by collecting several crystallographic data sets at increasingly higher temperatures. In a review article, Lanyi (2004b) described how this strategy has been used to map out a large part of the bacteriorhodopsin photocycle.

In practical terms, the freeze-then-activate method (Genick *et al.*, 1998) involves the following steps. A photoreceptor crystal is adapted to the dark, flash-frozen to liquid nitrogen temperature, and a data set for the dark-adapted state of the protein is collected. The crystal is then exposed to a light beam of moderate intensity, after which a second data set is recorded for the structure of the light-activated state. It should be noted that this scheme allows the collection of dark and light data sets on the same crystal without thawing, which is not possible for the activate-then-freeze approach. Collecting both data sets on the same crystal without intermittent thawing avoids a myriad of crystal-to-crystal variations or subtle differences introduced by changes in the crystal mounting and freezing procedures. As a result, in the freeze-then-activate method, differences between the two data sets are dominated by the light-driven structural changes in the photoreceptor structure.

To a noncrystallographer the freezing process may sound exotic, but the freezing of protein crystal to liquid nitrogen temperatures is a routine process and virtually all X-ray data collection systems are now equipped with cryostats that maintain crystals in a stream of cryogenic gas. The reason >90% of all new protein structures are now determined at cryogenic temperatures has nothing to do with cryotrapping. Low temperatures techniques are instead employed to minimize radiation damage from the ever-more powerful synchrotron X-ray beam used for X-ray data collection.

However, with the facilities for the freezing of protein crystals widely available, the only "experimental resource" required to perform freeze-then-activate experiments is a simple light source for crystal activation (see later).

Structural Interpretation of Kinetic Crystallography Results

Once kinetic crystallography data have been collected successfully, the next challenge lies in the interpretation. While structure determination by X-ray crystallography is now considered routine, the following factors conspire to make the interpretation of light-triggered structural changes in a protein much more challenging. First, photoactivation is virtually never complete. Typically only 25–50% of all molecules in a crystal will become activated even when activation conditions have been optimized. Experiments on PYP performed both in solution (Imamoto et al., 1996) and, more recently, in protein crystals (P.-D. Coureux and U. K. Genick, unpublished results) suggest that this limited photo conversion is inherent to cryo-cooled photoreceptor samples. Similar limits on achievable activation levels seem to also exist, at least for PYP, when activation is performed with ultrashort laser pulses (Anfinrud, personal communication). The net effect of the limited conversion is that the quality of the information for the light-activated state is substantially lower than would be expected for a fully light-activated crystal. Results on PYP indicated that the uncertainty in atomic position is inversely proportional to the occupancy of the atom. For example, if 25% of all molecules in a crystal have been activated, the accuracy with which the atomic coordinates of this activated state are known decreases fourfold. Second, structural changes during the photocycle take place in a confined environment so that the dark-adapted and light-activated molecules occupy the same physical space and the corresponding electron densities overlap. Third, the initial photoproducts retain a substantial portion of the absorbed photon energy in the form of conformational strain so that the chromophores can be expected to adopt noncanonical, distorted geometries. Therefore, stereochemical restraints (Engh and Huber, 1991), even when they are available for a chromophore, will most likely not be appropriate for refinement. Finally, the data quality from kinetic crystallography experiments is often considerably worse than that for a static structure. In the case of the Laue method, photons from the entire polychromatic beam will contribute to the background while individual reflections are only stimulated by one of the wavelengths. In addition, many high-resolution reflections stimulated by short wavelengths will fall on the detector in areas that are dominated by solvent scattering of longer wavelength X-rays, again contributing to background noise. As a result, useful Laue data are typically available to less than half of the maximal resolution obtained for the best monochromatic data collected for a static structure. Even for cryotrapping experiments the data quality for the light-activated intermediate is often less than optimal. Activation levels and overall diffraction quality will vary from crystal to crystal so that optimal activation and optimal diffraction quality rarely coincide.

All these factors make the structure determination of photocycle intermediates more difficult than the interpretation of static crystallographic data. At the same time the structural changes of interest are often very small and differences in atomic coordinates can have large implications for the functional interpretation of the structural result. It is therefore advisable that the interpretation of electron density maps or the refinement of atomic coordinates against crystallographic data of photocycle intermediates should be performed with the utmost care and with a healthy dose of skepticism. Unless ultrahigh resolution data (i.e., better than ~1.2 Å) are available and the intermediate has been accumulated to a high degree (i.e., >~30%), it will be very difficult to perform a meaningful refinement of the molecular geometry of photocycle intermediates. In particular it will be very difficult, both for the experimenter and for the reader of published reports, to assess the reliability of those refinement results. The absolute crystallographic R factor, which is often used as an indicator for the reliability of the structure of a protein, is virtually useless for assessing the reliability of the refined structure of a photocycle intermediate. The number of atoms involved in the structural changes during a protein photocycle and the occupancy of the intermediates are usually so low that the difference between the "correct" and a purposefully wrong model will scarcely change the R factor by 1/10th of a percent. The real-space correlation coefficient (Reddy *et al.*, 2003) provides a quantitative measure of the local agreement between model and experimental data and will therefore be much more sensitive than the R factor. In many cases it may be best to think of time-resolved crystallography not as a method to determine the structure of an intermediate *de novo*, but as a means to distinguish between a limited number of alternative models, where those models have been built based on prior knowledge, chemical intuition, and so on. The agreement of those models with experimental data can then be assessed by comparing different electron density maps that are experimentally observed with maps that are calculated from the model—ideally using the real-space correlation coefficient as a way to quantify this agreement. Readers who deem our assessment of time-resolved crystallography's powers to determine structures of photocycle intermediates *de novo* to be overly cautionary are urged to consult Lanyi (2004a), who described the case of the L-state photocycle intermediate in bacteriorhodopsin. Three research groups determined the structure of this photocycle intermediate independently. Taken by themselves the three structures appeared convincing, but they disagreed with each other dramatically.

Optical Properties of Protein Crystals

One property inherent to crystalline protein samples is their high protein concentration, which commonly ranges from ~300 to 800 mg/ml (i.e., tens of mM). This high-protein concentration and the efficiency with which photoreceptors capture photons make photoreceptor crystals very optically dense. The following equation relates the familiar molar extinction coefficient (M^{-1} cm^{-1}) to the specific optical density of a crystal in units of μm^{-1}.

$$OD\big/_{d} = \varepsilon \frac{n}{V} \cdot 0.166 \ \mu m^{-1} \tag{1}$$

where d is the crystal thickness in micrometers, the unit cell volume V is given by

$$V = a \cdot b \cdot c \sqrt{1 - \cos^2\alpha - \cos^2\beta - \cos^2\gamma + 2 \cos\alpha \cos\beta \cos\gamma} \ \overset{\circ}{A}^{3} \tag{2}$$

and n is the number of molecules per unit cell.

With molar extinction coefficients in the range of 10,000–60,000 M^{-1} cm^{-1} the specific optical density of photoreceptor crystals often approaches or exceeds 0.1 μm^{-1} so that crystals of the size (>100 μm diameter) typically used in crystallographic experiments can display optical densities in excess of 10 OD units. Using Beer's law $-\log(I/I_0) = OD$ we can see that under these conditions only 1 in 10^{10} of the photons that enter a crystal will be transmitted, making reliable absorption measurements impossible. Practical experience shows that measurements of ODs greater than 2 become unreliable. This then requires crystals to be no thicker than 5–10 μm.

Another unique property of protein crystals is their optical anisotropy. The efficiency of photon absorption depends on the proper alignment of the light's electromagnetic field vectors with the transition dipole moment of the chromophore. In solution, because photoreceptor molecules are free to rotate, the transition dipole moments are oriented randomly and the optical density is independent of the light's polarization. In contrast, in a protein crystal intra- and inter-molecular forces fix the orientation of the chromophore relative to the crystal axes. As a result, the efficiency of photon absorption will depend on the orientation of the light's plane of polarization relative to the crystal axes. Depending on the crystal's space group and on the orientation of chromophores in the unit cell the resulting anisotropy can be quite pronounced. For example, in PYP crystals grown in space group P6$_3$ the optical density for light with the plane of polarization perpendicular to the crystal's sixfold axis is five times higher than for light with the plane of polarization parallel to that axis (Ng *et al.*, 1995). Another effect of this anisotropy is that Beer's law does not hold strictly when measurements are performed with unpolarized light.

Mounting Crystals

Crystals mounted for X-ray data collection in kinetic crystallography experiments are mounted much the same way crystals are mounted for conventional crystallography. However, when crystals are mounted for spectroscopic measurements, a few special considerations apply. For measurements at cryogenic temperatures we employ the same cryoloops (Hampton Research, Aliso Viejo, CA) used for routine crystallographic measurements. The main difference is the size of the loop. While the rule of thumb for mounting crystals for X-ray data collection is to select loops that match the size of the crystal, this often produces less than desirable results when mounting crystals for spectroscopic measurements. A close match between crystal and loop size usually results in a near-spherical drop of mother liquor/cryoprotectant mixture around the crystal. Because of its curvature the drop acts as a small lens that redirects the measurement beam—the greater the curvature, the more pronounced the effect. To avoid formation of this mother liquor bead it helps to mount the crystal in a loop that is much larger than the crystal so that the mother liquor forms a flat film, in which the crystal is suspended. Alternatively, a loop much smaller than the crystal can be employed so that the crystal protrudes from the loop. When flash freezing is performed quickly, this latter strategy, in our hands, showed no deleterious effect from salt crystal formation or crystal drying and even PYP crystals grown from near-saturated $(NH_4)_2SO_4$ were perfectly preserved using this method. An added advantage of "dry" mounting a crystal in this fashion is that it allows the sample to be warmed to temperatures above the point where the flash-frozen vitreous mother liquor undergoes a phase transition to form cubic or hexagonal ice. Because the microscopic ice crystals generated in this transition scatter light strongly, the mother liquor drop turns from transparent to opaque and the phase transition can be observed readily by the naked eye. Usually this phase transition in the mother liquor is accompanied by a loss of diffraction from the protein crystal. However, the disruption of the protein crystal does not appear to originate from a phase transition of water inside the solvent channels of the crystal. Instead the disruption appears to be caused by the compression/shear forces exerted by the volume expansion associated with the ice crystal formation in the mother liquor drop. This interpretation is supported by our observation on rod-shaped PYP crystals that were mounted such that part of the crystal extended outside the loop and beyond the mother liquor drop. When controlled warming of a cryogenically frozen crystal triggered the phase transition of the mother liquor drop, diffraction from the area of the protein crystal surrounded by the mother liquor was lost simultaneously. Translation of the crystal so that the "dry" portion of the crystal was exposed to the X-ray beam showed that this portion of the crystal had survived the phase transition in the mother liquor unscathed.

For experiments at or near room temperature, evaporation of the mother liquor becomes a problem so crystals need to be mounted in an environment with controlled humidity. This is achieved most easily by mounting crystals in a sealed capillary containing a large drop of mother liquor placed a few millimeters from the crystal. The capillaries commonly used for X-ray data collection offer an attractive choice. While curved, the walls of these capillaries are so thin that they do not significantly distort the measurement beam. Choosing a large-diameter capillary further mitigates this already subtle distortion.

The real problem—once again—is the mother liquor. Any excess mother liquor surrounding the crystal will have a tendency to bead up and these beads will act as lenses to misdirect the measurement beam. An alternative method for mounting crystals at room temperature is the use of so-called microslides. Microslides are capillaries with two parallel walls and a rectangular cross section (Vitro Dynamics, Inc., Rockaway, NJ). These slides are often made from quartz or other UV-transparent glasses and have excellent optical qualities. The disadvantage of these slides is their high wall thickness, which makes them sturdy and attractive for spectroscopic work, but entirely unsuitable for X-ray diffraction work. When working with microslides we usually select slides with inner dimensions of $100-300$ μm by 3 mm. We mount many crystals together using capillary action to soak up some mother liquor and then soak up an entire crystallization drop containing multiple crystals followed by additional mother liquor. The capillary is then sealed with epoxy glue. Using this strategy, plate-shaped crystals often align themselves with the largest face of the crystal parallel to the optical walls of the slide—ideally positioned for measurement. By storing the

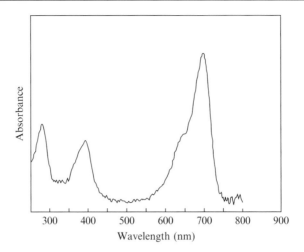

FIG. 3. Single crystal absorption spectrum recorded in a nanodrop (Nanodrop.com) spectrometer. The spectrum was recorded on a plate-shaped $200 \times 200 \times 5$-μm^3 crystal of a bacterial phytochrome and was provided courtesy of J. Wagner and K. Forest.

microslides flat on one side, crystals tend to adhere to one of the walls of the capillary so that the slide can then be rotated for measurement without dislodging the crystal from the capillary wall.

An interesting alternative to these mounting methods is to forego mounting of the crystal altogether. The nanodrop spectrometer (Nanodrop.com) allows transmission measurements on microliter-sized drops by positioning two optical fibers so closely that surface tension causes the formation of a liquid column between the two windows. This commercially available spectrometer can be used to record spectra on single crystals by pipetting a droplet containing the crystal into the measurement position. The spectra Jeremiah Wagner and Kathrina Forest recorded with this instrument (Fig. 3) on crystals of a bacterial phytochrome (Wagner *et al.*, 2005) suggest that for static absorption measurements at room temperature, this instrument may be an interesting, simple-to-use, low-cost alternative to the microscope-objective based spectrometer described below.

Design of a Single Crystal Microspectrophotometer

A spectrophotometer consists of three primary components: a light source, optics to deliver and collect the light, and a detector. The main differences between a conventional spectrometer and its single crystal equivalent are the requirements of the optical elements that deliver and collect the light. Standard spectrometers for solution measurements employ measurement beams with diameters of a few millimeters. In a single crystal spectrophotometer the beam size needs to be compressed to a few micrometers. This essentially requires the use of specialized microscope objectives. This section provides a description of our own microspectrophotometer design (Fig. 4), which builds on a previous design developed by Hajdu and colleagues (1993). The heart of both designs is a pair of mirror-based 15× objectives with 0.28 numerical aperture (Ealing; Rocklin, CA), which are focused on the sample from opposite directions. Light is delivered to and collected from these objectives by optical fibers. Advantages of these objectives are (i) their ability to deliver light across the 200- to 1000-nm wavelength range without chromatic or spherical aberration; (ii) the long working distance (25 mm), which provides ample room to

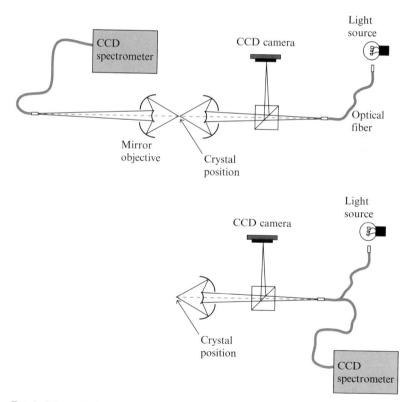

Fig. 4. Schematic drawings of a simple microspectrophotometer, which allows observation of the target along the optical axis of the measurement beam. This design can be converted easily between absorption and fluorescence mode.

deliver a cryogenic cooling jet and activating light beams to the sample; and (iii) the ability to produce extremely small focal spots (<3 μm diameter).

In our design the collimating lenses of the earlier design (Hadfield and Hajdu, 1993) are eliminated and the light diverging from the fiber end is collected directly by the mirror objective. The fiber and microscope objective are held 2.5 in. apart by mounting hardware (micro cage by Thorlabs Inc., Newton, NJ) with the fibers held in two-dimensional microtranslation stages. One of the objective/fiber assemblies contains a beam-splitting cube that delivers part of the light collected by the objective onto a low-cost "bullet-style" charge-coupled device (CCD) camera. This camera allows observation of the crystal target coaxially to the measurement beam, greatly aiding in the alignment of the crystal in the measurement beam and in determining the size of the measurement spot. The crystal is mounted in the focal spot of the spectrophotometer via a goniometer head that is attached to a three-axis translation stage. In our experience it is desirable to have a mounting stage with at least 1-in. travel in all three directions and to invest into a high-quality micrometer driven stage for this purpose.

Fiber-coupled light sources that produce continuous output from the near UV to the near IR are available commercially from Ocean Optics (oceanoptics.com) and a variety of other vendors. Reasonably priced detectors are available from the same firm and also from a range of other vendors often marketed under the label "hand-held spectrometers." The basic choice is between CCD- and diode array-based detectors. In our experience we would suggest a CCD detector because of the higher sensitivity of this detector technology (see section on activation by measurements light later) and would avoid diode array detectors, which produce spectra with higher signal-to-noise ratios, but require much higher light doses. In our experience the relatively low-cost Ocean Optics CCD detectors have performed so well that the considerable expense of a high-performance detector system would only be justified in very unusual cases.

Challenges of Recording UV-Visible Absorption Spectra in Crystals

The light used to record absorbance in crystals will always and inevitably trigger the light-sensing reaction in some of the molecules in the sample. So no matter how careful the experimenter, the act of monitoring the state of a photoreceptor sample by optical spectroscopy will always alter the state of that sample as well. For measurements in solution this is usually not a major concern. However, photoreceptors are very efficient at capturing photons and at translating photon absorption events into photochemical and protein structural changes. Photoreceptor crystals also contain very high protein concentrations and the measurement light has to be focused through a very small sample volume so that activation by the measurement beam becomes a serious concern.

Here is an example calculation. The type of CCD array detector in common use with microspectrophotometers (e.g., the USB2000 manufactured by Ocean Optics) requires measurement light with an I_0 of ~10 pJ^{-1} to obtain an absorption spectrum with an OD of 2 and a noise rmsd of ~10%. At this OD 99% of the I_0 photons will be absorbed in the sample. Because most photoreceptor absorption bands are 20–40 nm (FWHM) wide, the total dose of photons absorbed during collection of a medium-quality absorption spectrum is on the order of 300 pJ. Assuming a rod- or cube-shaped crystal with a specific OD of 0.25 μm^{-1} (e.g., PYP crystals), a crystal with an OD of 2 will be just 8 μm in diameter and to avoid artifacts caused by measurement light bleeding around the crystal (see later) the measurement spot diameter cannot be larger than 5 μm. Recording a spectrum will then deposit ~300 pJ of photon energy into a 157-μm^3 volume (8 μm long, 5 μm diameter) of the crystal. For the case of a PYP crystal this corresponds to ~6 × 10^8 photons deposited into a volume containing ~6 × 10^9 molecules or approximately one photon for every 10 molecules. Taking into account PYP's quantum yield of photo activation of around 30%, a series of 30 consecutive spectra will, on average, trigger the light response reaction in each molecule in the illuminated volume. Because microspectrophotometery of such small protein crystals requires the collection of multiple spectra alternating with careful alignment of the crystal in the measurement beam and subtle adjustments in the fiber position, a large number of protein molecules in the measurement volume may have been exposed to a photon before the desired spectrum has been recorded.

The easiest way to mitigate the activation problem is to increase the area over which the measurement light is distributed. Increasing the measurement spot from 5 to 50 μm in diameter reduces the number of photons per molecule by a factor of 100. In the example calculation given earlier, this would mean that not 1/10 but only 1/1000 molecules in the crystals would be activated, at which point light activation by the measurement beam is essentially negligible. Crystals in the shape of large thin plates are therefore the ideal specimens for absorption measurements. Such crystals allow the use of a large measurement beam while keeping sample thickness—and with it sample OD—in check. Unfortunately, such plate-shaped crystals are often not well suited for diffraction measurements, and the compact, "chunky" crystals that perform well in diffraction experiments are not suitable for spectroscopic measurement. Fortunately, every now and then crystallization trials during the optimization of crystallization conditions will yield drops in which crystals of the same microscopic crystal form will grow in distinctly different macroscopic crystal morphologies. These trials often generate the flattened crystal shapes so desirable for spectroscopic measurements.

Another way to mitigate the problem of high optical density is to take advantage of the pronounced optical anisotropy (see earlier discussion) of some photoreceptor crystals. Insertion of a polarizer into the light path of the microspectrophotometer between the delivery fiber and the first objective then allows measurements to be taken with light that is polarized to be minimally absorbed by the crystal. In the case of PYP this allows the use of crystals that are substantially larger (e.g., $\sim$40 μm instead of 8 μm thickness) so that the photon dose can be spread out over a volume containing more than 100 times the number of molecules.

Leaking Light Introduces Spectral and Kinetic Artifacts

While a large measurement beam is desirable to distribute the photon dose of the measurement beam over as large an area as possible, the beam size has to be chosen such that no measurement light can leak around the crystal. Such light leaks, that is, measurement light that bypasses the crystal, can lead to unexpected artifacts in both static and kinetic absorption measurements. In static measurements, leaking light emphasizes weak spectral features, whereas in kinetic measurements, it causes an apparent acceleration or slow down of reaction kinetics and may give the impression of multiexponential kinetics. Similar effects will also occur when the thickness of a crystal varies across the measurement beam, as may be the case in measurements on wedge-shaped crystals.

The source of these artifacts can be understood by considering that when a light leak is present, the light hitting the detector is the sum of two components: light that passed through the crystal and light that bypassed the crystal. Recalling Beers' law

$$OD = -\log\frac{I}{I_0} \tag{3}$$

we see that a sample with an OD of 4 absorbs all but 0.01% of the photons in the measurement beam. If we now consider an absorption measurement on a crystal with OD 4 but assume that 1% of the measuring light leaks around the crystal, then the detector will encounter 1.0099% of the original light intensity: 0.01% of the light that entered the crystal (i.e., 0.01% of 99% = 0.0099%) plus the 1% of the original light intensity contributed by the leak. The result is an apparent OD of 1.996. Generally, if 1% of light is leaking around the sample, the apparent OD will not exceed 2 and any measurement on crystals with a true OD approaching 2 will become seriously flawed. Also, the higher the true OD of the sample, the more pronounced the effect of even moderate light leaks. Besides resulting in flawed readings of absolute ODs, which are usually of little interest in single crystal absorption measurements, leaking light produces artifacts both when monitoring kinetics and when measuring absorption spectra. The net effect of light leaks on absorption spectra is a decrease in the apparent OD of strong absorbance bands, thus resulting in the relative amplification of weak spectral features. Figure 5 shows a simulation of how a light leak will affect a hypothetical absorption spectrum with a strong and a weak band. The leak dramatically reduces the apparent extinction coefficient of the strong (400 nm) band while the weak (500 nm) band is barely affected. Let us now assume that the 400- and 500-nm bands shown in Fig. 5 correspond to the dark-adapted and a light-activated form of a photoreceptor and that the extinction coefficients for the two states at their respective spectral peaks are identical. The true spectrum would then indicate that 25% of the molecules have accumulated in the intermediate state, while the spectrum affected by the light leak suggests 40% accumulation. In other words, the light leak causes a 60% overestimate in the intermediates occupancy. The higher the OD of the dominant peak, the stronger this effect will be.

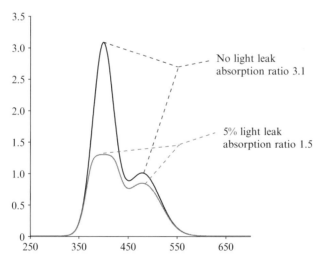

FIG. 5. Effect of light-leak artifact on absorption spectra. The simulation shows the effect of a 5% light leak on the apparent relative intensities of two absorption bands. The "true spectrum" is shown in black and the spectrum under the effect of the light leak in gray. The light-leak artifact drastically attenuates the apparent intensity of the 400-nm absorption band relative to the 500-nm band, leading to a dramatic overestimate of the contribution the 500-nm-absorbing species makes to the overall absorption spectrum.

In addition to these spectral artifacts, leaking light can also affect the measurement of reaction kinetics. In general, leaking light will appear to accelerate the kinetics, with which a spectral band builds up, while it will appear to slow down the decay of a spectral band. Figure 6 shows a simulation of how leaking light leads to the apparent acceleration in the buildup of a spectral band. The simulation assumes a crystal with OD = 3 and light leaks of 0.1 to 10%. In this case even a modest light leak of 1% doubles the apparent rate of the reaction. Disconcertingly, light leak artifacts are hard to spot. Shapes of time traces obtained from "leaky" measurements closely resemble single exponential decays and any deviation is easily masked by noise. If low noise data have been obtained, these residuals will become

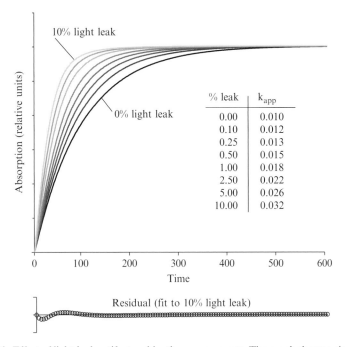

FIG. 6. Effect of light-leak artifact on kinetic measurements. The graph shows a simulation of the effect of successively larger light leaks on the apparent recovery kinetics when following the reappearance of a "bleached" absorption band. The amplitude of the time traces has been normalized. This normalization helps emphasize the effect of the light leak on the kinetics, but hides the fact that the light leak alters the amplitude as well as the kinetics of the recovery signal. Below the main panel is a separate graph showing the residual of a single-exponential fit to the recovery curve affected by a 10% light leak. The residual plot is shown on the same vertical scale as the main plot and displays the hallmark shape of an additional exponential component.

apparent. Unfortunately, the shape of the residual generated when a single-exponential decay is affected by a light leak (Fig. 6 lower panel) resembles the residuals generated when a multiexponential decay is fitted by a single exponential. The light-leak artifact may then be misinterpreted as evidence for multiexponential kinetics. Therefore, if multiexponential kinetics are observed in a single crystal absorption experiment, great care must be taken to ensure that such kinetics are not light-leak artifacts.

In photoreceptors the decay of one spectral feature is very often coupled to the buildup of a new spectral feature. If light leakage is not a problem, the time courses for the buildup and decay of those features should follow the same time course (Fig. 7). If light leakage is a problem, the kinetics of the two features will disagree. Specifically the time course of the buildup will appear to be faster than that of the decay. If the transition involves only the decay or buildup of one spectral feature, then a similar test can be performed by fitting time traces recorded at the peak and the shoulder of the absorption band. Again, the two time constants should agree; if not, leakage is a likely culprit.

Fluorescence Measurements

The high chromophore concentration that makes absorption measurements in protein crystals so difficult may often aid fluorescence measurements. Specifically, the high concentration of chromophores/fluorophores in protein crystals can result in a robust signal, even if the fluorescence efficiency of the fluorophores is so low that it would make the equivalent measurements in solution unattractive. Fluorescence measurements are also quite forgiving regarding the optical perfection of the crystal surface, size of the crystal, and leaking light and can be performed on both very large and very small crystals. In our opinion, single crystal microspectrofluorimetry with either endogenous or exogenous fluorophores (Weik *et al.*, 2004) is a highly underused technique.

Because conversion of our microspectrophotometer design for fluorescence measurements is rather easy, fluorescence measurements can be attempted without the need for additional hardware purchases. Figure 4 shows how the spectrophotometer can be adapted for fluorescence measurements in "back-scattering" geometry. The key to this assembly is a 2-to-1 fiber-optical cable (Ocean optics, oceanoptics.com) in which three fibers are spliced together in a Y configuration so that both excitation and emitted light are delivered/collected through the same optical fiber. This fiber assembly completely eliminates the need to align excitation and detection optics. Because the concentration of fluorophores in the crystal is so high, the intensity of the excitation light can be relatively low so that backscattered

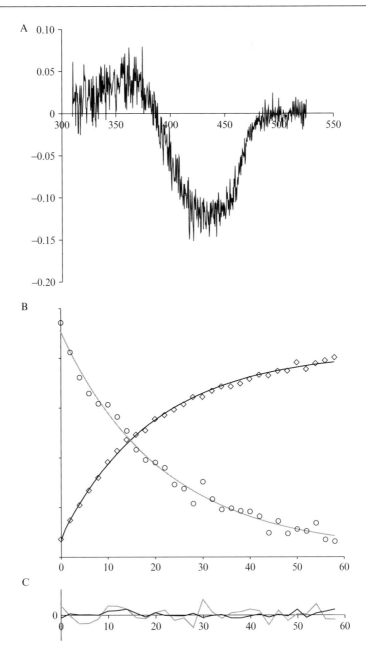

excitation light does not "bleed over" into the region of the spectrum where the fluorescence signal is detected. For the same reasons the requirements for the detector are not very stringent, and the same linear CCD array detector (USB2000, Ocean Optics, ocean optics.com) used for absorption measurements performs admirably for fluorescence measurements.

However, fluorescence measurements are not entirely without their own pitfalls. The main concern is the absorption of both excitation and emission photons. Because the excitation and detection wavelengths generally differ by a few tens of nanometers, optical densities of a crystal for excitation and emission light may differ substantially. The worst-case scenario combines a low OD at the excitation wavelength with a high OD at the monitoring wavelength. In this case the excitation light will penetrate deep into the crystal, where it will stimulate the emission of fluorescence photons. Many of the emitted fluorescence photons are then reabsorbed before they can escape the crystal and reach the detector. The result of this effect, commonly referred to as the "inner filter effect," is the appearance of a lowered fluorescence yield at the affected wavelengths. In many cases it would therefore be advantageous to choose an excitation wavelength at the peak of the excitation spectrum of the fluorophore, in which case penetration of the excitation light into the crystal and reabsorption of the emitted fluorescence photons will be minimized.

The presence of an inner filter effect can be detected in scatter plots of pairs of fluorescence emission intensities observed at the same wavelength, but stimulated by excitation with two different wavelengths (Fig. 8). If there is no internal filter effect and the two excitation wavelengths stimulate the same spectral transition, then the two emission spectra should be identical in shape and differ only by one overall scale factor. The scatter plot should therefore take the form of a straight line, with the slope of the line giving the scale factor. If internal filtering occurs, then the plot will take the appearance of an arrowhead pointing diagonally across the plot away from the origin. The two legs of the arrowhead will correspond to the two shoulders of the emission

---

Fig. 7. Difference spectrum (A) and recovery kinetics (B) of PPR-PYP crystals following light activation. To record these spectra, the intensity of the measurement beam was reduced until further reductions had no effect on the apparent rate of recovery. The spectrum is shown without any smoothing or other manipulation, giving an indication of the signal-to-noise ratio of the raw absorption signal. To determine the rate of recovery, the area under the 350- and 440-nm peaks was integrated and plotted as a function of time. Rates obtained for the recovering and decaying spectral bands agree to within their experimental error, indicating that the acceleration of the kinetics in PPR-PYP crystal relative to those in solution (data not shown) is not a measurement artifact caused by light leaks but rather a genuine effect of the crystal environment.

A

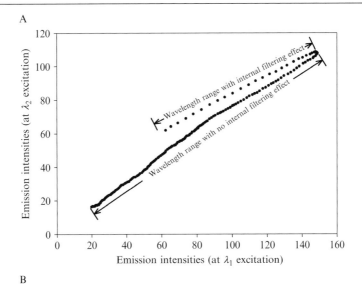

B

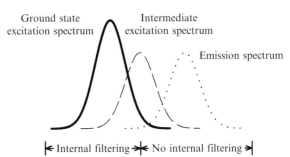

FIG. 8. Scatter plot used to detect the presence of an internal filtering effect on single crystal fluorescence spectra. This plot (A) is generated by taking emission spectra generated by stimulation at two different excitation wavelengths and generating a scatter plot of the intensities observed in these spectra for the same emission wavelength. The presence of the upper leg of the arrowhead shape indicates that internal filtering affects the measurement of these wavelengths. (B) A schematic with absorption spectra of a dark-adapted, nonfluorescent state and a light-activated, fluorescent intermediate as well as the fluorescence emission spectrum of the intermediate. Areas of the spectrum, which are noticeably affected by internal filtering, are indicated on the graph. Those areas should be omitted when fluorescence data are used to determine excitation spectra or quantify the buildup of intermediates. (A) Data were recorded on a cryogenically frozen light-activated 300-μm-diameter PYP crystal.

spectrum. The reason for this appearance is that internal filtering is bound to affect one of the shoulders of the spectrum more than the other, that is, fluorescent light with wavelengths in one of the shoulders will be attenuated more severely than light with wavelengths in the other shoulder.

By comparing these plots for multiple wavelengths, it will be possible to determine which portions of the emission spectrum are affected by internal filtering. These portions of the spectrum can then be ignored when excitation or emission spectra are calculated.

Microspectrophotometry: Summary and Warning

When interpreting any kind of microspectrophotometry data a healthy dose of skepticism is helpful—maybe even required. The basic fact is that protein crystals make poor spectroscopy samples. The protein crystal samples are generally very small, optically dense, have irregular surfaces, and, because they contain photoreceptors, are very light sensitive. As a result it is rarely possible to record microspectrophotometry data that are completely free of artifacts. It is, however, possible to detect many of these artifacts and to avoid potential misinterpretations caused by them. Diligence in ferreting out these potential problems is paramount when observed data are to be subjected to sophisticated data analysis algorithms, such as global analysis or singular value decomposition. Therefore, if a microspectrophotometry experiment indicates that the reaction kinetics in the crystal are accelerated or retarded relative to solution values, it may be helpful to think "measurement artifact" first and "exciting functional difference between solution and crystal environment" second.

Light Activation of Photoreceptor Crystals

Light activation of crystals can be divided into three strategies: pump probe, freeze-then-activate, and steady-state illumination (possibly followed by freeze trapping). The first strategy, usually employed for true time-resolved crystallography or for time-resolved spectroscopic measurements, usually relies on pulsed laser sources. Compact turnkey systems are now available in the form of OPOs pumped by nanosecond Q-switched Nd:YAG lasers (Opotek, opotek.com), but the cost of those systems (~$100K for a system powerful enough to saturate activation of a typical protein crystal and ~$50k for a system to perform kinetic measurements) will rarely be justified except for a specialist laboratory and we will therefore not discuss the details of choosing and using such lasers further. For steady-state experiments on photoreceptors that recover from light activation on the timescale of seconds or slower and for freeze-then-activate experiments, much simpler light sources will suffice. For preliminary experiments, a general-purpose light source with a light guide, such as those often used for the illumination of samples under stereomicroscopes, will be sufficient. Specific wavelength bands can be selected with the help of narrow band-pass interference filters.

Such filters are available commercially for a wide range of peak wavelengths and band path widths from companies such as CVI Laser Corp. or Newport Corp. (newport.com). However, in our experience the cost of those filters (~$300 each) quickly adds up and a more flexible light source may be desirable. Investment in a simple light source/monochromator combination will afford continuous wavelength tunability across most of the UV to near IR range, as well as considerable control over the width of the selected wavelength band. Simple light source/monochromator combinations are available from companies such as PTI Inc. (New Jersey), Newport Corp. (www. newport.com) and Jobin Yvon (www.jobinyvon) for ~ $7500.

Because the specific configurations of light bulbs, monochromator grating, and so on with which these systems are offered can be overwhelming, it is important to keep two facts in mind when making these choices. First, the absorption bands of virtually all biologically relevant chromophores are quite broad (FWHM $\gg$ 10 nm) so that the ability to select a very narrow band of wavelength has little importance. One of the key tradeoffs in monochromator design is between spectral resolution and throughput; for the purposes discussed here, the emphasis should be firmly on throughput. Second, because protein crystals are very small, what really matters is not the total amount of light a light source produces, but the ability to deliver as much light as possible into a very small spot; the technical term for this property is brilliance. It turns out that the brilliance of an optical system is determined by the light source. Downstream optical elements can degrade the brilliance of the light beam, but never improve it. As a result, select a light source that produces as much light as possible in a spot that is comparable to the size of a protein crystal. A low-power (e.g., 75 W) xenon arc lamp turns out to be a very good choice for the activation of photoreceptor crystals: The actual light source in these lamps is a very small, very bright plasma arc, and the spectrum emitted by this arc covers the near-UV to near-IR region and is devoid of sharp spectral lines in the UV-visible region. Also, because the total power of the light source is low, these light sources can be operated without the water cooling of ozone elimination systems required by higher wattage arc lamps.

For most applications it will be useful to deliver the light from the monochromator exit slit to the sample area via an optical fiber, not a set of mirrors. In our experience, fibers with a diameter of 400 μm are large enough to facilitate ease of alignment while retaining enough flexibility to guide the fibers through the crowded experimental setups. In our setup, the light emitted from the fiber end is imaged 1:1 on the sample area with a simple quartz lens. In our experience, a 2-in.-long 1-in.-diameter lens tube fitted with a 1-in.-diameter lens with a focal length of 25 mm images most of the light that exits the fiber to the sample, provides a small well-defined focal spot size (i.e., comparable in size to the fiber diameter), and provides a lense-to-sample distance sufficient to avoid interference with the components of the diffractometer and cryo-cooling device mounted around the crystal. The lens tube system offered by Thorlabs Inc. (Newton, NJ) provides a convenient and inexpensive way to integrate the focusing lens, a fiber holder, and a compact translation stage for the fiber, as well as a hardware mount into one compact assembly.

The assembly just described will deliver on the order of 1 nW/μm^2 of light with a 5-nm band pass to the crystal. This light intensity is, for example, sufficient to saturate the light response in a typical frozen crystal of PYP on the timescale of 100 s. For experiments on photoreceptor states with lifetimes shorter than 100 s, the band pass can be widened to the full width of the photoreceptors absorption band and so flux at the crystal can be increased by a factor of $\sim$10. For those cases where more rapid light activation is needed, it will be necessary to use more powerful light sources, in particular continuous wave lasers, such as argon ion, HeNe, or diode lasers equipped with mechanical shutters or pulsed lasers. These lasers, especially the diode lasers, are relatively inexpensive and powerful, but emit light at only a single fixed wavelength so any change in wavelength requires the purchase of a separate laser system.

Aligning the Activating Light Beam and the Crystal Position

For many experiments it is desirable to activate protein crystals while they are mounted for data collection. For those experiments it is necessary that the X-ray beam and the beam of the activating light intersect at the position of the crystal. To facilitate this alignment we have built the following home-built alignment tool. Despite, or maybe because of, its simplicity we have found this tool extremely useful. The tool consists of a 10×10-mm^2 sheet of thin brass with a 300-μm hole drilled into the center. (We drilled this hole using a very fine drill bit fitted to a Dremmel multitool.) This "pinhole" was then mounted on a standard crystallographic cryoloop base using a copper paperclip wire so that the hole in the brass sheet occupies the position normally occupied by the sample loop and the plane of the brass sheet is angled at 45° to the rotation axis of the pin base.

For the alignment procedure, the tool is mounted on the goniometer head via its cryoloop base. The pinhole is then centered relative to the beam position in the same way one centers a crystal for diffraction experiments. The position of the activation beam is then adjusted to maximize the intensity of the light transmitted through the hole. Coarse alignments are performed by adjustments of the mounting hardware, and the final fine adjustment is performed with the two-dimensional translation stage that mounts the optical fiber to the lens tube. A calibrated light meter can then be used to determine the photon flux through the pinhole, and with knowledge of the size of the hole, one can calculate the photon flux per unit area delivered to the crystal. We found that this procedure is much more reliable than commonly used alignment procedures based on the amount of light scattered from a needle tip, cryoloop, or mounted crystal. Having an actual measurement of the photon flux at the crystal position is also very useful in obtaining a rough estimate of the efficiency with which the activating light generates structural changes in the protein.

Summary and Outlook

The combination of light-activated histidine kinases and kinetic crystallography offers exciting opportunities for in-depth mechanistic studies of the molecular processes underlying biological sensing and signal transmission. Much of this chapter has been dedicated to the potential pitfalls of kinetic crystallography. Our goal in presenting those pitfalls was not to dissuade potential users from attempting their own kinetic crystallography experiments. Instead we hope to instill a healthy dose of skepticism that might help avoid overinterpretation of both spectroscopic and crystallographic results. We further hope that we have conveyed that this young field still offers much room for improvement in experimental and data processing procedures and therefore provides a fertile ground for researchers interested in developing as well as applying experimental techniques. Finally, we wish to encourage readers who are interested in further details to contact us directly.

References

Arndt, U. W., and Wonacott, A. J. (1977). "The Rotation Method in Crystallography." North-Holland Elsevier, Amsterdam.

Baca, M., Borgstahl, G. E., Boissinot, M., Burke, P. M., Williams, D. R., Slater, K. A., and Getzoff, E. D. (1994). Complete chemical structure of photoactive yellow protein: Novel thioester-linked 4-hydroxycinnamyl chromophore and photocycle chemistry. *Biochemistry* **33,** 14369–14377.

Bergo, V., Spudich, E. N., Scott, K. L., Spudich, J. L., and Rothschild, K. J. (2000). FTIR analysis of the SII540 intermediate of sensory rhodopsin II: Asp73 is the Schiff base proton acceptor. *Biochemistry* **39,** 2823–2830.

Bilwes, A. M., Alex, L. A., Crane, B. R., and Simon, M. I. (1999). Structure of CheA, a signal-transducing histidine kinase. *Cell* **96,** 131–141.

Bilwes, A. M., Quezada, C. M., Croal, L. R., Crane, B. R., and Simon, M. I. (2001). Nucleotide binding by the histidine kinase CheA. *Nat. Struct. Biol.* **8,** 353–360.

Birge, R. R., Einterz, C. M., Knapp, H. M., and Murray, L. P. (1988). The nature of the primary photochemical events in rhodopsin and isorhodopsin. *Biophys. J.* **53,** 367–385.

Bogomolni, R. A., and Spudich, J. L. (1987). The photochemical reactions of bacterial sensory rhodopsin-I. Flash photolysis study in the one microsecond to eight second time window. *Biophys. J.* **52,** 1071–1075.

Bourgeois, D., and Royant, A. (2005). Advances in kinetic protein crystallography. *Curr. Opin. Struct. Biol.* **15,** 538–547.

Christie, J. M., Salomon, M., Nozue, K., Wada, M., and Briggs, W. R. (1999). LOV (light, oxygen, or voltage) domains of the blue-light photoreceptor phototropin (nph1): Binding sites for the chromophore flavin mononucleotide. *Proc. Natl. Acad. Sci. USA* **96,** 8779–8783.

Crosson, S., Rajagopal, S., and Moffat, K. (2003). The LOV domain family: Photoresponsive signaling modules coupled to diverse output domains. *Biochemistry* **42,** 2–10.

Engh, R. A., and Huber, R. (1991). Accurate bond and angle parameters for X-ray protein structure refinement. *Acta Cryst. A* **47,** 392–400.

Essen, L. O. (2002). Halorhodopsin: Light-driven ion pumping made simple? *Curr. Opin. Struct. Biol.* **12,** 516–522.

Fankhauser, C. (2000). Phytochromes as light-modulated protein kinases. *Semin. Cell Dev. Biol.* **11,** 467–473.

Genick, U. K., Soltis, S. M., Kuhn, P., Canestrelli, I. L., and Getzoff, E. D. (1998). Structure at 0.85 A resolution of an early protein photocycle intermediate. *Nature* **392,** 206–209.

Gordeliy, V. I., Labahn, J., Moukhametzianov, R., Efremov, R., Granzin, J., Schlesinger, R., Buldt, G., Savopol, T., Scheidig, A. J., Klare, J. P., and Engelhard, M. (2002). Molecular basis of transmembrane signalling by sensory rhodopsin II-transducer complex. *Nature* **419,** 484–487.

Hadfield, A., and Hajdu, J. (1993). A fast and portable microspectrophotometer for protein crystallography. *J. Appl. Cryst.* **26,** 839–842.

Harigai, M., Yasuda, S., Imamoto, Y., Yoshihara, K., Tokunaga, F., and Kataoka, M. (2001). Amino acids in the N-terminal region regulate the photocycle of photoactive yellow protein. *J. Biochem. (Tokyo)* **130,** 51–56.

Harper, S. M., Christie, J. M., and Gardner, K. H. (2004). Disruption of the LOV-Jalpha helix interaction activates phototropin kinase activity. *Biochemistry* **43,** 16184–16192.

Harper, S. M., Neil, L. C., and Gardner, K. H. (2003). Structural basis of a phototropin light switch. *Science* **301,** 1541–1544.

Hurley, J. H. (2003). GAF domains: Cyclic nucleotides come full circle. *Sci. STKE* **2003,** PE1.

Imamoto, Y., Kataoka, M., and Tokunaga, F. (1996). Photoreaction cycle of photoactive yellow protein from *Ectothiorhodospira halophila* studied by low-temperature spectroscopy. *Biochemistry* **35,** 14047–14053.

Jiang, Z., Swem, L. R., Rushing, B. G., Devanathan, S., Tollin, G., and Bauer, C. E. (1999). Bacterial photoreceptor with similarity to photoactive yellow protein and plant phyto-chromes. *Science* **285,** 406–409.

Jung, K. H., Trivedi, V. D., and Spudich, J. L. (2003). Demonstration of a sensory rhodopsin in eubacteria. *Mol. Microbiol.* **47,** 1513–1522.

Kyndt, J. A., Meyer, T. E., and Cusanovich, M. A. (2004). Photoactive yellow protein, bacteriophytochrome, and sensory rhodopsin in purple phototrophic bacteria. *Photochem. Photobiol. Sci.* **3,** 519–530.

Lamparter, T., Carrascal, M., Michael, N., Martinez, E., Rottwinkel, G., and Abian, J. (2004). The biliverdin chromophore binds covalently to a conserved cysteine residue in the N-terminus of Agrobacterium phytochrome Agp1. *Biochemistry* **43,** 3659–3669.

Lanyi, J. K. (2004a). What is the real crystallographic structure of the L photointermediate of bacteriorhodopsin? *Biochim. Biophys. Acta* **1658,** 14–22.

Lanyi, J. K. (2004b). X-ray diffraction of bacteriorhodopsin photocycle intermediates. *Mol. Membr. Biol.* **21,** 143–150.

Lanyi, J. K., and Schobert, B. (2004). Local-global conformational coupling in a heptahelical membrane protein: Transport mechanism from crystal structures of the nine states in the bacteriorhodopsin photocycle. *Biochemistry* **43**, 3–8.

Losi, A. (2004). The bacterial counterparts of plant phototropins. *Photochem. Photobiol. Sci.* **3**, 566–574.

Losi, A., Braslavsky, S. E., Gartner, W., and Spudich, J. L. (1999). Time-resolved absorption and photothermal measurements with sensory rhodopsin I from *Halobacterium salinarum*. *Biophys. J.* **76**, 2183–2191.

Luecke, H., Schobert, B., Lanyi, J. K., Spudich, E. N., and Spudich, J. L. (2001). Crystal structure of sensory rhodopsin II at 2.4 angstroms: Insights into color tuning and transducer interaction. *Science* **293**, 1499–1503.

Luecke, H., Schobert, B., Richter, H. T., Cartailler, J. P., and Lanyi, J. K. (1999). Structural changes in bacteriorhodopsin during ion transport at 2 angstrom resolution. *Science* **286**, 255–261.

Moffat, K. (2003). The frontiers of time-resolved macromolecular crystallography: Movies and chirped X-ray pulses. *Faraday Discuss* **122**, 65–77; discussion 79–88.

Moukhametzianov, R., Klare, J. P., Efremov, R., Baeken, C., Goppner, A., Labahn, J., Engelhard, M., Buldt, G., and Gordeliy, V. I. (2006). Development of the signal in sensory rhodopsin and its transfer to the cognate transducer. *Nature* **440**, 115–119.

Nagel, G., Szellas, T., Kateriya, S., Adeishvili, N., Hegemann, P., and Bamberg, E. (2005). Channelrhodopsins: Directly light-gated cation channels. *Biochem. Soc. Trans.* **33**, 863–866.

Ng, K., Getzoff, E. D., and Moffat, K. (1995). Optical studies of a bacterial photoreceptor protein, photoactive yellow protein, in single crystals. *Biochemistry* **34**, 879–890.

Rasmussen, B. F., Stock, A. M., Ringe, D., and Petsko, G. A. (1992). Crystalline ribonuclease A loses function below the dynamical transition at 220 K. *Nature* **357**, 423–424.

Reddy, V., Swanson, S. M., Segelke, B., Kantardjieff, K. A., Sacchettini, J. C., and Rupp, B. (2003). Effective electron-density map improvement and structure validation on a Linux multi-CPU web cluster: The TB Structural Genomics Consortium Bias Removal Web Service. *Acta Crystallogr. D Biol. Crystallogr.* **59**, 2200–2210.

Ren, Z., Perman, B., Srajer, V., Teng, T. Y., Pradervand, C., Bourgeois, D., Schotte, F., Ursby, T., Kort, R., Wulff, M., and Moffat, K. (2001). A molecular movie at 1.8 A resolution displays the photocycle of photoactive yellow protein, a eubacterial blue-light receptor, from nanoseconds to seconds. *Biochemistry* **40**, 13788–13801.

Schobert, B., Brown, L. S., and Lanyi, J. K. (2003). Crystallographic structures of the M and N intermediates of bacteriorhodopsin: Assembly of a hydrogen-bonded chain of water molecules between Asp-96 and the retinal Schiff base. *J. Mol. Biol.* **330**, 553–570.

Schotte, F., Lim, M., Jackson, T. A., Smirnov, A. V., Soman, J., Olson, J. S., Phillips, G. N., Jr., Wulff, M., and Anfinrud, P. A. (2003). Watching a protein as it functions with 150-ps time-resolved X-ray crystallography. *Science* **300**, 1944–1947.

Semenza, G. L. (1999). Regulation of mammalian O2 homeostasis by hypoxia-inducible factor 1. *Annu. Rev. Cell Dev. Biol.* **15**, 551–578.

Spudich, J. L. (2006). The multitalented microbial sensory rhodopsins. *Trends Microbiol.* **14**, 480–487.

Swartz, T. E., Szundi, I., Spudich, J. L., and Bogomolni, R. A. (2000). New photointermediates in the two photon signaling pathway of sensory rhodopsin-I. *Biochemistry* **39**, 15101–15109.

Taylor, B. L., and Zhulin, I. B. (1999). PAS domains: Internal sensors of oxygen, redox potential, and light. *Microbiol. Mol. Biol. Rev.* **63**, 479–506.

van der Horst, M. A., van Stokkum, I. H., Crielaard, W., and Hellingwerf, K. J. (2001). The role of the N-terminal domain of photoactive yellow protein in the transient partial unfolding during signalling state formation. *FEBS Lett.* **497**, 26–30.

Vierstra, R. D., and Davis, S. J. (2000). Bacteriophytochromes: New tools for understanding phytochrome signal transduction. *Semin. Cell Dev. Biol.* **11,** 511–521.

Vreede, J., van der Horst, M. A., Hellingwerf, K. J., Crielaard, W., and van Aalten, D. M. (2003). PAS domains: Common structure and common flexibility. *J. Biol. Chem.* **278,** 18434–18439.

Wagner, J. R., Brunzelle, J. S., Forest, K. T., and Vierstra, R. D. (2005). A light-sensing knot revealed by the structure of the chromophore-binding domain of phytochrome. *Nature* **438,** 325–331.

Watts, K. J., Sommer, K., Fry, S. L., Johnson, M. S., and Taylor, B. L. (2006). Function of the N-terminal cap of the PAS domain in signaling by the aerotaxis receptor Aer. *J. Bacteriol.* **188,** 2154–2162.

Weik, M., Vernede, X., Royant, A., and Bourgeois, D. (2004). Temperature derivative fluorescence spectroscopy as a tool to study dynamical changes in protein crystals. *Biophys. J.* **86,** 3176–3185.

Wuichet, K., Alexander, R. P., and Zhulin, I. B. (2007). Comparative genomic and protein sequence analyses of a complex system controlling bacterial chemotaxis. *Methods Enzymol.* **422**(1), (this volume).

[16] Synthesis of a Stable Analog of the Phosphorylated Form of CheY: Phosphono-CheY

By CHRISTOPHER J. HALKIDES, CORY J. BOTTONE, ERIC S. CASPER, R. MATTHEW HAAS, and KENNETH MCADAMS

Abstract

The chemical modification of a cysteinyl residue of D57C CheY by the addition of a phosphonomethyl group, $(HO)_2P(O)\text{-}CH_2\text{-}$, is described. This modification produces a nonlabile analog of an aspartyl phosphate residue in the active form of CheY. The chemically modified protein, phosphono-CheY, is suitable for structural and functional studies. An extensive discussion of the synthetic methodology and purification strategy is presented. A detailed protocol is given.

Introduction

Bacterial signal transduction takes place largely through the use of two component systems, one of which is a histidine autokinase and the other is the response regulator (Stock *et al.*, 2000). The histidine kinase phosphorylates itself on a catalytic histidine and then transfers the phosphoryl group to an aspartyl residue on a response regulator. The phosphorylated response regulator is the active or signaling state, and the dephosphorylated response regulator is quiescent. The phosphorylated aspartyl residue has a large range of labilities, ranging from roughly 1 s to 10 h. The lability of the phosphorylated response regulator makes it challenging to perform structural studies on the active state (Lowry *et al.*, 1994).

The purpose of phosphonomethylation is to produce an analog of the phosphorylated state of a response regulator that is indefinitely stable. Phosphonomethylated proteins are suitable for crystallographic and other structural studies. We have solved the structure of phosphono-CheY, the protein that controls the direction of flagellar rotation in *Escherichia coli* (Halkides *et al.*, 2000). Phosphonomethylation has allowed us to perform structure/function studies on mutant forms of CheY in its active state, as opposed to the inactive state (submitted for publication).

The production of phosphonomethylated response regulators such as phosphono-CheY begins with production of a suitable mutant in which the nucleophilic aspartyl residue (Asp57 in CheY from *E. coli*) is mutated to a cysteinyl residue to produce D57C CheY (Fig. 1). It may be prudent to mutate other cysteinyl residues to serinyl, threoninyl, or alanyl residues

METHODS IN ENZYMOLOGY, VOL. 422
0076-6879/07 $35.00
DOI: 10.1016/S0076-6879(06)22016-0

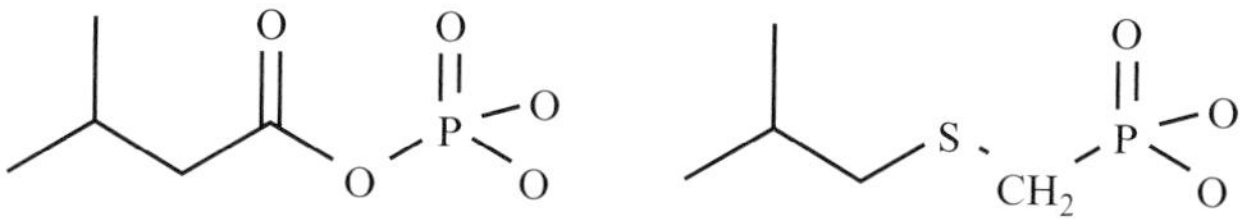

CheY~P   Phosphono-CheY

FIG. 1. Comparison between phosphorylated CheY (CheY~P) and phosphono-CheY. CheY~P is created by the phosphorylation of residue Asp57 of CheY. It is the active, signaling state of this protein, and the phosphoryl group is very labile. Phosphono-CheY is created when D57C CheY is modified chemically by the addition of a phosphonomethyl group onto Cys57. It is a nonlabile analog of CheY~P.

so that no unwanted phosphonomethylation takes place. This eliminates the other cysteinyl residue as a site of modification and also simplifies the quantitation of free thiol groups. In the case of nonconserved cysteine residues, one might change these into residues found in different species. In cases where site-directed mutagenesis of other cysteinyl residues is not possible, phosphonomethylation may still be successful, provided that the other cysteinyl residues are less nucleophilic than the active-site cysteinyl residue and that a suitable assay for the degree of phosphonomethylation exists.

Once the mutant protein has been expressed, it must be purified, either from the soluble fraction or from inclusion bodies. The majority of D57C CheY from E. coli is found in inclusion bodies in the two expression systems we have used (Halkides et al., 1998). Inclusion bodies have been observed previously in other cysteinyl mutants of CheY (Krueger et al., 1990). When we purify D57C CheY from inclusion bodies, we use an anion-exchange chromatography column, a Cibacron-blue column, which was not discussed in our previous purification protocol (Halkides et al., 1998), and sometimes a size-exclusion column. When we purify from the soluble fraction, we use the same columns as previous workers (Lowry et al., 1994). We have modified the purification of wild-type E. coli CheY in several ways to take into account the presence of the cysteinyl residue and the apparently lower solubility and higher pI value of the mutant as opposed to wild-type CheY (Lowry et al., 1994). We add 5–10 mM 2-mercaptoethanol (2-ME) to most buffers just prior to use. We include 1 mM EDTA in most buffers to sequester metal ions that would catalyze oxidation of the cysteinyl residue. D57C CheY from E. coli precipitates below pH values of about 6.0; therefore, we use MES at this pH, as opposed to 5.3 when purifying wild-type CheY. We typically substitute dithiothreitol (DTT) for 2-ME in the final storage buffer. We use a shallower gradient in the DE52 ion-exchange chromatography step.

In some cases, it is necessary to reduce the protein prior to phosphono-methylation (see later). When the protein is in an alkaline buffer (pH 8–9), we use DTT for 1–15 h. However, when the protein is in a buffer near neutral pH, bis(2-mercaptoethyl) sulfone (BMS) is considerably faster than DTT (Singh et al., 1995). In contrast to its low solubility in water, we have found that BMS dissolves in 0.5 M acetonitrile; therefore, we make additions of BMS in acetonitrile to the protein. We have also used tris(2-carboxyethyl)phosphine (TCEP)(Getz et al., 1999). However, TCEP is not compatible with phosphate buffers.

We chose phosphonomethyltriflate (PMT) for the modification of response regulators because the trifluoromethanesulfonate (triflate) leaving group is much more reactive than leaving groups such as iodide and tosylate. We have synthesized related compounds with leaving groups having reactivities that are intermediate between triflate and iodide, specifically trifluoroethanesulfonate and pentafluorobenzenesulfonate. Based on preliminary experiments with these leaving groups, we believe that they are insufficiently reactive for the modification of D57C CheY but are useful for modifying other proteins.

We have made some modifications to the synthesis of PMT. We purify di-tert-butylhydroxymethylphosphonate by recrystallization or with silica gel chromatography. We often distill trifluoromethanesulfonic anhydride on the same day it is used. We keep the temperature at which we add trifluoro-methanesulfonic anhydride to di-tert-butylhydroxymethylphosphonate as low as possible, typically −75 to −80°. These modifications tend to minimize the appearance of a dark brown colored impurity that appears in some syntheses of PMT or to remove some of the color. Other workers have suggested replacing triethylamine with more sterically hindered bases to minimize possible side reactions (Netscher and Bohrer, 1996). We have very limited experience with these other bases.

We flow N_2 gas over PMT after it has been opened and store it desiccated at −20 to −80°. We weigh small portions of PMT into microcentrifuge tubes quickly, as it is hygroscopic. The PMT is mixed first with ethanol and then triethylamine immediately prior to addition to protein. We estimate the half-life of PMT in alkaline solution to be 5 min. Therefore, we terminate phosphonomethylation reactions after 45–60 min.

The solution conditions profoundly affect the yield of phosphonometh-ylated protein and may influence how selectively PMT modifies the response regulator. As pH increases, the nucleophilicity of the target cysteinyl residue increases until the pK_a of the residue is reached (this pK_a value is not known but is assumed to be 8–10). As the pH is raised above the pK_a of the cysteinyl residue, there is little further gain in reactivity. However, lysyl residues typically have higher pK_a values and would continue to increase in reactivity in this range of pH values (Kyte, 1995). Therefore, we select the pH that produces an acceptable yield with little evidence of modification at other sites.

We choose a buffer with a pK_a value near the optimal pH (typically 8.5–9.0) and should be nonnucleophilic (typically a sterically hindered secondary or tertiary amine). We have had success with AMPSO, TAPS, and CHES buffers in a concentration range of 250 to 750 mM. PMT is a diprotic acid, and we commonly employ concentrations of 100–150 mM. We often find that the fall in pH when PMT is added is greater than we predict. We use 3–4 equivalents of triethylamine to neutralize PMT. The concentration of the buffer must be high enough to counteract any acidity not neutralized by the triethylamine. The number of equivalents of triethylamine and the concentration of buffer must be adjusted empirically and jointly. We find that spotting about 1 μl of reaction mixture onto finely graded pH paper gives useful pH information.

We presume that the metal ion is necessary to give PMT some affinity for the active site of the response regulator. We have the greatest success with three alkaline earth metal chlorides: barium, calcium, and strontium ions. These three ions often give yields that are within experimental error of each other; therefore, it is difficult to choose a single best alkaline earth ion. Zinc ion gives a lower yield of *E. coli* phosphono-CheY, possibly because the lower solubility of $Zn(OH)_2$ necessitates working at lower concentrations of this metal ion. We typically use a final concentration of 150–250 mM of the metal chloride salt. We make up stock solutions of each metal chloride that are at or near the limit of solubility, about 3.2 M for $SrCl_2$. Experiments with trivalent ions did not produce better results than what was obtained with the alkaline earth ions. Other metal ions might give better results with other response regulators.

Three assays are used routinely to judge the success of phosphonomethylation reactions: a protein assay, a DTNB assay, and a chromatographic assay, usually reversed-phase, high-performance liquid chromatography (RP-HPLC). The protein assay is used both to assess whether protein is lost in the phosphonomethylation or biotinylation steps and to calculate the percentage of free thiol groups, in conjunction with the DTNB assay (Riddles *et al.*, 1983; Wright and Viola, 1998).

Phosphonomethyltriflate imparts a light brown color to the protein solution, which would interfere with measurements of protein concentration by UV absorbance unless a buffer exchange is done after the phosphonomethylation reaction. We have found both the Bradford assay and the bicinchoninic acid (BCA) assay to be especially useful when buffer exchange cannot be performed immediately after phosphonomethylation (see later); however, the BCA assay can only be employed in the absence of the thiol-based reducing agents discussed earlier.

The reaction between DTNB and the active-site cysteinyl group requires 20–40 min to come to completion, unlike small molecule thiol groups, which react almost as soon as the solution is mixed. Therefore, we modified the standard DTNB assay in the following ways: We often run the assay thermostatted at 25.0°, and we check that the absorbance of DTNB in buffer is constant in time for at least 45 s. If we do not obtain a stable reading, we discard this solution and cuvette. If this fails to produce a stable reading, we prepare a fresh solution of DTNB. We add protein to the cuvette, mix, and take absorbance readings for at least 20 min and the reaction is essentially complete. We extrapolate back to time zero to obtain the initial absorbance. We ignore any instantaneous increase in absorbance because it is most likely caused by the incomplete removal of DTT or 2-ME. Small amounts of precipitate sometimes form after the addition of protein, possibly due to the formation of a phosphate salt of the alkaline earth metal ion (the DTNB assay is buffered with phosphate). Thorough mixing minimizes this problem. One could also substitute a buffer of Tris-HCl, pH 7.5–8.0, for the phosphate buffer, pH 7.28 (Wright and Viola, 1998).

We have found that RP-HPLC detects a shortening of the retention time for phosphono-CheY versus CheY when we employ a C18 column and elute with a shallow gradient of acetonitrile (about 7% change in acetonitrile per hour) with 0.1% (v/v) trifluoroacetic acid (TFA) as the organic ion-pairing reagent. We note that C4 is a possible alternative to C18, based on the phosphorylation of OmpR (Head et al., 1998). Although we have employed other separation methods less frequently, we believe that hydrophobic interaction chromatography (HIC) and ion-exchange chromatography will prove useful in some cases. Indeed, HIC of D54C/C81S CheY from *Thermotoga maritima* is able to separate phosphono-CheY from CheY (unpublished experiments). We speculate that native gel electrophoresis might also be able to detect phosphonomethylation of response regulators, based on results with SpoOF and SpoOF~P (Zapf et al., 1996).

We initially search for conditions using parallel reactions. We employ a 2.1-mm RP-HPLC column for this phase of the work to minimize sample size. We subject a portion of protein to buffer exchange, assay for protein and free thiol groups, and divide into three to four aliquots. We add metal chloride to each of the aliquots. We then mix the PMT with ethanol and triethylamine and rapidly add aliquots of this mixture to the protein, mixing quickly. After a reaction time of 45–60 min, we remove portions for protein, DTNB, and HPLC assays. We vary the concentration or identity of metal ion, or the number of equivalents of base within a set of reactions, and we vary the pH and the identity of the buffer among sets of reactions for reasons discussed earlier.

These reactions are judged by several comparisons of the sample before and after phosphonomethylation. For the unmodified protein we typically observe that the percentage of free thiol groups to protein concentration is 80% or higher. If this ratio is too low, we discontinue the phosphonomethylation reaction and attempt to reduce the protein as discussed previously. We stress that calculation of the percentage of free thiol groups depends on an *accurate* as well as precise protein concentration determination. If protein concentration can be determined precisely but not accurately, one expects that the percentage of free thiol groups will be reproducibly above or below the stoichiometric ratio. We also typically see only one peak by RP-HPLC when the protein is pure and completely reduced; however, the converse is not always true.

The percentage of free thiol groups should decrease to half or less of its initial value in favorable cases. After the phosphonomethylation reaction is finished, we change buffers into one suitable for the biotinylation reaction. We sometimes observe a fall in the amount of protein concentration at this point; presumably protein has been lost in the phosphonomethylation reaction and small amounts have been removed for assays. The RP-HPLC chromatogram shows peaks corresponding to phosphono-CheY, followed by unreacted CheY. The ratio of peak areas should be consistent with the percentage of free thiol groups, within experimental error.

The strategy for purification of phosphono-CheY is to modify the unreacted cysteine residue with a biotin-containing reagent (to biotinylate) unreacted CheY and to purify over immobilized avidin. Phosphono-CheY passes through the column, but biotinylated CheY is retained. We then change buffers to remove the unreacted biotinylating reagent and DTT, prior to chromatography over avidin.

The reaction between CheY and the biotinylating reagent may be followed by DTNB or HPLC assays. We typically use roughly a 10-fold excess of biotinylation reagent over unmodified protein. This creates a situation in which the loss of the free thiol groups on CheY is approximately first-order kinetically. We establish conditions of pH and time that give >90% conversion of the cysteine residue into its biotinylated form. We have had the most consistent results with a biotin reagent with a water-soluble linker terminating in an iodoacetamide group, PEO-iodoacetyl biotin. Our initial experiments in which we reacted CheY with a biotinylation reagent bearing a maleimide group (PEO-maleimide activated biotin) at pH ~7 were promising, but we often observed a large, unexplained loss of protein. Therefore, we do not routinely use this reagent. Once conditions for biotinylation have been firmly established, we do not typically follow this reaction by taking time points, but rather terminate the reaction by adding excess DTT after a set time period.

In experiments with D54C/C81S CheY from *T. maritima*, we have used a biotin reagent in the form of a mixed disulfide, biotin-HPDP, at pH 8.0–8.5. This reagent is expected to have very high selectivity for modifying thiol groups. Moreover, upon reaction with protein, this reagent releases a chromophore, pyridine-2-thione, that absorbs light at 343 nm, $\varepsilon = 8080\ M^{-1}cm^{-1}$. Therefore, the biotinylation reaction can be followed to completion by direct spectrophotometry. This reagent has two disadvantages: (1) it must be prepared as a concentrated stock in dimethyl sulfoxide (DMSO) or dimethylformamide and (2) this biotinylated protein must be kept apart from reducing agents. Neither of these two disadvantages is serious.

We use immobilized monomeric avidin to separate biotinylated CheY from phosphono-CheY. We chose monomeric avidin over other forms because binding of biotin derivatives is reversible at pH 2.8. Therefore, we regenerate the column using a glycine buffer at this pH.

In RP-HPLC experiments, biotinylated CheY elutes with a similar retention time to phosphono-CheY, which can complicate the interpretation. We advocate spiking HPLC samples of presumed phosphono-CheY with a similar concentration of biotinylated CheY to resolve ambiguities in assigning the chromatographic peaks. Preliminary experiments with HIC chromatography suggest that this method may give more discrimination between phosphono-CheY and biotinylated CheY.

The purified phosphono-CheY gives a single major peak by RP-HPLC with a retention time that is about 12 min less than D57C CheY. We often observe a few small peaks with retention times that are close to the major peak. We have occasionally used assays for inorganic phosphate to verify the presence of the phosphonomethyl group. When phosphorus-containing proteins are assayed, the ashing step must be sufficient to oxidize considerably more organic material than when phosphorus-containing small molecules are assayed. We have used magnesium nitrate as the oxidant (Buss and Stull, 1983), and others have used hydrogen peroxide in sulfuric acid (Bailey and Colman, 1987). We have observed at least 20% deviation of the concentration of phosphate from that expected on the basis of protein assays. We verify the presence of the phosphonomethyl group by mass spectrometry.

There are three buffer exchange steps in the overall phosphonomethylation process. To affect removal of small molecules we use both commercial gel filtration columns and spin columns (Penefsky, 1979). When we use spin columns, we typically use two columns in tandem to improve the completeness of exchange. When we use a gel filtration column prior to phosphonomethylation, we next concentrate the protein in Centricon-3 centrifugation devices if the protein concentration is below our preferred range. We have not observed oxidation of the protein during the concentration step, possibly due to the presence of EDTA in our buffers. The spin columns have the advantage of speed, whereas the gel filtration columns tend to give slightly higher recovery of protein.

Many methods may be used to assess how well a phosphonomethylated response regulator mimics the properties of the labile phosphorylated response regulator. We have shown that phosphono-CheY binds both immobilized FliM and CheZ, two proteins that interact more strongly with phosphorylated than unphosphorylated CheY (Halkides et al., 1998). We have also shown that phosphono-CheY binds to peptides derived from FliM and CheZ with much higher affinity than unphosphorylated CheY binds them (Halkides et al., 2000).

Protocols

Synthesis of Phosphonomethyltriflate

Triethylamine and dichloromethane are stirred with calcium hydride for about 1 day and distilled under dry conditions. Glassware is oven dried, and reactions are performed under nitrogen. Trifluoromethanesulfonic (triflic) anhydride is from Aldrich or TCI Chemicals. Di-*tert*-butylhydroxymethylphosphonate is synthesized as described previously with minor modifications (Halkides et al., 1998). Impure preparations of this compound are recrystallized from *n*-heptane or are purified over silica gel with up to 8% methanol in CHCl$_3$ as the solvent. The purified compound is stored at $-80°$ in a desiccator. Phosphonomethyltrifiate is synthesized from di-*tert*-butylhydroxymethylphosphonate as described previously with certain modifications (Halkides et al., 1998). All glassware is oven dried, and dry N$_2$ gas is the atmosphere in all reactions. Triethylamine and dichloromethane are stirred with CaH$_2$ and distilled. Trifluoromethanesulfonic anhydride is distilled freshly before use. The solution of trifluoromethanesulfonic anhydride (2.20 g, 1.2 equivalents) in 5 ml of dry CH$_2$Cl$_2$ is added dropwise to the solution of di-*tert*-butylhydroxymethylphosphonate (1.46 g, 1 equivalent) and triethylamine (1.27 ml, 1.4 equivalents) in 5 ml of CH$_2$Cl$_2$ at approximately $-75°$ using a dry ice/acetone bath. The temperature is slowly raised to $-30°$ and returned to about $-75°$. After two washes of the solution with cold, saturated sodium bicarbonate, some preparations of the intermediate di-*tert*-butylphosphonomethyltrifluoromethanesulfonate are purified by vacuum silica chromatography to reduce the amount of a brown impurity. The column is prepared with 15 g of silica over a bed of diatomaceous earth in a sintered glass funnel. The column is eluted with 20 ml of CH$_2$Cl$_2$, followed by 25% ethyl acetate in CH$_2$Cl$_2$ as the solvent. The first

20 ml of effluent is discarded; the subsequent effluent is dried with $MgSO_4$, and the volatiles are removed with rotary evaporation. A ^{31}P NMR spectrum of di-*tert*-butylphosphonomethyltrifluoromethanesulfonate displays a single peak ($\delta_p = 4.06$ ppm) in $CDCl_3$. A ^1H nuclear magnetic resonance (NMR) spectrum in $CDCl_3$ displays peaks at $\delta_H = 4.46$ ppm (doublet, $^2J_{PH} = 8.5$ Hz) and 1.55 ppm (singlet) that integrate to a ratio of approximately 1:9. The *tert*-butyl groups are removed by stirring the intermediate di-*tert*-butylphosphonomethyltrifluoromethanesulfonate with 1 ml of fresh 4 M HCl in dioxane for 4 h. The solvent is removed with rotary evaporation, followed by high vacuum for about 15 h. Phosphonomethyltrifluoromethanesulfonate is stored at $-80°$ in a desiccator. When dissolved in d_6-DMSO, this compound displays two doublets by ^1H NMR, at 4.70 and 4.45 ppm, whose ratio changes over time. The ^{31}P NMR spectra of this compound likewise display two signals, at 12.1 and 8.0 ppm, whose ratio changes with time. We tentatively ascribe this time-dependent change in NMR spectra to the nucleophilic attack of DMSO on PMT. ^1H NMR spectra of PMT sometimes display a signal at 3.5 ppm that arises from an impurity.

Purification of D57C CheY

All steps are carried out near 5°, except the adjusting of buffers to the desired pH. Buffers are adjusted to the desired pH at room temperature, taking into account the temperature dependence of their pK_a values. Site-directed mutagenesis is used on plasmid pRL22 to produce a gene coding for D57C CheY. *E. coli* strains B and K12 are transformed with this plasmid by electroporation. *E. coli* D57C/T87I CheY and D57C/T87I/Y106W CheY cloned into separate pet24a(+) plasmid vectors (Novagen) are from Phil Matsumura. Each plasmid is transformed by standard electroporation methods into *E. coli* strain B834(DE3).

Cells are grown in Luria broth at 37° with shaking. Typically, 1.0 mM IPTG is added to induce protein production when the cells achieve an optical density of approximately 0.5 at 600 nm. The growth is continued into stationary phase, and the cells are centrifuged for approximately 20 min at 4000 rpm in a Beckman JA-10 rotor. The cells are resuspended in 50 mM Tris-HCl (pH 7.50), 1 mM EDTA, and 0.2 mM phenylmethylsulfonyl fluoride (PMSF). The concentration of cells is generally 0.2 g wet cell paste per milliliter of sonication buffer. Cells are lysed with a Fisher 550 sonic dismembrator using multiple 30-s bursts, interspersed with cooling. Lysozyme to a final concentration of 0.2 mg/ml is occasionally added to assist in breaking the cell walls. The cell debris is centrifuged at approximately 16,000 rpm for 30–45 min in a JA-20 rotor. The pellet is homogenized in 10 mM EDTA (pH 7.5), 30% (weight to volume of solution) sucrose using a Dounce tissue homogenizer. The mixture is spun at approximately 17,000 rpm for 45–60 min, and the pellet is homogenized in 50 mM Tris-HCl (pH 7.5), 5 mM EDTA, and 1% (volume to final volume) Triton X-100. The homogenate is spun at approximately 16,000 rpm for 45–60 min, and the pellet is resuspended in 6 M urea, 50 mM Tris-HCl (pH 7.5), 10 mM DTT, and 1 mM EDTA and incubated for at least 90 min. The urea had been prepared as a stock solution of 8 M and deionized by passing through a column of AG 501-X8(D) (Bio-Rad). When the 8 M urea stock solution is prepared well in advance, it is frozen as 25 ml-aliquots in plastic centrifuge tubes that are kept horizontal during freezing and stored at −80°. The translucent mixture is centrifuged at approximately 17,000 rpm, homogenized with a second portion of buffered urea, and centrifuged again. All of the supernatants are checked for protein using SDS-PAGE. Generally the first two urea supernatants form the inclusion-body purification. In some cases the sonication supernatant is also used to begin a separate, soluble-fraction purification.

The urea supernatants are combined and diluted to a final protein concentration of approximately 0.3 mg/ml. In our hands the Bradford assay sometimes overestimates the concentration of protein at this stage. The protein is dialyzed three to four times against 5 mM Tris-HCl (pH 7.5), 0.5–1.0 mM EDTA, 7.1 mM 2-ME, and 0.02% sodium azide. The first portion of dialysis buffer also includes 0.2 mM PMSF. The protein solution is applied to a 2.5 by 11-cm column of DE-52 (Whatman) in the chloride form. The column is rinsed with starting buffer to elute unbound protein, and the protein elutes with a 0–300 mM, 1-liter gradient of NaCl at 1 ml/min. Fractions are pooled on the basis of A_{280} readings and SDS-PAGE. Pooled fractions are dialyzed into 50 mM Tris-HCl (pH 7.9), 1 mM EDTA, 7.1 mM 2-ME, and 0.02% sodium azide. The pool is applied to a 2.5 by 12-cm column of Affigel Blue (Bio-Rad). After unbound proteins are eluted from the column with starting buffer, CheY is eluted with a step gradient of 2 M NaCl in the same buffer. In some cases a 2.5 by 54-cm column of Sephadex G-50–150 (Sigma) is used to remove higher molecular weight impurities that remain after the DE-52 anion-exchange column and the Affigel Blue column. The protein is dialyzed into 20 mM BES·KOH (pH 7.0–7.5), 0.1 mM EDTA, 5 mM DTT, and 0.02% sodium azide. Small portions of the protein are frozen and stored at −80°.

The soluble fraction is dialyzed three times into 50 mM MES·KOH (pH 6.0), 1 mM EDTA, 7.1 mM 2-ME, and 0.02% sodium azide. The first portion of dialysis buffer has 0.2 mM PMSF. The protein is applied to a 2.5 by 12-cm column of DE-52 in the chloride form. After the column is rinsed with starting buffer, a 600-ml gradient of 0–50 mM NaCl in the same buffer is used to elute the protein at 1 ml/min. The protein is collected in 6-ml fractions. Fractions are pooled on the basis of A_{280} values and SDS-PAGE. The protein is then purified with DE-52 and Affigel Blue as described earlier for the purification from inclusion bodies and frozen as small aliquots.

Phosphonomethylation

Phosphonomethylation is typically carried out on 3- to 15-mg portions of D57C CheY. If the ratio of free thiol groups to protein is less than about 75%, the aliquot of protein is reduced with DTT, BMS, or TCEP overnight. The following protocol assumes a fully reduced, 4-mg aliquot of CheY in $400\,\mu l$ of buffer after buffer exchange. The protein is exchanged into 500 mM AMPSO·KOH (pH 9.0–9.1), 1 mM EDTA and samples are taken for the protein assay, DTNB assay, and HPLC assay. The first two assays are completed and the percentage of free thiol groups is calculated prior to phosphonomethylation. Typically, a $\sim$20-μl aliquot of 3.2 M SrCl$_2$ is added (CaCl$_2$ and BaCl$_2$ are approximately as effective, as discussed earlier). A 12-mg portion of PMT (MW $\sim$250) is weighed into a small, plastic centrifuge tube and the tube is capped. A 10-μl aliquot of anhydrous ethanol is used to dissolve the PMT, a 20-μl aliquot of triethylamine (3 equivalents relative to PMT) is added, and the mixture is quickly added to a small, conical centrifuge tube containing CheY. The final concentration of PMT is about 120 mM. A precipitate sometimes develops (possibly the strontium salt of hydroxymethylphosphonate); therefore, the reaction mixture is centrifuged at the end of the reaction, typically about 1 h. Then the crude phosphono-CheY is applied to a desalting column equilibrated in a buffer appropriate for biotinylation, and the product is assayed for protein and for free thiol groups and by RP-HPLC.

RP-HPLC

Two Vydac 218TP columns each with a pore size of 300 Å are used. The flow rates are 0.2 ml/min for a 2.1 by 250-mm column and 1.0 ml/min for a 4.6 by 250-mm column, respectively. Mobile phase A is 5% acetonitrile with 0.1% TFA (v/v), and mobile phase B is 80% acetonitrile with TFA added to give approximately the same absorbance at 215 nm as mobile phase A, roughly 0.08% TFA. The gradient is 45.5–52.5% MPB over 45–60 min. We prepare the protein in a mixture of mobile phases A and B such that the percentage of acetonitrile is about 5% lower than its concentration at the start of the gradient. We load 5 μg of CheY on the 2.1-mm column, and on the 4.6-mm column, we load 15 μg. When the gradient is 45 min long, phosphono-CheY elutes at 15.8 min and CheY elutes at 28.5 min. The retention times vary slightly with different columns, and the gradient is adjusted accordingly.

DTNB Assay

A buffer of 0.10 M potassium phosphate, 1 mM EDTA is prepared at pH 7.3. A stock solution of 20 mM DTNB in this buffer is prepared without further adjustment of the pH. In a typical assay, 950 μl of buffer and 50 μl of DTNB are combined, and the absorbance is monitored at 410 nm. A 10-μg sample of CheY is added, and the absorbance is monitored for about 40 min to obtain a final absorbance reading. The initial value of absorbance is found by extrapolation of the progress curve back to time zero. In some cases, the reaction is run at 25.0° to obtain a stable reading prior to addition of protein. The concentration of free thiol groups calculated from this assay is divided by the protein concentration to obtain a percentage of free thiol groups.

Biotinylation

The concentration of biotinylating reagent is typically 2.5 mM, and the concentration of D57C CheY is about 250 μM. For reactions using PEO-iodoacetyl biotin (Pierce), the buffer is 50 mM AMPSO·KOH (pH 8.5), 1 mM EDTA, or 50 mM AMPSO·KOH (pH 9.0), 1 mM EDTA. The reaction is essentially complete in 4.5 h when assayed by DTNB or by RP-HPLC. However, the reaction appears to be slightly slower at pH 9.0, requiring more than 7 h. Based on these results, we use a reaction time of 12–18 h. For reactions using PEO-maleimide activated biotin (Pierce) the buffer is 50 mM BES·KOH (pH 7.0), 50 mM NaCl, and 1 mM EDTA. The reaction appears to be 80% complete after 2 h. However, longer reaction times lead to protein precipitation in several cases. For reactions with biotin-HPDP (Pierce or G Biosciences) and D54C/C81S CheY from *T. maritima*, the buffer is Tris-HCl (pH 8.0), 1 mM EDTA, and 0.02% sodium azide. A stock solution of 50 mM biotin-HPDP is prepared in DMSO with slight warming. It is added to the protein to produce a final concentration of 2.5 mM. The reaction is followed at 343 nm, and it appears to be 95% complete in 60 min at 25°. For some reactions employing PEO-iodoacetyl biotin, the reaction is terminated with the addition of DTT or 2-ME. After the biotinylation reaction is complete, the protein is exchanged into 50 mM BES·KOH (pH 7.1), 1 mM EDTA, and 0.02% sodium azide to remove the unreacted biotinylation reagent.

A 5-ml column of monomeric avidin (Pierce) is prepared essentially according to the manufacturer's directions, except that BES buffer is substituted for phosphate-buffered saline in the two buffers that are used near neutral pH. The mixture of phosphono-CheY and biotinylated CheY in a volume of approximately 2.5 ml is applied to the column by letting the solution into the column bed ~1 ml at a time, with incubations of at least several minutes to allow biotinylated CheY to diffuse to avidin. The effluent is collected starting at the beginning of the time the protein solution is applied to the column. About 2.5 additional column volumes (12.5 ml) of BES buffer (see earlier discussion) are used to elute the unbound protein (phosphono-CheY). This ~15-ml fraction is collected and concentrated in a CentriPrep, a Centricon-3 (Millipore), and then analyzed by HPLC. The exact capacity of monomeric avidin for biotinylated CheY is not known, but we have estimated it to be 0.2 mg/ml of column bed. If there is still biotinylated CheY present, the solution is passed over regenerated and reequilibrated avidin a second time and concentrated again. To regenerate and reequilibrate the monomeric avidin, the column is rinsed with 30 ml of glycine-HCl (pH 2.8) and then with 20 ml of BES buffer.

Buffer Exchange

When the volume of protein solution is 50 μl, a single Penefsky column is used. When the volume is 50–300 μl, two Penefsky columns are used sequentially. When the volume is 300–1500 μl, a commercial G25 column (Pharmacia PD-10) is used. Penefsky (spin) columns are prepared by filling a plastic cylinder having a plastic frit with about 3 ml of coarse Sephadex G25. Two column volumes of buffer are passed through the gel. The columns are placed in 12-ml plastic tubes and are spun in a clinical centrifuge (IEC) at 90% of maximal rotation speed for 60 s. The tops of 1.5-ml conical, plastic centrifuge tubes are cut off, and the conical tubes are placed beneath the column. The protein solution is applied to the now-opaque gel, and the columns are spun for 90 s. The volume of the protein solution may increase by about 15% for each run.

Acknowledgments

We acknowledge technical assistance from John Iacuzio and Tyler Davis and helpful conversations with Dennis Phillion. Support for this work was provided by North Carolina Biotechnology Center ARIG Grant 9905ARG0026 and by National Institutes of Health (NIH) Grant 1R15GM063514-01A1. Its contents are solely the responsibility of the authors and do not necessarily represent the official views of the North Carolina Biotechnology Center or the NIH.

References

Bailey, J. M., and Colman, R. F. (1987). Isolation of the glutamyl peptide labeled by the nucleotide analogue 2-(4-bromo-2,3-dioxobutylthio)-1,N(6)-ethenoadenosine 2′,5′-biphosphate in the active site of NADP+-specific isocitrate dehydrogenase. *J. Biol. Chem.* **262,** 12620–12626.

Buss, J. E., and Stull, J. T. (1983). Measurement of chemical phosphate in proteins. *Methods Enzymol.* **99,** 7–14.

Getz, E. B., Xiao, M., Chakrabarty, T., Cooke, R., and Selvin, P. R. (1999). A comparison between the sulfhydryl reductants tris(2-carboxyethyl)phosphine and dithiothreitol for use in protein biochemistry. *Anal. Biochem.* **273**, 73–80.

Halkides, C. J., McEvoy, M. M., Casper, E., Matsumura, P., Volz, K., and Dahlquist, F. W. (2000). The 1.9 Å resolution crystal structure of phosphono-CheY, an analogue of the active form of the response regulator, CheY. *Biochemistry* **39**, 5280–5286.

Halkides, C. J., Zhu, X., Phillion, D. P., Matsumura, P., and Dahlquist, F. W. (1998). Synthesis and biochemical characterization of an analog of CheY-phosphate, a signal transduction protein in bacterial chemotaxis. *Biochemistry* **37**, 13674–13680.

Head, C. G., Tardy, A., and Kenney, L. J. (1998). Relative binding affinities of OmpR and OmpR-phosphate at the ompF and ompC regulatory sites. *J. Mol. Biol.* **281**, 857–870.

Krueger, J. M., Stock, A. M., Schutt, C. E., and Stock, J. B. (1990). Inclusion bodies from proteins produced at high levels in *Escherichia coli*. In "Protein Folding: Deciphering the Second Half of the Genetic Code" (L. Gierasch and J. King, eds.), p. 334. American Association for the Advancement of Science, Washington, DC.

Kyte, J. (1995). "Structure in Protein Chemistry." Garland, New York.

Lowry, D. F., Roth, A. F., Rupert, P. B., Dahlquist, F. W., Moy, F. J., Domaille, P. J., and Matsumura, P. (1994). Signal transduction in chemotaxis. *J. Biol. Chem.* **269**, 26358–26362.

Netscher, T., and Bohrer, P. (1996). Formation of sulfinate esters in the synthesis of triflates. *Tetrahedron Lett.* **37**, 8359–8362.

Penefsky, H. S. (1979). A centrifuged-column procedure for the measurement of ligand binding by beef heart F_1. *Methods Enzymol.* **56**, 527–530.

Riddles, P. W., Blakeley, R. L., and Zerner, B. (1983). Reassessment of Ellman's reagent. *Methods Enzymol.* **91**, 49–60.

Singh, R., Lamoureux, G. V., Lees, W. J., and Whitesides, G. M. (1995). Reagents for rapid reduction of disulfide bonds. *Methods Enzymol.* **251**, 167–173.

Stock, A. M., Robinson, V. L., and Goudreau, P. N. (2000). Two-component signal transduction. *Annu. Rev. Biochem.* **69**, 183–215.

Wright, S. K., and Viola, R. E. (1998). Evaluation of methods for the quantitation of cysteines in proteins. *Anal. Biochem.* **265**, 8–14.

Zapf, J. W., Hoch, J. A., and Whiteley, J. M. (1996). A phosphotransferase activity of the *Bacillus subtilis* sporulation protein Sop0F that employs phosphoramidate substrates. *Biochemistry* **35**, 2926–2933.

[17]   Application of Fluorescence Resonance Energy Transfer to Examine EnvZ/OmpR Interactions

By S. Thomas King and Linda J. Kenney

Abstract

The EnvZ/OmpR two-component regulatory system is best known for regulating the porin genes *ompF* and *ompC* in response to changes in the osmolarity of the growth medium. In response to an unknown signal, EnvZ is autophosphorylated by ATP on a histidine residue. The phosphoryl group is subsequently transferred to a conserved aspartate residue on OmpR. Phosphorylation of OmpR increases its affinity for the regulatory regions of the porin genes, altering their expression. Phosphorylation also alters the interaction with EnvZ and OmpR. In order to study the interactions of EnvZ and OmpR, we employed a full-length EnvZ construct fused to the green fluorescent protein (GFP) that was overexpressed and targeted to the inner membrane. Spheroplasts were prepared and lysed in microtiter plates containing purified, fluorescent-labeled OmpR protein. Fluorescence resonance energy transfer (FRET) from the GFP donor to fluorescein- or rhodamine-conjugated OmpR acceptor occurred, indicating that the two proteins interact. We then used FRET to further characterize the effect of phosphorylation on the interaction parameters. Results indicate that the full-length EnvZ behaves similarly to the isolated cytoplasmic domain EnvZc alone. Furthermore, the phospho-OmpR protein has a reduced affinity for the EnvZ kinase. This chapter describes general considerations regarding such experiments and provides detailed protocols for quantitatively measuring them.

Introduction

Two-component signal transduction systems are the primary means by which prokaryotic organisms sense and respond to changing environmental conditions (Hoch and Silhavy, 1995). In their simplest form, these systems consist of a membrane-bound sensor kinase, which senses environmental conditions, and a cytoplasmic response regulator, which mediates the appropriate intracellular response. In the EnvZ/OmpR system, EnvZ is a histidine kinase that spans the inner membrane (Forst *et al.*, 1987; Liljestrom, 1986). In response to an unknown signal, EnvZ is phosphorylated by ATP at histidine 243 (Roberts *et al.*, 1994). The phosphoryl group is then transferred to aspartic acid 55 of the OmpR response regulator (Delgado *et al.*, 1993). This signaling pathway controls expression of

METHODS IN ENZYMOLOGY, VOL. 422
0076-6879/07 $35.00
DOI: 10.1016/S0076-6879(06)22017-2

the genes encoding the outer membrane proteins OmpF and OmpC (Fig. 1). At low osmolarity, OmpF is expressed, and at high osmolarity, *ompF* transcription is repressed and *ompC* is activated (van Alphen and Lugtenberg, 1977). In *Salmonella enterica* serovar Typhimurium, OmpR activates the SsrA/B two-component regulatory system responsible for turning on genes located on *Salmonella* pathogenicity island 2 to establish a systemic infection (Feng *et al.*, 2003, 2004; Lee *et al.*, 2000).

In addition to the aforementioned autophosphorylation and phospho-transfer activities, the EnvZ kinase can stimulate dephosphorylation of phospho-OmpR (OmpR~P). In a signaling pathway, dephosphorylation of the response regulator can be a key element, one that might be regulated to govern the duration or gradation of the response. At low osmolarity,

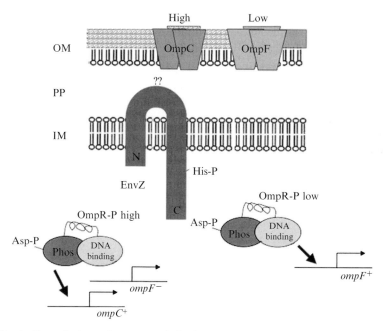

FIG. 1. General scheme for osmoregulation in *E. coli*. The sensor kinase, EnvZ, is an inner membrane protein. The response regulator OmpR consists of two domains, an N-terminal phosphorylation domain and a C-terminal DNA-binding domain, separated by a flexible linker. The phosphorylation site is indicated by Asp~P. At low osmolarity, OmpF is the major porin in the outer membrane ("OmpR~P low"). At high osmolarity, the concentration of OmpR~P increases ("OmpR~P high"), activating *ompC* transcription and repressing *ompF* transcription. OmpC is the major porin in the outer membrane at high osmolarity. OM, outer membrane; PP, periplasm; IM, inner membrane; high, high osmolarity; low, low osmolarity. (See color insert.)

OmpR~P levels are presumed to be low because the kinase activity of EnvZ is low or the phosphatase activity is high (Fig. 1). At high osmolarity, OmpR~P levels are thought to increase either by increasing the kinase or by decreasing the phosphatase activity of EnvZ.

Previous studies used a chimera of the transmembrane and periplasmic domains of the aspartate chemoreceptor (Tar) fused to the cytoplasmic domain of EnvZ (Taz) and proposed that the phosphatase activity of EnvZ was the regulated step in response to an osmotic signal (Yang and Inouye, 1993). As a first test of this model, we measured the cytoplasmic domain of EnvZ (EnvZc) binding to OmpR and OmpR~P; OmpR binds to EnvZc with high affinity, whereas OmpR~P has reduced affinity for EnvZc. These differences in affinity would promote OmpR binding to EnvZ~P, whereupon OmpR would become phosphorylated and its lower affinity for EnvZ would promote its dissociation and enhanced DNA binding (Mattison and Kenney, 2002). Therefore, it seems less likely that the EnvZ phosphatase is important *in vivo* for controlling OmpR~P levels. However, because EnvZc is not in its native membrane environment and lacks the transmembrane segments and periplasmic domain, it was of interest to develop an assay that employed the full-length EnvZ in its native cytoplasmic membrane.

The low copy number of EnvZ (approximately 100 molecules/cell) (Cai and Inouye, 2002) and its membrane location have made its characterization difficult. To circumvent these problems, a full-length protein was fused to the green fluorescent protein (GFP; kindly provided by Mark Goulian, University of Pennsylvania) and overexpressed. We then employed this construct and examined its interactions with purified, fluorescent-labeled OmpR using fluorescence resonance energy transfer (FRET) *in vitro*. Results indicate that phosphorylation of OmpR lowers its affinity for the kinase EnvZ and raise doubts as to whether the phosphatase activity of EnvZ is relevant *in vivo*. Other results also suggest that the isolated cytoplasmic domain of EnvZc behaves similarly to the full-length protein.

Overexpression of EnvZ and Preparation of Spheroplasts
(Based on Osborn *et al.*, 1972)

1. Innoculate 250 ml LB containing 50 μg/ml ampicillin with 2.5 ml of an overnight culture of *Escherichia coli* TOP10 containing the plasmid pSL5. The culture is grown with aeration to an OD_{600} of 0.6. Add isopropyl-β-D-thiogalactoside (IPTG) to a final concentration of 1 mM. At OD_{600} of 0.8, pellet the cells for 5 min at 8000 rpm.

2. Resuspend the pellet in 14 ml cold 0.75 M sucrose, 10 mM Tris-HCl, pH 7.8, and add 1.4 μl 10 μg/ml lysozyme and stir on ice. Slowly add 1.5 mM EDTA, pH 7.5, over 10 min.

A   B

C

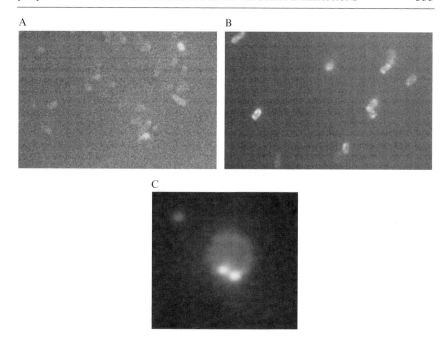

FIG. 2. EnvZ-GFP-expressing cells were suspended in phosphate-buffered saline and added to the cover glass in the TIRF microscope and allowed to sediment onto the coated surface (for image collection details, see Dubin-Thaler *et al.*, 2004). (A) Uninduced culture of *E. coli* Top10 containing plasmid pSL5. (B) Induced with 1 m*M* IPTG for 3 h. (C) Magnified view from the TIRF image in B.

3. Pellet the cells at 4000 rpm for 20 min at 4°. Pour off the supernatant and resuspend in 5 ml cold 0.25 *M* sucrose 3 m*M* Tris-HCl, pH 7.8, 1 m*M* EDTA. Verify spheroplast formation by light microscopy (see Fig. 2).

Protein Purification and Fluorescent Labeling of OmpR

Express and purify OmpR as described previously (Head *et al.*, 1998). To label OmpR with fluorescein or rhodamine maleimide, employ the following protocol.

1. Label OmpR at its native lone cysteine residue, C67, although many engineered unique cysteines are also readily labeled using this protocol (Maris *et al.*, 2005). React OmpR (25 μM) with a 10-fold molar excess of fluorescein-5-malemide or tetramethylrhodamine-5-maleimide (Invitrogen) for 2 h at 4° in Tris-buffered saline (TBS), pH 7.2 (Hermanson, 1996).

2. Separate the labeled protein from the unreacted probe by dialysis in 10 liter TBS overnight at $4°$ using SLIDE-A-Lyzer dialysis kits with a 10,000 molecular weight cutoff (Pierce). Change the dialysis solution two times, recover, and store the sample at $4°$.

3. Phosphorylate OmpR with the small phosphodonor phosphorami- date (PA). Synthesize ammonium hydrogen phosphoramidate according to Sheridan *et al.* (1971). Dissolve the resulting crystalline product to 0.5 *M* in water and use to phosphorylate OmpR. Carry out phosphorylation reac- tions in 200-μl volumes containing 50 m*M* Tris (pH 7.5), 50 m*M* KCl, 20 m*M* MgCl$_2$, 25 mM PA, and 10 μM OmpR. Incubate reactions for 1 h at room temperature. To measure the concentration of OmpR~P, inject samples onto a C4 HPLC column and elute using reversed phase chroma- tography as described (Tran *et al.*, 2000). Titrate fluorescein-labeled OmpR or OmpR~P into wells of a microtiter plate at increasing concentrations for the FRET assay described later.

4. To determine the percentage of protein labeled, measure a 1:10 dilution of dialyzed OmpR at three wavelengths, 280, 492, and 540 nm, for OmpR, fluorescein, and rhodamine using the following molar extinction coefficients 13,490, 83,000, and 95,000, respectively (Head *et al.*, 1998; Hermanson, 1996). In the experiment shown in Fig. 3, 90% of the OmpR protein is fluorescein labeled.

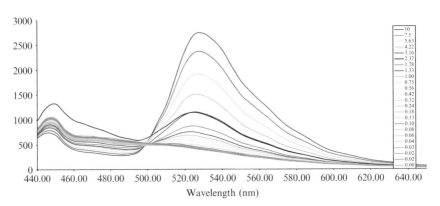

FIG. 3. Fluorescence resonance energy transfer (FRET) with EnvZ-GFP to fluorescein- conjugated OmpR. EnvZ-GFP was overexpressed and spheroplasts were prepared and lysed in cold H$_2$O according to Osborn *et al.* (1972). Fluorescent-labeled OmpR (fluorescein or rhoda- mine) ranged from 0 to 10 μM in the presence of 250 nM EnvZ-GFP. A control experiment was performed with 0 to 1000 nM unconjugated rhodamine (or fluorescein) in the presence of 250 nM EnvZ-GFP to measure the amount of nonspecific interaction of the donor (GFP) and acceptor (rhodamine or fluorescein) fluorophores. The OmpR concentration is shown in the boxed inset. (See color insert.)

Fluorescence Resonance Energy Transfer

1. Based on a molar extinction coefficient of 30,000 (A_{395}) for EnvZ-GFP, adjust the concentration to 250 nM and aliquot 15 μl into wells of a 384 white microtiter plate. Use 250 nM as a receptor concentration so that greater than 10-fold molar excess of OmpR or OmpR~P can be measured.

2. Add 30 μl 10 μM fluorescent-OmpR or OmpR~P to the first wells of the dilution series. Serially dilute OmpR into the 30 μl of EnvZ-GFP and incubate at room temperature for 30 min. In separate wells, also include unlabeled fluorescein or rhodamine to measure the amount of random association of EnvZ-GFP with the free fluorescent probe (Hicks, 2002). Measure four replicates at each dilution.

3. Measure fluorescence intensity in a Molecular Devices Spectra Max Gemini with the instrument set on autogain, 5-nm slit widths, and the following excitation and emission wavelengths: 425/450 nm (GFP), 425/530 and 480/530 for fluorescein, and 425/590 and 560/590 for rhodamine. Collect data in relative fluorescent units (RFU). Subtract the RFU for the GFP transfer to fluorescein (425/530) or rhodamine (425/590) from the samples, measuring the interaction of the free probe and EnvZ-GFP. For FRET measurements, it is advantageous to measure both the decrease in donor emission (EnvZ-GFP) and the increase in acceptor emission (rhodamine- or fluorescein-OmpR).

Figure 3 shows that excitation of the GFP at 425 nm in the presence of OmpR leads to fluorescein emission at 530 nm. Clearly there is energy transfer, as evident by both the decrease in the GFP emission (donor) at 450 nm and an increase in fluorescein emission (acceptor) at 530 nm. The efficiency of the EnvZ-GFP emission transfer to OmpR-fluorescein or OmpR-rhodamine is calculated to be 21 and 18%, respectively, based on the relative decrease of the GFP emission with OmpR-conjugated rhodamine or fluorescein. The change in relative fluorescence units as a function of OmpR and OmpR~P is plotted in Fig. 4. Data are fit to the equation shown in the legend to Fig. 4 using the sum of least squares, and both the binding affinity (K_d) and the Hill coefficient (H) were determined.

OmpR Has a Higher Affinity for EnvZ Than OmpR~P

The K_d for unphosphorylated OmpR interacting with EnvZ-GFP via FRET is 519 nM. This is in good agreement with a K_d determined for OmpR and EnvZc using fluorescence anisotropy (Mattison and Kenney, 2002). Our observation that two different constructs of EnvZ exhibit an identical affinity for OmpR suggests that the transmembrane and periplasmic domains do not influence the interaction between EnvZ and OmpR. A comparison of

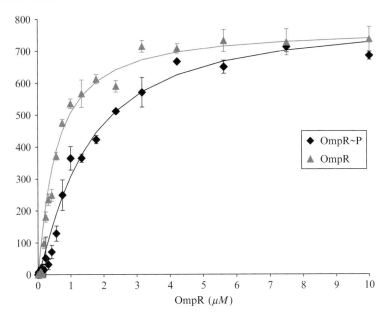

FIG. 4. The binding curve was a result of the OmpR-fluorescein and EnvZ-GFP FRET signal at 530 nm subtracted from the control titration of EnvZ-GFP and the fluorescein signal of the unconjugated probe and gave a similar value to a binding curve resulting from the decrease in GFP emission at 450 nm subtracted from the control titration of EnvZ-GFP and the rhodamine signal of the unconjugated probe. Data obtained were fit to the equation $F = F_{max}[OmpR]^H/K_d + [OmpR]H$ using the sum of least squares. The K_d for OmpR and for OmpR~P is 519 nM and 1.6 μM, respectively. The Hill coefficient was 1.17. (See color insert.)

the results of a FRET experiment with OmpR and OmpR~P is shown in Fig. 4. It is evident from this experiment that phosphorylation of OmpR reduces its affinity for the EnvZ kinase by about threefold. The K_d for OmpR~P/EnvZ-GFP is 1.6 mM. In a previous study using fluorescence anisotropy, it was demonstrated the OmpR~P had a lower affinity for EnvZc, but the K_d was not determined (Mattison and Kenney, 2002). The Hill coefficient was 1.17, indicating that the EnvZ/OmpR interaction was not cooperative. This is in contrast to a previous study that used SDS native PAGE to isolate complexes of OmpR and EnvZc that reported that OmpR binding to EnvZc was cooperative, although no estimates of Hill coefficients were presented (Yoshida et al., 2002). An important difference between the two studies is that the FRET experiment reported here is a solution-based assay allowing for the determination of the dissociation constants, whereas the previous study was not.

Concluding Remarks

The present study used an EnvZ-GFP fusion protein to study the interaction between OmpR and OmpR~P with its cognate kinase in its native membrane environment. These interactions were easily measured using fluorescence resonance energy transfer (Fig. 3). The observation that OmpR~P had a lower affinity for EnvZ was suggested by previous experiments (Mattison and Kenney, 2002), but the application of FRET enabled us to determine the K_d. Furthermore, the fact that EnvZc and EnvZ-GFP interact similarly with OmpR suggests that EnvZ/OmpR interactions are not affected by the transmembrane or periplasmic domains of EnvZ. It will now be of interest to employ EnvZ mutants that are reported to be biased in their kinase and phosphatase activities to see how the substitutions affect the interaction with OmpR and OmpR~P.

Acknowledgments

We are grateful to Dr. Mark Goulian, University of Pennsylvania, for his generous gift of EnvZ-GFP and for stimulating discussions and to Dr. Richard Gemeinhart (Department of Pharmaceutics and Bioengineering, University of Illinois-Chicago) for use of the fluorometer. We also thank Dr. Art Johnson (Texas A&M University) for helpful discussions regarding the FRET experiments. This work was supported by grants from the National Science Foundation (MCB-0613014) and National Institutes of Health (GM-058746) to L.J.K.

References

Cai, S. J., and Inouye, M. (2002). EnvZ-OmpR interaction and osmoregulation in *Escherichia coli. J. Biol. Chem.* **277,** 24155–24161.

Delgado, J., Forst, S., Harlocker, S., and Inouye, M. (1993). Identification of a phosphorylation site and functional analysis of conserved aspartic acid residues of OmpR, a transcriptional activator for *ompF* and *ompC* in *Escherichia coli. Mol. Microbiol.* **10,** 1037–1047.

Dubin-Thaler, B. J., Giannone, G., Dobereiner, H. G., and Sheetz, M. P. (2004). Nanometer analysis of cell spreading on matrix-coated surfaces reveals two distinct cell states and STEPs. *Biophys. J.* **86,** 1794–1806.

Feng, X., Oropeza, R., and Kenney, L. J. (2003). Dual regulation by phospho-OmpR of *ssrA/B* gene expression in *Salmonella* pathogenicity island 2. *Mol. Microbiol.* **48,** 1131–1143.

Feng, X., Walthers, D., Oropeza, R., and Kenney, L. J. (2004). The response regulator SsrB activates transcription and binds to a region overlapping OmpR binding sites at *Salmonella* pathogenicity island 2. *Mol. Microbiol.* **54,** 823–835.

Forst, S., Comeau, D., Norioka, S., and Inouye, M. (1987). Localization and membrane topology of EnvZ, a protein involved in osmoregulation of OmpF and OmpC in *Escherichia coli. J. Biol. Chem.* **262,** 16433–16438.

Head, C. G., Tardy, A., and Kenney, L. J. (1998). Relative binding affinities of OmpR and OmpR-phosphate at the ompF and ompC regulatory sites. *J. Mol. Biol.* **281,** 857–870.

Hermanson, G. T. (1996). "Bioconjugate Techniques." Academic Press, San Diego, CA.

Hicks, B. W. (2002). "Green Fluorescent Protein Applications and Protocols." Humana Press, Totowa, NJ.

Hoch, J. A., and Silhavy, T. J. (1995). "Two-Component Signal Transduction." ASM Press, Washington, DC.

Lee, A. K., Detweiler, C. S., and Falkow, S. (2000). OmpR regulates the two-component system SsrA-SsrB in *Salmonella* pathogenicity island 2. *J. Bacteriol.* **182,** 771–781.

Liljestrom, P. (1986). "Structure and Expression of the *ompB* Operon of *Salmonella typhimurium* and *Escherichia coli.*" Ph.D. Thesis, University of Helsinki, Helsinki, Finland.

Maris, A., Walthers, D., Mattison, K., Byers, N., and Kenney, L. J. (2005). The response regulator OmpR oligomerizes via β-sheets to form head-to-head dimers. *J. Mol. Biol.* **350,** 843–856.

Mattison, K., and Kenney, L. J. (2002). Phosphorylation alters the interaction of the response regulator OmpR with its sensor kinase EnvZ. *J. Biol. Chem.* **277,** 11143–11148.

Osborn, M. J., Gander, J. E., Parisi, E., and Carson, J. (1972). Mechanism of assembly of the outer membrane of *Salmonella typhimurium*: Isolation and characterization of cytoplasmic and outer membrane. *J. Biol. Chem.* **247,** 3962–3972.

Roberts, D. L., Bennett, D. W., and Forst, S. A. (1994). Identification of the site of phosphorylation on the osmosensor, EnvZ, of *Escherichia coli*. *J. Biol. Chem.* **269,** 8728–8733.

Sheridan, R. C., McCullough, J. F., and Wakefield, Z. T. (1971). Phosphoramidic acid and its salts. *Inorg. Synth.* **13,** 23–26.

Tran, V. K., Oropeza, R., and Kenney, L. J. (2000). A single amino acid substitution in the C-terminus of OmpR alters DNA recognition and phosphorylation. *J. Mol. Biol.* **299,** 1257–1270.

van Alphen, W., and Lugtenberg, B. (1977). Influence of osmolarity of the growth medium on the outer membrane protein pattern of *Escherichia coli*. *J. Bacteriol.* **131,** 623–630.

Yang, Y., and Inouye, M. (1993). Requirement of both kinase and phosphatase activities of an *Escherichia coli* receptor (Taz1) for ligand-dependent signal transduction. *J. Mol. Biol.* **231,** 335–342.

Yoshida, T., Qin, L., and Inouye, M. (2002). Formation of the stoichiometric complex of EnvZ, a histidine kinase, with its response regulator, OmpR. *Mol. Microbiol.* **46,** 1273–1282.

[18] Gene Promoter Scan Methodology for Identifying and Classifying Coregulated Promoters

By IGOR ZWIR, OSCAR HARARI, and EDUARDO A. GROISMAN

Abstract

A critical challenge of the postgenomic era is to understand how genes are differentially regulated. Genetic and genomic approaches have been used successfully to assign genes to distinct regulatory networks in both prokaryotes and eukaryotes. However, little is known about what determines the differential expression of genes within a particular network, even when it involves a single transcription factor. The fact that coregulated genes may be differentially expressed suggests that subtle differences in the shared *cis*-acting regulatory elements are likely to be significant. This chapter describes a method, termed gene promoter scan (GPS), that discriminates among coregulated promoters by simultaneously considering a variety of *cis*-acting regulatory features. Application of this method to the PhoP/PhoQ two-component regulatory system of *Escherichia coli* and *Salmonella enterica* uncovered novel members of the PhoP regulon, as well as regulatory interactions that had not been discovered using previous approaches. The predictions made by GPS were validated experimentally to establish that the PhoP protein uses multiple mechanisms to control gene transcription and is a central element in a highly connected network.

Introduction

The two-component system constitutes a major form of bacterial signal transduction. Typically, a two-component system consists of a sensor kinase that responds to a specific signal by modifying the phosphorylated state of a cognate response regulator. The majority of response regulators are DNA-binding proteins that modulate gene transcription. Because the phosphorylated form of the response regulator binds to target promoters with higher affinity than the unphosphorylated one, sensor-promoted changes in the phosphorylated state of a response regulator can have a profound impact in the gene expression profile of an organism.

Genomic analysis revealed that there is a direct correlation between genome size and the number of two-component systems present in a given bacterial species. In addition, organisms that live in varied environments tend to have a larger number of two-component systems than those that occupy a single environment. For example, the aphid endosymbiont

METHODS IN ENZYMOLOGY, VOL. 422
Copyright 2007, Elsevier Inc. All rights reserved.

Buchnera aphidicola has a genome size of approximately 640 kb that does not encode two-component systems (Shigenobu *et al.*, 2000). In contrast, *Escherichia coli* has a genome size of 4.5 Mb encoding 30 such systems (Blattner *et al.*, 1997), and the environmental microbe and opportunistic pathogen *Pseudomonas aeruginosa*, with a genome size of 6.3 Mb, harbors 118 two-component system proteins (Stover *et al.*, 2000).

The number of targets that a response regulator controls varies among the different systems found in a given bacterial species and between homologous systems in related bacterial species. In *E. coli*, for example, the response regulator KdpD appears to govern transcription of a single promoter, whereas the response regulator ArcA modulates expression of >30 operons (Georgellis *et al.*, 1999; Salgado *et al.*, 2004). Because the products encoded by the multiple targets of regulation of a response regulator such as ArcA are likely required in different amounts and/or for different extents of time, the corresponding genes must differ in their *cis*-acting promoter sequences responsible for the distinct gene expression patterns of individual members of a regulon (i.e., a group of genes that is coordinately regulated by a regulatory protein). The analysis of coregulated genes is complicated by the fact that two-component systems can control gene expression indirectly by modulating the expression and/or activity of other two-component systems, transcriptional regulators, and sigma factors. Moreover, the targets of regulation of orthologous response regulators overlap only partially in closely related species such as *Salmonella* and *E. coli*, suggesting that small changes in the amino acid sequence of a response regulator and/or in *cis*-acting promoter features can have a big impact on gene regulation. Cumulatively, these issues highlight the need for methods that identify the critical elements of a promoter determining gene expression and that are not heavily dependent on sequence conservation such as phylogenetic footprinting methods (Manson McGuire and Church, 2000).

The material required for analyzing the promoter features governing bacterial gene expression is widely available. It consists of genome sequences (often of multiple isolates of a given bacterial species), genome-wide transcription data (typically obtained using microarrays), and biological databases containing examples of previously explored cases. However, it is not yet possible to scan a bacterial genome sequence and readily predict the expression behavior of genes belonging to a regulon. In principle, coregulated genes could be differentiated by incorporating into the analysis quantitative and kinetic measurements of gene expression (Ronen *et al.*, 2002) and/or considering the participation of other transcription factors (Bar-Joseph *et al.*, 2003; Beer and Tavazoie, 2004; Conlon *et al.*, 2003). However, there are constraints in such analyses due to systematic errors in microarray experiments, the extra work required to obtain kinetic data, and the missing information about additional signals impacting on gene expression. These constraints hitherto only allow a relatively crude classification of gene expression patterns into a limited number of classes (e.g., up- and downregulated genes [Oshima *et al.*, 2002; Tucker *et al.*, 2002]).

This chapter discusses a methodology designed to identify and classify promoters that are coregulated by a bacterial transcriptional regulator, such as the response regulators of two-component systems. This methodology, termed gene promoter scan (GPS), groups promoters sharing distinct sets of promoter features to generate groupings that may reflect biological properties of a system under investigation such as the time and place that a promoter is activated or silenced.

We have applied the GPS method to investigate the targets of regulation of the response regulator PhoP, which together with the sensor kinase PhoQ form a two-component system that is a major regulator of virulence and of the adaptation to low Mg^{2+} environments in several gram-negative species (Groisman, 2001). The PhoQ protein responds to the levels of extracytoplasmic Mg^{2+} by modifying the phosphorylated state of the DNA-binding protein PhoP (Castelli *et al.*, 2000; Chamnongpol *et al.*, 2003; Montagne *et al.*, 2001). The PhoP/PhoQ system is a particularly interesting case study because (1) it controls the expression of a large number of genes, amounting to approximately 3% of the genes in the case of *Salmonella* (Zwir *et al.*, 2005). (2) Promoters harboring a binding site for the PhoP protein may differ in the distance and orientation of the PhoP box relative to the RNA polymerase-binding site, as well as in other promoter features. (3) PhoP also controls gene expression indirectly by regulating the expressionand/ or activity of other two-component systems at the transcriptional (e.g., RstA/ RstB) (Minagawa *et al.*, 2003), posttranscriptional (e.g., SsrB/SpiR) (Bijlsma and Groisman, 2005), and posttranslational (e.g., PmrA/PmrB) (Kato and Groisman, 2004) levels. In addition, PhoP regulates the levels of the alternative σ factor RpoS (Tu *et al.*, 2006) and participates in a feed-forward loop with the regulatory protein SlyA (Shi *et al.*, 2004) (Fig. 1).

Challenge of Identifying Promoter Features Governing
Gene Transcription

Identification of the promoter features that determine the distinct expression behavior of coregulated genes is a challenging task because of the difficulty in ascertaining the role that subtle differences in shared *cis*-acting regulatory elements of coregulated promoters play in gene transcription. Therefore, approaches that homogenize features among promoters (e.g., relying on consensuses to describe the various promoter features)

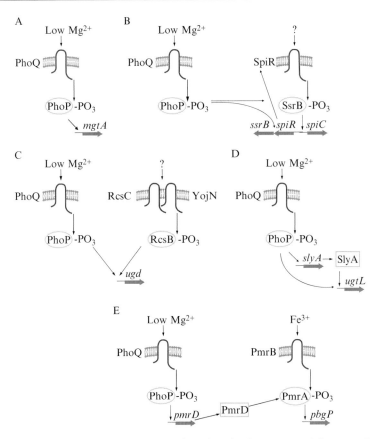

FIG. 1. The PhoP/PhoQ system uses a variety of mechanisms to control the expression of a large number of genes in a direct or indirect fashion. (A) The PhoP protein recognizes a direct hexanucleotide repeat separated by five nucleotides, which has been termed the PhoP box, activating the *mgtA* promoter of *Salmonella*. (B) The PhoP/PhoQ uses a transcriptional cascade mediated by the SsrB/SpiR two-component system to regulate the *spiC* promoter. (C) The PhoP/PhoQ system works cooperatively with the RcsB/RcsC system to activate the *ugd* promoter. (D) The PhoP/PhoQ system utilizes a feed-forward loop mediated by the SlyA protein to activate the *ugtL* promoter in *Salmonella*. (E) The PhoP/PhoQ system controls the *pbgP* promoter at the posttranslational level, where the PhoP-dependent PmrD protein activates the regulatory protein PmrA. (See color insert.)

and even across species can hamper the discovery of key differences that distinguish promoters coregulated by the same transcriptional regulator. For example, methods that look for matching of a sequence to a consensus have been successfully used to identify promoters controlled by particular transcription factors (Bailey and Elkan, 1995; Martinez-Antonio and

Collado-Vides, 2003; Stormo, 2000). Although these methods often increase specificity, their strict cutoffs decrease sensitivity (Hertz and Stormo, 1999; Stormo, 2000), which makes it difficult to detect binding sites with a weak resemblance to a consensus sequence. The complexity of the analysis is exacerbated by the need to consider other sequence elements relevant to differential expression patterns, including the class and location of the RNA polymerase, the presence of binding sites for other transcription factors, and their topological location in the DNA (Beer and Tavazoie, 2004; Pritsker et al., 2004). Indeed, similar expression patterns can be generated from different features or a mixture of multiple underlying features, thus making it more difficult to discern the molecular basis for analogous gene expression.

GPS Methodology as an Integrated Algorithm

The increased availability of biological information, such as genome sequences, microarray gene expression, as well as text data stored in public databases, and knowledge-discovery techniques (or data mining) is used to currently generate hypotheses that need to be evaluated. For example, when groups of coregulated transcripts are identified by clustering the expression patterns generated by a series of microarray experiments, the promoter sequences for each transcript in a cluster may be fed to a motif discovery algorithm to find common elements implicated in transcriptional regulation among coexpressed genes. These approaches incorporate knowledge in a decision-making cascade that can be summarized as follows: find genes with similar expression patterns and then see if they have similar promoters (Holmes and Bruno, 2000).

Most of the available algorithms implemented by the approaches described earlier base their decision in cutoffs that constrain one analysis stage on the previous one. Therefore, the analysis is hampered by the need to decide whether to consider first gene expression data or promoter features. In addition, the analysis is complicated because of the noisy nature of microarray data, the possibility of cryptic promoter elements contributing to gene expression, the potential for interaction among regulatory proteins, and the existence of alternative modes of transcription regulation, which remain poorly understood.

There are simple algorithms that ignore the constraints just listed, which appear to generate interesting results and be of practical benefits (Tavazoie et al., 1999). However, identification of the promoter features that determine the distinct expression behavior of coregulated genes within a regulon requires a more detailed and integrated analysis of the regulatory features. Why is it useful to have an integrated model? One reason is that cascade algorithms and integrated algorithms are solving subtly different problems. In contrast to cascade algorithms, integrated algorithms can be summarized as finding clusters of genes that have (a) similar expression patterns *and* (b) similar promoters (Holmes and Bruno, 2000).

The gene promoter scan is a machine learning method (Cheeseman and Oldford, 1994; Cook *et al.*, 2001; Cooper and Herskovits, 1992) that identifies, differentiates, and groups sets of coregulated promoters by simultaneously considering multiple *cis*-acting regulatory features and gene expression (Fig. 2). GPS carries out an exhaustive description of *cis*-acting regulatory features, including the orientation, location, and number of binding sites for a regulatory protein, the presence of binding site submotifs, and the class and number of RNA polymerase sites (Fig. 3).

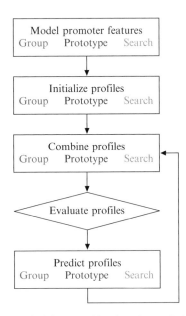

FIG. 2. The GPS method. GPS is a machine learning technique that *models* promoter features as well as relations between them, uses them to describe promoters, *combines* such characterized promoters into groups termed profiles, *evaluates* the resulting profiles to select the most significant ones, and performs genome-wide *predictions* based on such profiles. To accomplish this task, GPS carries out three basic operations: *grouping* observations from the data set; *prototyping* such groups into their most representative elements (centroid); and *searching* in the set of optimal solutions (i.e., Pareto optimal frontier) to retrieve the most relevant profiles, which are used to describe and identify new objects by similarity with the prototypes. (See color insert.)

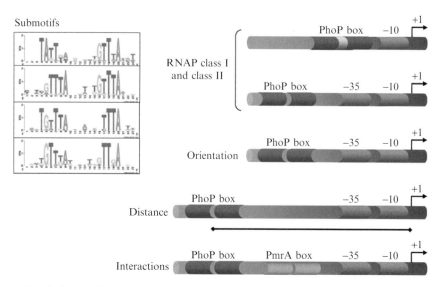

FIG. 3. Schematics of PhoP-regulated promoters harboring different features analyzed by GPS. GPS performs an integrated analysis of promoter regulatory features, initially focusing on six types of features for describing a training set of promoters: *submotifs*, which model the studied transcription factor-binding motifs; *RNA pol sites*, which characterize the RNA polymerase motif, the class of $\sigma70$ promoter that differentiates *class I* from *class II* promoters, and the distance distributions (*close, medium,* and *remote*) between RNA polymerase and transcription factor-binding sites in activated and repressed promoters; *activated/repressed*, where we learn activation and repression distributions by compiling distances between binding sites for RNA polymerase and a transcription factor; *interactions,* where we evaluate motifs for several transcription factor-binding sites and model the distance distributions between motifs colocated in the same promoter regions; and *expression,* which considers gene expression levels. (See color insert.)

The GPS method is specifically aimed at handling the variability in sequence, location, and topology that characterize gene transcription. Instead of using an overall consensus model for a feature, where potentially relevant differences are often concealed because of intrinsic averaging operations between promoters and even across species, we decompose a feature into a family of models or building blocks. This approach maximizes the sensitivity of detecting those instances that weakly resemble a consensus (e.g., binding site sequences) without decreasing the specificity. In addition, features are considered using fuzzy assignments (i.e., not precisely defined) instead of categorical entities (Bezdek, 1998; Gasch and Eisen, 2002; Ruspini and Zwir, 2002), which allow us to encode how well a particular sequence matches each of the multiple models for a given promoter feature.

Individual features are then linked into more informative composite models that can be used to explain the kinetic expression behavior of genes.

It should be noted that GPS treats each of the promoter features with equal weight because it is not known beforehand which features are important. To circumvent limitations imposed by relatively few classes of gene expression levels to *cis*-acting features, the GPS method treats gene expression data as one feature among many (as opposed to testing it as a dependent vaiable). The various features are analyzed concurrently and recurrent relations are recognized to generate profiles, which are groups of promoters having features in common. GPS uses an unsupervised strategy (i.e., preexisting examples are not required), as well as multiobjective optimization techniques, which enhance the likelihood of recovering all optimal feature associations rather than potentially biased subsets (Deb, 2001; Ruspini and Zwir, 2002). The resulting profiles group promoters that may share underlying biological properties.

Exploring Targets of Regulation of a Response Regulator Using GPS

GPS Built-in Features

The GPS method performs an integrated analysis of promoter regulatory features to identify profiles, which are sets of promoters described by common sets of features. We initially focused on six types of features for describing a training set of promoters (Bar-Joseph *et al.*, 2003; Beer and Tavazoie, 2004; Li *et al.*, 2002; Zwir *et al.*, 2005b): *submotifs*, which model the studied transcription factor-binding motifs; *orientation*, which characterizes the binding boxes as either in direct or opposite orientation relative to the open reading frame; *RNA pol sites*, which characterize the RNA polymerase motif (Cotik *et al.*, 2005), the class of $\sigma70$ promoter (Romero Zaliz *et al.*, 2004) that differentiates *class I* from *class II* promoters, and distance distributions (*close, medium,* and *remote*) between RNA polymerase and transcription factor-binding sites in activated and repressed promoters (Salgado *et al.*, 2004); *activated/repressed*, where we learn activation and repression distributions by compiling distances between binding sites for RNA polymerase and a transcription factor; *interactions*, where we evaluate motifs for several transcription factor-binding sites and model the distance distributions between motifs colocated in the same promoter regions; and *expression*, which considers gene expression levels.

GPS Initialization Strategy

The GPS method takes a list of candidate genes obtained from the literature, gene expression experiments (e.g., microarray, ChiP, or RT-PCR), or user-based hypothesis and generates initial profiles of each individual type of feature. The generation of initial profiles increases the sensitivity of a feature without decreasing its specificity (Zwir *et al.*, 2005a). This distinguishes GPS from methods relying on a single consensus, which often fail to describe and retrieve potentially interesting candidates that exhibit a weak resemblance to an average consensus pattern, which may be construed as a gene being indirectly regulated by a transcription factor (Zwir *et al.*, 2005b).

Input data should be specified according to each type of feature, that is, DNA sequences for the *submotifs* and, if available, gene expression levels for the *expression* (see online manual at http://gps-tools.wustl.edu). The initial models for each feature can also be provided by the user. For example, GPS uses position weight matrices generated by the Consensus/Patser method (Stormo, 2000), but also can accept any other built-in matrix generated from other methods (Tompa *et al.*, 2005). Indeed, the number of profiles can be a priori specified or calculated automatically using the Xie-Beni index (Zwir *et al.*, 2005a). Although the specifications of these initial conditions are crucial for clustering algorithms (Bezdek *et al.*, 1992), they are not critical for GPS and can be solved later by the dynamic approach followed by the method (Zwir *et al.*, 2005a).

Two or more promoter regions containing different binding sites for a given transcription factor are considered as distinct instances, which can be later associated by the method as more features become incorporated into the analysis. Indeed, GPS considers promoters independently of phylogenetic conservation. Therefore, after dissecting direct and indirect regulation, each instance in the database is constrained to a promoter region where a binding site motif of the studied transcription factor is found. Several features describe promoters exhibiting a binding site of the studied transcription factor. For example, one or more RNA polymerase-binding sites can be predicted around that site, which can be *class I* or *II* and located at different distances termed *close, medium,* or *remote.* The subjacent models for these features were learned from experimental examples provided by the RegulonDB database (Salgado *et al.*, 2004). However, we generated predictions from raw data rather than using original data present in RegulonDB. Additionally, GPS characterizes binding sites for a transcription factor as participating in either activation or repression by evaluating their distance from the RNA polymerase site. In this way, GPS establishes relationships between the different binding sites and their topology in a promoter region. Finally, because the expression is considered as one feature among many, promoters can be analyzed in the absence of expression data.

One of the most salient properties of the strategy followed by GPS to encode features is the use of metadata. Thus, GPS can encode features as fuzzy data (i.e., not precisely defined) instead of categorical entities

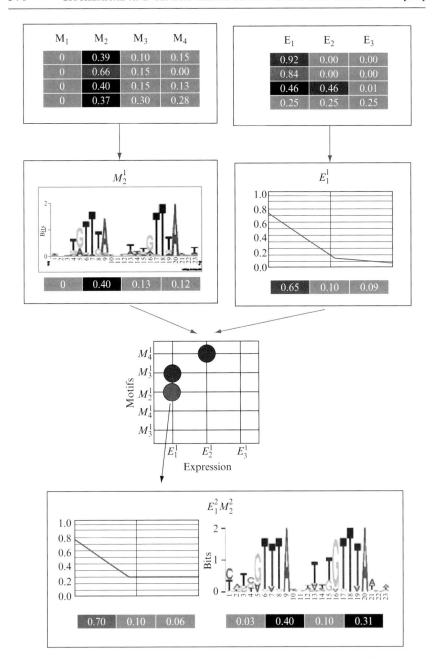

FIG. 4. Using GPS to build promoter profiles. GPS generation of the profile $E_1^2 M_2^2$ is shown here. It partially corresponds to the highlighted substructure of the lattice shown in Fig. 5. GPS starts by using information from databases and microarray data to construct a family of models

(Bezdek, 1998; Gasch and Eisen, 2002; Ruspini and Zwir, 2002), where a promoter instance can be related to more than one model. This captures the variability that exists in biological systems and delays the grouping of promoters until more information (i.e., features) is added. For example, a sequence corresponding to a transcription factor-binding site can be initially similar to both *submotif* M_1 and *submotif* M_2. Later, it could be assigned to a profile containing M_1 after adding the *orientation* and the *RNA pol site* features. Moreover, new intermediate profiles can be generated by taking advantage of the implementation of the profiles as dynamic fuzzy clusters (Bezdek, 1998).

The GPS method also uses metadata to analyze composite features. It could be the case, for example, that two or more features would not be independent of each other. Thus, GPS joins them by using fuzzy predicates [i.e., such as P(A and B) in a probabilistic interpretation]. Indeed, the distance between binding sites for RNA polymerase and for a transcription factor is meaningless if one does not consider the occurrences of the sites.

GPS Grouping Strategy

The GPS method performs an exhaustive combination of the features, which are dynamically rediscretized at each level of a lattice searching space (Fig. 4). The method allows reassignments of observations between sibling profiles, thereby solving initial misspecifications and allowing to identify cohesive sets of promoters in environments with reduced data sets

for each feature (e.g., expression levels E_1 to E_3, PhoP box submotif M_1 to M_4, as well as other features [not shown]). The promoters are described using the modeled features, the degree of matching between features and promoters being encoded as a vector of independent values, where 1 (red color) corresponds to maximum matching and 0 (green color) corresponds to the absence of the feature. For each feature, the promoters are then grouped into subsets that share similar patterns using fuzzy clustering. Each subset shown in the initial panel is prototyped by locating the centroid that best represents the group to generate the initial, level 1 profiles (e.g., $E_1^1 M_2^1$ and I_3^1). The centroids are encoded as a vector and also visualized by graphical plots for the "expression" and the "interactions" features and by a sequence logo (Crooks *et al.*, 2004) for the "submotifs" feature. These level 1 profiles are combined to generate level 2 profiles (e.g., $E_1^2 M_2^2$ and $M_2^2 I_3^2$) by the intersection of the ancestor profiles and then prototyped. (Blue circles represent profiles containing other subsets of promoters. The absence of a circle signifies that no promoters are classified into these profiles.) Further navigation through the feature-space lattice generates the level 3 profiles (for example, $E_1^3 M_2^3 I_3^3$) after incorporating the "interactions" feature (Fig. 5). Note that the vectors of the daughter profiles are built anew from the constituent promoters and are slightly different than those of their ancestors because of the refinement that takes place during the profile learning process. (See color insert.)

and high levels of uncertainty. For example, a promoter that initially resembles both M_2^1 and M_3^1 *submotifs* and is initially assigned to M_2^1 can later be reassigned to M_3^3 in a level 3 profile E_1^3 M_3^3 I_3^3. This happens because, unlike hierarchical clustering, GPS allows movement between sibling profiles E_1^3 M_2^3 I_3^3 → E_1^3 M_3^3 I_3^3 and dynamically reformulates the initial discretization of the profiles (Zwir *et al.*, 2005a) (Fig. 5).

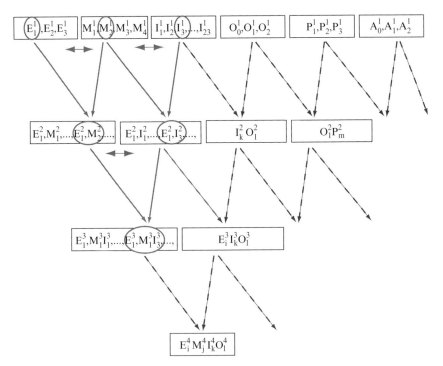

FIG. 5. GPS navigates through the feature-space lattice, generating and evaluating profiles. For analysis of promoters regulated by the PhoP protein, we identified up to five models for each type of feature, which are used to describe the promoters. Then, GPS generates profiles, which are groups of promoters sharing common sets of features. (Subscripts denote the different profiles for each feature, whereas superscripts denote the level in the lattice of the profile.) For example, E_1^1 is a particular *expression* profile that differs from E_2^1 and E_3^1. These level 1 profiles of each feature are combined to identify level 2 profiles; similarly, level 2 profiles are combined to create level 3 profiles. In addition, because of the fuzzy formulation of the clustering, any promoter that was initially assigned to a specific profile E_i^t can participate in profile of level t (i.e., indicated as a double-headed arrow). Thus, observations can migrate from parental to offspring clusters (i.e., hierarchical clustering) and among sibling clusters (i.e., optimization clustering). Here we show a small part of the complete lattice, where the part that is highlighted in red is also described in Fig. 4. (See color insert.)

GPS Evaluation Strategy

The profile searching and evaluation process is carried out as a multi-objective optimization problem (Deb, 2001; Rissanen, 1989; Ruspini and Zwir, 2002), which must consider conflicting criteria: the extent of the profile, the quality of matching among its members and the corresponding features, and its diversity (Cook *et al.*, 2001; Ruspini and Zwir, 2002). This strategy allows the identification of sets of optimal, instead of single or maximum estimated, profiles as models of alternative hypotheses describing distinct regulatory scenarios.

GPS Validation Strategy

GPS is an unsupervised method that does not need the specification of output classes, which is in contrast to supervised approaches (Zwir *et al.*, 2005a). Thus, the discovered profiles can be used for independently explaining external classes as a process often termed labeling (Mitchell, 1997). These classes can be introduced as a control in GPS, which automatically correlates them with the obtained profiles. For example, GPS uses the expression as one feature among many often derived from a constrained microarray gene expression experiment that just distinguishes between up- and downregulated genes (e.g., mutant vs wild-type conditions). However, the posterior availability of more discriminating classes, such as those derived from time-dependent ChiP experiments, can be used as an external phenomenon to be explained by the learned profiles.

Technical Specifications of GPS

Programming Resources

The GPS system has been implemented to be a platform-independent method, with a flexible and fast performance. It combines various machine learning techniques, implemented in cohesive programming languages and frameworks, to satisfy these nonfunctional requirements. The software consists of a core application, which executes sequentially as well as in parallel fashion on a cluster of computers, and two remote interfaces: a light web front end user interface developed in php, which accepts user's input and e-mails results, and a web service interface coded in java.

User Interface

Data definition and parameters are specified to the system as a single XML document (Wang *et al.*, 2005). This standard provides the required flexibility and readability to allow specification of the database, features, and initial profiles. An XML schema is provided (http://gps-tools.wustl.edu/gps/gps.xsd) to verify the document and to facilitate its editing. Apache Tomcat and Apache Axis are used to provide Web service interface, allowing application-to-application interaction in a standardized fashion (http://gps-tools.wustl.edu:8080/gps).

The *core system* is coded in a java-independent platform. Advanced java virtual machines with adaptive and just-in-time compilation and other techniques now typically provide performance up to 50 to 100% the speed of C++ programs (Lindsey *et al.*, 2005). We also encapsulated the execution of existing position weight matrice software by developing ad-hoc scripts in perl scripting language.

Parallel Execution

Components that require a large amount of processing power are executed in parallel in a high-performance computing environment provided by the Condor High throughput computing workload management system (Basney and Livny, 1999), which administers batch jobs on clusters of dedicated computing resources.

GPS Input

The GPS method captures the input specifications by an XML file that contains two parts. The first part corresponds to the specifications of the features, whereas the second corresponds to the database composed of the promoter values for the features.

Feature Specifications

Here we describe examples of several features. The complete manual is online at http://gps-tools.wustl.edu.

Purpose: representing DNA-binding site submotifs
Syntax

<Feature type="sequence" name="submotif" >

Indicates that input data can be a DNA sequence or a position weight matrix containing a motif, and its name, which must be unique.

<Bin name="M_1" membershipFile="gps_data/M_1.mat" file Type ="mat" />

Specifies one input bin (i.e., *submotif*) that was previously clustered and preprocessed as a position weight matrix and stored in a file termed M_1 with extension "mat" located in a user-defined directory.

```
<Bin name="M_2">
<InitialMember name="mgtC_681"/>
<InitialMember name="mgtC_718"/>
<InitialMember name="mgtC_925"/>
</Bin>
```

Specifies a bin containing the name of promoters belonging to a desired submotif, which are used to automatically calculate the initial matrices if they are not available.

```
<Bin cluster="n">
```

If a single bin is proposed, GPS automatically clusters the instances into ("n") bins. If the number of clusters is null (""), GPS uses the Xie-Beni index to calculate the initial number of clusters.

Description: GPS takes initial lists of candidate promoter sequences for a specific transcription factor and clusters them using Fuzzy C-Means algorithm into bins. Each of these bins is further encoded as position weight matrices using the Consensus/Patser method and used as single-type initial profiles. If the number of clusters is not specified, GPS uses the Xie-Beni index to provide the corresponding number. Matrices provided by other methods (e.g., MEME) can be also directly incorporated as input data. The initial bins would be dynamically reformulated when new features were aggregated.

Purpose: representing microarray gene expression
Syntax

```
<Feature type="expression" name="Expression" >
```

Indicates that input data can be a vector of continuous values corresponding to levels of gene expression resulting from one or more experiments. The name must be unique.

```
<Bin  name="E_1"  membershipFile="gps_data/E_1.exp"  fileType
  ="exp" />
```

Specifies one input expression bin, where columns are distinct experimental or time conditions that were previously clustered and preprocessed as a prototype (i.e., centroid or array of real numbers) and stored in a file termed E_1 with extension "exp" located in a user-defined directory.

```
<Bin name="E_2"/>
<InitialMember name="mgtC_681"/>
<InitialMember name="mgtC_718"/>
<InitialMember name="mgtC_925"/>
</Bin>
```

Specifies a bin containing the name of promoters belonging to a desired expression profile, which is used to automatically calculate the corresponding initial prototype.

<Bin cluster="n">

If a single bin is proposed, GPS automatically clusters the instances into (*"n"*) profiles. If the number of clusters is null (*""*), GPS uses the Xie-Beni index to calculate the initial number of profiles.

Description: GPS takes lists of candidate genes, where columns indicate different or time-dependent experiments. GPS clusters them using Fuzzy C-Means algorithm into bins. Each of these bins is further encoded as a centroid and used as single-type initial profiles. If the number of clusters is not specified, GPS uses the Xie-Beni index to provide the corresponding number. The initial profiles would be dynamically reformulated when new features were aggregated.

Purpose: representing the orientation or topological order of a regulatory element
Syntax

<Feature type="value" name="Orientation" deviation_ factor="0.5">

Indicates that input data can be a continuous/integer value, which represents continuous or discrete events, respectively. For example, the orientation of a binding site relative to the open reading frame (e.g., direct or opposite) or the topological order of a regulatory element regarding another (e.g., in front of or behind). The prototypes are uniformly discretized according to the *deviation factor*. For example, choosing a partition with three values P_0, P_1, and P_2, GPS establishes that P_1 will be the central value and $P_{0,2} = P_1 \pm df \times stdex$.

Description: GPS takes lists of promoters characterized by a discrete or continuous values (e.g., direct = 0 and indirect = 1; repressed = 0, activated = 1, and fuzzy activated or repressed = 0.5). The method clusters them using Fuzzy C-Means algorithm into bins. Each of these bins is further encoded as prototypes calculated as specified in the syntax section.

Purpose: representing fuzzy features
Syntax

<Feature type="fuzzy" name="Fuzzy_Motif" input type="sequence"
interpretation="possibilistic|fuzzy">

Indicates metadata that encode the degree of matching between an instance and several profiles (i.e., the similarity between instances and the prototypes that represent the profiles). The input can accept different types of features: "sequence," "expression," etc. The encoding method could be fuzzy or possibilistic (i.e., membership to all profiles do not have to sum 1).

> <*Bin name="MF_1" centroid="[0.652 0.036 0.160 0.528]" wi=" 0.5339" (default=1)/>*
>
> <*Bin name="MF_2" centroid="[0.808 0.284 0.348 0.174]" wi= "0.1264"/>*
>
> <*Bin name="MF_3" centroid="[0.433 0.229 0.422 0.625]" wi= "0.1015"/>*

The initial profiles (e.g., submotifs) are encoded as vectors of continuous values (*centroids*) that represent the averaged similarity of their members to a feature submodel (e.g., the position weight matrix of submotif M_1). Indeed, the vector contains the similarity values of the profile members to all other single-type profiles. The *wi* values correspond to the amplitude of the fuzzy clusters defined for each centroid.

> <*Bin name="MF_4" centroid="" wi=" membershipFile= "gps_data/ M_1.mat" ... membershipFile= "gps_data/M_4.mat"/>*

If the centroids are not specified, GPS calculates them based on the profiles defined in *membershipFile* and adjusts the amplitude of the fuzzy cluster based on the *wi* parameter.

Description: GPS allows each promoter instance to belong to multiple profiles in parallel by encoding into a metadata its degree of similarity to all profile prototypes. Therefore, these membership values can be considered by GPS, instead of original data, during the learning phase of the method. This codification allows representing different types of input data (e.g., expression, sequences) into the same framework composed of numeric vectors. Moreover, this approach allows encoding intermediate classes that were not initially specified (e.g., the expression class representing the concept "between high and medium" corresponding to those genes where expression is consistent with both levels of expression: high and medium).

GPS Output

The output of the program is composed of four main sections: the XML file submitted by the user, the list of explored profiles, the selected nondominated profiles, and a snapshot matrix designed to export the results into a typical spreadsheet or into the Spotfire environment (Wilkins, 2000).

List of Profiles

This section is identified by the tag "+*Profiles:*" and enumerates all profiles in the lattice of potential hypothesis. The name of a profile corresponds to the abbreviated names of the features contained in that profile (e.g., the profile named *orientation_i.expression_j.motif_k* is composed of the rediscretized version of the original features and *i*, *j*, and *k* correspond to the initial single-type profiles). Values corresponding to the evaluation of the profiles are dumped as the probability of intersection (i.e., profile extent evaluated by the probability of the features intersection [PI]) and similarity degree of matching between promoters and the prototypes of the profile (SI). Finally, we describe each profile by listing its features, its prototype or centroid, and the recovered promoters, indicating name, feature values, and evaluation score.

Dominance Relationship

The start of this section is indicated by the "+*Dominance* tag," where each profile is described by name, PI and SI scores, and a tag indicating if it is either dominated or nondominated. Profiles containing unique promoters are not considered. Dominated profiles also contain a list of their dominating profiles.

Snapshot Matrix

This section is identified by the "+*Matrix* tag," where columns represent profiles and rows correspond to the profile name, the domination status, the PI and SI values, the number of promoters recovered by the profile, the number of features that characterize the profile, and, finally, the membership value of all of the promoters to the profile.

Uncovering Promoter Profiles Regulated by Response Regulator
PhoP Using GPS

We examined the genome-wide transcription profile of wild-type and *phoP E. coli* strains experiencing low Mg^{2+}, and identified genes whose expression differed statistically between the two strains (Li and Wong, 2001; Tusher *et al.*, 2001). We used these genes, as well as *Salmonella enterica* promoters suspected to be regulated by PhoP, which were provided from our own laboratory knowledge and the literature to generate the initial list of promoter candidates.

We utilized this list to make the initial models of the features, which were used with relaxed thresholds (Hertz and Stormo, 1999) to describe promoters with weak matching to consensus. For example, GPS clustered these genes by their expression similarity: E_1 and E_2, consisting of upregulated genes, and E_3, harboring downregulated genes. Then we classified all candidates based on the similarity of their expression to that of models built for each of the three expression groups, permitting individual genes to belong to more than one group (i.e., E_1 and/or E_2) (Bezdek, 1998; Gasch and Eisen, 2002). This enabled us to recover weakly expressed genes that would have otherwise gone undetected using strict statistical filters (Li and Wong, 2001; Tusher *et al.*, 2001). GPS applied the same strategy to the other features. For example, the initial submotifs corresponding to the PhoP-binding site were dissected by GPS, allowing the recovery of PhoP-regulated promoters with weak matching to the PhoP box consensus, such as the *Salmonella pmrD* promoter, that could not be detected using consensus cutoffs (Hertz and Stormo, 1999; Stormo 2000), despite being regulated and footprinted by the PhoP protein (Kato *et al.*, 2003; Kox *et al.*, 2000).

We used several features for the initial profiles, including discrimination of PhoP box submotifs (M_1-M_4), the orientation (O_1-O_2) and distance of the PhoP box relative to the RNA polymerase site (P_1-P_3), the class of $\sigma70$ promoter (because $\sigma70$ is responsible for the transcription of PhoP-regulated genes [Yamamoto *et al.*, 2002]) (P_1-P_3), the presence of potential binding sites for 60+ transcription factors (Salgado *et al.*, 2001) (I_0-I_4), and whether the position of the PhoP box suggests that a promoter is activated or repressed (A_1-A_3). Then, GPS applied its grouping, prototyping, and searching strategy and uncovered several optimal profiles, which were validated experimentally (Zwir *et al.*, 2005b).

One of the profiles identifies canonical PhoP-regulated promoters. This profile, $P_1^4 E_1^4 M_2^4 I_3^4$ (PI = 0.39, SI = 0.07), encompasses promoters (e.g., those of the *phoP, mgtA, ybcU,* and *yhiW* genes of *E. coli* and the *slyB* gene of *Salmonella*) that share the same RNA polymerase sites, expression patterns, PhoP box submotif, and the same pattern for other transcription factor-binding sites. The profile includes not only the prototypical *phoP* and *mgtA* promoters (Minagawa *et al.*, 2003), but also the promoters of the *yhiW* gene, which was not known to be under PhoP control.

Another profile describes promoters with PhoP boxes in the opposite orientation of the canonical PhoP-regulated promoters. This profile, $P_3^2 O_1^2$, (PI = 0.07, SI = 0.17), includes promoters also with the PhoP box in the opposite orientation (e.g., those of the *slyB* and *yhiW* genes of *E. coli* and the *ybjX, mig-14, virK, mgtC,* and *pagC* genes of *Salmonella*) but differs from the former profile in that the PhoP box is located further upstream from the RNA polymerase site than the typical PhoP-regulated gene. Notably, these promoters could be assigned to a profile even in the absence of expression data. Despite the unusual orientation of the PhoP box in these promoters, the identified PhoP boxes are bona fide PhoP-binding sites (Shi *et al.*, 2004; Shin and Groisman, 2005; Zwir *et al.*, 2005b). Curiously, it had been suggested that PhoP regulates these genes of *Salmonella* indirectly because a PhoP-binding site could not be identified at a location typical of other PhoP-activated genes (Lejona *et al.*, 2003).

By using gene expression as one feature among many, GPS could distinguish between promoters of the acid resistance genes (Masuda and Church, 2003; Tucker *et al.*, 2002) that otherwise would have stayed undifferentiated within the same expression group. These promoters were found to belong to one of three distinct profiles: $E_2^3 \, M_0^3 \, I_1^3$ (PI = 0.11, SI = 0.03), includes promoters for acid resistance structural genes lacking a recognizable PhoP box (e.g., those of the *dps* and *gadA* genes of *E. coli*); $E_2^2 \, M_4^2$ (PI = 0.25, SI = 0.10), comprises promoters of a different set of structural genes that include *hdeD* and *hdeAB*; and $E_2^2 \, P_3^2$ (PI = 0.419, SI = 0.185), harbors promoters of the acid resistance regulatory genes *yhiE* and *yhiW* (also termed *gadE* and *gadW*, respectively). Promoters in the latter two profiles harbor PhoP boxes but these profiles differ in the RNA polymerase sites and their distance to the PhoP box. These findings enabled the prediction that PhoP uses at least two modes of regulation to control transcription of acid resistance genes: a feed-forward loop and classical transcriptional cascade (Zwir *et al.*, 2005b).

Conclusions

We have described an unsupervised machine learning method, termed GPS, that discriminates among coregulated promoters by simultaneously considering both *cis*-acting regulatory features and gene expression. The GPS method encodes regulatory features specifically aimed at handling the variability in sequence, location, and topology that characterize gene transcription. Then, the method uses an integrated approach for discovering promoter profiles, thereby uncovering an unsuspected complexity in the regulatory targets that are under direct and indirect transcriptional control of the regulatory protein.

Several characteristics of GPS contribute to its power. First, it considers gene expression as one feature among many, thereby allowing classification of promoters even in its absence (Beer and Tavazoie, 2004; Conlon *et al.*, 2003). Particularly, GPS differs from supervised learning methods (Mitchell, 1997) that group features and observations based on explicitly defined dependent variables (Beer and Tavazoie, 2004; Conlon *et al.*, 2003; Quinlan, 1993). Second, GPS performs a local feature selection for each profile because not every feature is relevant for all profiles (Kohavi and John, 1997), and, a priori, we do not know which feature is biologically meaningful for a given promoter. This is in contrast to approaches that filter or reduce features for all possible clusters (Yeung and Ruzzo, 2001). Third, GPS finds all optimal solutions among multiple criteria (Pareto optimality) (Deb, 2001), which avoids the biases that might result from using any specific weighing scheme (Rissanen, 1989). This can detect cohesion within a small number of promoters that would remain undetected by methods that emphasize the number of promoters in a profile (Agrawal and Shafer, 1996). Fourth, GPS has a multimodal nature that allows alternative descriptions of a system by providing several adequate solutions (Deb, 2001; Ruspini and Zwir, 2002), thus recovering locally optimal solutions, which have been shown to be biologically meaningful (Azevedo et al., 2005; Cotik et al., 2005). This differentiates GPS from methods that focus on a single optimum (Gutierrez-Rios et al., 2003; Martinez-Antonio and Collado-Vides, 2003). Finally, GPS allows promoters to be members of more than one profile by using fuzzy clustering (Bezdek, 1998; Cordon et al., 2002; Gasch and Eisen, 2002), thus explicitly treating the profiles as hypotheses, which are tested and refined during the analysis (Mitchell, 1997). This distinguishes GPS from clustering approaches that prematurely force promoters into disjointed groups (Qin et al., 2003). In addition, GPS recognizes that not every profile is meaningful (Bezdek, 1998), which avoids the constraints of methods that force membership even to uninteresting groups because the sum of membership is required to be one (Cooper and Herskovits, 1992).

The GPS method can be generalized to a method for grouping, prototyping, and searching in the lattice space of hypotheses, which can be used in different structural domains. For example, we have described the analysis of the targets of regulation of the response regulator PhoP (Zwir et al., 2005b) and it is now being applied to describe other two-components systems (e.g., PmrA/PmrB) and general regulators (e.g., CRP) in different genomes (e.g., Yersinia pestis and Vibrio cholerae). Moreover, it is being used to mine the Gene Ontology database (Ashburner et al., 2000) to discover and annotate profiles across biological processes, cellular components, and molecular functions and to identify molecular pathways that provide insight into the host response over time to systemic inflammatory insults (Calvano et al., 2005).

Acknowledgments

Our research is supported, in part, by grants from the National Institutes of Health to E.A.G., who is an Investigator of the Howard Hughes Medical Institute. I.Z. is also a member of the Computer Science Department at the University of Granada, Spain, and is supported, in part, by the Spanish Ministry of Science and Technology under project BIO2004-0270E and TIN2006-12879.

References

Agrawal, R., and Shafer, J. C. (1996). Parallel mining of association rules. *IEEE Trans. Knowledge Data Engineer.* **8,** 962–969.

Ashburner, M., Ball, C. A., Blake, J. A., Botstein, D., Butler, H., Cherry, J. M., Davis, A. P., Dolinski, K., Dwight, S. S., Eppig, J. T., Harris, M. A., Hill, D. P., *et al.* (2000). Gene ontology: Tool for the unification of biology. The Gene Ontology Consortium. *Nat. Genet.* **25,** 25–29.

Azevedo, R. B., Lohaus, R., Braun, V., Gumbel, M., Umamaheshwar, M., Agapow, P. M., Houthoofd, W., Platzer, U., Borgonie, G., Meinzer, H. P., and Leroi, A. M. (2005). The simplicity of metazoan cell lineages. *Nature* **433,** 152–156.

Bailey, T. L., and Elkan, C. (1995). The value of prior knowledge in discovering motifs with MEME. *Proc. Int. Conf. Intell. Syst. Mol. Biol.* **3,** 21–29.

Bar-Joseph, Z., Gerber, G. K., Lee, T. I., Rinaldi, N. J., Yoo, J. Y., Robert, F., Gordon, D. B., Fraenkel, E., Jaakkola, T. S., Young, R. A., and Gifford, D. K. (2003). Computational discovery of gene modules and regulatory networks. *Nat. Biotechnol.* **21,** 1337–1342.

Basney, J., and Livny, M. (1999). Deploying a high throughput computing cluster. *In* "High Performance Cluster Computing" (R. Buyya, ed.). Prentice Hall, New York.

Beer, M. A., and Tavazoie, S. (2004). Predicting gene expression from sequence. *Cell* **117,** 185–198.

Bezdek, J. C. (1998). Pattern analysis. *In* "Handbook of Fuzzy Computation" (W. Pedrycz, P. P. Bonissone, and E. H. Ruspini, eds.), pp. F6.1.1–F6.6.20. Institute of Physics, Bristol.

Bezdek, J. C., Pal, S. K., and IEEE Neural Networks Council (1992). "Fuzzy Models for Pattern Recognition: Methods That Search for Structures in Data." IEEE Press, New York.

Bijlsma, J. J., and Groisman, E. A. (2005). The PhoP/PhoQ system controls the intramacrophage type three secretion system of *Salmonella enterica. Mol. Microbiol.* **57,** 85–96.

Blattner, F. R., Plunkett, G., 3rd, Bloch, C. A., Perna, N. T., Burland, V., Riley, M., Collado-Vides, J., Glasner, J. D., Rode, C. K., Mayhew, G. F., Gregor, J., Davis, N. W., *et al.* (1997). The complete genome sequence of *Escherichia coli* K-12. *Science* **277,** 1453–1474.

Calvano, S. E., Xiao, W., Richards, D. R., Felciano, R. M., Baker, H. V., Cho, R. J., Chen, R. O., Brownstein, B. H., Cobb, J. P., Tschoeke, S. K., Miller-Graziano, C., Moldawer, L. L., *et al.* (2005). A network-based analysis of systemic inflammation in humans. *Nature* **437,** 1032–1037.

Castelli, M. E., Garcia Vescovi, E., and Soncini, F. C. (2000). The phosphatase activity is the target for Mg^{2+} regulation of the sensor protein PhoQ in *Salmonella. J. Biol. Chem.* **275,** 22948–22954.

Chamnongpol, S., Cromie, M., and Groisman, E. A. (2003). Mg^{2+} sensing by the Mg^{2+} sensor PhoQ of *Salmonella enterica. J. Mol. Biol.* **325,** 795–807.

Cheeseman, P., and Oldford, R. W. (1994). "Selecting Models from Data: Artificial Intelligence and Statistics IV." Springer-Verlag, New York.

Conlon, E. M., Liu, X. S., Lieb, J. D., and Liu, J. S. (2003). Integrating regulatory motif discovery and genome-wide expression analysis. *Proc. Natl. Acad. Sci. USA* **100,** 3339–3344.

Cook, D. J., Holder, L. B., Su, S., Maglothin, R., and Jonyer, I. (2001). Structural mining of molecular biology data. *IEEE Eng. Med. Biol. Mag.* **20,** 67–74.

Cooper, G. F., and Herskovits, E. (1992). A Bayesian method for the induction of probabilistic networks from data. *Machine Learn.* **9,** 309–347.

Cordon, O., Herrera, F., and Zwir, I. (2002). Linguistic modeling by hierarchical systems of linguistic rules. *IEEE Trans. Fuzzy Syst.* **10,** 2–20.

Cotik, V., Zaliz, R. R., and Zwir, I. (2005). A hybrid promoter analysis methodology for prokaryotic genomes. *Fuzzy Sets Syst.* **152,** 83–102.

Crooks, G. E., Hon, G., Chandonia, J. M., and Brenner, S. E. (2004). WebLogo: A sequence logo generator. *Genome Res.* **14,** 1188–1190.

Deb, K (2001). "Multi-objective Optimization Using Evolutionary Algorithms." Wiley, New York.

Gasch, A. P., and Eisen, M. B. (2002). Exploring the conditional coregulation of yeast gene expression through fuzzy k-means clustering. *Genome Biol.* **3,** RESEARCH0059.

Georgellis, D., Kwon, O., and Lin, E. C. (1999). Amplification of signaling activity of the arc two-component system of *Escherichia coli* by anaerobic metabolites: An *in vitro* study with different protein modules. *J. Biol. Chem.* **274,** 35950–35954.

Groisman, E. A. (2001). The pleiotropic two-component regulatory system PhoP-PhoQ. *J. Bacteriol.* **183,** 1835–1842.

Gutierrez-Rios, R. M., Rosenblueth, D. A., Loza, J. A., Huerta, A. M., Glasner, J. D., Blattner, F. R., and Collado-Vides, J. (2003). Regulatory network of *Escherichia coli*: Consistency between literature knowledge and microarray profiles. *Genome Res.* **13,** 2435–2443.

Hertz, G. Z., and Stormo, G. D. (1999). Identifying DNA and protein patterns with statistically significant alignments of multiple sequences. *Bioinformatics* **15,** 563–577.

Holmes, I., and Bruno, W. J. (2000). Finding regulatory elements using joint likelihoods for sequence and expression profile data. *Proc. Int. Conf. Intell. Syst. Mol. Biol.* **8,** 202–210.

Kato, A., and Groisman, E. A. (2004). Connecting two-component regulatory systems by a protein that protects a response regulator from dephosphorylation by its cognate sensor. *Genes Dev.* **18,** 2302–2313.

Kato, A., Latifi, T., and Groisman, E. A. (2003). Closing the loop: The PmrA/PmrB two-component system negatively controls expression of its posttranscriptional activator PmrD. *Proc. Natl. Acad. Sci. USA* **100,** 4706–4711.

Kohavi, R., and John, G. H. (1997). Wrappers for feature subset selection. *Artificial Intelligence* **97,** 273–324.

Kox, L. F., Wosten, M. M., and Groisman, E. A. (2000). A small protein that mediates the activation of a two-component system by another two-component system. *EMBO J.* **19,** 1861–1872.

Lejona, S., Aguirre, A., Cabeza, M. L., Garcia Vescovi, E., and Soncini, F. C. (2003). Molecular characterization of the Mg^{2+}-responsive PhoP-PhoQ regulon in *Salmonella enterica*. *J. Bacteriol.* **185,** 6287–6294.

Li, C., and Wong, W. H. (2001). Model-based analysis of oligonucleotide arrays: Expression index computation and outlier detection. *Proc. Natl. Acad. Sci. USA* **98,** 31–36.

Li, H., Rhodius, V., Gross, C., and Siggia, E. D. (2002). Identification of the binding sites of regulatory proteins in bacterial genomes. *Proc. Natl. Acad. Sci. USA* **99,** 11772–11777.

Lindsey, C. S., Tolliver, J. S., and Lindblad, T. (2005). "JavaTech: An Introduction to Scientific and Technical Computing with Java." Cambridge University Press, New York.

Manson McGuire, A., and Church, G. M. (2000). Predicting regulons and their cis-regulatory motifs by comparative genomics. *Nucleic Acids Res.* **28,** 4523–4530.

Martinez-Antonio, A., and Collado-Vides, J. (2003). Identifying global regulators in transcriptional regulatory networks in bacteria. *Curr. Opin. Microbiol.* **6,** 482–489.

Masuda, N., and Church, G. M. (2003). Regulatory network of acid resistance genes in *Escherichia coli*. *Mol. Microbiol.* **48,** 699–712.

Minagawa, S., Ogasawara, H., Kato, A., Yamamoto, K., Eguchi, Y., Oshima, T., Mori, H., Ishihama, A., and Utsumi, R. (2003). Identification and molecular characterization of the Mg^{2+} stimulon of *Escherichia coli*. *J. Bacteriol.* **185,** 3696–3702.

Mitchell, T. M. (1997). "Machine Learning." McGraw-Hill, New York.

Montagne, M., Martel, A., and Le Moual, H. (2001). Characterization of the catalytic activities of the PhoQ histidine protein kinase of *Salmonella enterica* serovar Typhimurium. *J. Bacteriol.* **183,** 1787–1791.

Oshima, T., Aiba, H., Masuda, Y., Kanaya, S., Sugiura, M., Wanner, B. L., Mori, H., and Mizuno, T. (2002). Transcriptome analysis of all two-component regulatory system mutants of *Escherichia coli* K-12. *Mol. Microbiol.* **46,** 281–291.

Pritsker, M., Liu, Y. C., Beer, M. A., and Tavazoie, S. (2004). Whole-genome discovery of transcription factor binding sites by network-level conservation. *Genome Res.* **14,** 99–108.

Qin, Z. S., McCue, L. A., Thompson, W., Mayerhofer, L., Lawrence, C. E., and Liu, J. S. (2003). Identification of co-regulated genes through Bayesian clustering of predicted regulatory binding sites. *Nat. Biotechnol.* **21,** 435–439.

Quinlan, J. R. (1993). "C4.5: Programs for Machine Learning." Morgan Kaufmann, San Mateo, CA.

Rissanen, J. (1989). "Stochastic Complexity in Statistical Inquiry." World Scientific, Singapore.

Romero Zaliz, R., Zwir, I., and Ruspini, E. H. (2004). Generalized analysis of promoters: A method for DNA sequence description. *In* "Applications of Multi-Objective Evolutionary Algorithms" (C. A. Coello Coello and G. B. Lamont, eds.), pp. 427–450. World Scientific, Singapore.

Ronen, M., Rosenberg, R., Shraiman, B. I., and Alon, U. (2002). Assigning numbers to the arrows: Parameterizing a gene regulation network by using accurate expression kinetics. *Proc. Natl. Acad. Sci. USA* **99,** 10555–10560.

Ruspini, E. H., and Zwir, I. (2002). Automated generation of qualitative representations of complex objects by hybrid soft-computing methods. *In* "Pattern Recognition: From Classical to Modern Approaches" (S. K. Pal and A. Pal, eds.), pp. 454–474. World Scientific, New Jersey.

Salgado, H., Gama-Castro, S., Martinez-Antonio, A., Diaz-Peredo, E., Sanchez-Solano, F., Peralta-Gil, M., Garcia-Alonso, D., Jimenez-Jacinto, V., Santos-Zavaleta, A., Bonavides-Martinez, C., and Collado-Vides, J. (2004). RegulonDB (version 4.0): Transcriptional regulation, operon organization and growth conditions in *Escherichia coli* K-12. *Nucleic Acids Res.* **32,** D303–D306.

Salgado, H., Santos-Zavaleta, A., Gama-Castro, S., Millan-Zarate, D., Diaz-Peredo, E., Sanchez-Solano, F., Perez-Rueda, E., Bonavides-Martinez, C., and Collado-Vides, J. (2001). RegulonDB (version 3.2): Transcriptional regulation and operon organization in *Escherichia coli* K-12. *Nucleic Acids Res.* **29,** 72–74.

Shi, Y., Latifi, T., Cromie, M. J., and Groisman, E. A. (2004). Transcriptional control of the antimicrobial peptide resistance ugtL gene by the *Salmonella* PhoP and SlyA regulatory proteins. *J. Biol. Chem.* **279,** 38618–38625.

Shigenobu, S., Watanabe, H., Hattori, M., Sakaki, Y., and Ishikawa, H. (2000). Genome sequence of the endocellular bacterial symbiont of aphids *Buchnera* sp. APS. *Nature* **407,** 81–86.

Shin, D., and Groisman, E. A. (2005). Signal-dependent binding of the response regulators PhoP and PmrA to their target promoters *in vivo. J. Biol. Chem.* **280,** 4089–4094.

Stormo, G. D. (2000). DNA binding sites: Representation and discovery. *Bioinformatics* **16,** 16–23.

Stover, C. K., Pham, X. Q., Erwin, A. L., Mizoguchi, S. D., Warrener, P., Hickey, M. J., Brinkman, F. S., Hufnagle, W. O., Kowalik, D. J., Lagrou, M., Garber, R. L., Goltry, L., *et al.* (2000). Complete genome sequence of *Pseudomonas aeruginosa* PA01, an opportunistic pathogen. *Nature* **406,** 959–964.

Tavazoie, S., Hughes, J. D., Campbell, M. J., Cho, R. J., and Church, G. M. (1999). Systematic determination of genetic network architecture. *Nat. Genet.* **22,** 281–285.

Tompa, M., Li, N., Bailey, T. L., Church, G. M., De Moor, B., Eskin, E., Favorov, A. V., Frith, M. C., Fu, Y., Kent, W. J., Makeev, V. J., Mironov, A. A., *et al.* (2005). Assessing computational tools for the discovery of transcription factor binding sites. *Nat. Biotechnol.* **23**, 137–144.

Tu, X., Latifi, T., Bougdour, A., Gottesman, S., and Groisman, E. A. (2006). The PhoP/PhoQ two-component system stabilizes the alternative sigma factor RpoS in *Salmonella enterica*. *Proc. Natl. Acad. Sci. USA* **103**, 13503–13508.

Tucker, D. L., Tucker, N., and Conway, T. (2002). Gene expression profiling of the pH response in *Escherichia coli. J. Bacteriol.* **184**, 6551–6558.

Tusher, V. G., Tibshirani, R., and Chu, G. (2001). Significance analysis of microarrays applied to the ionizing radiation response. *Proc. Natl. Acad. Sci. USA* **98**, 5116–5121.

Wang, X., Gorlitsky, R., and Almeida, J. S. (2005). From XML to RDF: How semantic web technologies will change the design of 'omic' standards. *Nat. Biotechnol.* **23**, 1099–1103.

Wilkins, C. L. (2000). Data mining with Spotfire Pro 4.0. *Anal. Chem.* **72**, 550a.

Yamamoto, K., Ogasawara, H., Fujita, N., Utsumi, R., and Ishihama, A. (2002). Novel mode of transcription regulation of divergently overlapping promoters by PhoP, the regulator of two-component system sensing external magnesium availability. *Mol. Microbiol.* **45**, 423–438.

Yeung, K. Y., and Ruzzo, W. L. (2001). Principal component analysis for clustering gene expression data. *Bioinformatics* **17**, 763–774.

Zwir, I., Huang, H., and Groisman, E. A. (2005a). Analysis of differentially-regulated genes within a regulatory network by GPS genome navigation. *Bioinformatics* **21**, 4073–4083.

Zwir, I., Shin, D., Kato, A., Nishino, K., Latifi, T., Solomon, F., Hare, J. M., Huang, H., and Groisman, E. A. (2005b). Dissecting the PhoP regulatory network of *Escherichia coli* and *Salmonella enterica. Proc. Natl. Acad. Sci. USA* **102**, 2862–2867.

[19]   Targeting Two-Component Signal Transduction: A Novel Drug Discovery System

By Ario Okada, Yasuhiro Gotoh, Takafumi Watanabe, Eiji Furuta, Kaneyoshi Yamamoto, and Ryutaro Utsumi

Abstract

We have developed two screening systems for isolating inhibitors that target bacterial two-component signal transduction: (1) a differential growth assay using a temperature-sensitive *yycF* mutant (CNM2000) of *Bacillus subtilis*, which is supersensitive to histidine kinase inhibitors, and (2) a high-throughput genetic system for targeting the homodimerization of histidine kinases essential for the bacterial two-component signal transduction. By using these methods, we have been able to identify various types of inhibitors that block the autophosphorylation of histidine kinases with different modes of actions.

Introduction

The rapid emergence of antibiotic resistance in pathogenic bacteria has underscored the need for an accelerated approach to the discovery of new antibacterial agents. In the past decade, the developments of bacterial genomics, bioinformatics, and gene manipulation have led to the discovery of many novel protein targets for antibacterial agents (Moir *et al.*, 1999). For instance, the two-component signal transduction systems (TCSs) of bacteria, which consist of two proteins, histidine kinase (HK) and response regulators (RR), have received increasing attention for their potential as novel antibacterial drug targets for the following reasons (Barrette and Hoch, 1998; Macielag and Goldschmidt, 2000; Matsushita and Janda, 2002). First, TCSs are essential for coordinated expression of stress-response genes, including those for virulence factors. Second, some TCSs regulate the expression of antibiotic resistance determinants, including drug-efflux pumps (Eguchi *et al.*, 2003; Hirakawa *et al.*, 2003; Kato *et al.*, 2000). Third, protein-histidine phosphorylation in the signal transduction pathway in bacteria is distinct from serine/threonine and tyrosine phosphorylation in higher eukaryotes. Finally, the high degree of structural homology in the catalytic domain of HKs and in the receiver domain of RRs suggests that multiple TCSs within a single bacterium could be

METHODS IN ENZYMOLOGY, VOL. 422
0076-6879/07 $35.00
DOI: 10.1016/S0076-6879(06)22019-6

inhibited simultaneously, thereby lowering the frequency of drug-resistant strains.

It should also be noted that a small number of TCS-encoding genes are essential in both gram-negative and gram-positive bacteria (Beir and Frank, 2000; Fabret and Hoch, 1998; Hancock and Perego, 2002; Lange *et al.*, 1999; Martin *et al.*, 1999; Throup *et al.*, 2000). In *Bacillus subtilis*, there are 36 HKs and 34 RRs, although only YycG(HK)/YycF(RR) is essential for growth (Fabret and Hoch, 1998; Fukuchi *et al.*, 2000). Homologs of the *B. subtilis* YycG/YycF pair have also been identified in *Staphylococcus aureus, Streptococcus pneumoniae,* and *Enterococcus faecalis* (Hancock and Perego, 2002; Lange *et al.*, 1999; Martin *et al.*, 1999; Throup *et al.*, 2000). *yycF* temperature-sensitive (TS) strains were isolated in *S. aureus* as well as in *B. subtilis* (Fabret and Hoch, 1998; Martin *et al.*, 1999; Watanabe *et al.*, 2003b). Since these point mutations in *yycF* resulted in severe defects in bacterial growth at nonpermissive temperatures, the YycG/YycF system is an excellent target for novel antibiotics (Barrette and Hoch, 1998; Yamamoto *et al.*, 2001). Based on these observations, we have developed two different screening methods for inhibitors targeting YycG/YycF two-component signal transduction, and this chapter describes the principles and detailed methods.

Differential Growth Assay

Principle

Bacillus subtilis strain CNM2000 carries the H215P mutation in YycF (the response regulator of YycG/YycF TCS) (Fabret and Hoch, 1998; Watanabe *et al.*, 2003a,b) and shows a temperature-sensitive (ts) phenotype in growth. The H215P mutation is located on the loop connecting the $\alpha 3$ helix to the $\beta 6$ and $\beta 7$ C-terminal strands of YycF. Substitution of a proline for a histidine at the site perturbs the DNA-binding properties at elevated temperatures (Watanabe *et al.*, 2003a). Upon a shift to 47°, the CNM2000 strain stopped growing after approximately 30 min and then cell turbidity at OD_{600} decreased. We examined the sensitivity of the CNM2000 to cefazolin, amikacin, vancomycin, erythromycin, ofloxacin, and a histidine kinase inhibitor, NH127 (3-benzyl-1-lauryl-2-methyl imidazolium iodide) (Yamamoto *et al.*, 2000) at 37° (Fig. 1). As a result, the growth of CNM2000 was found to be more sensitive than that of the *B. subtilis* 168 to NH127. Therefore, based on this property of CNM2000, which is highly sensitive to the HK inhibitor, YycG inhibitors should be selectively screened.

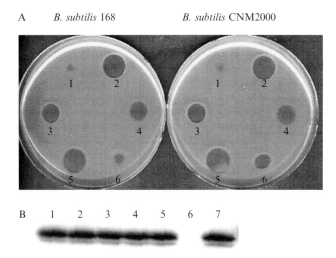

FIG. 1. Sensitivity of CNM2000 to the histidine kinase inhibitor NH127. (A) Sensitivity of *B. subtilis* 168 and CNM2000 to antibacterial agents. Cefazolin (1), amikacin (2), vancomycin (3), erythromycin (4), ofloxacin (5), and NH127 (6) were bioassayed using 168 (left) and CNM2000 (right) as described in the text. One microliter of each antibacterial agent (50 μg/ml) in DMSO was spotted on trypticase soy agar plates containing 168 or CNM2000. (B) Effect of the antibacterial agents on YycG autophosphorylation. Autophosphorylation of YycG purified from *B. subtilis* was performed in the presence or absence of various antibacterial agents (50 μg/ml); lanes 1–7 contain cefazolin, amikacin, vancomycin, erythromycin, ofloxacin, NH127, and DMSO (control), respectively.

Bacterial Strains

Bacillus subtilis 168 (*trpC2*) and its temperature-sensitive mutant CNM2000 (*trpC2, yycF[H215P],Cm^r*) were used in a differential growth assay (Watanabe *et al.*, 2003b).

Growth Media (Per Liter of Water)

Tryptic soy broth (TSB): 30 g Trypticase soy broth (BBL)
Bottom agar: 7.5 g Trypticase soy broth (BBL), 15 g agar powder (Nacalai Tesque, Inc.).
Top agar: 7.5 g Trypticase soy broth (BBL), 5.0 g agar powder (Nacalai Tesque, Inc.).

Other Reagents

10× kinase buffer: 500 mM Tris-HCl (pH 7.5), 500 mM KCl, 100 mM MgCl$_2$

ATP solution (20 μl): 1.0 μl of [γ-^{32}P]ATP (370 kBq/μl, Amersham Bioscience Corp.), 10 μl of 25 μM ATP, 9.0 μl of distilled water

Screening Procedure

1. Prepare a polypropylene culture tube (50 ml, IWAKI) with 10 ml of TSB or TSB containing chloramphenicol (5 μg/ml).
2. Inoculate a single colony (*B. subtillis* 168 and CNM2000) and culture at 37° overnight.
3. Preheat a small test tube at 55°.
4. Place 3 ml of top agar melted by a microwave oven into the tube.
5. Add 30 μl of overnight culture to the tube and pour on a bottom agar.
6. Cool the plate for 30 min at room temperature to harden the top agar layer.
7. Add 1 μl of screening sample to the top agar and leave for 30 min.
8. Incubate at 37° overnight (see Figs. 1 and 2).

A

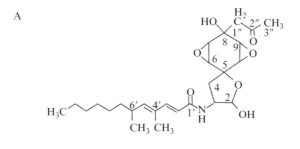

B   *B. subtilis* 168   *B. subtilis* CNM2000

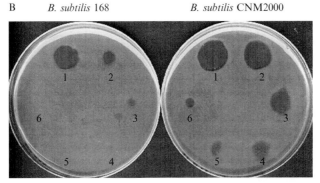

FIG. 2. Bioassay of the histidine kinase inhibitor aranorosinol B screened by the differential growth assay. (A) Chemical structure of aranorosinol B. (B) Purified aranorosinol B was bioassayed using 168 (left) and CNM2000 (right) as described in the text. One microliter of aranorosinol B at six concentrations (μg/ml) was spotted on trypticase soy agar plates containing 168 or CNM2000: 10.0 (1), 5.00 (2), 2.50 (3), 1.25 (4), 0.625 (5), and 0.313 (6).

Autophosphorylation Assay for YycG

1. Dispense 7 μl of YycG solution (2.5 μl of YycG [2 pmol/μl], 1.0 μl of 10× kinase buffer, 3.5 μl of distilled water) into the tube.
2. Add 1 μl of a screened sample.
3. After 5 min, add 2 μl of ATP solution.
4. Let stand for 10 min at room temperature.
5. Add 3 μl of 5× SDS sample buffer, load it on an SDS polyacrylamide gel (Silhavy *et al.*, 1984), and run electrophoresis at 20 mA.
6. Immerse the gel in a solution containing 45% methanol and 10% acetic acid.
7. After 5 min, wrap the gel onto a piece of Whatman 3MM paper in Saran Wrap and dry the gel using a gel dryer (Bio-Rad, Model 583).
8. Expose the gel to an imaging plate (BAS-IPMS2040, Fuji Photo Film Co., Ltd.) for 3 h and quantitate with BAS 1000 (Fuji Photo Film Co., Ltd.).
9. Calculate IC_{50} from the intensity of the bands of each phosphorylated product after subtraction of the background.

Results and Discussion

We performed a differential growth assay to identify YycG inhibitors by screening samples of acetone extracts from 4000 microbes. A total of 11 samples showed greater activity against CNM2000 than strain 168. Seven of those samples significantly inhibited the autophosphorylation activity of YycG HK, while the other four samples were only weakly active. Starting with the most potent extract from the microbe WF140196, identified as *Gymnascella dankaliensis* (*Pseudomonas roseus*), an active compound was purified that was more potent to CNM2000 than to strain 168, and it was identified as aranorosinol B by ^{13}C nuclear magnetic resonance (NMR) analysis (Watanabe *et al.*, 2003b). The purified aranorosinol B inhibited the autophosphorylation of YycG from both *B. subtilis* and *S. aureus* with IC_{50} values of 232 and 211 μM, respectively. As expected, the mutant CNM2000 was more sensitive to the purified aranorosinol B than the wild-type strain 168 (Fig. 2). Aranorosinol B did not show antibacterial activity against gram-negative bacteria, including *Escherichia coli* and *Pseudomonas aeruginosa*, while MICs against *B. subtilis* and *S. aureus* were 31.25 and 15.62 μg/ml, respectively. The method described here is presently being used in our laboratory for high-throughput screening for the identification of HK inhibitors and is of value in the subsequent stages of lead compound identification involving iterative rounds of chemical refinement.

High-Throughput Genetic System

Principle

Cytoplasmic or truncated forms of HKs have been known to dimerize *in vitro* (Hidaka *et al.*, 1997). Solution-state NMR (Tomomori *et al.*, 1999) indeed revealed dimerization of the homodimeric core domain in *E. coli* EnvZ HK, which belongs to the same Pho subfamily of HKs as YycG. Dimerization of YycG is considered an essential step in autophosphorylation of this enzyme (Watanabe *et al.*, 2003a). Therefore, inhibitors targeting the homodimerization of YycG are assumed to be a novel class of antibiotics. To identify such drugs, we have developed a high-throughput genetic system for targeting the homodimerization of HK, which is based on the dimerization properties of the IclR repressor of *E. coli* using green fluorescent protein (GFP) as the reporter gene (Furuta *et al.*, 2005).

In order to quantify the ability of dimerization of proteins by fluorescent intensity, a homodimerization assay vector, pFI003, was constructed in which the enhanced GFP (EGFP) gene was fused directly downstream of the *iclR* promoter (Fig. 3). The leucine zipper of yeast GCN4 and the homodimerized C-terminal domain of the FadR of *E. coli* were fused to the N-terminal (100 amino acids: N-100) portion (DNA-binding domain) of the IclR repressor in pFI003 and expressed in *E. coli* JM109. The chimeric repressors (N100-zip, N100-CFad) repressed the expression of EGFP (Fig. 4; Furuta *et al.*, 2005). The chimeric repressors, in which N100 of pFI003 was fused to functional homodimerization domains of HK containing YycG, can also suppress EGFP expression under the *iclR* promoter. In fact, N100 was fused to the YycG cytoplasmic domain (CYycG) to

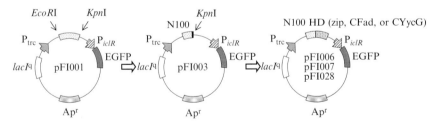

FIG. 3. Construction of pFI006, pFI007, pFI028, and pFI001. The leucine zipper (zip, 33 amino acid sequence) of yeast GCN4, the dimerization domain of FadR of *E. coli* (CFad, 83–239 amino acids from N-terminal), and the cytoplasmic domain of YycG of *S. aureus* (CYycG, 204–608 amino acids from N-terminal) are fused at the *Kpn*I site of pFI003 to construct pFI006, pFI007, and pFI028, respectively. Homodimerization domain (HD): pFI006, leucine zipper; pFI007, CFad; pFI028, CYycG. In pFI001, the full length of the *iclR* gene is cloned. N100: DNA-binding domain (1–100 amino acids from N-terminal) of IclR of *E. coli*.

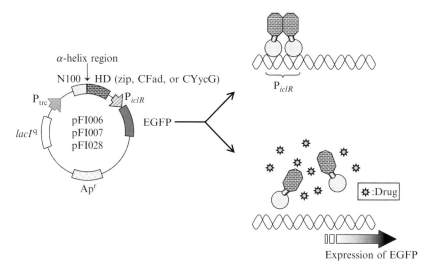

FIG. 4. Schematic diagram of homodimerization system.

construct pFI028 (Figs. 3 and 4). JM109/pFI028 showed a substantial decrease in fluorescent activity compared to that of N100. This system identifies and localizes the dimerization domain of virtually any protein capable of forming a homodimer. Therefore, it is useful for screening inhibitors of HK dimerization (Fig. 4).

Bacterial Strains and Plasmids

Escherichia coli JM109 [*endA1, recA1, gyrA96, thi, hsdR17, relA1, supE44, Δ(lac-proAB), (F', traD36, proAB, lacIqZ ΔM15*)] containing pFI028 or pFI001 (Fig. 3) was cultured in Luria Bertani broth (LB) containing ampicillin (100 μg/ml) to screen a chemical library.

Screening Procedure

1. Prepare a culture with 5 ml LB containing ampicillin (100 μg/ml) in a test tube.
2. Inoculate a single colony of *E. coli* JM109/pFI028 or JM109/pFI001 (Fig. 3) and culture at 37° overnight.
3. Dispense 194 μl of LB, 1 μl of overnight culture, and 5 μl of sample or dimethyl sulfoxide (DMSO) into each well of a 96-well tissue culture-treated microplate (flat bottom, black/clear, with lids installed, Matrix Technologies Corp.).

4. Shake or rotate (200 rpm) at 37° for 20 h.
5. Measure optical density at 595 nm (OD_{595}) and fluorescence levels at an excitation of 485 nm (ex_{485}) and an emission of 535 nm (em_{535}) with a spectrophotometer (Model 3550 microplate reader Bio-Rad) and the Wallac 1420 ARVOsx (Perkin Elmer Life Science), respectively.
6. Normalize the fluorescent intensity of each sample according to the following formula:

$$F_s = (E_{535} \text{ sample} - E_{535} \text{ LB})/(OD_{595} \text{ sample} - OD_{595} \text{ LB})$$
$$F_c = (E_{535} \text{ DMSO} - E_{535} \text{ LB})/(OD_{595} \text{ DMSO} - OD_{595} \text{ LB})$$
$$\text{Relative fluoresence} = F_s/F_c$$

Results and Discussion

To identify YycG inhibitors, we screened a chemical library against JM109/pFI028 and JM109/pFI001 in 96-titer wells. Through this screening process, we identified a compound named I-8-15 (1-dodecyl-2-isopropylimidazole), which significantly increased the fluorescent intensity of JM109/pFI028, while no change in fluorescent intensity was observed in JM109/pFI001 in the

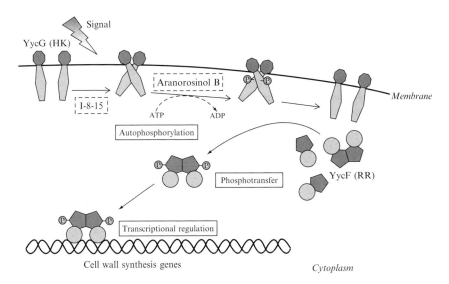

FIG. 5. A new class of antibacterial agents targeting a two-component system, YycG/YycF. Aranorosinol B and I-8-15 were newly screened with the differential growth assay and the high-throughput genetic system, respectively.

presence or absence of I-8-15. In fact, I-8-15 inhibited autophosphorylation of CYycG with an IC_{50} of 76.5 μM, and it also significantly inhibited the growth of methicillin-resistant *S. aureus* and vancomycin-resistant *E. faecalis* with MICs at 25 and 50 $\mu g/ml$, respectively. However, when aranorosinol B, which was isolated using the differential growth assay, was added to the culture of JM109/pFI028, no increase in fluorescent intensity was observed (Furuta *et al.*, 2005). I-8-15 acted on CYycG in a different manner than that of aranorosinol B and exerted its bactericidal activity by inhibiting the autophosphorylation of CYycG. Thus, by using both a differential growth assay and a high-throughput genetic system, it is possible to isolate various types of HK inhibitors with different modes of actions (Fig. 5).

Acknowledgments

The authors are grateful to Dr. K. Watabe for critical reading of this chapter. This work was supported by the Research and Development Program for New Bio-industry Initiatives (2006–2010) from the Bio-oriented Technology Research Advancement Institution and the Academic Frontier Project for Private Universities through a matching funds subsidy from the Ministry of Education, Culture, Sports, Science, and Technology (2004–2008). It was also supported in part by Prefecture Collaboration of Regional Entities for the Advancement of Technological Excellence, JST (2005–2009).

References

Barrette, J. F., and Hoch, J. A. (1998). Two-component signal transduction as a target for microbial anti-infective therapy. *Antimicrob. Agents Chemother.* **42,** 1529–1536.

Beir, D., and Frank, R. (2000). Molecular characterization of two-component systems of *Helicobacter pylori*. *J. Bacteriol.* **182,** 2068–2076.

Eguchi, Y., Oshima, T., Mori, H., Aono, R., Yamamoto, K., Ishihama, A., and Utsumi, R. (2003). Transcriptional regulation of efflux genes by EvgAS, a two-component system in *Escherichia coli*. *Microbiology* **149,** 2819–2828.

Fabret, C., and Hoch, J. A. (1998). A two-component signal transduction system essential for growth of *Bacillus subtilis*: Implications for anti-infective therapy. *J. Bacteriol.* **180,** 6375–6383.

Fukuchi, K., Kasahara, Y., Asai, K., Kobayashi, K., Moriya, S., and Ogasawara, N. (2000). The essential two-component regulatory system encoded by *yycF* and *yycG* modulates expression of the *ftsAZ* operon in *Bacillus subtilis*. *Microbiology* **146,** 1573–1583.

Furuta, E., Yamamoto, K., Tatebe, D., Watabe, K., Kitayama, T., and Utsumi, R. (2005). Targeting protein homodimerization: A novel drug discovery system. *FEBS Lett.* **579,** 2065–2070.

Hancock, L., and Perego, M. (2002). Two-component signal transduction in *Enterococcus faecalis*. *J. Bacterial.* **184,** 5819–5825.

Hidaka, Y., Park, H., and Inouye, M. (1997). Demonstration of dimer formation of the cytoplasmic domain of a transmembrane osmosensor protein, EnvZ of *Escherichia coli* using N-histidine tag affinity chromatography. *FEBS Lett.* **400,** 238–243.

Hirakawa, H., Nishino, K., Hirota, Y., and Yamaguchi, A. (2003). Comprehensive studies of drug resistance mediated by over-expression of response regulators of two-component signal transduction system in *Escherichia coli*. *J. Bacteriol.* **185,** 1851–1856.

Kato, A., Ohnishi, H., Yamamoto, K., Furuta, E., Tanabe, H., and Utsumi, R. (2000). Transcription of *emrKY* is regulated by the EvgA-EvgS two-component system in *Escherichia coli* K-12. *Biosci. Biotechnol. Biochem.* **64,** 1203–1209.

Lange, R., Wagner, C., de Saizieu, A., Flint, N., Molnos, J., Stieger, M., Caspers, P., Kamber, M., Keck, W., and Amrein, K. E. (1999). Domain organization and molecular characterization of 13 two-component systems identified by genome sequencing of *Streptococcus pneumoniae*. *Gene* **237,** 223–234.

Macielag, M. J., and Goldschmidt, R. (2000). Inhibitors of bacterial two-component signaling systems. *Exp. Opin. Invest. Drugs* **9,** 2351–2369.

Martin, P. K., Li, T., Sun, D., Biek, D. P., and Schmid, M. B. (1999). Role in cell permeability of an essential two-component system in *Staphylococcus aureus*. *J. Bacteriol.* **181,** 3666–3673.

Matsushita, M., and Janda, K. D. (2002). Histidine kinases as targets for new antimicrobial agents. *Bioorg. Med. Chem.* **10,** 855–867.

Moir, D. T., Shaw, K. J., Hare, R. S., and Voris, G. F. (1999). Genomics and antimicrobial drug discovery. *Antimicrob. Agents Chemother.* **43,** 439–446.

Silhavy, T. J., Berman, M. L., and Enquist, L. W. (1984). Preparation of SDS protein extracts. *In* "Experiments with Gene Fusions," pp. 208–212. Cold Spring Harbor Laboratory, Cold Spring Harbor, NY.

Throup, J. P., Koretke, K. K., Bryant, A. P., Ingraham, K. A., Chalker, A. F., Ge, Y., Marra, A., Wallis, N. G., Brown, J. R., Holmes, D. J., Rosenberg, M., and Burnham, M. K. (2000). A genomic analysis of two-component signal transduction in *Streptococcus pneumoniae*. *Mol. Microbiol.* **35,** 566–576.

Tomomori, C., Tanaka, T., Dutta, R., Park, H., Saha, S. K., Zhu, Y., Ishima, R., Liu, D., Tong, K. I., Kurokawa, H., Qian, H., Inouye, M., and Ikura, M. (1999). Solution structure of the homodimeric core domain of *Escherichia coli* histidine kinase EnvZ. *Nat. Struct. Biol.* **6,** 729–734.

Watanabe, T., Hashimoto, Y., Umemoto, Y., Tatebe, D., Furuta, E., Fukamizo, T., Yamamoto, K., and Utsumi, R. (2003a). Molecular characterization of the essential response regulator protein YycF in *Bacillus subtilis*. *J. Mol. Microbiol. Biotechnol.* **6,** 155–163.

Watanabe, T., Hashimoto, Y., Yamamoto, K., Hirao, K., Ishihama, A., Hino, M., and Utsumi, R. (2003b). Isolation and characterization of inhibitors of the essential histidine kinase, YycG in *Bacillus subtilis* and *Staphylococcus aureus*. *J. Antibiot.* **56,** 1045–1052.

Yamamoto, K., Kitayama, T., Ishida, N., Watanabe, T., Tanabe, H., Takatani, M., Okamoto, T., and Utusmi, R. (2000). Identification and characterization of a potent antibacterial agent, NH125 against drug-resistant bacteria. *Biosci. Biotechnol. Biochem.* **64,** 912–923.

Yamamoto, K., Kitayama, T., Minagawa, S., Watanabe, T., Sawada, S., Okamoto, T., and Utsumi, R. (2001). Antibacterial agents that inhibit histidine protein kinase YycG of *Bacillus subtilis*. *Biosci. Biotechnol. Biochem.* **65,** 2306–2310.

[20]   The Essential YycFG Two-Component System of *Bacillus subtilis*

By Hendrik Szurmant, Tatsuya Fukushima, and James A. Hoch

Abstract

The YycFG two-component system, highly conserved in the low G+C gram positives, is essential for cell viability in most organisms in which it has been studied. The system is organized within an operon that includes at least one but often three to four other genes. Products of two of these genes, *yycH* and *yycI*, have been shown to have a regulatory role on this two-component system. Immunofluorescent studies identified YycG kinase localization at the cell division sites consistent with its role in regulating cell divisional processes. The essential nature and operon organization of this system commanded special requirements in studying this system genetically. This chapter presents methods utilized in identifying the regulatory circuit that controls the activity of the YycG kinase in *Bacillus subtilis*. Most aspects of our approaches are applicable to other two-component systems in *B. subtilis* and the gram positives. Some are limited to essential systems, such as the YycFG system.

Introduction

The YycFG two-component system is arguably the most intriguing signal transduction system in *Bacillus subtilis* and other gram-positive bacteria due to its essential role for cell viability and its high amino acid conservation (Fabret and Hoch, 1998; Martin *et al.*, 1999). Since our discovery of its essential nature, this system has been studied in some detail in various organisms, and several laboratories have contributed nicely, clarifying the important role this system plays (Fukuchi *et al.*, 2000; Howell *et al.*, 2006; Ng *et al.*, 2003). The regulon controlled by this two-component system among different organisms is diverse, but a common theme is the control of genes for cell wall metabolic processes, cell membrane composition, and cell division (Dubrac and Msadek, 2004; Fukuchi *et al.*, 2000; Howell *et al.*, 2003; Mohedano *et al.*, 2005).

Two important questions to answer when studying a novel two-component system are what are the input signals feeding into the system and what is the output regulated by this system? The input is defined as the signals sensed by the histidine kinase, which can be as diverse as a nutrient, pH, temperature, or interaction with other proteins (Kaspar and Bott, 2002;

METHODS IN ENZYMOLOGY, VOL. 422   0076-6879/07 $35.00
                DOI: 10.1016/S0076-6879(06)22020-2

Mansilla *et al.*, 2005; Neiditch *et al.*, 2006; Tiwari *et al.*, 1996). The output is defined as the genes controlled by this system (the regulon) for the standard DNA-binding response regulator. In well-studied organisms the regulon is identified by microarray analysis of a deletion strain in comparison to the wild-type strain. The procedure is more complicated when the two-component system is essential. Because the system cannot be inactivated, a technique has to be designed allowing activity control of the two-component system to change expression levels of the regulon. The most straightforward approach is either overexpression of the two-component system of interest or construction of a strain depleted for the sensor kinase or response regulator (Mohedano *et al.*, 2005). Particularly, the overexpression of the signaling proteins can lead to secondary effects and complicate the analysis.

Howell and colleagues (2003) designed an interesting approach that led to the identification of the consensus DNA-binding sequence for *B. subtilis* YycF. This approach utilized the fact that *B. subtilis* expresses a phylogenetically closely related two-component system to YycFG, the PhoPR system. This system has been well studied by Hulett and colleagues (1996). Activity of the kinase PhoR is induced under phosphate limitation conditions (Hulett *et al.*, 1994). Construction of a hybrid response regulator consisting of the PhoP response regulator domain and the YycF DNA-binding domain allowed for the phosphate-dependent regulation of YycF-dependent gene expression.

Had a signal been known for the YycFG system, identification of the regulon would have been simpler. Unfortunately, the identification of input signals for two-component systems has been difficult and is not as straightforward as identification of the regulon. Indeed, signals controlling histidine kinase activity remain unknown for most two-component systems currently under investigation. With some exceptions, well-defined signals are available only for systems responsible for utilization of nutrient sources.

Our studies, described here, were designed to help the identification of signals feeding into the YycFG two-component system. Some of these methods are necessary because of the essentiality of the YycFG system, whereas others are widely applicable to the study of many two-component systems. Certainly, these studies are complementary to those identifying the output of a two-component system and present an alternative approach when first studying a novel two-component system.

Construction of Conditional Mutants

The first indication that a protein is essential for cell viability is usually the observation that no insertion mutants can be obtained. Of course this mere inability cannot be considered proof of essentiality. Usually, conditional mutants are constructed that allow the control of expression or activity of the system. If cells do not grow under conditions in which the gene is not expressed, a given system is considered essential.

For the YycFG system, two such conditional mutants have been obtained. One is a temperature-sensitive *yycF* mutant and the other is an insertion mutant that places the *yycFG* operon under control of the isopropyl-β-D-thiogalactoside-inducible P_{spac} promoter (Fabret and Hoch, 1998). Construction of the latter is trivial utilizing the pMUTIN plasmid and is therefore not discussed further (Vagner *et al.*, 1998). The temperature-sensitive *yycF* mutant has proven useful, as it can be used in transposon mutagenesis to identify negative regulators of the YycFG two-component system. A detailed protocol describing the required steps in obtaining such a mutant is therefore described.

The method is a multistep process (Fig. 1). Steps include identification and construction of a suitable mutation delivery system, construction of a mutation library, delivery and screening of mutants for the desired thermosensitive phenotype, and finally identification of the mutation in a positive clone followed by confirmation that the mutation is responsible for the phenotype.

Choosing a Suitable Delivery System

To construct a temperature-sensitive strain in the gene of choice a suitable method to replace the wild-type copy with the mutant version has to be identified. This method has to be highly efficient as it takes numerous strains to identify one temperature-sensitive mutant. The *yycFG* operon is closely linked to the adenylosuccinate synthetase gene *purA* involved in adenine biosynthesis (Kunst *et al.*, 1997). Therefore, *yycF* or *yycG* mutants could be introduced utilizing this gene as a nutritional marker. A *purA* strain was required as recipient allowing for the delivery of *yycFG* mutations by selecting for Ade$^+$ on a minimal plate. In the absence of a suitable nutritional marker it is possible to link the chromosome to an antibiotic resistance marker and deliver the mutants that way.

Construction of a Mutant Library

Following the identification of a suitable system able to introduce mutations, a mutant library has to be constructed. In the case of the *yycFG* system, a 5.4-kbp fragment coding for the *purA* gene, several tRNA genes, *yycF* and *yycG,* was amplified utilizing the Expand polymerase chain reaction (PCR) kit (Boehringer Mannheim) and standard PCR conditions. We assumed that the fidelity of the polymerase was low enough to introduce sufficient mutations. In hindsight, temperature-sensitive mutants

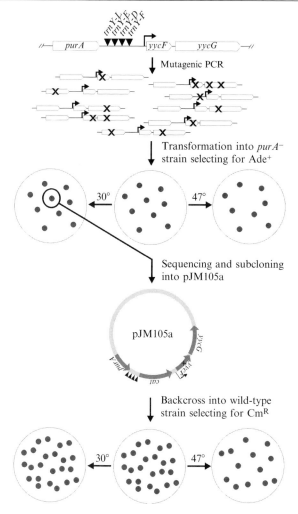

Fig. 1. Isolation of a temperature-sensitive YycF mutant. A temperature-sensitive mutant was created in a multistep process. In the first step a suitable delivery system is identified. For YycF the adjacent nutritional marker *purA* responsible for adenine auxotrophy (Ade⁻) was a selective marker in transformation. Second, the region of interest is amplified by mutagenic PCR creating a mutant library. The mutant library is directly transformed into an Ade⁻ *purA B. subtilis* strain selecting for Ade⁺. Colonies are replica plated onto two plates that are incubated at 30 or 47°. Colonies growing at the low but not at the high temperature are potential candidates. Screening of 3200 colonies resulted in one temperature-sensitive strain. From this strain, *yycF* and *yycG* were PCR amplified and cloned into the vector pJM105a downstream of a *cm^R* cassette, and the region upstream of the *yyc* operon was cloned upstream of a *cm^R* cassette. *yycF* and *yycG* genes were sequenced. The vector was backcrossed into the wild-type *B. subtilis* strain selecting for Cm^R followed by screening for the temperature-sensitive phenotype. Sequencing revealed a single base pair change in *yycF* causing a H215P mutation. The backcross resulted in 30% temperature-sensitive colonies, confirming that the mutation was responsible for this phenotype.

might have been obtained at higher frequency if error-prone PCR conditions (such as $MnCl_2$ or disproportional dNTP concentrations) would have been applied. The optimal error rate to maximize the chance of identifying a thermosensitive mutant has to be determined empirically. The natural competence of *B. subtilis* allowed for direct transformation of the PCR fragment and therefore eliminated the need of cloning the mutant library.

Delivery of Mutants and Screening

The resulting PCR fragment could be transformed into the *purA B. subtilis* strain Mu8u5u16 selecting for Ade^+ by plating the transformation on glucose minimal plates supplemented with required nutrients except adenine. To identify a temperature-sensitive strain, colonies are transferred to two LB plates by replica plating. One plate is incubated at 30° and one at 47°. Colonies that do not grow at 47° but do so at 30° are potential candidates. In our approach, 3200 colonies had to be screened in order to observe a single temperature-sensitive strain.

Identification of Mutations

Once a mutant has been obtained it is necessary to confirm that the temperature-sensitive phenotype is due to a mutation in the appropriate gene. The first step is to amplify the genes of interest by PCR followed by sequencing. In the case of the *yycF* and *yycG* genes, only a single base pair substitution was observed in *yycF* that introduced a H215P mutation.

Following its identification the responsibility of the mutation for causing the temperature-sensitive phenotype has to be assured by a genetic backcross. For *yycF* this was accomplished utilizing the vector pJM105a (Perego, 1993). This vector has a cm^R cassette flanked by up- and downstream multiple cloning sites. The mutated *yycF* gene, including the promoter region, was cloned downstream of the cm^R marker, and the region upstream of the *yyc* operon was cloned upstream of the cm^R marker. The resulting vector was transformed into a wild-type *B. subtilis* strain confirming linkage between the cm^R marker and the temperature-sensitive phenotype and demonstrating that the mutation in the *yycF* gene was responsible. It should be mentioned that this construct could have been used to introduce the mutant library to begin with eliminating the necessity of a nearby nutritional marker.

Transposon Mutagenesis to Identify Regulatory Elements

The aforementioned temperature-sensitive strain allowed for a transposon mutagenesis study aimed at revealing negative control elements of the YycFG system. The rationale was that at the nonpermissive temperature the levels of phosphorylated $YycF_{H215P}$ protein drop below a critical threshold. Elimination of any process that regulates the YycFG system negatively, either on an activity basis or on a transcriptional basis, might complement the temperature-sensitive phenotype by raising levels of phosphorylated YycF above that critical threshold (Fig. 2). This process successfully identified YycH and later YycI as negative regulators of YycG activity (Szurmant *et al.*, 2005; Szurmant *et al.*, 2007). Of course such a transposon mutagenesis relies on the ability to select for deletion of a negative control gene. This type of selection can only be achieved for two-component systems that are essential and therefore the described approach is limited to such systems.

Transposition

Once a temperature-sensitive strain has been obtained the strain can be the subject of a transposon mutagenesis. We utilized the mini-*Tn10* delivery plasmid pIC333 (Steinmetz and Richter, 1994). This plasmid features an *ery*^{R} marker, a thermosensitive gram-positive origin of replication, and the transposase gene. The 2.4-kb large transposable element includes a *spc*^{R} marker and the *Escherichia coli* ColE1 origin. The usefulness of these features will become clear in the following paragraphs. The only disadvantage of a mini-*Tn10*-based system is that insertion into the genome shows preference for specific sequences and is therefore not entirely random. Our laboratory has now available a new Mariner-based transposon system that eliminates this problem and can be used in future transposon mutagenesis projects (Wilson and Hoch, unpublished result).

To perform a transposon mutagenesis the plasmid pIC333 is first transformed into the strain of interest—in our case the temperature-sensitive *yycF* mutant strain—and plated on LB medium selecting for Ery^R at the permissive temperature of 30°. A single colony is initially grown at 30° in a 1-ml LB culture for 16 h. To initiate the transposition event the strain is subcultured (1:500) into fresh LB medium supplemented with spectinomycin. The temperature is raised to 37° and the culture is grown for an additional 16 h. At 37° the gram-positive origin of replication in pIC333 is inactive. Therefore the plasmid including the Spc^R is lost unless the transposable element is inserted into the chromosome. Following the incubation period the culture is plated on LB supplemented with spectinomycin and grown at the nonpermissive temperature of 47°, thereby selecting for transposition events that repair the temperature-sensitive *yycF* phenotype. Thirty-two colonies are obtained relatively easily by this method.

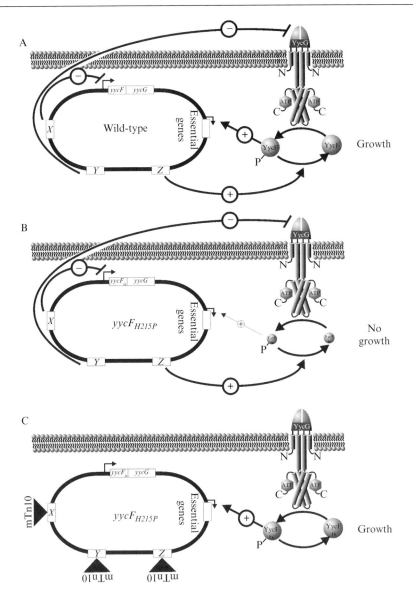

Fig. 2. Rationale for the identification of a negative regulator of the YycFG two-component system by transposon mutagenesis. (A) In a wild-type strain expression of the *yyc* operon, activity of YycG, or stability of YycF-P is likely controlled by unknown negative regulators X, Y, and Z, respectively. Expression of essential YycF-dependent genes is fine-tuned to allow for growth. (B) In the temperature-sensitive strain at the nonpermissive temperature levels of YycF-P are low, likely due to decreased protein stability. Expression of essential genes drops below the critical threshold, thereby not allowing for growth. (C) Disruption of any of the negative regulators X, Y,

Identification of Mutants

Two steps are necessary to identify mutations responsible for suppression of the temperature-sensitive phenotype. The first is a backcross that confirms linkage between the spc^R marker and the repair of the temperature-sensitive phenotype. The second is identification of the position of the transposon in the genome.

The first step provided an unexpected problem in our study. The temperature-sensitive *yycF* strain showed low competence in DNA uptake. Only 4 of the 32 colonies were successfully backcrossed into the $yycF_{ts}$ strain. Ultimately the result obtained from the four colonies resulted in sufficient data to follow up on and the other 28 colonies were ignored. Otherwise measures would have had to be taken to confirm results of the other strains.

In order to identify the genes interrupted by the transposon and confirmed by the backcross, the *E. coli* origin of replication found within the transposable element comes into play. Genomic DNA was prepared from individual colonies and subjected to an *Eco*RI digest followed by ligation. The ligation reaction was transformed into *E. coli* strain DH5α selecting for Spc^R. A diagnostic restriction digest was performed to eliminate possible siblings that can be numerous during a transposon mutagenesis. Unique plasmids were sequenced, identifying the location of insertion.

The gene in which the transposon is inserted does not necessarily have to be the one responsible for rescue of the temperature-sensitive phenotype. If the gene is part of a larger operon, any downstream gene is a potential candidate as the transposable element asserts a polarity effect. The transposon study presented here identified multiple insertions in the *yycH* gene found immediately downstream of the *yycG* kinase gene (Szurmant *et al.*, 2005). In-frame deletions of *yycH* and the other downstream genes later identified both YycH and YycI to have a negative effect on YycG activity (Szurmant *et al.*, 2007). A method to produce markerless in-frame deletions that do not exhibit such a polarity effect is therefore presented next.

Constructing In-Frame Deletions in the *yyc* Operon

The *yycFGHIJK* operon is one of many two-component system operons that contain additional genes. In *B. subtilis* some of these genes have been shown to regulate the two-component system that they are associated with by genomic context (Dartois *et al.*, 1997; Jordan *et al.*, 2006; Szurmant *et al.*, 2005). When investigating the role of these individual genes through

---

or Z by transposon mutagenesis results in the rescue of the temperature-sensitive phenotype by raising YycF-P levels above the critical threshold. This method identified YycH and YycI as negative regulators of YycG activity (type Y).

deletion analysis it is important to design deletion strains to not induce a polarity effect on downstream genes in the operon. The purest way of achieving this is the design of markerless in-frame deletions of the genes to be studied. This has been somewhat difficult in *B. subtilis*, as no widely available counterselection system has been described.

In lieu of a counterselection system, genes have been successfully deleted in-frame by a gene conversion methodology (Szurmant *et al.*, 2003). This is a tedious two-step process. The simple first step involves replacement of the gene of interest with an antibiotic marker via double crossover homologous recombination. In the tedious second step the antibiotic marker is removed by double crossover recombination from a plasmid with a gram-positive origin of replication containing regions flanking the gene, which is the subject for deletion. This is a low-frequency event (1:200–1:500), which has to be screened for instead of allowing for selection. Applying this approach to delete *yycH*, *yycI*, and *yycJ* and screening in excess of 3000 colonies did not result in antibiotic-sensitive colonies. This might have been due to the close neighborhood of the *yyc* operon to the *B. subtilis* origin of replication (Kunst *et al.*, 1997). We argue that during transformation most cells will be arrested in a state that has the *yyc* operon already duplicated in the cell. If a double crossover recombination event occurs in these cells, the result would be mixed colonies. Because we are screening for the loss of the antibiotic marker, this event would be difficult to identify.

A much more efficient method utilizes the *Saccharomyces cerevisiae* homing endonuclease I-*Sce*I and has been described for *Bacillus anthracis* (Janes and Stibitz, 2006). The authors suggested that their method might be applicable to achieve gene deletions in all firmicutes. We adopted this system to construct deletion within the *B. subtilis yyc* operon, thereby confirming that this method is suitable for organisms other than *B. anthracis*. I-*Sce*I is a restriction enzyme with an 18-bp recognition sequence. The method involves introduction of a I-*Sce*I recognition site into the genome along with the deletion construct via a single-crossover recombination. This is followed by expression of the restriction enzyme to induce a double-stranded DNA break, resulting in DNA repair via homologous recombination and inclusion of the desired mutation in 50% of all colonies (Fig. 3). In principle, the only requirement for this approach to work is that the genome of interest is devoid of a native I-*Sce*I site, which is highly likely.

Plasmid Construction

The method requires two plasmids, one that introduces a I-*Sce*I restriction site into the chromosome along with the desired deletion and a second one that expresses the I-*Sce*I restriction enzyme in order to cut the

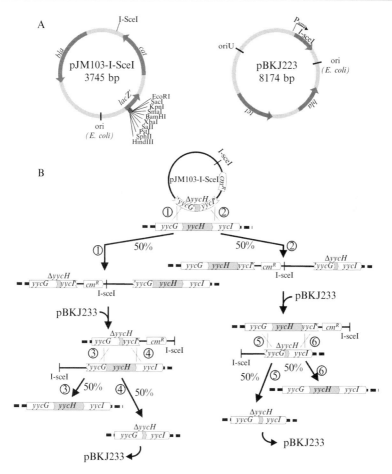

FIG. 3. I-*Sce*I strategy to construct in-frame deletions of genes in *B. subtilis*. (A) Two plasmids are required to delete a gene from the *B. subtilis* chromosome by the I-*Sce*I method. The first plasmid, pJM103-I-SceI, is a suicide vector featuring the cm^R marker *cat* for selection in *B. subtilis*, the amp^R gene *bla* for selection in *E. coli*, the *E. coli* pMB1 origin, and a I-*Sce*I restriction site. The multiple cloning site is located in the *lacZ* α-peptide gene, allowing for a blue/white screen. The second plasmid, pBKJ223, contains the gram-positive oriU origin, the gram-negative pMB1 origin, an amp^R gene *bla* for selection in *E. coli*, and a tet^R gene *tet* for selection in *B. subtilis* and expresses *I-sce*I under control of the amylase promoter (Janes and Stibitz, 2006). (B) The deletion of genes is a multistep process. First, 500 bp of up- and downstream regions to the desired gene (shown here for *yycH*) is introduced by single crossover recombination utilizing the pJM103-I-*Sce*I vector. Two possible strains can result from this recombination event, potentially resulting in different colony phenotypes. A DNA double-stranded break is introduced in either strain by expressing I-*Sce*I from the pBKJ223 vector. The break is repaired by homologous recombination, resulting in either the desired in-frame deletion strain or a wild-type strain at roughly a 1:1 distribution, detected by PCR analysis. The pBKJ223 plasmid is lost from the desired deletion strain by repeated subculturing in LB broth.

chromosome and induce homologous recombination. The second plasmid, pBKJ223 constructed by Janes and Stibitz (2006), could be used in *B. subtilis* without modification and it is likely suitable for use in most gram-positive organisms. For the first one we modified the suicide integration vector pJM103 to include an I-*Sce*I site (Perego, 1993). Any suicide integration vector can be modified for use in this approach by introduction of an I-*Sce*I site. Introduction of an I-*Sce*I site can be achieved by annealing two complementary oligonucleotides coding for the I-*Sce*I site with overhangs complementary to a suitable restriction site in the vector of choice. For pJM103, the restriction site was *Bgl*II and the oligonucleotide sequences were

5′-GATC**TAGGGATAACAGGGTAAT**-3′ and
5′-GATC**ATTACCCTGTTATCCCTA**-3′

with the I-*Sce*I restriction site in bold. The *Bgl*II-digested plasmid was ligated with the annealed oligonucleotides and transformed into *E. coli* TG1, resulting in pJM103-I-*Sce*I. Digesting the plasmid with the commercially available I-*Sce*I restriction endonuclease identified positive clones easily.

To delete the desired genes from the chromosome, up- and downstream regions are cloned into the suicide integration vector. This is achieved most easily with two subsequent PCR steps. The gene of choice, including about 500-bp up- and downstream regions, is cloned into the pJM103-I-*Sce*I plasmid. The gene is deleted by a second PCR with inverted primers that anneal at the 5′ and 3′ prime ends of the gene. This second PCR then amplifies the entire vector and the flanking regions but removes the gene of choice. Amplifying long DNA fragments can be tricky. We had good success when using iProof high-fidelity DNA polymerase (Bio-Rad) for this step. The to-be deleted areas have to be chosen carefully. Ideally, most of the gene is removed in-frame; however, overlapping genes and the position of ribosome-binding sites have to be taken into account. For *yycH*, *yycI*, and *yycJ* deletion strains, oligonucleotides were designed to remove all but 23, 18, or 9 N-terminal and 17, 8, or 16 C-terminal amino acid codons, respectively. Oligonucleotides were designed to also introduce a unique restriction site on each end (*Msc*I for YycH and *Eco*RV for YycI and YycJ). This allowed for cutting the PCR fragment and subsequent ligation. This is an efficient way of assuring that truncated primers often found in unpurified samples do not prevent the deletion from being in-frame. Furthermore, the primers do not have to be phosphorylated first, and we have observed that the ligation step is more efficient. Finally, double and triple deletion constructs for adjacent genes can be derived by an easy subcloning step if compatible restriction sites were used.

Introduction of the Suicide Plasmid

The deletion plasmid is transformed into *B. subtilis* selecting for Cm[R]. Single-crossover recombination can occur either upstream or downstream of the gene chosen for deletion, resulting in two possible genotypes. If the deletion of the gene results in a colony phenotype, two different types of colonies might be observed. In our experience this is irrelevant for future steps. We also noticed (by PCR) that more often than not multiple plasmids are inserted into the chromosome. This is likely because of the fact that competent *B. subtilis* preferentially, if not exclusively, takes up concatemeric plasmid DNA. Again this is irrelevant for future steps. Therefore, colonies do not have to be investigated further at this point.

Introduction of the I-SceI Expressing Plasmid

To introduce a double-stranded break in the DNA and homologous recombination, a culture grown from an individual Cm[R] colony is transformed with the pBKJ223 plasmid and plated on medium supplemented with 10 μg/ml tetracycline. Tetracycline is light sensitive and therefore plates need be fresh (less than 1 week old) and kept in the dark. A single Tet[R] colony is streaked twice over LB supplemented with tetracycline but not chloramphenicol. Following at least 10 individual colonies are screened for a Cm[S] phenotype, indicating that homologous recombination has occurred. At this point we usually observed more than 50% of colonies to have lost the chloramphenicol resistance. Cm[S] colonies will have lost the chloramphenicol resistance by homologous recombination, resulting in either wild type or the desired deletion strain. Deletion strains can be identified by PCR. To exclude mixed colonies it is advisable to perform two different PCR reactions, one with a set of primers that anneals within the deleted gene and one that anneals in regions flanking the deleted gene. In our attempts to delete *yycH*, *yycI*, and *yycJ*, roughly 50% of all screened colonies contained the desired deletion. We therefore advise screening about 5 colonies by PCR.

Removal of the I-SceI Expressing Plasmid

Following the identification of an in-frame deletion strain the final step is to cure the strain from the *I-sceI* expressing plasmid. pBKJ223 was unstable in *B. anthracis*. Restreaking a strain twice in the absence of an antibiotic easily caused spontaneous loss of the plasmid, resulting in Tet[S] colonies (Janes and Stibitz, 2006; Saile and Koehler, 2002). The plasmid is maintained more stably in *B. subtilis*. Mere restreaking did not result in the identification of cured strains. Instead strains were cured by repeated growth in liquid LB for at least 2 days, subculturing (1:500) every 8–14 h. Following Tet[S] colonies were identified by replica plating.

Studying Interactions between the YycG Kinase and Its
  Regulatory Proteins

The majority of two-component sensor kinases are transmembrane proteins and a subset of these receive input by interactions with other proteins. The transmembrane architecture makes *in vitro* interaction studies difficult. It is impossible to predict whether truncated fragments devoid of the transmembrane helices are able to fold correctly and form physiologically relevant quaternary structures. For example, overexpression and purification of the periplasmic sensing domain of YycG resulted in the formation of inclusion bodies. The little amount of protein observed in solution tended to aggregate and was highly susceptible to proteolytic cleavage, suggesting incorrect folding. Because of these problems, *in vitro* interaction studies between this kinase fragment and regulatory proteins YycH and YycI were inconclusive.

Problems can be avoided by utilizing *in vivo* interaction systems. These have the advantage that interaction of full-length proteins can be investigated. A disadvantage is that the strength of interaction can only be addressed on a comparative basis. The yeast two-hybrid system has been used traditionally (Miller and Stagljar, 2004). More recently, bacterial two-hybrid systems have been developed (Karimova *et al.*, 2002). One of these systems relies on reconstitution of adenylate cyclase activity, which in turn relaxes catabolite repression and induces expression of *lacZ* among others (Karimova *et al.*, 1998). Unlike in the yeast two-hybrid system, fusion proteins are not required to bind DNA, a fact that renders this system more sensitive and less susceptible to false negatives.

To study the interaction between YycG and its control proteins YycH and YycI, we employed the bacterial hybrid system, which utilizes the *Bordatella pertussis* adenylate cyclase and is performed in an adenylate cyclase-deficient *cya E. coli* strain. This system has been well described elsewhere (Karimova *et al.*, 2000). It is explained here as applied to the study of the YycGHI complex.

Plasmid Construction

The *B. pertussis* adenylate cyclase gene features two domains: T18 and T25 (Ladant and Ullmann, 1999). Unfused fragments do not afford activity; however, activity can be reconstituted by fusing these fragments to interacting proteins (Fig. 4). Bacterial two-hybrid plasmids available for cloning are pUT18, pUT18C, and pKT25 (Karimova *et al.*, 2001). The pUT18 plasmids feature a high-copy number ColEI origin, the ampicillin resistance gene, and express the T18 fragment with a multiple cloning site located either at the N-term (pUT18) or at the C-term (pUT18C). Plasmid pKT25

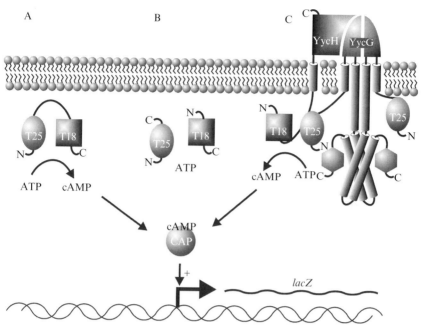

FIG. 4. The bacterial two-hybrid system identifies interactions of transmembrane proteins. The bacterial two-hybrid system relies on the *B. pertussis* adenylate cyclase enzyme. This enzyme contains two domains, T25 and T18. (A) In an adenylate cyclase-deficient *E. coli* strain, *B. pertussis* adenylate cyclase activates the CAP protein through generation of cAMP, resulting in induction of *lacZ* expression and production of blue colonies on a plate supplemented with X-gal. (B) When the two domains are expressed individually the protein is inactive. No cAMP is produced and the expression of *lacZ* remains repressed, resulting in white colonies. (C) If the fragments are fused to interacting proteins shown here for YycG and YycH, activity is achieved, resulting in blue colonies.

features the low-copy number 15A origin, a kanamycin resistance gene, and a multiple cloning site at the C-term of fragment T25. For our interaction studies among YycG, YycH, and YycI we utilized vectors pKT25 and pUT18C. This assured—assuming that all constructs are properly integrated into the membrane—that the fragments located N-terminal to our proteins of interest are facing the cytoplasm. Of note is that pKT25 and pUT18C plasmids contain similar multiple cloning sites. Several restriction sites allow for easy subcloning between the two vectors. In our studies we chose sites *Xba*I and *Bam*HI. The entire coding sequences for YycG, YycH, and YycI only excluding the start codons were amplified, introducing a 5′ *Xba*I site and a 3′ *Bam*HI site, and the fragments were cloned into the same sites of vector pUT18C and pKT25.

Cotransformation and Screening

Following the construction of these plasmids, all possible combinations of pUT18C and pKT25 vectors (16 strains including negative controls) are cotransformed into an adenylate cyclase-deficient *E. coli* strain. Two strains are available. BTH101 is a fast-growing Rec[+] strain with the disadvantage of occasional plasmid instability, whereas DHM1 is a slower growing *recA* strain (Karimova *et al.*, 2005). For studying YycGHI interactions, BTH101 is utilized. Cotransformations are straightforward with this strain and easily achieved if the strain is of sufficient competence. We have had good experience with electrocompetent cells prepared by standard techniques. Transformations are plated on LB plates supplemented with 40 μg/ml X-gal, 30 μg/ml kanamycin, and 100 μg/ml ampicillin and are incubated at 37° for 14 h, at which point sufficient blue color is observed to identify interacting proteins, whereas negative control strains show little color.

Quantification of Interactions

As mentioned earlier, one disadvantage of *in vivo* interaction systems is that no dissociation constants are forthcoming. Nevertheless an attempt can be made to quantify the interactions. Daniel *et al.* (2006) suggested that plating strains on minimal plates with X-gal and comparing the intensity of blue color formed after an 14-h incubation at 30° resulted in the most reliable and reproducible measure of interaction strength. We observed reproducible results when performing a liquid assay of β-galactosidase activity according to Miller (1972). The advantage is that one can put a number on the interaction, which saves the cost of publication of a color figure. However, it is not clear how dissociation constant and β-galactosidase activity are correlated. One gets another comparative measure of interaction strength instead.

The liquid assay is performed according to Miller (1972). At least three individual colonies of every strain are grown overnight (14 h) in 2 ml LB supplemented with 100 μg/ml ampicillin and 30 μg/ml kanamycin. The OD_{600} nm is determined. Cells are collected by centrifugation and resuspended in 1 ml Z buffer (100 mM $Na_XH_YPO_4$, pH 7.0, 10 mM KCl, 1 mM $MgSO_4$, 50 mM β-mercaptoethanol). Permeabilization of cells is achieved by the addition of 10 μl toluene and 10 μl of a 0.1% SDS solution and vortexing. Shaking cell suspensions for 30 min at 37° evaporated the toluene. β-Glactosidase activity is determined in a microtiter plate. In a pilot assay, 100 μl of cell suspensions is combined with 50 μl Z buffer. Reactions are started by the addition of 20 μl of a 4.5-mg/ml aqueous 2-nitrophenyl-β-D-galactopyranoside substrate solution. After 15 min reactions are stopped by the addition of 30 μl of a 1.2 M Na_2CO_3 solution. Absorbance at 420 and 550 nm is determined in a microplate reader. The assay is then repeated, adjusting cell volumes to receive an A_{420} nm between 0.2 and 0.6. The product absorbance is corrected for scattered light in the following Eq. (1):

$$A_{420-550} = A_{420} \text{ nm} - 1.4 \times A_{550} \text{ nm} \qquad (1)$$

β-Galactosidase activity is calculated in Miller units in the following Eq. (2):

$$\text{Activity} = (1000 \times A_{420-550})/(\text{t} \times \text{OD}_{600} \text{ nm} \times \text{V}) \qquad (2)$$

where t is the reaction time in minutes and V is the volume of cells used in milliliters.

Subcellular Localization Studies

The discovery that chemotaxis receptors localize to the cell poles in *Caulobacter crescentus* and subsequent similar observations in *E. coli*, *B. subtilis,* and other bacteria led to a new line of studies (Alley *et al.*, 1992; Kirby *et al.*, 2000; Maddock and Shapiro, 1993). In the absence of a known regulon or signal of a two-component system, cellular localization might give a hint to functions. Questions that can be answered by such an approach are where does this system function and therefore where does it receive its signal input? For the YycG kinase, localization was observed at the cell division sites consistent with its role in controlling cell division processes (Fukuchi *et al.*, 2000).

Subcellular localization can be visualized by immunofluorescent staining or by tagging the protein of interest with the green fluorescent protein (GFP) or any of its derivatives. The latter has the advantages that it can be performed in living cells and that the movement of a protein can be studied in addition to its localization. However, we observed for several *B. subtilis* histidine kinases, including YycG, that native expression levels were too low to visualize the localization of GFP-tagged proteins. Over-expression of a GFP-tagged protein can lead to artifactual results and we therefore recommend immunofluorescent staining when first studying localization of a novel two-component kinase. A detailed protocol is described next.

Antibody Preparation

A high-quality antibody devoid of cross-reacting material is a requirement for immunofluorescent localization studies. Most antibodies require purification before they can be used in subsequent studies. We outline an affinity purification protocol applied to YycG and YycF antibodies.

The antibody is precipitated from 1 ml YycG or YycF antiserum derived from New Zealand white rabbits by the addition of ammonium sulfate to a final saturation of 45%, and the mixture is shaken gently for 14 h at 4°. The precipitated antibody is pelleted by centrifugation (13,000 rpm, 30 min at 4°). The pellet is dissolved in phosphate-buffered saline (PBS) and then dialyzed against three changes of PBS. Note that antibody derived from different animals precipitates at different ammonium sulfate saturations (Harlow and Lane, 1988).

In the meantime, YycG or YycF protein is immobilized on a matrix to generate an affinity column. Purified proteins are dialyzed against 20 mM sodium borate buffer, pH 9.0, which can be supplemented with NaCl if necessary. As a matrix we use Reacti-Gel CDI Supports (Pierce). One milliliter of gel slurry is placed in an Eppendorf tube and washed three times with cold H_2O by centrifugation (1000g, 1 min). The washed matrix is mixed with 2–5 mg dialyzed protein at room temperature for a minimum of 12 h. At lower temperatures, covalent linking of protein to the matrix is less efficient. To assure that sufficient protein has been linked to the matrix, the supernatant can be subjected to a protein concentration assay. The matrix is collected by centrifugation and resuspended in 1 ml 50 mM Tris-HCl, pH 9.0, and shaken at room temperature for a minimum of 4 h to inactivate the remaining protein-binding sites. Finally, the matrix is placed into a disposable column and washed with 100 column volumes PBS.

Dialyzed antiserum and protein matrix are added into an Eppendorf tube and shaken gently at 4° for 14 h. The slurry is placed into a disposable column and washed with 100 ml PBS, collecting the flow through. The antibody is eluted with 10 ml 0.2 M glycine, pH 2.2. The eluate is immediately neutralized by the addition of an equal volume of 1 M Tris-HCl, pH 8.0. Finally, the matrix is again washed with 100 ml PBS, collecting the flow through. Additional antibody can be extracted from the flow-through fractions by repeating the steps two more times.

Eluted antibody from the three steps (60 ml) is concentrated down to 1 ml using a Centricon plus-20 with a molecular mass cutoff of 5000 Da (Millipore). Following this, the antibody is dialyzed into PBS and stored at −20°. The purity of the antibody can then be addressed by immunoblotting. No cross-reacting material should be present at this point.

Slide Preparation

Prior to immunofluorescent studies, microscope slides have to be prepared to adsorb cells efficiently. A multiwell microscope slide (we use either 12-well diagnostic printed slides by Erie Scientific or 15-well multi-test slides by MP Biomedical) is treated with 10 or 20 μl poly-L-lysine solution per well (Sigma-Aldrich) for 30 min at room temperature. The solution is removed with a Pipetman and the slide is washed once with H_2O and air dried at room temperature.

For clarity in all subsequent steps involving the microscope slide, solutions are added and aspirated with a Pipetman. The volume of solutions depends on the size of the wells on the slide and a range is given that reflects the two different slides used in our experiments. Because cell adsorption capability declines quickly, slides should be prepared on the sampling day.

Cell Growth and Fixation

Prior to cell sampling, a formaldehyde solution has to be prepared. Paraformaldehyde (0.16 g) is suspended in 1.2 ml of 0.166 M $Na_XH_YPO_4$, pH 7.4, and then dissolved by maintaining the mixture at 65°, resulting in a formaldehyde concentration of 13.33%. When necessary, 1.5 μl of a 25% glutaraldehyde solution is added. This can help cell fixation, but can raise the level of background immunofluorescence. Ready-made formaldehyde solutions oxidize over time. We recommend preparing the formaldehyde fresh from solid paraformaldehyde.

To sample cells the appropriate strain is grown in the medium of choice. Aliquots (500 μl) are removed at appropriate times, and 120 μl of the formaldehyde solution is added to the cell suspensions, resulting in a final concentration of 2.58%. Cells are fixed by keeping the mixture at room temperature for 20 min. It is important that *B. subtilis* cells are not centrifuged prior to the addition of formaldehyde as this can lead to artifactual fluorescence localization (Harry *et al.*, 1995). Following this, the fixed cells are collected by centrifugation (14,000 rpm, 2 min, room temperature). The pellet is washed twice with PBS and suspended in PBS to an OD_{525} nm of less than 0.4. Ten to 20 μl of cell suspension is added to a poly-L-lysine-treated multiwell slide and kept at room temperature for 30 min. The suspension is removed carefully, and the plate is washed three times by the addition and removal of 15–30 μl PBS. Finally, 10–20 μl of solution I (25 mM Tris-HCl, pH 8.0, 10 mM EDTA, 50 mM glucose) is added to the wells, and the slide is kept at room temperature for 10 min.

Cell Permeablization

To permeabilize the cells, 10 or 20 μl solution I supplemented with 2 mg/ml lysozyme is added to the wells. To assure incomplete lysis, the lysozyme is allowed to act for no longer than 4 min. The solution is removed, and the slide is washed 10 to 16 times with PBS to assure complete removal of lysozyme. Following washing, the slide glass is submerged into ice-cold methanol for no longer than 5 min and subsequently dried.

Finally, 10 to 20 μl of PBS supplemented with 2% BSA (PBS-B) is added to the wells and kept at room temperature for 20 min, preparing the slides for antibody staining.

Fluorescence Staining and Visualization

The purified primary antibody is diluted in PBS-B. The optimal dilution has to be determined empirically. For reference, the YycG antibody is used at a 1:50 dilution in contrast to a 1:10,000 dilution required for immunoblotting. Ten to 20 μl antibody solution is added to the wells and allowed to equilibrate at 4° overnight. The slides are then washed 10–16 times with PBS-B. Ten to 20 μl PBS-B is applied to the wells, and the slide glass is kept at room temperature for at least 10 min. If a second protein is to be visualized in the same cells (e.g., performed to observe colocalization of YycG with the cell divisional master regulator FtsZ) the primary antibody against that protein (FtsZ) is applied following overnight incubation with the first antibody, and cells are equilibrated an additional night and washed as for the initial primary antibody. Secondary antibody is then diluted in PBS-B, and 10 to 20 μl of secondary antibody solution is added to the wells and the slide is kept in the dark for 1 to 3 h at room temperature. If two secondary antibodies are required, a mixture of the two antibodies is applied. To visualize YycG and YycF, we use Alexa Fluor 488 (green)-conjugated antirabbit antibody at a typical dilution of 1:500 (Molecular Probes). To confirm colocalization of YycG with FtsZ, we utilize Alexa Fluor 546-conjugated antisheep antibody. Washes are as described earlier. DNA is visualized by the addition of PBS supplemented with 3 μg/ml 4',6-diamidino-2-phenylindole (DAPI). Following a 5-min incubation time, DAPI is removed and the cells are washed once with PBS. An antifade solution (Immuno-Fluore Mounting Medium, MP Medicals) is added to the wells and the wells are concealed with a cover glass. The edges are sealed with nail polish, and the slide glass is kept at 4° until use in fluorescent microscopy.

To visualize localization of the protein(s), a high-quality fluorescent microscope equipped with suitable filters has to be available. Given the cost of such equipment, investigators usually have to deal with whatever equipment is available at their institute. For reference we obtained excellent results with the Bio-Rad Radiance 2100 Rainbow laser-scanning confocal microscope (Zeiss) when studying colocalization of YycG and FtsZ (Fukushima and Hoch, unpublished). The confocal component has seven available laser lines for excitation and a wide range of adjustable band-pass filters, which permit one to determine optimal bands to perform spectral separation.

Concluding Remarks

When studying a novel two-component system, even in the absence of knowledge about the genes under transcriptional control of this system, experiments can be designed that help identify the role that the signal transduction plays. Some of the experiments presented here were designed to utilize the essentiality of the YycFG system and can be applied to any essential signaling system. Others are universally suitable to study two-component systems and their regulation.

Acknowledgments

Some of the presented methods either utilized tools constructed by others or were adopted with only slight modification from methods developed by others and we thank the authors of these tools and methods for their effort. Our studies benefited greatly from their work. Most noteworthy are Michael Steinmetz and Rosalyne Richter who constructed the pIC333 transposon delivery vector, Scott Stibitz who established the I-SceI method in B. anthracis, Marta Perego who constructed the pJM103-I-SceI vector, and Daniel Ladant and Gouzel Karimova who developed the bacterial two-hybrid system. The authors acknowledge funding by grant GM019416 from the National Institute of General Medicine Sciences and grant AI055860 from the National Institute of Allergy and Infectious Diseases, National Institutes of Health, USPHS.

References

Alley, M. R., Maddock, J. R., and Shapiro, L. (1992). Polar localization of a bacterial chemoreceptor. Genes Dev. 6, 825–836.

Daniel, R. A., Noirot-Gros, M. F., Noirot, P., and Errington, J. (2006). Multiple interactions between the transmembrane division proteins of Bacillus subtilis and the role of FtsL instability in divisome assembly. J. Bacteriol. 188, 7396–7404.

Dartois, V., Djavakhishvili, T., and Hoch, J. A. (1997). KapB is a lipoprotein required for KinB signal transduction and activation of the phosphorelay to sporulation in Bacillus subtilis. Mol. Microbiol. 26, 1097–1108.

Dubrac, S., and Msadek, T. (2004). Identification of genes controlled by the essential YycG/YycF two-component system of Staphylococcus aureus. J. Bacteriol. 186, 1175–1181.

Fabret, C., and Hoch, J. A. (1998). A two-component signal transduction system essential for growth of Bacillus subtilis: Implications for anti-infective therapy. J. Bacteriol. 180, 6375–6383.

Fukuchi, K., Kasahara, Y., Asai, K., Kobayashi, K., Moriya, S., and Ogasawara, N. (2000). The essential two-component regulatory system encoded by yycF and yycG modulates expression of the ftsAZ operon in Bacillus subtilis. Microbiology 146(Pt. 7), 1573–1583.

Harlow, E., and Lane, D. P. (1988). Storing and purifying antibodies. In "Antibodies, a Laboratory Manual," pp. 285–318. Cold Spring Harbor Laboratory, Cold Spring Harbor, NY.

Harry, E. J., Pogliano, K., and Losick, R. (1995). Use of immunofluorescence to visualize cell-specific gene expression during sporulation in Bacillus subtilis. J. Bacteriol. 177, 3386–3393.

Howell, A., Dubrac, S., Andersen, K. K., Noone, D., Fert, J., Msadek, T., and Devine, K. (2003). Genes controlled by the essential YycG/YycF two-component system of Bacillus subtilis revealed through a novel hybrid regulator approach. Mol. Microbiol. 49, 1639–1655.

Howell, A., Dubrac, S., Noone, D., Varughese, K. I., and Devine, K. (2006). Interactions between the YycFG and PhoPR two-component systems in *Bacillus subtilis*: The PhoR kinase phosphorylates the non-cognate YycF response regulator upon phosphate limitation. *Mol. Microbiol.* **59,** 1199–1215.

Hulett, F. M. (1996). The signal-transduction network for Pho regulation in *Bacillus subtilis*. *Mol. Microbiol.* **19,** 933–939.

Hulett, F. M., Lee, J., Shi, L., Sun, G., Chesnut, R., Sharkova, E., Duggan, M. F., and Kapp, N. (1994). Sequential action of two-component genetic switches regulates the PHO regulon in *Bacillus subtilis*. *J. Bacteriol.* **176,** 1348–1358.

Janes, B. K., and Stibitz, S. (2006). Routine markerless gene replacement in *Bacillus anthracis*. *Infect. Immun.* **74,** 1949–1953.

Jordan, S., Junker, A., Helmann, J. D., and Mascher, T. (2006). Regulation of LiaRS-dependent gene expression in *Bacillus subtilis*: Identification of inhibitor proteins, regulator binding sites, and target genes of a conserved cell envelope stress-sensing two-component system. *J. Bacteriol.* **188,** 5153–5166.

Karimova, G., Dautin, N., and Ladant, D. (2005). Interaction network among *Escherichia coli* membrane proteins involved in cell division as revealed by bacterial two-hybrid analysis. *J. Bacteriol.* **187,** 2233–2243.

Karimova, G., Ladant, D., and Ullmann, A. (2002). Two-hybrid systems and their usage in infection biology. *Int. J. Med. Microbiol.* **292,** 17–25.

Karimova, G., Pidoux, J., Ullmann, A., and Ladant, D. (1998). A bacterial two-hybrid system based on a reconstituted signal transduction pathway. *Proc. Natl. Acad. Sci. USA* **95,** 5752–5756.

Karimova, G., Ullmann, A., and Ladant, D. (2000). A bacterial two-hybrid system that exploits a cAMP signaling cascade in *Escherichia coli*. *Methods Enzymol.* **328,** 59–73.

Karimova, G., Ullmann, A., and Ladant, D. (2001). Protein-protein interaction between *Bacillus stearothermophilus* tyrosyl-tRNA synthetase subdomains revealed by a bacterial two-hybrid system. *J. Mol. Microbiol. Biotechnol.* **3,** 73–82.

Kaspar, S., and Bott, M. (2002). The sensor kinase CitA (DpiB) of *Escherichia coli* functions as a high-affinity citrate receptor. *Arch. Microbiol.* **177,** 313–321.

Kirby, J. R., Niewold, T. B., Maloy, S., and Ordal, G. W. (2000). CheB is required for behavioural responses to negative stimuli during chemotaxis in *Bacillus subtilis*. *Mol. Microbiol.* **35,** 44–57.

Kunst, F., Ogasawara, N., Moszer, I., Albertini, A. M., Alloni, G., Azevedo, V., Bertero, M. G., Bessieres, P., Bolotin, A., Borchert, S., Borriss, R., Boursier, L., *et al.* (1997). The complete genome sequence of the gram-positive bacterium *Bacillus subtilis*. *Nature* **390,** 249–256.

Ladant, D., and Ullmann, A. (1999). *Bordatella pertussis* adenylate cyclase: A toxin with multiple talents. *Trends Microbiol.* **7,** 172–176.

Maddock, J. R., and Shapiro, L. (1993). Polar location of the chemoreceptor complex in the *Escherichia coli* cell. *Science* **259,** 1717–1723.

Mansilla, M. C., Albanesi, D., Cybulski, L. E., and de Mendoza, D. (2005). Molecular mechanisms of low temperature sensing bacteria. *Ann. Hepatol.* **4,** 216–217.

Martin, P. K., Li, T., Sun, D., Biek, D. P., and Schmid, M. B. (1999). Role in cell permeability of an essential two-component system in *Staphylococcus aureus*. *J. Bacteriol.* **181,** 3666–3673.

Miller, J., and Stagljar, I. (2004). Using the yeast two-hybrid system to identify interacting proteins. *Methods Mol. Biol.* **261,** 247–262.

Miller, J. H. (1972). *In* "Experiments in Molecular Genetics," pp. 352–355. Cold Spring Harbor Laboratory, Cold Spring Harbor, NY.

Mohedano, M. L., Overweg, K., de la Fuente, A., Reuter, M., Altabe, S., Mulholland, F., de Mendoza, D., Lopez, P., and Wells, J. M. (2005). Evidence that the essential response regulator YycF in *Streptococcus pneumoniae* modulates expression of fatty acid biosynthesis genes and alters membrane composition. *J. Bacteriol.* **187,** 2357–2367.

Neiditch, M. B., Federle, M. J., Pompeani, A. J., Kelly, R. C., Swem, D. L., Jeffrey, P. D., Bassler, B. L., and Hughson, F. M. (2006). Ligand-induced asymmetry in histidine sensor kinase complex regulates quorum sensing. *Cell* **126,** 1095–1108.

Ng, W. L., Robertson, G. T., Kazmierczak, K. M., Zhao, J., Gilmour, R., and Winkler, M. E. (2003). Constitutive expression of PcsB suppresses the requirement for the essential VicR (YycF) response regulator in *Streptococcus pneumoniae* R6. *Mol. Microbiol.* **50,** 1647–1663.

Perego, M. (1993). Integrational vectors for genetic manipulation in *Bacillus subtilis.* *In* "*Bacillus subtilis* and Other Gram-Positive Bacteria: Biochemistry, Physiology, and Molecular Genetics" (A. L. Sonenshein, J. A. Hoch, and R. Losick, eds.), pp. 615–624. American Society for Microbiology, Washington, DC.

Saile, E., and Koehler, T. M. (2002). Control of anthrax toxin gene expression by the transition state regulator abrB. *J. Bacteriol.* **184,** 370–380.

Steinmetz, M., and Richter, R. (1994). Easy cloning of mini-Tn10 insertions from the *Bacillus subtilis* chromosome. *J. Bacteriol.* **176,** 1761–1763.

Szurmant, H., Bunn, M. W., Cannistraro, V. J., and Ordal, G. W. (2003). *Bacillus subtilis* hydrolyzes CheY-P at the location of its action, the flagellar switch. *J. Biol. Chem.* **278,** 48611–48616.

Szurmant, H., Mohan, M. A., Imus, P. M., and Hoch, J. A. (2007). YycH and YycI interact to regulate the essential YycFG two-component system in *Bacillus subtilis. J. Bacteriol.* **189,** in press.

Szurmant, H., Nelson, K., Kim, E. J., Perego, M., and Hoch, J. A. (2005). YycH regulates the activity of the essential YycFG two-component system in *Bacillus subtilis. J. Bacteriol.* **187,** 5419–5426.

Tiwari, R. P., Reeve, W. G., Dilworth, M. J., and Glenn, A. R. (1996). Acid tolerance in *Rhizobium meliloti* strain WSM419 involves a two-component sensor-regulator system. *Microbiology* **142**(Pt. 7), 1693–1704.

Vagner, V., Dervyn, E., and Ehrlich, S. D. (1998). A vector for systematic gene inactivation in *Bacillus subtilis. Microbiology* **144**(Pt. 11), 3097–3104.

Section III

Physiological Assays and Readouts

[21] Isolation and Characterization of Chemotaxis Mutants of the Lyme Disease Spirochete *Borrelia burgdorferi* Using Allelic Exchange Mutagenesis, Flow Cytometry, and Cell Tracking

By Md. A. Motaleb, Michael R. Miller, Richard G. Bakker, Chunhao Li, and Nyles W. Charon

Abstract

Constructing mutants by targeted gene inactivation is more difficult in the Lyme disease organism, *Borrelia burgdorferi*, than in many other species of bacteria. The *B. burgdorferi* genome is fragmented, with a large linear genome and 21 linear and circular plasmids. Some of these small linear and circular plasmids are often lost during laboratory propagation, and the loss of specific plasmids can have a significant impact on virulence. In addition to the unusual structure of the *B. burgdorferi* genome, the presence of an active restriction-modification system impedes genetic transformation. Furthermore, *B. burgdorferi* is relatively slow growing, with a 7- to 12-h generation time, requiring weeks to obtain single colonies. The beginning part of this chapter details the procedure in targeting specific *B. burgdorferi* genes by allelic exchange mutagenesis. Our laboratory is especially interested in constructing and analyzing *B. burgdorferi* chemotaxis and motility mutants. Characterization of these mutants with respect to chemotaxis and swimming behavior is more difficult than for many other bacterial species. We have developed swarm plate and modified capillary tube assays for assessing chemotaxis. In the modified capillary tube chemotaxis assay, flow cytometry is used to rapidly enumerate cells that accumulate in the capillary tubes containing attractants. To assess the swimming behavior and velocity of *B. burgdorferi* wild-type and mutant cells, we use a commercially available cell tracker referred to as "Volocity." The latter part of this chapter presents protocols for performing swarm plate and modified capillary tube assays, as well as cell motion analysis. It should be possible to adapt these procedures to study other spirochete species, as well as other species of bacteria, especially those that have long generation times.

Introduction

Borrelia burgdorferi is a spirochete that causes Lyme disease. It is transmitted to humans and other animals by the bite of a tick and is the most prevalent arthropod borne infection in the United States. As with

METHODS IN ENZYMOLOGY, VOL. 422
0076-6879/07 $35.00
DOI: 10.1016/S0076-6879(06)22021-4

other spirochete species, *B. burgdorferi* has a unique structure. Outermost is an outer membrane sheath. This membrane is devoid of lipopolysaccharides. Within the outer membrane sheath is the cell cylinder. Residing in the periplasmic space are 14 to 22 periplasmic flagella, with 7 to 11 periplasmic flagella inserted near each end of the cell that extend toward the cell body center (Charon and Goldstein, 2002; Rosa *et al.*, 2005).

Genetic manipulations are especially difficult in virulent, low-passage *B. burgdorferi*. The *B. burgdorferi* genome is composed of a large linear chromosome approximately 960 kb, with 12 linear plasmids (lp) and 9 circular plasmids totaling approximately 600 kb and ranging in size from 5 to 56 kb (Fraser *et al.*, 1997). This notably fragmented genome with so many plasmids is relatively unique among bacterial species. During propagation in the laboratory, some plasmids are spontaneously lost, resulting in diminished infectivity using animal models of Lyme disease (Elias *et al.*, 2002; Norris *et al.*, 1995; Schwan *et al.*, 1988). Therefore, even after successful transformation, during clonal selection, many isolates lose essential virulence plasmids and cannot be used in studies related to pathogenesis. In addition, restriction-modification genes that encode proteins that inhibit genetic transformation by degrading introduced foreign DNA are located on plasmids lp25 and lp56. Inactivation or loss of one of the restriction genes located in the lp25 (locus BBE02) results in a higher efficiency of transformation (Kawabata *et al.*, 2004). However, this inactivation results in cells with a slightly compromised infectivity (Kawabata *et al.*, 2004).

In contrast, genetic manipulation is relatively easy in high-passage, avirulent *B. burgdorferi*. Avirulent strain B31-A retains approximately 12 of the 21 natural plasmids (unpublished observation). The fact that this strain lacks lp25 and lp56 may explain, in part, its easier genetic manipulation. The complexity of genetic manipulations in *B. burgdorferi* is illustrated by the report that, for unknown reasons, it is extremely difficult to complement cells even with a restriction system that is altered (Shi *et al.*, 2006).

Borrelia burgdorferi Mutagenesis

Samuels *et al.* (1994) first described the successful genetic transformation of *B. burgdorferi* using a coumermycin-resistant cassette that was isolated from a spontaneously mutated *B. burgdorferi gyrB* gene. However, because the gene within this cassette contained a native, mutated *gyrB* gene, constructs containing this cassette and genes of interest were found to inefficiently recombine with the targeted genes: This inefficiency is due to recombination of the cassette at the chromosomal *gyrB* gene instead of at the targeted gene site (Rosa *et al.*, 1996). A kanamycin-resistant gene (*kan*) derived from *Escherichia coli* Tn903 has been developed in which *B. burgdorferi* flagella *flgB* or *flaB* promoters were fused to the *kan* gene (Bono *et al.*, 2000). The development of this *kan* cassette has permitted relatively efficient gene targeting. Several other antibiotic resistant cassettes (e.g., streptomycin, gentamycin, erythromycin, and synthesized coumermycin cassettes) have since been developed to target *B. burgdorferi* genes, allowing the construction of mutant strains harboring multiple mutations (Elias *et al.*, 2003; Frank *et al.*, 2003; Sartakova *et al.*, 2000, 2003). We are using both high- and low-passage *B. burgdorferi* cells (B31A and B31-A3, respectively) (Elias *et al.*, 2002; Samuels *et al.*, 1994) to inactivate motility and chemotaxis genes using allelic exchange mutagenesis. While inactivation in high-passage *B. burgdorferi* is relatively easy with the cassettes described earlier, targeted mutagenesis in low-passage, virulent cell remains a difficult endeavor.

Chemotaxis and Motility Analysis

We are interested in investigating the effect of specific targeted mutations on *B. burgdorferi* chemotaxis and motility. In traditional capillary tube chemotaxis assays, bacteria that accumulate in capillary tubes containing attractants after incubation are enumerated by viable plate counts. However, because of the long generation time of *B. burgdorferi*, this approach is not feasible, and counting cells manually in a Petroff-Hausser chamber is laborious, time-consuming, and of low throughput. We developed a modified capillary tube assay to quantitate *B. burgdorferi* that has a considerably higher throughput (Bakker *et al.*, 2007; Motaleb *et al.*, 2005). The main modifications of this assay were use of larger capillary tubes and rapid enumeration of cells by flow cytometry. Swarm plates are also a conventional means of characterizing chemotaxis and motility mutants (Adler, 1974); however, the swarm assay had to be modified for analyzing *B. burgdorferi*. We found that diluting the growth medium enhanced swarm ring migration (Li *et al.*, 2002; Motaleb *et al.*, 2000, 2005).

Motility and chemotaxis mutants of *B. burgdorferi* have altered swimming behaviors as viewed by light microscopy (Li *et al.*, 2002; Motaleb *et al.*, 2000, 2005). Critical analysis of *B. burgdorferi* movement is essential for understanding the impact of inactivating specific genes on cell motility and chemotaxis. Our initial studies used the "Hobson BacTracker" to track mutant and wild-type *B. burgdorferi* cells (Li *et al.*, 2002). However, more recently we have found that the "Volocity" system from Improvision Inc. is more convenient (M. Motaleb and N. Charon-unpublished; Bakker *et al.*, 2007). The Volocity system has also been used for other species of bacteria (Butler and Camilli, 2004). Details of our modified swarm plate and capillary tube chemotaxis assays, as well as an analysis of cell swimming behavior, are described in detail in this chapter.

Materials and Methods

Allelic Exchange Mutagenesis in B. burgdorferi

B. burgdorferi (avirulent B31-A and virulent B31-A3) cells are grown in liquid Barbour-Stonner-Kelly II media (BSK-II, see later) at 33° in a humidified chamber with 3% CO_2 (Barbour, 1984).

Materials

Protocols for growing *B. burgdorferi*, plating, and electroporation have been described in detail elsewhere (Barbour, 1984; Samuels *et al.*, 1994).

Culture and Plating Media

Liquid BSK-II: To prepare 1 liter BSK-II, dissolve the following in ~800 ml H_2O. Bovine serum albumin (BSA, Celliance Inc.) 50.0 g, Bacto Neopeptone (BD & Co.) 5.0 g, Bacto TC Yeastolate (BD & Co.) 2.5 g, HEPES (Sigma-Aldrich) 6.0 g, glucose (Fisher Scientific) 5.0 g, Na-citrate (Fisher Scientific) 0.7 g, Na-pyruvate (Sigma-Aldrich Co., St. Louis, MO) 0.8 g, *N*-acetylglucosamine (Sigma-Aldrich) 0.4 g, Na-bicarbonate (Fisher Scientific) 2.2 g. Stir for 2–3 h, adjust pH to 7.6 with fresh NaOH solution, and add H_2O to 1 liter. Filter (0.22 μM) sterilize and store at −20°. To test sterility, transfer ~4.0 ml of media into a culture tube aseptically and incubate the tube in a 33° incubator for several days and examine media microscopically for contamination.

10× CMRL-1066 (US Biologicals): Prepare and adjust a 10× solution of CMRL-1066 to pH 7.6 with Na-bicarbonate according to the manufacturer's protocol. After filter sterilizing, store 10× CMRL-1066 at 4°, protected from light (wrapped in foil or placed within a closed box).

Complete BSK-II medium: To complete BSK-II medium, mix 100.0 ml BSK-II with 9.1 ml of 10× CMRL-1066 and 5.8 ml of rabbit serum. Complete BSK-II medium should be stable for at least a month at 4°. Prolonged storage (>2 months) results in loss of growth potential.

Plating BSK medium (P-BSK): To prepare 1 liter P-BSK, mix in 800 ml H_2O: 83 g BSA, 8.3 g Bacto Neopeptone, 4.2 g Bacto TC Yeastolate, 10.0 g HEPES, 8.3 g glucose, 1.2 g Na-citrate, 1.3 g Na-pyruvate, 0.7 g *N*-acetylglucosamine, 3.7 g Na-bicarbonate. Stir, pH, filter, and check for contamination as described earlier for preparing BSK-II. Store at −20°.

Plating B. burgdorferi

a. Mix 240.0 ml of P-BSK, 38.0 ml of 10× CMRL-1066, 12.0 ml of rabbit serum, and 20.0 ml of freshly prepared, 0.22 μM filtered 5% sodium bicarbonate solution. Warm the mixture at 50° in a water bath. Add 200.0 ml of autoclaved 1.7% agarose (SeaKem LE Agarose, Cambrex BioScience Rockland, Inc.) equilibrated to 50° and keep the mixture warm to avoid solidification.

b. Pipette 17.0 ml of the molten medium into 12 plates (100 × 15-mm sterile dishes, Fisher Scientific or any brand) and allow the agar to solidify inside a laminar flow hood. Return the rest of the medium to a 50° water bath.

c. Pipette desirable amount of *B. burgdorferi* (e.g., electroporated cells, see later) into a 50-ml plastic tube. Add 16–17.0 ml of the medium containing agar (medium should be warm, ~45°) to the cells, and add the cell-medium mixture onto the solidified bottom agarose slowly. Avoid forming bubbles and allow the plate to solidify at room temperature inside a hood.

d. Invert solidified plates (upside down) and incubate at 33° in a 3% CO_2 humidified incubator until colonies appear (10–30 days depending on the strain: wild-type cells should appear in 10–14 days, whereas motility and chemotaxis mutant cells may take 20–30 days).

e. Isolate single colonies by picking with a 0.2- or 1.0-ml micropipette tip and transfer into 3 ml of liquid BSK-II media in a plastic tube (Falcon 16-ml tubes with caps, or any brand) with the cap loose. Add antibiotics if appropriate. Check cell growth under a microscope periodically; some mutants (motility mutants) may accumulate at the bottom of the tube.

Comments on BSK-II and P-BSK Media. Both BSA and rabbit serum exhibit tremendous differences in the ability of different lots (preparations) to support *B. burgdorferi* growth. Therefore, it is necessary to screen several lots of albumin and serum for *B. burgdorferi* growth in both broth (BSK-II) and plates (P-BSK) before purchasing large amounts of either medium component.

Alternative of Plating B. burgdorferi: *96-Well Plate*

After electroporation and overnight incubation (see later), add 100–200 ml of fresh BSK-II media (add antibiotic if appropriate) to the electroporated cells. Mix and then add 0.1 ml of the mixture containing electroporated cells into each well of 96-well plates and then incubate at 33° as described previously.

Comments on Using 96-Well Plates. The 96-well plate is a convenient approach to obtain clonal isolates; however, it also has limitations. For example, plating *B. burgdorferi* using the P-BSK approach usually permits isolation of single colonies and, consequently, a homogeneous population of cells. However, when cells are distributed in a small volume in 96-well plates, it is possible that a mixture of cells will be deposited in a single well. In our experience, we found that when cells were electroporated to inactivate the *flaB* gene (Motaleb *et al.*, 2000) and plated into 96-well plates, many individual wells contained both mutant cells (nonmotile and rod shaped) and cells that had not undergone allelic exchange mutagenesis. Because *flaB* mutant cells were rod shaped and nonmotile, the mutant and wild-type cells were clearly differentiated by dark-field microscopy.

Preparing DNA for Targeted Gene Inactivation

The genome sequence of *B. burgdorferi* is published (Fraser *et al.*, 1997) and available at http://cmr.tigr.org. While constructing a vector to inactivate the gene of interest in *B. burgdorferi*, a few points that have been discussed briefly (Motaleb *et al.*, 2000) are important to consider: (1) Delete a segment of DNA from the gene to be inactivated and insert an antibiotic resistance cassette (kanamycin, gentamycin, erythromycin, or streptomycin/spectinomycin cassette). Deleting part of the targeted gene has advantages in that addition of the respective antibiotic to the growth medium is not required to keep the mutation intact. This is also important if the mutants are planned to be used for animal studies. However, deletion within very small genes is problematic. (2) Allow at least 1 kb of flanking DNA on each side of the antibiotic-resistant cassette to be inserted within the targeted gene. This will maximize the chance of recombination of the electroporated DNA (targeted DNA plus antibiotic resistance cassette) with the homologous DNA in the *B. burgdorferi* genome. If, for any reason, this is problematic, 400 bp of flanking DNA on each side of the antibiotic resistance cassette will work, but less efficiently. (3) When inactivating a gene that is within an operon, consider inserting the cassette in the same direction of transcription as the targeted gene to minimize the chance of polar effects on downstream gene expression.

Targeted Mutagenesis: Construction of DNA Vectors

a. Design polymerase chain reaction (PCR) primers to amplify a segment of DNA containing the targeted gene (the primers can be outside of the gene to be inactivated) and then clone the PCR DNA into any cloning vector (e.g., pGEM-T vectors from Promega or Topo cloning vectors from Invitrogen) (Bono *et al.*, 2000; Motaleb *et al.*, 2000).

b. Identify unique restriction site(s) *within* the targeted gene. In the case of small genes, two restriction sites may not be available within the targeted gene.

c. PCR amplify an antibiotic-resistant cassette with PCR primers engineered to contain a specific restriction site(s) that is identified within the target gene. Clone the PCR DNA into a cloning vector.

d. Digest both the vectors containing the target gene and the antibiotic-resistant cassette with the same restriction enzyme(s) and ligate the cassette within the targeted gene. Recombinant DNAs should be restriction digested and sequenced to confirm proper insertion of the cassette within the desired restriction site(s).

e. PCR amplify the targeted gene plus the antibiotic cassette with the primers initially used to amplify the targeted gene plus flanking DNA. Alternatively, restriction digest the vector and then isolate the *B. burgdorferi* DNA with the inserted antibiotic resistance cassette. Purify DNA to be electroporated *without any traces of salt*. Dissolve DNA in small volume of sterile water. Electroporate enough DNA into competent *B. burgdorferi*. For high-passage B31A and low-passage B31-A3 strains of *B. burgdorferi*, 4–5 and 20–30 μg of DNA, respectively, should achieve targeted mutagenesis if all other conditions are met.

Preparation of Competent B. burgdorferi *and Electroporation*

The following procedures are based on those described by Samuels *et al.* (1994). Sterile materials needed: pipettes (10 and 25 ml), micropipette tips, buffers to make competent cells, and Eppendorf tubes to store competent cells. A centrifuge with a precooled (4°) rotor should be available.

Reagents and Buffers

Wash buffer: Dulbecco's phosphate-buffered saline (PBS without calcium)
Electroporation solution (EPS): 9.3 g sucrose and 15 ml of glycerol in a total volume of 100 ml and filter sterilize. EPS should be made fresh. Keep both EPS and PBS on ice.

Preparation of Competent Cells

1. Grow 100–200 ml of *B. burgdorferi* in a humidified incubator as described earlier. Cells should be very active and in log growth phase (2–5×10^7 cells/ml)—do not use cells approaching stationary phase.

2. Transfer cells into sterile centrifuge tubes and close lids. Place the tubes on ice for 15–20 min. Collect cells by centrifuging at 5000g for 20 min at 4° in a Sorvall centrifuge (SS34).

3. Carefully decant the medium inside a hood. Add 30 ml of cold PBS into the tubes and incubate on ice for 5 min to allow the pellet to loosen. Resuspend the cell pellet by pipetting up and down with precooled 10- or 25-ml pipettes very gently, as *B. burgdorferi* is very fragile.

4. Centrifuge at 4000*g* for 10 min at 4°. Wash again with PBS as before.

5. Add 20 ml of cold EPS to the pellet, resuspend and centrifuge cells as before. Repeat this step twice.

6. Resuspend the final pellet with sufficient volume of EPS to make 10^9 cells/ml. Transfer 70–80 μl of resuspended competent cells into a precooled Eppendorf tube. Keep tubes on ice until they are ready for electroporation or freeze cells in liquid N_2 for storage. The transformation efficiency will decline as competent cells are stored at $-80°$, but should be suitable for at least a month. We recommend using freshly prepared low-passage (B31-A3) *B. burgdorferi* rather than frozen competent cells.

Electroporation

Keep the electroporation cuvette (0.2-cm gap, Bio-Rad) and DNA to be electroporated on ice. Add 4 ml of BSK-II medium in a sterile, plastic culture tube and warm the medium to room temperature.

1. Add up to 15 μl of targeted gene cassette containing DNA (4–5 μg for high-passage or 20–30 μg for low-passage *B. burgdorferi*) into a tube containing competent cells. Mix by flicking the tube gently. Incubate on ice for 10–15 min and then transfer the entire content into a cold electroporation cuvette. Bring the cell–DNA mixture into the bottom of the cuvette by shaking or gently dropping on a solid surface. Wipe off any fluids from outside of the cuvette.

2. Place the cuvette in the gene pulser (Gene pulser II, Bio-Rad) and allow a single exponential decay pulse of 2.5 KV, capacitance of 25 μF, and resistance of 200 Ω. This should produce a time constant of 4–5 ms. The pulser will spark if there is any trace of salt present in the DNA.

3. Place the cuvette inside a laminar hood and add 1 ml of room temperature BSK-II, mix very gently, and transfer the entire mixture into the tube containing 4 ml of BSK-II. Do not add any antibiotic at this step. Incubate the tube at 33° in the incubator overnight.

4. After incubating for ~20 h, plate cells out as described earlier (dish or 96 well). Add antibiotic (for *kan*, we use 200 μg/ml for B31-A3 and 200–400 μg/ml for B31-A) during plating and incubate plates until colonies appear on the plate. If using the 96-well procedure, after diluting cells with 100–200 of fresh BSK-II, pipette 0.1 ml diluted cells into the plates with antibiotics, if appropriate; change of medium color from red to yellowish

is an indication that cells are growing, which is confirmed by microscopic examination of cells in the wells.

5. Pick single colonies from plates (single well in case of 96-well plate), grow in BSK-II, with antibiotic if appropriate, and analyze the clones for insertion of the antibiotic resistance cassette within the desired site by PCR and Southern blotting. Western blotting with wild-type and mutant cells should be done if the respective antibody is available.

Characterization of Chemotaxis Mutants

B. burgdorferi cells are chemotactic and contain multiple copies of chemotaxis genes (e.g., 2-*che*As, 3-*che*Ys, 5-*mcp*s). Swarm plate and capillary tube chemotaxis assays were utilized to assess the effects of specific *B. burgdorferi* mutations on chemotaxis genes (Motaleb *et al.*, 2000, 2005).

Swarm Plate Chemotaxis Assay

1. Media for swarm plate assay. Mix 34.2 ml plating BSK (P-BSK) with 1.7 ml of rabbit serum. Transfer 8 ml of the P-BSK-rabbit serum mixture in a fresh bottle and then add 3.13 ml fresh 5% Na-bicarbonate, 5 ml 10 × CMRL-1066, and 48.0 ml Dulbecco's PBS into the mixture. Warm this solution to 50° and then add 16.0 ml of softened 1.7% agarose adjusted to 50°. After mixing, transfer 36.0 ml of this swarm plate mixture into a plate. Allow the media in plates to solidify inside a hood. Scale up accordingly.

2. Grow *B. burgdorferi* to the late logarithmic growth phase. Collect cells by slow centrifugation, decant medium carefully, and gently resuspend the cell pellet such that 5 μl contains approximately 1×10^6 cells.

3. Inoculate 5 μl cells just under the agarose surface using a micropipette; the pipette tip should be carefully inserted such that the tip does not touch the bottom surface of the plate. Keep the plate lids open inside a laminar flow hood for 5–10 min. Carefully transfer the plates inside an incubator and incubate for 3–4 days. Document the swarm plate assay and measure swarm diameters.

Modified Capillary Tube Chemotaxis Assay

Counting solution: 0.01 M HEPES, 0.15 M NaCl, pH 7.4, 10 nM fresh Syto61 (Molecular Probes Inc., Eugene, OR)

Motility buffer: 136.9 mM NaCl, 8.10 mM Na$_2$HPO$_4$, 2.7 mM KCl, 1.47 mM KH$_2$PO$_4$, 2% recrystallized BSA (Sigma-Aldrich Co.), 0.1 mM EDTA, pH 7.4

2% methylcellulose: methylcellulose (Fisher), 400 mesh, is added to water to create a 2% (w/v) solution in a bottle with a stir bar. This solution is autoclaved and then stirred for several days until the solution is homogeneous.

Flow Cytometry Enumeration of B. burgdorferi *(Bakker* et al., *2007; Motaleb* et al., *2005)*

All solutions, including cell culture medium, must be 0.1 μm filtered to remove debris that would otherwise interfere with flow cytometric analysis—the only exception is methylcellulose, which cannot be 0.1 μm filtered. Dilute *B. burgdorferi* into counting solution with or without 6-μm-diameter polystyrene beads ($3–5 \times 10^3$/ml final concentration, Duke Scientific Co., Palo Alto, CA). Syto61 is a membrane-permeable, nucleic acid-binding, fluorescent dye excitable at 635 nm with an emission at 647 nm. We have enumerated avirulent *B. burgdorferi* cells using a Becton-Dickinson FACS Calibur with 15-mW air-cooled argon and red diode lasers operating at 488 and 635 nm, respectively, and a flow rate of 12 or 16 μl/min for 60 or 120 s to ensure at least 100 events in the gate where cells are observed. Compensation is unnecessary, as there was no spectral overlap between detectors. We use CellQuestPro (Becton-Dickinson, San Jose, CA) for all data acquisition (Bakker *et al.*, 2007).

Capillary Tube Chemotaxis Assay *(Bakker* et al., *2007)*

Prepare chemotaxis chambers before collecting cells for flow cytometry. Eppendorf tubes (2 ml) with a Parafilm sheet closed under a perforated cap or two 96-well plates "sandwiched" face to face with tape and with holes on one side for inserting capillary tubes can be used as chemotaxis chambers (Fig. 1; Bakker *et al.*, 2007). The capillary tube assay developed by Pfeffer and described by Adler (1974) was optimized to measure *B. burgdorferi* chemotaxis. Grow *B. burgdorferi* to late logarithmic phase ($\sim1 \times 10^8$ cells/ml) and collect cells by centrifuging at 1800g for 15 min at room temperature. Gently resuspend the cell pellet in motility buffer. Count cells using flow cytometry as described earlier. These cells are then mixed gently with 2% methylcellulose (400 mesh, Sigma-Aldrich Co.) in 0.1 M HEPES, 0.15 M NaCl, 0.01 mM EDTA and with motility buffer to obtain a final solution of 1×10^7 cells/ml, 1% (final) methylcellulose. Add the cell suspension into each of the Eppendorf tube chemotaxis chambers (0.3 ml) with a perforated lid or into wells of 96-well plates (0.2 ml/well). For Eppendorf tubes, place a paraffin sheet under the perforated cap and then close the cap; for 96-well plates, invert and tape another 96-well plate with holes in the bottoms of wells to the plate containing cells (Fig. 1).

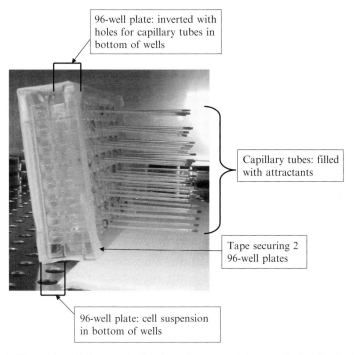

96-well plate: inverted with
holes for capillary tubes in
bottom of wells

Capillary tubes: filled
with attractants

Tape securing 2
96-well plates

96-well plate: cell suspension
in bottom of wells

FIG. 1. Ninety six-well chemotaxis chambers. One type of chemotaxis chamber is shown in which the cell suspension is placed in the wells of one 96-well plate; holes are created in the corresponding wells of another plate, and the plates are taped together with the bottom of both sets of wells facing outward. Capillary tubes containing attractants are placed through the holes into the cell suspension. The chamber is then placed in an incubator, positioned so that capillary tubes are approximately horizontal. (See color insert.)

Prepare concentrated test attractant solutions (glucosamine or *N*-acetyl-glucosamine, Sigma-Aldrich Co.) in motility buffer and 0.1-μm filter. Serially dilute attractant solutions with motility buffer and then mix with an equal volume of 2% methylcellulose before filling 75-μl capillary tubes. Seal one end of each capillary tube with silicone grease, wipe off any excess fluids outside the tubes, and then insert the open end of the tubes into cells in Eppendorf tubes or 96-well plates (Fig. 1). We found that four to five replicates are sufficient for generating reliable results; controls containing motility buffer, 1% methylcellulose with no potential attractant are required to provide background or random cell entry values; positive control attractants must be included. Incubate tubes or plates containing capillary tubes at 33° for 2 h. Remove capillary tubes, carefully wipe fluids outside the capillary tubes with a paper towel, and expel the contents into

Eppendorf tubes by centrifugation at 1000g for 3–4 s. Mix 0.01 ml of the expelled capillary tube solutions with 0.5 ml of counting solution and then enumerate with flow cytometry as described previously. Flow cytometry analysis can be automated, facilitating rapid analysis of many samples with minimal investigator "down time." Figure 2 shows a representative result of flow cytometry analysis, with few cells entering the tube containing no attractant (Fig. 2A) and many more cells entering the tube containing 0.1 M N-acetylglucosamine (Fig. 2B), which elicits a positive reaction and can be used as a positive control.

Calculate the mean and standard deviations for replicate capillary tubes and apply statistical analysis to determine if known or potential attractants significantly increase the relative chemotactic response, which is the ratio of mean number of cells in capillary tubes containing attractant to the mean number of cells in negative control tubes. Chemicals that meet the following criteria are considered attractants: (1) ≥twofold more cells enter capillary tubes containing test compounds, relative to control capillary tubes containing buffer, (2) the response is concentration dependent (Fig. 3), (3) the test compound does not elicit a positive response in chemotaxis mutant cells, and (4) the test compound does not elicit a response in the absence of a concentration gradient, that is, when the cell suspension

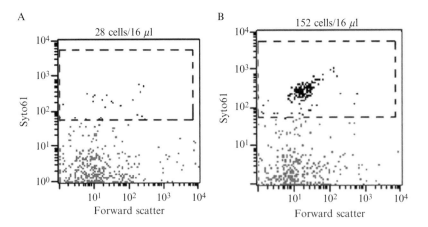

Fig. 2. Flow cytometry analysis. Following incubation in a chemotaxis chamber with *B. burgdorferi*, capillary tubes containing no attractant (A) or 0.1 *M* N-acetylglucosamine (B) were removed and the contents prepared for flow cytometry as described. The *X* axis is forward scattering, a measure of particle complexity, and the *Y* axis is Syto61 fluorescence intensity. The dashed line indicates regions "gated" or regions in which events (*B. burgdorferi* cells) were counted during a 16-μl sampling period. Approximately five times more cells entered the capillary tube containing N-acetylglucosamine than the tube containing no attractant (negative control). (See color insert.)

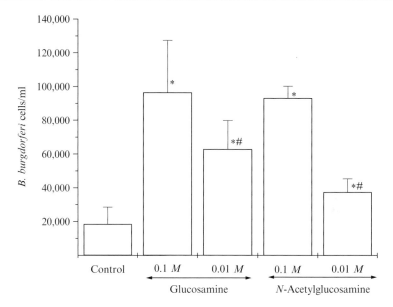

FIG. 3. Chemotaxis response is dependent on the concentration of attractants. The modified chemotaxis assay was performed with *B. burgdorferi* exposed to capillary tubes containing no attractants (control) or different concentrations of glucosamine or *N*-acetylglucosamine. Fewer cells enter capillary tubes containing *N*-acetylglucosamine or glucosamine as the concentration of these chemicals is decreased. An asterisk indicates significantly greater ($p < 0.05$) than control, and a number sign indicates that 0.01 *M* concentrations of attractants are significantly lower ($p > 0.05$) than 0.1 *M* concentrations of attractants.

contains the same concentration of test compound that is in the capillary tubes (Kelly-Wintenberg and Montie, 1994; Kennedy and Yancey, 1996; Lukat *et al.*, 1991; Motaleb *et al.*, 2005; Silversmith *et al.*, 2001).

Comments. We have identified chitosan dimers (0.01 *M*), *N*-acetylglucosamine (0.1 *M*), and glucosamine (0.1 *M*) as attractants that can be used as positive controls in the modified capillary tube assay. The pH of attractant solutions should be checked and readjusted to 7.6 if necessary. Care must be taken in centrifuging and resuspending *B. burgdorferi*; centrifuging too hard and/or resuspending too vigorously will result in cell clumping, damaged cells, and/or poor motility. General cell "health" and motility should be verified microscopically at the beginning and end of the assay—cells should be moving well and appear healthy when the assay is finished.

The relative chemotactic response exhibited by *B. burgdorferi* for attractants identified to date is ~2–5, which is in the lower range of that reported for other bacteria, for example, 10–72 for *E. coli* (Adler, 1974; Mesibov *et al.*, 1973). However, the modified capillary tube assay described here uses larger capillary tubes (70 μl) than those used in the classical capillary tube assay (1 μl). The larger capillary tubes produce a shallower gradient of attractants and may contribute to an apparently lower chemotactic response (Adler, 1974). For the compounds identified as attractants to date, relatively high concentrations (0.1 *M*) often elicited maximum chemotaxis. The utilization of relatively large capillary tubes, producing shallower concentration gradients, may also require high concentrations of attractants to elicit a positive chemotactic response. Note that 0.1 *M* concentrations of attractants elicited the maximal chemotactic response for *E. coli* and *Salmonella enterica* serovar Typhimurium (Hedblom and Adler, 1983; Melton *et al.*, 1978; Mesibov and Adler, 1972) to some chemoattractants.

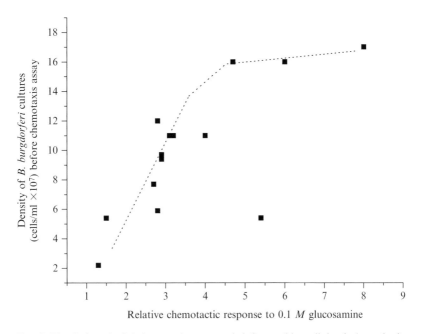

Fig. 4. The *B. burgdorferi* chemotaxis response is influenced by cell density/growth phase. Compiling the response of *B. burgdorferi* to 0.1 *M* N-acetylglucosamine over many different experiments indicated that the chemotactic response increased as the density of cell cultures increased and was generally highest as cells approached stationary phase ($\sim$2 × 10^8 cells/ml). The relative chemotactic response is the ratio of cells that migrate into attractant-filled vs buffer-filled capillary tubes and is used to normalize day-to-day variability (Kelly-Wintenberg and Montie, 1994; Kennedy and Yancey, 1996; Mazumder *et al.*, 1999; Moulton and Montie, 1979).

In our hands the relative chemotactic response of avirulent *B. burgdorferi* B31-A is somewhat dependent on cell density or cell growth phase (Fig. 4). Therefore, it is important to assess the response to potential attractants at different cell densities or phases of growth.

Cell Tracking

Grow and pass cells daily for several days to ensure that the cells are very active and in log-phase growth. Gently mix growing cells and 2% methylcellulose (1:1, v/v) to obtain a homogeneous suspension—*B. burgdorferi* swim best in somewhat viscous, gel-like media (Goldstein *et al.*, 1994). Alternatively, collect cells by slow centrifugation (1800*g* for 15 min at room temperature) and gently resuspend the cells in motility buffer. Mix the cell suspension with 2% methylcellulose (1:1, v/v) and resuspend the cells as described previously. Pipette 50 μl of the mixture on a microscope slide. Lightly apply vacuum grease to all four sides of a cover glass (Fisher Scientific or any brand) and then place on top of the mixture on the slide. Dilute cells if necessary to ensure few cells (two to five good swimming cells) per microscope filled. Capture cells using a dark-field microscope equipped with a 35° heated stage and a charged-couple device (CCD) camera. Track cells using a bacterial cell tracker and determine the velocity of the swimming cells.

Our equipment and settings are the followings: Zeiss Axioskop 2 under dark-field illumination (Carl Zeiss Inc., Oberkochen, Germany) equipped with a 35° heated stage (Physitemp Inc., Clifton, NJ) coupled with a CCD camera and a software package "Volocity"* (Improvision Inc., Coventry, UK). Using Volocity, video sequences of swimming cells are captured with iMovie on a Apple PowerMac G4 using a Scion LG-5 (Scion Inc., Fredrick, MD) frame grabber card and a Dage MTI (Dage-MTI Inc., Michigan City, IN) black-and-white video camera. Videos are exported as QuickTime movies and imported into OpenLab (Improvision Inc.) where the frames are cropped, calibrated using a stage micrometer, and saved as LIFF files. Volocity is then used to track the LIFF files. Determine the swimming velocity from the software.

Acknowledgments

This work was supported by U.S. Public Health Service Grant AI29743 to N. W. Charon, and NIH grant RR16440 to the West Virginia University Flow Cytometric Core Facility. We thank C. Cunningham for assistance with flow cytometry.

---

* Note, the new volocity software eliminates the need for importation into OpenLab.

References

Adler, J. (1974). A method for measuring chemotaxis and use of the method to determine optimum conditions for chemotaxis by *Escherichia coli*. *J. Gen. Microbiol.* **74**, 77–91.

Bakker, R. G., Li, C., Miller, M. R., Cunningham, C., and Charon, N. W. (2007). Identification of specific chemoattractants and genetic complementation of a *Borrelia burgdorferi* chemotaxis mutant: A flow cytometry-based capillary tube chemotaxis assay. *Appl. and Environ. Microbiol.* **73**, 1180–1188.

Barbour, A. G. (1984). Isolation and cultivation of Lyme disease spirochetes. *Yale J. Biol. Med.* **57**, 521–525.

Bono, J. L., Elias, A. F., Kupko, J. J., III, Stevenson, B., Tilly, K., and Rosa, P. (2000). Efficient targeted mutagenesis in *Borrelia burgdorferi*. *J. Bacteriol.* **182**, 2445–2452.

Butler, S. M., and Camilli, A. (2004). Both chemotaxis and net motility greatly influence the infectivity of *Vibrio cholerae*. *Proc. Natl. Acad. Sci. USA* **101**, 5018–5023.

Charon, N. W., and Goldstein, S. F. (2002). Genetics of motility and chemotaxis of a fascinating group of bacteria: The spirochetes. *Annu. Rev. Genet.* **36**, 47–73.

Elias, A. F., Bono, J. L., Kupko, J. J., III, Stewart, P. E., Krum, J. G., and Rosa, P. A. (2003). New antibiotic resistance cassettes suitable for genetic studies in *Borrelia burgdorferi*. *J. Mol. Microbiol. Biotechnol.* **6**, 29–40.

Elias, A. F., Stewart, P. E., Grimm, D., Caimano, M. J., Eggers, C. H., Tilly, K., Bono, J. L., Akins, D. R., Radolf, J. D., Schwan, T. G., and Rosa, P. (2002). Clonal polymorphism of *Borrelia burgdorferi* strain B31 MI: Implications for mutagenesis in an infectious strain background. *Infect. Immun.* **70**, 2139–2150.

Frank, K. L., Bundle, S. F., Kresge, M. E., Eggers, C. H., and Samuels, D. S. (2003). aadA confers streptomycin resistance in *Borrelia burgdorferi*. *J. Bacteriol.* **185**, 6723–6727.

Fraser, C. M., Casjens, S., Huang, W. M., Sutton, G. G., Clayton, R., Lathigra, R., White, O., Ketchum, K. A., Dodson, R., Hickey, E. K., Gwinn, M., Dougherty, B., *et al.* (1997). Genomic sequence of a Lyme disease spirochaete, *Borrelia burgdorferi*. *Nature* **390**, 580–586.

Goldstein, S. F., Charon, N. W., and Kreiling, J. A. (1994). *Borrelia burgdorferi* swims with a planar waveform similar to that of eukaryotic flagella. *Proc. Natl. Acad. Sci. USA* **91**, 3433–3437.

Hedblom, M. L., and Adler, J. (1983). Chemotactic response of *Escherichia coli* to chemically synthesized amino acids. *J. Bacteriol.* **155**, 1463–1466.

Kawabata, H., Norris, S. J., and Watanabe, H. (2004). BBE02 disruption mutants of *Borrelia burgdorferi* B31 have a highly transformable, infectious phenotype. *Infect. Immun.* **72**, 7147–7154.

Kelly-Wintenberg, K., and Montie, T. C. (1994). Chemotaxis to oligopeptides by *Pseudomonas aeruginosa*. *Appl. Environ. Microbiol.* **60**, 363–367.

Kennedy, M. J., and Yancey, R. J., Jr. (1996). Motility and chemotaxis in *Serpulina hyodysenteriae*. *Vet. Microbiol.* **49**, 21–30.

Li, C., Bakker, R. G., Motaleb, M. A., Sartakova, M. L., Cabello, F. C., and Charon, N. W. (2002). Asymmetrical flagellar rotation in *Borrelia burgdorferi* nonchemotactic mutants. *Proc. Natl. Acad. Sci. USA* **99**, 6169–6174.

Lukat, G. S., Lee, B. H., Mottonen, J. M., Stock, A. M., and Stock, J. B. (1991). Roles of the highly conserved aspartate and lysine residues in the response regulator of bacterial chemotaxis. *J. Biol. Chem.* **266**, 8348–8354.

Mazumder, R., Phelps, T. J., Krieg, N. R., and Benoit, R. E. (1999). Determining chemotactic responses by two subsurface microaerophiles using a simplified capillary assay method. *J. Microbiol. Methods* **37**, 255–263.

Melton, T., Hartman, P. E., Stratis, J. P., Lee, T. L., and Davis, A. T. (1978). Chemotaxis of *Salmonella typhimurium* to amino acids and some sugars. *J. Bacteriol.* **133**, 708–716.

Mesibov, R., and Adler, J. (1972). Chemotaxis toward amino acids in *Escherichia coli*. *J. Bacteriol.* **112**, 315–326.

Mesibov, R., Ordal, G. W., and Adler, J. (1973). The range of attractant concentrations for bacterial chemotaxis and the threshold and size of response over this range: Weber law and related phenomena. *J. Gen. Physiol* **62**, 203–223.

Motaleb, M. A., Corum, L., Bono, J. L., Elias, A. F., Rosa, P., Samuels, D. S., and Charon, N. W. (2000). *Borrelia burgdorferi* periplasmic flagella have both skeletal and motility functions. *Proc. Natl. Acad. Sci. USA* **97**, 10899–10904.

Motaleb, M. A., Miller, M. R., Li, C., Bakker, R. G., Goldstein, S. F., Silversmith, R. E., Bourret, R. B., and Charon, N. W. (2005). CheX is a phosphorylated CheY phosphatase essential for *Borrelia burgdorferi* chemotaxis. *J. Bacteriol.* **187**, 7963–7969.

Moulton, R. C., and Montie, T. C. (1979). Chemotaxis by *Pseudomonas aeruginosa*. *J. Bacteriol.* **137**, 274–280.

Norris, S. J., Howell, J. K., Garza, S. A., Ferdows, M. S., and Barbour, A. G. (1995). High- and low-infectivity phenotypes of clonal populations of *in vitro*-cultured *Borrelia burgdorferi*. *Infect. Immun.* **63**, 2206–2212.

Rosa, P., Samuels, D. S., Hogan, D., Stevenson, B., Casjens, S., and Tilly, K. (1996). Directed insertion of a selectable marker into a circular plasmid of *Borrelia burgdorferi*. *J. Bacteriol.* **178**, 5946–5953.

Rosa, P. A., Tilly, K., and Stewart, P. E. (2005). The burgeoning molecular genetics of the Lyme disease spirochaete. *Nat. Rev. Microbiol.* **3**, 129–143.

Samuels, D. S., Mach, K. E., and Garon, C. F. (1994). Genetic transformation of the Lyme disease agent *Borrelia burgdorferi* with coumarin-resistant *gyrB*. *J. Bacteriol.* **176**, 6045–6049.

Sartakova, M., Dobrikova, E., and Cabello, F. C. (2000). Development of an extrachromosomal cloning vector system for use in *Borrelia burgdorferi*. *Proc. Natl. Acad. Sci. USA* **97**, 4850–4855.

Sartakova, M. L., Dobrikova, E. Y., Terekhova, D. A., Devis, R., Bugrysheva, J. V., Morozova, O. V., Godfrey, H. P., and Cabello, F. C. (2003). Novel antibiotic-resistance markers in pGK12-derived vectors for *Borrelia burgdorferi*. *Gene* **303**, 131–137.

Schwan, T. G., Burgdorfer, W., and Garon, C. F. (1988). Changes in infectivity and plasmid profile of the Lyme disease spirochete, *Borrelia burgdorferi*, as a result of *in vitro* cultivation. *Infect. Immun.* **56**, 1831–1836.

Shi, Y., Xu, Q., Seemanapalli, S. V., McShan, K., and Liang, F. T. (2006). The dbpBA locus of *Borrelia burgdorferi* is not essential for infection of mice. *Infect. Immun.* **74**, 6509–6512.

Silversmith, R. E., Smith, J. G., Guanga, G. P., Les, J. T., and Bourret, R. B. (2001). Alteration of a nonconserved active site residue in the chemotaxis response regulator CheY affects phosphorylation and interaction with CheZ. *J. Biol. Chem.* **276**, 18478–18484.

[22] Phosphorylation Assays of Chemotaxis Two-Component System Proteins in *Borrelia burgdorferi*

By Md. A. Motaleb, Michael R. Miller, Chunhao Li, and Nyles W. Charon

Abstract

Borrelia burgdorferi has a complex chemotaxis signal transduction system with multiple chemotaxis gene homologs similar to those found in *Escherichia coli* and *Bacillus subtilis*. The *B. burgdorferi* genome sequence encodes two *cheA*, three *cheY*, three *cheW*, two *cheB*, two *cheR*, but no *cheZ* genes. Instead of *cheZ*, *B. burgdorferi* contains a different CheY-P phosphatase, referred to as *cheX*. The multiple *B. burgdorferi* histidine kinases (CheA1 and CheA2) and response regulators (CheY1, CheY2, and CheY3) possess all the domains and functional residues found in *E. coli* CheA and CheY, respectively. Understanding protein phosphorylation is critical to unraveling many biological processes, including chemotaxis signal transduction, motility, growth control, metabolism, and disease processes. *E. coli*, *Salmonella enterica* serovar Typhimurium, and *B. subtilis* chemotaxis systems have been studied extensively, providing models to understand chemotaxis signaling in the Lyme disease spirochete *B. burgdorferi*. Both genetic approaches and biochemical analyses are essential in understanding its complex two-component chemotaxis systems. Specifically, gene inactivation studies assess the importance of specific genes in chemotaxis and motility under certain conditions. Furthermore, biochemical approaches help determine the following *in vitro* reactions: (1) the extent that the histidine kinases, CheA1 and CheA2, are autophosphorylated using ATP; (2) the transfer of phosphate from CheA1-P and CheA2-P to each CheY species; and (3) the dephosphorylation of each CheY-P species by CheX. We hypothesize that characterizing protein phosphorylation in the *B. burgdorferi* two-component chemotaxis system will facilitate understanding of how the periplasmic flagellar bundles located near each end of *B. burgdorferi* cells are coordinately regulated for chemotaxis. During chemotaxis, these bacteria run, pause (stop/flex), and reverse (run again). This chapter describes protocols for assessing *B. burgdorferi* CheA autophosphorylation, transfer of phosphate from CheA-P to CheY, and CheY-P dephosphorylation.

METHODS IN ENZYMOLOGY, VOL. 422   0076-6879/07 $35.00
DOI: 10.1016/S0076-6879(06)22022-6

Introduction

Bacterial chemotaxis systems govern how bacterial cells swim after sensing environmental signals. For example, *Escherichia coli, Salmonella enterica* serovar Typhimurium, and *Bacillus subtilis* alternate their swimming behavior in response to chemical stimuli in the environment, such that they move toward chemical attractants and away from chemical repellents (Sourjik, 2004; Szurmant and Ordal, 2004; Wadhams and Armitage, 2004). In *Borrelia burgdorferi*, chemotaxis is hypothesized to play a major role in the spirochete life cycle, whereby it facilitates migration of the spirochetes from tick guts into the mammalian hosts during tick feeding. Chemotaxis is also believed essential for facilitating the migration of spirochetes from infected mammals into uninfected ticks, which then act as new vectors for disease spread. In *E. coli* and *S. enterica*, the movement of cells is determined by the direction flagella rotate: Rotation of flagella counter-clockwise (CCW) produces smooth swimming, or a "run," whereas when one or more flagella rotate clockwise (CW), cells "tumble," which allows them to change direction (Parkinson *et al.*, 2005; Szurmant and Ordal, 2004). Swimming toward attractants occurs via a "biased random walk," in which runs are extended when cells move toward increasing concentrations of an attractant (tumbling is suppressed during runs). Many proteins are involved in bacterial chemotactic signal transduction pathways, including, but not limited to, the following: (1) membrane-bound receptor proteins, or methyl-accepting chemotaxis proteins (MCPs) that bind to chemical attractants or repellants. (2) CheA histidine kinases that "sense" the binding of the ligand to MCPs. CheA is autophosphorylated via ATP, and upon binding ligand, MCPs either decrease (i.e., *E. coli*) or increase (i.e., *B. subtilis*) CheA autophosphorylation, depending on the bacterial species. (3) The "response regulator" CheY accepts phosphate from phosphorylated CheA. CheY-P binds to the flagellar motor protein FliM, causing one or more flagella to rotate CW, which results in a tumble. (4) CheZ dephosphorylates CheY-P efficiently. The concentration of CheY-P in a cell determines whether the flagella rotate CCW or CW; hence, whether the cell runs (low CheY-P) or tumbles (high CheY-P), respectively.

Regulation of CheY-P

The concentration of CheY-P is regulated primarily by the rate of phosphate transfer from CheA-P to CheY and by the rate of CheY-P dephosphorylation. In *E. coli* and *S. enterica*, binding of an attractant to an MCP reduces CheA autophosphorylation, which in turn reduces the level of CheY-P. This results in an increase in CCW rotation of flagella and extended runs (Macnab and Ornston, 1977; Turner *et al.*, 2000). Inactivating *cheA* or *cheY* results in a constantly running phenotype and the loss of chemotaxis (Alon *et al.*, 1998; Boesch *et al.*, 2000; Parkinson, 1978). However, binding of attractants to *B. subtilis* receptors leads to the activation of CheA and therefore increased levels of CheY-P. In *B. subtilis,* interaction of CheY-P with FliM increases the probability of CCW rotation for smooth swimming, and these cells tumble when their flagella rotate CW (Szurmant and Ordal, 2004).

CheY-P can be dephosphorylated in multiple ways (Szurmant and Ordal, 2004). CheY-P can autodephosphorylate, with half-lives of CheY-P ranging from ∼1 to 2 s for $CheY_6$ in *Rhodobacter sphaeroids* (Porter and Armitage, 2002) to ∼1.5 to 2 min in *B. subtilis* (Szurmant *et al.*, 2004). However, additional, more efficient means of removing phosphate from CheY-P are required to achieve the observed rapid response to chemotactic stimuli. The most extensively studied means of removing phosphate from CheY-P is via CheZ, a CheY-P phosphatase found in *E. coli* and other γ- and β-proteobacteria (Szurmant and Ordal, 2004). The inactivation of *cheZ* results in nonchemotactic cells that tumble constantly. Other organisms, including Archaea, *Thermotoga* sp., spirochetes, gram-positive bacteria, and some proteobacteria, for example, *Vibrio cholerae* and *Myxococcus,* utilize recently discovered phosphatases called CheC-CheD/ CheX/FliY (Motaleb *et al.*, 2005; Park *et al.*, 2004; Szurmant and Ordal, 2004; Yang and Li, 2005). *V. cholerae* is the only known organism that possesses both CheZ and CheC-like phosphatases (Heidelberg *et al.*, 2000; Szurmant and Ordal, 2004). However, in some species containing multiple CheY proteins, for example, *Sinorhizobium meliloti, Rhodobacter sphaeroides*, and other α-proteobacteria, at least one CheY protein appears to act as a "phosphate sink," sequestering phosphate from another CheY protein that regulates the direction of flagellar motor rotation (Wadhams *et al.*, 2004).

While the phosphorylation pathway in the two-component chemotaxis system has been studied extensively in *E. coli* and some other bacteria, relatively little is known about the *B. burgdorferi* chemotaxis system and phosphorylation pathways (Li *et al.*, 2002; Szurmant and Ordal, 2004). Furthermore, it is not yet clear how these spirochetes coordinate their two bundles of periplasmic flagella, attached to opposite ends of the cell, to run, flex (tumble), and reverse in a manner that permits chemotactic movement (Charon and Goldstein, 2002). By inactivating specific *B. burgdorferi* genes, we have demonstrated essential roles for *cheA2, cheY3*, and *cheX* in chemotaxis: All three of these mutants are nonchemotactic, and while *cheA2* or *cheY3* mutants constantly run in one direction without flexing, *cheX* mutants constantly flex and are unable to translate (Li *et al.*, 2002; Motaleb *et al.*, 2005; M. Motaleb and N. Charon, in preparation).

The biochemical analysis of *B. burgdorferi* CheA1 and CheA2 auto-phosphorylation and the transfer of phosphate from these CheA proteins to the three CheY's are being studied in our laboratory. In addition, we are characterizing the autodephosphorylation of CheY-P, as well as CheX-mediated dephosphorylation. The following sections outline the biochemical procedures that will enable us to better understand the regulation of *B. burgdorferi* motility and chemotaxis (Motaleb *et al.*, 2005).

Overview

In order to study CheA autophosphorylation, transfer of phosphate to CheY, and dephosphorylation of CheY, it is essential to express and purify these proteins in their native (active) state. Basic molecular biology approaches are often successful, but for some proteins this can be one of the most difficult steps. Generally, CheA proteins are active as homo-dimers, whereby each subunit (auto)phosphorylates the other using ATP. CheA is incubated with $[\gamma^{-32}P]ATP$ and reaction conditions ($[MgCl_2]$, [KCl], pH, etc.) are optimized to affect maximum formation of CheA-^{32}P (Fig. 1A). If CheA autophosphorylation is the only reaction being studied, reactions can be stopped by the addition of stop buffer, and CheA-^{32}P is separated from unreacted $[\gamma^{-32}P]ATP$ by sodium dodecyl sulfate polyacryl-amide gel electrophoresis (SDS-PAGE). Gels are dried and exposed to X-ray film or a phosphorimager screen to visualize CheA-^{32}P, and relative intensities of CheA-^{32}P are determined using a phosphorimager screen

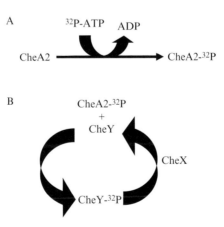

FIG. 1. Bacterial phosphorylation pathways. (A) CheA autophosphorylation with ATP. (B) Transfer of phosphate from phosphorylated CheA (CheA-P) to CheY, forming CheY-P, and CheX-mediated dephosphorylation of CheY-P.

analyzer. When studying transfer of phosphate from CheA-^{32}P to CheY, CheA-^{32}P and unreacted [γ-^{32}P]ATP must be separated in a manner that preserves the native form of CheA-^{32}P. CheA-^{32}P is then incubated with CheY to analyze transfer of [^{32}P] to CheY, as well as CheY-P auto-dephosphorylation (Fig. 1B). CheA proteins are much larger than CheY proteins, and these [^{32}P]-containing proteins are readily resolved by SDS-PAGE and analyzed as described earlier. The addition of CheX or other potential CheY-P phosphatases to the reaction mixture allows analysis of phosphatase-mediated CheY-P dephosphorylation (Fig. 1B).

Materials and Methods

Solutions

> Lysis buffer: 50 mM NaH$_2$PO$_4$·H$_2$O, 300 mM NaCl, and 10 mM imidazole, pH adjusted to 8.0 with fresh NaOH
>
> Wash buffer: The wash buffer is the same as lysis buffer, but with 20 mM imidazole
>
> Elution buffer: The elution buffer is the same as lysis buffer, but with 25 mM imidazole. All buffers are stored at 4°.
>
> Dialysis buffer: 25 mM NaCl and 50 mM Tris-HCl, 10% glycerol (v/v), pH 8.5. Autoclave and store at 4°.
>
> TKM reaction buffer: 50 mM Tris-HCl, 50 mM KCl, and 5 mM MgCl$_2$, pH 8.5
>
> Stop buffer: 50 mM Tris-HCl, 100 mM dithiothreitol, 2% (w/v) SDS, 0.1% (w/v) bromophenol blue, 10% (v/v) glycerol, and 5% (v/v) 2-mercaptoethanol, pH 6.8
>
> ^{32}P-ATP: 0.3 mM [γ-^{32}P]ATP (5000 Ci/mmol, 10 mCi/ml, from MP Biomedicals)

Purification of Soluble, Recombinant Chemotaxis Proteins
*   Overexpressed in* E. coli

The pQE31 expression vector (Qiagen Inc.) is used to clone *cheA1*, and pCR T7/NT-TOPO (Invitrogen Inc.) is used for *cheA2*. The pQE30 vector (Qiagen Inc.) is used for cloning *cheY1, cheY2, cheY3,* and *cheX* (Li *et al.*, 2002; Motaleb *et al.*, 2005; M. Motaleb and N. Charon, in preparation). Vectors containing *cheA1, cheY1, cheY2, cheY3,* and *cheX* are expressed in M15(pREP4) cell lines as described in the manufacturer's (Qiagen Inc.) protocol. The *cheA2*-containing vector is expressed in the BL21(DE3) codon plus host cell line (Stratagene Inc.). All proteins are expressed with an N-terminal 6× His tag. We use the protocol of Qiagen to purify all chemotaxis proteins. *E. coli* cells expressing any of the chemotaxis genes are grown in the presence of antibiotics (ampicillin plus kanamycin for M15 or ampicillin plus chloramphenicol for codon plus cell) at 37° overnight. The following morning, overnight cultures are diluted 1:20 into fresh LB broth with appropriate antibiotics until OD_{600} becomes ∼0.6. Isopropyl-β-D-thiogalactoside is then added to 1 mM to induce expression, and cultures are incubated for another 4 to 5 h. Under these conditions, no proteins form inclusion bodies, and recombinant protein expression levels are at a satisfactory level. Cells are collected by centrifugation at 6000g for 20 min, and the supernatant fluid is decanted. The cell pellets are stored at −70°. Frozen cell pellets are resuspended with cold (4°) lysis buffer and kept on ice (resuspend pellets derived from 100 ml of culture media into 5 ml of lysis buffer). Cells are lysed by sonication with a microtip probe or via a French pressure cell. Protease inhibitor cocktail solution (Sigma-Aldrich Inc.) may be added at this stage, according to the manufacturer's instructions. Lysed cells are centrifuged at 12,000g for 30 min at 4°, and supernatant fluids are collected. Ni-NTA agarose (Qiagen Inc.) is added to the supernatant fluids according to the recommendation by Qiagen, and tubes are rotated gently at 4° for 1 h. The agarose slurry is poured into a column and allowed to flow by gravity. Agarose columns are washed with wash buffer at least two to three times, and then His-tagged proteins are eluted in a new tube with 2–5 ml of cold elution buffer. Eluted proteins are dialyzed at 4° for several hours, while carefully monitoring for protein precipitation during dialysis. Dialyzed proteins are electrophoresed on SDS-PAGE and Coomassie blue stained to determine integrity and purity; the amount of purified protein is estimated by comparison to different amounts of a standard protein run in the same gel. In addition, the Bio-Rad protein assay kit (Bio-Rad Inc.) is also used to measure protein concentration, where bovine serum albumin is used as a standard.

Notes

Expression. While we described protein expression and purification for *B. burgdorferi* CheAs, CheYs, and CheX, specific measures must be taken to optimize expression and purification for a given protein. For example, a given protein may not express in a given host cell line, but changing the host cell and/or the expression vector may improve expression. Furthermore, the expression of some proteins may be limited due to the rarity of certain tRNAs that are abundant in the organism from where the heterologous proteins are derived. When inducing protein expression in a different host, the pool of rare tRNAs may be depleted and translation halted. Codon plus cell lines provide a more abundant supply of these rare tRNAs for efficient induced expression of heterologous proteins, which is absent in conventional cell lines. Sometimes changing the tag (e.g., His tag to GST, FLAG, or other tags) and location (e.g., from N-terminal to C-terminal) may be helpful in achieving expression of active proteins. Recombinant CheA1 was readily expressed and overproduced in M15 (pREP4). However, although we used codon plus for obtaining recombinant CheA2, its overproduction is erratic, and thus other unknown variables are evidently involved in its synthesis in *E. coli*.

Lysis. We found that the lysis buffer alone is not enough to lyse many bacterial cells. Lysis buffer plus lysozyme and sonication improves lysis, but we found a significant percentage of cells were still intact. In our hands a French pressure cell produced most efficient cell lysis and retained activity of the expressed protein.

Purification. Each protein is unique, and occasionally proteins will precipitate during dialysis. In this case dialysis buffers should be modified to minimize precipitation. For example, if a protein contains cysteine residues, consider adding a reducing agent, such as dithiothreitol or Tris (2-carboxyethyl)phosphine hydrochloride. Sometimes adding 5–10% glycerol in the dialysis buffer may minimize precipitation. Additional information regarding expression vectors, host cell lines, expression, and purification can be obtained from specific manufacturer's protocols.

CheA Autophosphorylation

CheA proteins are a histidine kinase(s) that autophosphorylates in the presence of ATP and $MgCl_2$. Different CheA proteins may have somewhat different properties or preferences, and it is important to empirically determine optimum autophosphorylation reaction conditions for each CheA, including pH, $[MgCl_2]$, and [KCl].

Borrelia burgdorferi CheA1 and CheA2 autophosphorylation reactions utilize ~200 pmol of each protein, 0.3 mM [γ-^{32}P]ATP (5000 Ci/mmol, 10 mCi/ml) in a 0.1-ml reaction volume and are optimized with respect to [KCl], $[MgCl_2]$, and pH. Reactions are initiated by the addition of [γ-^{32}P] ATP, incubated for 30 min at room temperature, and stopped by the addition of 25 μl stop buffer. Reactions are then subjected to SDS-PAGE, and unincorporated [γ-^{32}P]ATP is electrophoresed off the gel, into the lower buffer reservoir. Gels are dried and subjected to phosphorimage analysis or autoradiography with X-ray film. For both CheA1 and CheA2, optimum autophosphorylation occurs with TKM buffer. The extent of CheA1 and CheA2 autophosphorylation increases rapidly during the first 30 min and is essentially complete by 60 min, indicating that both CheA1 and CheA2 are capable of autophosphorylation using ATP.

Phosphotransfer from CheA-^{32}P to CheY(s)

To study transfer of [^{32}P] from CheA-^{32}P to CheY, CheA-^{32}P is separated from unincorporated [γ-^{32}P]ATP using Micro Bio-spin 6 chromatography columns (Bio-Rad Inc.). CheA-^{32}P can then be used immediately or stored at $-70°$ for a week for phosphotransfer to CheY. *In vivo*, the ratio of histidine kinase, response regulator, and CheY-P phosphatase is generally 1:6:1; therefore, in phosphotransfer studies, CheY is six- to sevenfold molar excess of CheA. To study phosphate transfer from CheA-^{32}P to CheY, 20 pmol CheA-^{32}P is added to 150 pmol CheY in TKM buffer, with or without 20 pmol of CheX (see later) in 0.1 ml total volume. In the absence of CheX, a time course (10 s to several minutes) provides information about the rate of phosphate transfer to each CheY, as well as the stability of CheY-P (i.e., CheY-P autodephosphorylation). At each time point, 10-μl aliquots of the reaction are transferred to Eppendorf tubes containing 2 μl of stop buffer, followed by rapid mixing. CheA is separated from CheY by SDS-PAGE, and dried gels are subjected to phosphorimage analysis. We found that *B. burgdorferi* CheA1-P and CheA2-P are able to transfer phosphate to all three response regulators, CheY1, CheY2, and CheY3. However, CheA2-P transfers phosphate more rapidly to CheY1 and CheY3 than to CheY2. For most species, CheA-P is rapidly transferred to CheY, and CheY-P autodephosphorylates relatively rapidly (approximately 14 s for *E. coli*). However, we found that the half-lives of *B. burgdorferi* CheY1-P and CheY3-P are longer ($>$15 min) than any species reported to date (Porter *et al.*, 2002; Silversmith *et al.*, 2003; Sourjik and Schmitt, 1998; Szurmant *et al.*, 2004) (Fig. 2).

Dephosphorylation of CheY-P with the Phosphatase CheX

To study phosphatase-mediated CheY-P dephosphorylation, phosphotransfer reactions described earlier are performed in the presence of CheX or other potential CheY-P phosphatases. Depending on the stability of CheY-P, phosphatases can be incubated with CheY before the addition of CheA-^{32}P (if CheY-P autodephosphorylates rapidly) or after the addition of CheA-^{32}P (if CheY-P autodephosphorylates slowly). After adding CheX to the reaction mixture of CheA2-^{32}P and CheY3, 10 μl of the mixture is transferred at various times to Eppendorf tubes containing 2 μl stop buffer (see Fig. 2, last two lanes). We find that *B. burgdorferi* CheX dephosphorylates CheY1-P, CheY2-P, and CheY3-P efficiently (Fig. 2; unpublished observations).

Notes

The CheA and CheY proteins are phosphorylated on His and Asp residues, respectively, which are much more labile than phosphorylated Ser, Thr, or Tyr residues. Therefore, it is important that the phosphorylation

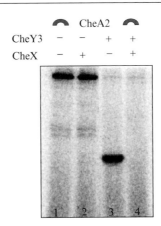

FIG. 2. *Borrelia burgdorferi* CheA2 was autophosphorylated with [γ-^{32}P]ATP as described in the text. After removing unreacted [γ-^{32}P]ATP, CheA-^{32}P was incubated without any additions (lane 1), with CheX alone (lane 2), with CheY3 alone for 5 min (lane 3), or with CheY3 for 5 min, followed by the addition of CheX for 10 s (lane 4).

reactions are not boiled prior to SDS-PAGE and that gels are not fixed with methanol/acetic acid (Hess *et al.*, 1991). *B. burgdorferi* CheY-Ps are small (~14 kDa); when performing SDS-PAGE, phosphotransfer and dephosphorylation reactions should be run in 15% SDS polyacrylamide gels, and a commercial set of visible (standard) proteins, at least one of which is ~14 kDa, should be electrophoresed in one gel lane to ensure that the small CheY-P does not migrate off the gel. Following electrophoresis, we dry gels with a heated, vacuum gel dryer, and dried gels are exposed to a phosphorimage screen (usually for 1 h). Exposed screens are analyzed with a Molecular Dynamics Storm 820 phosphoimager, and the intensity of phosphorylated proteins is quantitated using Image Quant v2003.02 software.

Acknowledgments

We thank R. B. Bourret and R. E. Silversmith for critical reading of this chapter and continuous support in our research. This work was supported by U.S. Public Health Service Grant AI29743 to Nyles W. Charon.

References

Alon, U., Camarena, L., Surette, M. G., Arcas, B. A. Y., Liu, Y., Leibler, S., and Stock, J. B. (1998). Response regulator output in bacterial chemotaxis. *EMBO J.* **17,** 4238–4248.
Boesch, K. C., Silversmith, R. E., and Bourret, R. B. (2000). Isolation and characterization of nonchemotactic CheZ mutants of *Escherichia coli. J. Bacteriol.* **182,** 3544–3552.

Charon, N. W., and Goldstein, S. F. (2002). Genetics of motility and chemotaxis of a fascinating group of bacteria: The spirochetes. *Annu. Rev. Genet.* **36,** 47–73.

Heidelberg, J. F., Eisen, J. A., Nelson, W. C., Clayton, R. A., Gwinn, M. L., Dodson, R. J., Haft, D. H., Hickey, E. K., Peterson, J. D., Umayam, L., Gill, S. R., Nelson, K. E., *et al.* (2000). DNA sequence of both chromosomes of the cholera pathogen *Vibrio cholerae. Nature* **406,** 477–483.

Hess, J. F., Bourret, R. B., and Simon, M. I. (1991). Phosphorylation assays for proteins of the two-component regulatory system controlling chemotaxis in *Escherichia coli. Methods Enzymol.* **200,** 188–204.

Li, C., Bakker, R. G., Motaleb, M. A., Sartakova, M. L., Cabello, F. C., and Charon, N. W. (2002). Asymmetrical flagellar rotation in *Borrelia burgdorferi* nonchemotactic mutants. *Proc. Natl. Acad. Sci. USA* **99,** 6169–6174.

Macnab, R. M., and Ornston, M. K. (1977). Normal-to-curly flagellar transitions and their role in bacterial tumbling: Stabilization of an alternative quaternary structure by mechanical force. *J. Mol. Biol.* **112,** 1–30.

Motaleb, M. A., Miller, M. R., Li, C., Bakker, R. G., Goldstein, S. F., Silversmith, R. E., Bourret, R. B., and Charon, N. W. (2005). CheX is a phosphorylated CheY phosphatase essential for *Borrelia burgdorferi* chemotaxis. *J. Bacteriol.* **187,** 7963–7969.

Park, S. Y., Chao, X., Gonzalez-Bonet, G., Beel, B. D., Bilwes, A. M., and Crane, B. R. (2004). Structure and function of an unusual family of protein phosphatases: The bacterial chemotaxis proteins CheC and CheX. *Mol. Cell* **16,** 563–574.

Parkinson, J. S. (1978). Complementation analysis and deletion mapping of *Escherichia coli* mutants defective in chemotaxis. *J. Bacteriol.* **135,** 45–53.

Parkinson, J. S., Ames, P., and Studdert, C. A. (2005). Collaborative signaling by bacterial chemoreceptors. *Curr. Opin. Microbiol.* **8,** 116–121.

Porter, S. L., and Armitage, J. P. (2002). Phosphotransfer in *Rhodobacter sphaeroides* chemotaxis. *J. Mol. Biol.* **324,** 35–45.

Silversmith, R. E., Guanga, G. P., Betts, L., Chu, C., Zhao, R., and Bourret, R. B. (2003). CheZ-mediated dephosphorylation of the *Escherichia coli* chemotaxis response regulator CheY: Role for CheY glutamate 89. *J. Bacteriol.* **185,** 1495–1502.

Sourjik, V. (2004). Receptor clustering and signal processing in *E. coli* chemotaxis. *Trends Microbiol.* **12,** 569–576.

Sourjik, V., and Schmitt, R. (1998). Phosphotransfer between CheA, CheY1, and CheY2 in the chemotaxis signal transduction chain of *Rhizobium meliloti. Biochemistry* **37,** 2327–2335.

Szurmant, H., Muff, T. J., and Ordal, G. W. (2004). *Bacillus subtilis* CheC and FliY are members of a novel class of CheY-P-hydrolyzing proteins in the chemotactic signal transduction cascade. *J. Biol. Chem.* **279,** 21787–21792.

Szurmant, H., and Ordal, G. W. (2004). Diversity in chemotaxis mechanisms among the bacteria and archaea. *Microbiol. Mol. Biol. Rev.* **68,** 301–319.

Turner, L., Ryu, W. S., and Berg, H. C. (2000). Real-time imaging of fluorescent flagellar filaments. *J. Bacteriol.* **182,** 2793–2801.

Wadhams, G. H., and Armitage, J. P. (2004). Making sense of it all: Bacterial chemotaxis. *Nat. Rev. Mol. Cell Biol.* **5,** 1024–1037.

Yang, Z., and Li, Z. (2005). Demonstration of interactions among *Myxococcus xanthus* Dif chemotaxis-like proteins by the yeast two-hybrid system. *Arch. Microbiol.* **183,** 243–252.

[23] Regulation of Respiratory Genes by ResD–ResE Signal Transduction System in *Bacillus subtilis*

By HAO GENG, PETER ZUBER, and MICHIKO M. NAKANO

Abstract

Successful respiration in *Bacillus subtilis* using oxygen or nitrate as the terminal electron acceptor requires the ResD–ResE signal transduction system. Although transcription of ResDE-controlled genes is induced at the stationary phase of aerobic growth, it is induced to a higher extent upon oxygen limitation. Furthermore, maximal transcriptional activation requires not only oxygen limitation, but also nitric oxide (NO). Oxygen limitation likely results in conversion of the ResE sensor kinase activity from a phosphatase-dominant to a kinase-dominant mode. In addition, low oxygen levels promote the production and maintenance of NO during nitrate respiration, which leads to elimination of the repression exerted by the NO-sensitive transcriptional regulator NsrR. ResD, after undergoing ResE-mediated phosphorylation, interacts with the C-terminal domain of the α subunit of RNA polymerase to activate transcription initiation at ResDE-controlled promoters.

Introduction

Two respiratory pathways are known in *Bacillus subtilis*; the bacterium undergoes respiration with oxygen as the terminal electron acceptor under aerobic conditions and with nitrate under anaerobic conditions. Unlike *Escherichia coli*, which carries two different quinones, ubiquinone and menaquinone, *B. subtilis* only contains menaquinone. Electrons transferred to the menaquinone pool are used to reduce oxygen by multiple membrane-bound menaquinol oxidases or by menaquinol:cytochrome *c* oxidoreductase. Under anaerobic conditions, reduced menaquinone is oxidized by a membrane-bound nitrate reductase when nitrate is present.

Expression of genes involved in aerobic and anaerobic respiration is dependent on the ResD response regulator and, to a lesser extent, on the ResE kinase (Sun *et al.*, 1996b). ResD and ResE are also required for full induction of the Pho (phosphate) regulon (Sun *et al.*, 1996a) as ResDE activates expression of aerobic terminal oxidase genes (Sun *et al.*, 1996b), which are needed to induce the *phoPR* operon encoding the Pho two-component regulatory proteins (Schau *et al.*, 2004).

METHODS IN ENZYMOLOGY, VOL. 422   0076-6879/07 $35.00

ResDE orthologs have been found in Bacilli, Listeria, and Staphylococci and were shown to function in virulence gene regulation. SrrAB (also called SrhSR) in *Staphylococcus aureus* regulates energy production (Throup *et al.*, 2001) and production of virulence factors (Pragman *et al.*, 2004; Yarwood *et al.*, 2001) in response to oxygen limitation. ResD from *Listeria monocytogenes* is required for repression of virulence gene expression in the presence of easily fermentable carbon sources (Larsen *et al.*, 2006). ResDE-dependent regulation of the enterotoxin gene was reported in *Bacillus cereus*. Curiously, the *resE* mutation showed a more adverse effect on enterotoxin production than the *resDE* mutation (Duport *et al.*, 2006).

ResD and ResE belong to the OmpR-EnvZ subfamily of two-component regulatory proteins. The ResE sensor kinase has two transmembrane domains with a fairly large extracytoplasmic domain. A HAMP linker and a PAS domain exist between the second transmembrane domain and the kinase domain. The PAS domain is required for ResE activity and probably functions in sensing a signal derived from oxygen limitation (Baruah *et al.*, 2004). ResD is a transcriptional regulator, which is active in its phosphorylated form. However, ResD carrying the phosphorylation-site mutation (D56A) can activate transcription both *in vivo* and *in vitro*, albeit to a lesser extent than phosphorylated protein (Geng *et al.*, 2004). Therefore, the role of phosphorylation in activation of ResD is somewhat mysterious. Previous studies showed that ResD is a monomer in either the unphosphorylated or the phosphorylated state (Zhang and Hulett, 2000), and the ResD monomer is likely able to bind at certain ResD-controlled promoters (Geng *et al.*, 2007). Direct interaction of ResD~P to regulatory regions of target genes was demonstrated in the intergenic region of the *ctaA* and the *ctaB* operons that function in cytochrome *caa*$_3$ oxidase assembly (Zhang and Hulett, 2000), as well as the promoter regions of the *nasD* operon encoding nitrite reductase (Geng *et al.*, 2004; Nakano *et al.*, 2000b); the flavohemoglobin gene, *hmp* (Geng *et al.*, 2004; Nakano *et al.*, 2000b); the *yclJ* operon encoding a pair of two-component regulatory proteins (Härting *et al.*, 2004); and a gene-encoding anaerobic transcriptional regulator, *fnr* (Geng *et al.*, 2007). This chapter focuses on ResDE-controlled transcription in response to oxygen limitation and nitric oxide (NO).

Oxygen Limitation and ResDE-Dependent Transcription

Overview

What is the signal perceived by ResE and how does ResE activity change in response to the signal? Given the conditions that activate expression of ResDE-controlled genes, such a signal could be generated both during the stationary phase of aerobic culture growth and more so upon oxygen limitation. Several metabolic parameters have been tested as a source of such a signal, such as NADH/NAD$^+$ ratio and the redox state of menaquinone (M. M. Nakano, unpublished results), yet the question remains unanswered. However, we know that ResE phosphatase activity is partly responsible for the repression of ResDE-dependent transcription under aerobic conditions. When phosphorylated ResD (ResD~P) is provided via cross talk with a kinase other than ResE, aerobic expression of ResDE-controlled genes such as *fnr* and *resA* increases sharply, but only when ResE is absent (Nakano *et al.*, 1999). This result suggests that ResD could be phosphorylated by another kinase (or with a small high-energy phosphate donor), but ResE is mainly, if not solely, responsible for the dephosphorylation of ResD~P under these conditions. It also suggests that ResE is in a phosphatase-dominant mode under aerobic conditions and kinase dominant under anaerobic conditions (Nakano and Zhu, 2001). This section describes *B. subtilis* culture conditions to use for investigating the role of oxygen limitation in the ResD–ResE signal transduction system. For anaerobic cultures on agar plates, the use of GasPak Plus Anaerobic System Envelopes (Becton Dickinson and Co.) is convenient. The system generates hydrogen and carbon dioxide. It was reported that the oxygen concentration decreases to less than 0.2% within 100 min after activation. As for liquid cultures, two methods are usually used. In the first method a serum bottle with a rubber septum is filled completely with cell suspension. Obtaining samples of the culture or addition of supplements is carried out using a Hamilton syringe. In the second method, cell suspension is transferred to a 2-ml microcentrifuge tube so that the tube is completely filled. The tube is tightly capped and a single tube is used only for one sample time point. In either case, oxygen remaining in starting cultures is gradually consumed during cell growth and the ResDE-dependent transcription starts to be activated around 2 h after inoculation. Here we describe the second culture method with rich medium used routinely for the measurement of β-galactosidase activities of *lacZ* fused to promoters activated by oxygen limitation. Readers can refer to a previous report for descriptions for assembling anaerobic cultures using defined media (Nakano *et al.*, 1997).

Culture of B. subtilis *under Anaerobic Conditions*

    1. Harvest *B. subtilis* cells from overnight cultures on a Difco sporulation medium (DSM) plate (Nakano *et al.*, 1988) by dispersing colonies and suspending cells with 2× YT liquid medium.

2. Inoculate cells at a starting optical density at 600 nm (OD_{600}) of 0.02 in 2× YT (Nakano *et al.*, 1988) supplemented with 1% glucose and 0.2% KNO_3 (for growth under nitrate respiration) or 2× YT supplemented with 0.5% glucose and 0.5% pyruvate (for fermentative growth).
3. Divide the cell suspension into 2-ml screw-cap tubes and close the cap tightly.
4. Incubate the tubes in a 37° water-bath incubator.
5. Mix gently by inverting tubes before taking samples for measurement of β-galactosidase activities (Miller, 1972).

Stimulatory Effect of NO on ResDE-Dependent Transcription

Overview

An implication that NO might stimulate expression of the ResDE regulon came from the observation that oxygen limitation alone is not sufficient to fully activate transcription. Although *B. subtilis* grows anaerobically by generating ATP via fermentation (Cruz Ramos *et al.*, 2000; Nakano *et al.*, 1997), expression of ResDE-controlled genes is much lower under these conditions compared to those of cells grown by nitrate respiration (LaCelle *et al.*, 1996; Nakano *et al.*, 1998, 2000a). This is particularly the case with *nasD* and *hmp*, which are not expressed under aerobic conditions even if ResD~P is present (M. M. Nakano, unpublished results), unlike the other genes described earlier. Later studies revealed that increases in transcription during nitrate respiration are caused by the stimulatory effect of NO (Nakano, 2002), which is produced from nitrite during nitrate respiration. A microarray analysis also confirmed that *nasD* and *hmp* are members of the NO stimulon (Moore *et al.*, 2004). Since *B. subtilis* is not a denitrifier, the mechanism by which NO is produced remains to be elucidated. However, three lines of evidence indicate that *B. subtilis* senses NO and responds by activating ResDE-dependent transcription. First, when cells undergo anaerobic fermentative growth where cells are unable to produce NO endogenously, addition of NO or nitrite can induce transcription (Nakano, 2002). Second, this positive effect of NO and nitrite is eliminated in the presence of an NO-specific scavenger (Nakano *et al.*, 2006). Third, we have discovered that the NO-sensitive transcription factor NsrR is involved in ResDE-dependent gene regulation (Nakano *et al.*, 2006). In the absence of NO, NsrR downregulates transcription by interacting with the promoter region of *nasD* and *hmp*. Expression of other

ResD-controlled genes is also stimulated by NO under anaerobic conditions, although the interaction of NsrR at these promoters has not been demonstrated. Hence, further investigation is required to fully uncover the role of NO in regulation of ResDE-dependent transcription.

We describe here how to examine the effect of NO on gene expression in *B. subtilis*. Many NO donors are available commercially. We usually use various NONOate compounds available from Cayman Chemical. These chemicals have advantages over other NO donors because (1) they are stable under alkaline pH and release NO spontaneously under physiological pH conditions, therefore handling of the compounds is simplified, and (2) NONOate compounds with a wide variety of half-lives are available and we can choose an appropriate NO donor depending on the experiment. The range of half-life is from 1.8 s under pH 7.4 at 37° (for PROLI NONOate) to 20 h (for DETA NONOate). As for NO scavengers, hemoglobin is often used, but 2-(4-carboxyphenyl)-4,4,5,5-tetramethylimidazoline-1-oxyl-3-oxide (carboxy-PTIO) (Molecular Probes) is a better choice to determine whether the observed stimulatory effect is attributable to NO or nitrite. Because carboxy-PTIO interacts with NO and generates nitrite (Akaike *et al.*, 1993; Goldstein *et al.*, 2003), if transcription in nitrate-respiring cells is decreased by the addition of carboxy-PTIO, it is safe to say that NO, but not nitrite, should be the direct effector. This strategy was used to draw the conclusion that *B. subtilis* senses NO and responds through activation of ResDE-dependent transcription (Nakano *et al.*, 2006).

Examination of Stimulatory Effect of NO on Transcription

1. Inoculate 2× YT supplemented with 0.5% glucose, 0.5% pyruvate, and appropriate antibiotics with *B. subtilis* carrying the *lacZ* gene fused to a ResDE-controlled promoter as described earlier.

2. Dissolve spermine NONOate in 10 mM NaOH and carboxy-PTIO in water each at a concentration of 100 mM. The half-life of spermine NONOate is 39 min at 37°, pH 7.4, and 2 mol NO is generated per mol donor.

3. At midlog phase of the growth curve, withdraw an aliquot of culture as a control. To the cultures, add spermine NONOate to a final concentration of 100 μM, and to one of the spermine NONOate-supplemented cultures, add carboxy-PTIO to 1 mM. As a control, the effect of carboxy-PTIO itself on expression of the *lacZ* fusion should be examined.

4. Take samples after 0.5 and 1 h for measurement of β-galactosidase activities.

Phosphorylation Assay Using Full-Length ResE

Overview

Although ResE is a membrane-bound sensor kinase, *in vitro* studies involving measurements of autophosphorylation and phosphorylation/dephosphorylation of its substrate ResD have been carried out using a truncated soluble form of ResE, as is often done with other membrane-bound kinases. We have shown previously that the cytoplasmic region of ResE senses oxygen limitation, but *in vivo* activity of the soluble ResE as measured by ResDE-dependent transcription is lower than that of the full-length protein (Baruah *et al.*, 2004). This result could be explained by the presence of a second sensing domain in the extracytoplasmic region of ResE or a higher accessibility of membrane-bound ResE to a signal ligand. In order to investigate if membrane localization of ResE is important for its activities, we have purified the His$_6$-tagged full-length ResE. A review is available on details of strategies for the reconstitution of membrane proteins into liposomes (Rigaud *et al.*, 1995). Briefly, the full-length ResE is overproduced in *E. coli* and extracted from the membrane with detergent. After purification using Ni^{2+}-nitrilotriacetic acid (NTA) column chromatography, ResE is incorporated into liposomes prepared from *E. coli* phospholipids. Figure 1 shows autophosphorylation of the full-length ResE. ResE in *E. coli* membrane fractions autophosphorylates in the presence of [γ-^{32}P]ATP and transfers the phosphoryl group to an unidentified *E. coli* protein. ResE, solubilized by detergent and purified by Ni^{2+}-NTA column chromatography, undergoes autophosphorylation

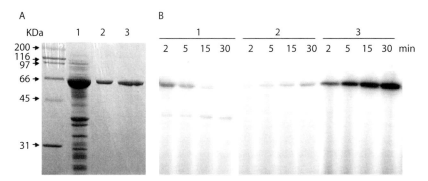

FIG. 1. Autophosphorylation of the full-length ResE. (A) Coomassie staining of ResE-overproduced *E. coli* membrane fraction (1), solubilized ResE (2), and ResE-proteoliposomes (3). (B) Autoradiography of ResE phosphorylation with the samples shown in A. Each sample was incubated in the presence of [γ-^{32}P]ATP at room temperature for the indicated times.

but the activity increases sharply after reconstitution into proteoliposomes. The reconstituted ResE was also able to phosphorylate purified ResD protein (data not shown). Although detailed characterization of the full-length ResE awaits future experiments, we outline the reconstitution method next.

Reconstitution Protocol of Full-Length ResE

1. Culture BL21(DE3)/pLysS carrying pMMN622 in Luria-Bertani (LB) medium supplemented with 25 μg/ml ampicillin and 5 μg/ml of chloramphenicol at 37°.
2. Add 0.5 mM isopropyl-β-D-thiogalactopyranoside (IPTG) when OD_{600} reaches around 0.5 and incubate further at room temperature for 4 h.
3. Harvest cells and suspend with buffer A (50 mM Tris-HCl, pH 7.7, 10 mM $MgCl_2$).
4. Break cells by passing through a French press.
5. Centrifuge the cell lysate at 15,000g for 20 min.
6. Centrifuge the supernatant at 200,000g for 65 min.
7. Wash the pellet (membrane fraction) twice with 1 mM Tris-HCl, pH 7.7, 3 mM EDTA.
8. Suspend the pellet (1.5 mg/ml) in buffer B (50 mM Tris-HCl, pH 7.7, 10 mM β-mercaptoethanol, 10% glycerol) and homogenize the suspension.
9. Add slowly lauryldimethylamine oxide (to final concentration of 2%, w/v) to the homogenate and gently stir for 30 min in an ice bath.
10. Centrifuge at 200,000g for 65 min.
11. Apply the supernatant to a Ni^{2+}-NTA column (Qiagen) equilibrated with buffer C (50 mM Tris-HCl, pH 7.7, 500 mM NaCl, 10% glycerol, 0.04% dodecyl maltoside) with 10 mM imidazole.
12. Wash the column with 20 volumes of buffer C plus 10 mM imidazole and elute with buffer C plus 100 mM imidazole.
13. Evaporate chloroform from *E. coli* phospholipids (25 mg) (polar lipid extract from Avanti Polar Lipids).
14. Add 2.5 ml of 20 mM potassium phosphate buffer (pH 7.2) containing 40 mg N-octyl-β-D-glucopyranoside to the *E. coli* phospholipid preparation.
15. Dialyze against 20 mM potassium phosphate buffer (pH 7.2) overnight.
16. Freeze-thaw phospholipids for three cycles in liquid nitrogen. Thaw slowly at 20°.
17. Add dodecyl maltoside to 0.58% (w/v).

18. Add 1 mg of purified ResE and stir at 20° for 10 min.
19. Add Bio-Beads SM-2 (Bio-Rad Laboratories Inc.) (the ratio of Bio-Beads to detergent is 20 [w/w]) and shake overnight at 4°. Bio-Beads should be prewashed with methanol, then with water. Treatment with Bio-Beads results in the insertion of ResE into the liposomes, thus generating proteoliposomes.
20. Remove Bio-Beads, add fresh Bio-Beads, and shake at room temperature for 1 h and then at 4° for 1 h.
21. Withdraw the solution and centrifuge at 200,000g for 1 h.
22. Suspend the pellet with buffer B.
23. Make aliquots, freeze in liquid nitrogen, and store at −80°.

In Vivo Effect of α-CTD Alanine Substitutions on ResDE-Dependent Transcription

Overview

Most of the response regulators of two-component signal transduction systems are known to be transcriptional regulators that often activate transcription by interacting with a subunit(s) of RNA polymerase (RNAP) at the target promoter. Transcription of some, if not all, of ResD-controlled genes requires interaction between ResD~P and the C-terminal domain of the RNAP α subunit (α-CTD; Geng *et al.*, 2007). A library of *E. coli* α-CTD mutations has been widely used to characterize the mechanism by which a variety of transcriptional activators recruit RNAP to promoter regions by interaction with RNAP. In order to investigate the mechanism of activator-stimulated transcription initiation in *B. subtilis*, we constructed a library of alanine substitutions in *B. subtilis* α-CTD (from amino acid residues 251 to 314) (Zhang *et al.*, 2006). A fragment carrying a 3′-half of *rpoA* and its downstream region was used for a polymerase chain reaction (PCR)-based targeted mutagenesis. Each base substitution, with the exception of E254A, Y263A, and R311A, was designed to result in generation or loss of a restriction enzyme site. The PCR fragments were cloned downstream of an IPTG-inducible promoter of pAG58-ble-1 (Youngman *et al.*, 1989) as described previously (Nakano *et al.*, 2000c). Integration of the recombinant plasmid into the *B. subtilis rpoA* locus could be achieved only in the presence of IPTG, which allows expression of essential downstream genes encoding ribosomal proteins. We took advantage of this feature to generate the mutant alleles of *rpoA*. Removal of IPTG from the plasmid transformants results in 100% recovery of plasmid-cured strains carrying either the wild-type or the mutant allele of *rpoA*. Introduction of the mutation is confirmed by restriction enzyme digestion or sequencing of a PCR product generated from chromosomal DNA of the plasmid-cured isolates.

Because the library was constructed by replacing the wild-type *rpoA* gene with a mutant allele, *B. subtilis* carrying some mutant alleles could not be recovered, probably due to the essential role of substituted amino acid residues. Five mutations, E255A, R261A, R268A, R289A, and G292A, could not be constructed in *B. subtilis*. E255 corresponds to D259 in the "261 determinant" of *E. coli* αCTD that interacts with the "593–604 determinant" of σ^{70} region 4 (Chen *et al.*, 2003; Ross *et al.*, 2003), and the structure-based model suggested that D259 and E261 contact R603 on σ^{70} (Ross *et al.*, 2003). Consistent with this finding, the E255A mutation was able to be constructed in *B. subtilis* cells carrying substitution R362A in σ^{A}, the residue corresponding to R603 in the *E. coli* σ (H. Geng and M. M. Nakano, unpublished result). The *B. subtilis* E255A mutant was also recovered when K356 or H359 in σ^{A} region 4 is substituted with glutamate or arginine, respectively. The *E. coli* residues R265 and G296 corresponding to R261 and G292, respectively, are in the "265 determinant" that is required for αCTD–DNA interaction (Busby and Ebright, 1999). Residues 269, 278, and 301 are originally alanine, and only A269 was substituted with isoleucine.

The effect of amino acid substitutions in α-CTD on ResDE-dependent transcription of *fnr, nasD*, and *hmp* was investigated. In order to confirm that the mutational effect is specific to ResDE-activated promoters, *rpsD* encoding ribosomal protein S4 was used as a positive control. Results showed that *hmp* expression was not significantly affected by any one amino acid substitution in α-CTD. In contrast, the Y263A, K267A, and A269I mutations reduced *fnr* and *nasD* expression to less than 20% of the wild-type level. The expression was also affected (20 to 50% of the wild-type level) by the E254A, V260A, and N290A mutations. Among the substitutions of the six amino acid residues, only the Y263A has a moderate negative effect on *rpsD* expression (unpublished results; Zhang *et al.*, 2006). All five residues except N290 are located in close proximity to one another, forming a surface-exposed patch (Fig. 2). It is likely that some of these residues make contact with ResD~P at the *fnr* and *nasD* promoter. N290 is located on a surface distinct from the other residues. The corresponding residue (N294) in *E. coli* α-CTD is in the "265 determinant" required for DNA interaction (Gaal *et al.*, 1996) as described earlier and it is likely that N290 is also involved in interaction with DNA.

Because this α-CTD library is the most comprehensive library in *B. subtilis* and is widely used for studying transcriptional activation in *B. subtilis*, we briefly summarize the mutant construction.

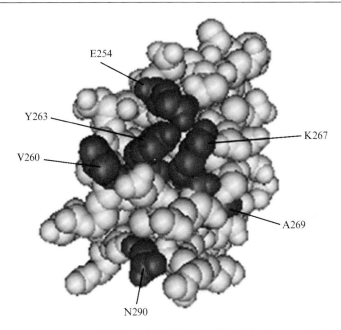

FIG. 2. The structural model of *B. subtilis* RNAP α-CTD (Newberry *et al.*, 2005). Amino acid residues identified as important for ResD-dependent activation of *fnr* and *nasD* are shown in black. Adapted from (Geng *et al.*, 2007) with permission.

Construction of α-CTD Mutants

1. Transform a *B. subtilis* strain with a pAG58-ble1 derivative carrying a mutant allele of *rpoA* and select on DSM agar supplemented with 5 μg/ml chloramphenicol and 1 mM IPTG.
2. Grow the transformant at 37° in 2 ml of 2× YT with chloramphenicol and IPTG.
3. Transfer 50 μl of the overnight cultures to 2 ml of 2× YT without chloramphenicol and IPTG and incubate at 37° until midlog phase.
4. Plate appropriate dilutions of cells on DSM agar.
5. Replica plate colonies onto DSM and DSM agar supplemented with chloramphenicol.
6. Prepare chromosomal DNA from chloramphenicol-sensitive isolates.
7. Amplify the *rpoA* region by PCR and digest with appropriate restriction enzymes or sequence directly to examine whether the desired codon substitution is introduced into the *rpoA*-coding sequence.

In Vitro Effect of α-CTD Alanine Substitutions on ResDE-Dependent Transcription

Overview

If certain amino acid substitutions of α-CTD affect transcription of ResDE-controlled genes but not *rpsD in vivo* as described previously, it is very likely that the affected amino acid positions are involved in direct interaction with ResD~P. This possibility is examined by *in vitro* analysis, such as *in vitro* runoff transcription and DNase I footprinting. Our DNase I footprinting analysis confirmed that α-CTD binds to a site adjacent to the ResD-binding site of the *fnr* promoter and facilitates formation of a ResD~P-RNAP-DNA ternary complex (Geng *et al.*, 2007).

For these experiments, *B. subtilis* RNAP is purified from strain MH5636 (constructed in the F. M. Hulett laboratory) carrying a His_{10}-tagged *rpoC* encoding the β' subunit of RNAP with a modification of the original method (Qi and Hulett, 1998). RNAP subunits, σ^A, α, and α-CTD, are overexpressed as self-cleavable intein-tag versions in *E. coli* and purified as detailed later. RNAP carrying a mutant allele of α-CTD can be purified from a strain constructed by transforming the His_{10}-tagged *rpoC* (carrying a chloramphenicol resistance marker) into *B. subtilis*, producing a mutant α-CTD.

Purification of RNAP

1. Culture MH5636 in $2\times$ YT supplemented with $5\,\mu g/ml$ of chloramphenicol.
2. Harvest cells at OD_{600} ~1.0–1.2.
3. Suspend harvested cells with buffer D ($50\,mM$ Tris-HCl, pH 8.0, $100\,mM$ NaCl, $5\,mM$ $MgCl_2$, 5% glycerol) and break cells by passing through a French press.
4. Centrifuge cell lysate at $15,000g$ for 20 min.
5. Apply the cleared lysate to a Ni^{2+}-NTA column equilibrated with buffer D and wash the column with buffer D containing $30\,mM$ imidazole.
6. Elute protein with buffer D plus $200\,mM$ imidazole.
7. Pool fractions containing RNAP and apply to a HiQ column (Bio-Rad Laboratories Inc.) equilibrated with buffer D.
8. After washing with buffer D, elute protein with NaCl gradient (200 to $500\,mM$) in buffer D.
9. Concentrate protein with a Centricon YM-10 (10 kDa cutoff) filter (Millipore Co.) and dialyze against buffer E ($10\,mM$ Tris-HCl, pH 8.0, $10\,mM$ $MgCl_2$, $100\,mM$ KCl, 50% glycerol).
10. Store at $-20°$.

Expression and Purification of α

1. Culture BL21(DE3)/pLysS carrying pSN28 at 37° in LB supplemented with 25 μg/ml ampicillin and 5 μg/ml of chloramphenicol.
2. At OD$_{600}$ ~0.4, add 0.5 mM IPTG and incubate at 30° for 3 h.
3. Suspend harvested cells with buffer F (25 mM Tris-HCl, pH 8.0, 500 mM NaCl, 5% glycerol) and break cells by passing through a French press.
4. Centrifuge cell lysate at 15,000g for 20 min.
5. Apply cleared cell lysate to a chitin resin (New England BioLabs) column equilibrated with buffer F and wash the column with buffer F.
6. Wash the column with 3 volumes of buffer G (25 mM Tris-HCl, pH 8.0, 100 mM NaCl, 5% glycerol).
7. Flush the column with 3 volumes of buffer G with 50 mM dithiothreitol (DTT).
8. Induce cleavage of the tag by incubating buffer G with 50 mM DTT.
9. Elute protein with buffer G.
10. Pool fractions containing α and apply to a HiQ column equilibrated with buffer G.
11. Wash the column with buffer G and elute with NaCl gradient (from 100 to 700 mM) in buffer G.
12. Pool fractions containing α, concentrate using a Centricon YM-7 (7 kDa cutoff) filter (Millipore Co.), and dialyze against buffer G containing 0.1 mM DTT. Divide into aliquots and freeze at −80°.

Expression and Purification of α-CTD

1. Culture BL21(DE3)/pLysS carrying pSN37 at 37° in LB supplemented with ampicillin and chloramphenicol.
2. At OD$_{600}$ ~0.4, add 0.5 mM IPTG, and incubate at 30° for 3 h.
3. Suspend harvested cells with buffer H (25 mM Tris-HCl, pH 8.0, 500 mM NaCl, 1 mM MgCl$_2$, 5% glycerol) and break cells by passing through a French press.
4. Centrifuge cell lysate at 15,000g for 20 min.
5. Apply cleared cell lysate to a chitin resin column equilibrated with buffer H and wash the column with buffer H.
6. Wash the column with buffer I (25 mM Tris-HCl, pH 8.0, 100 mM KCl, 1 mM MgCl$_2$, 5% glycerol).
7. Flush the column with 3 volumes of buffer I plus 50 mM DTT.
8. Induce cleavage of the tag by incubating the column overnight in buffer I with 50 mM DTT.
9. Elute protein with buffer I.

10. Pool fractions containing α-CTD and apply to a HiQ column equilibrated with buffer I.
11. Wash the column with buffer I and elute with KCl gradient (from 100 to 250 mM) in buffer I.
12. Pool fractions containing α-CTD, concentrate using a Centricon YM-7 filter, and dialyze against buffer J (25 mM Tris-HCl, pH 8.0, 100 mM KCl, 5 mM MgCl$_2$, 0.1 mM DTT, 10% glycerol). Divide into aliquots and freeze at $-80°$.

Expression and Purification of σ^A

1. Culture ER2566 carrying pSN64 at 28° in LB supplemented with ampicillin and chloramphenicol.
2. At OD$_{600}$ ~0.5, add 0.5 mM IPTG and incubate at 25° for 3 to 4 h.
3. Suspend harvested cells with buffer K (25 mM Tris-HCl, pH 8.0, 500 mM NaCl, 0.1 mM EDTA, 1 mM MgCl$_2$, 10% glycerol) containing 0.05% Triton X.
4. After breaking cells through a French press, collect cleared lysate by centrifugation as described earlier.
5. Apply the cleared lysate to a chitin column and wash with buffer K and then wash with 3 volumes of buffer L (25 mM Tris-HCl, pH 8.0, 100 mM KCl, 0.1 mM EDTA, 1 mM MgCl$_2$, 10% glycerol).
6. Flush the column with 3 volumes of buffer L with 50 mM DTT and incubate overnight at 4°.
7. Elute protein with buffer L and pool fractions containing σ^A.
8. Apply the pooled sample to a HiQ column.
9. After washing with buffer L, elute with KCl gradient (from 100 to 500 mM) in buffer L.
10. Pool fractions containing σ^A, concentrate using a Centricon YM-7, and dialyze against buffer L. Divide into aliquots and freeze at $-80°$.

In Vitro *Transcription Assay*

1. Mix ResD and ResE (each 1 μM) in 20 μl of transcription buffer (25 mM Tris-HCl, pH 7.5, 100 mM KCl, 0.1 mM EDTA, 0.5 mM DTT, 5 mM MgCl$_2$, 0.25 mM ATP, 50 μg of bovine serum albumin/ml, 10% glycerol, 0.4 U of RNasin [Promega]) and incubate at room temperature for 15 min.
2. Add purified RNAP and σ^A protein (both at 25 nM) and 5 nM template DNA generated by PCR. Incubate the reaction mixtures at room temperature for 10 min.

3. Add 50 μM of ATP, GTP, and CTP, 25 μM UTP, and 5 μCi of $[\alpha-^{32}P]$ UTP (800 Ci/mmol) to start transcription.
4. Incubate at 37° for 25 min, add 10 μl of stop solution (1 M ammonium acetate, 0.1 mg/ml yeast RNA, 0.03 M EDTA), and then precipitate with EtOH.
5. Run samples on 6% polyacrylamide–7 M urea gel and analyze with a PhosphorImager.

DNase I Footprinting Analysis

1. Mix 3 to 6 μM ResD with 2 μM ResE in 20 μl buffer (25 mM Tris-HCl, pH 7.5, 100 mM KCl, 0.1 mM EDTA, 0.5 mM DTT, 5 mM MgCl$_2$, 0.25 mM ATP) and incubate at room temperature for 15 min.
2. Label an oligonucleotide using T4 DNA polynucleotide kinase and $[\gamma-^{32}P]ATP$.
3. Use the labeled primer and a cold primer to amplify coding (or noncoding) strand by PCR.
4. Run the PCR product on a 6% nondenaturing polyacrylamide gel and expose the gel to X-ray film.
5. Cut the radioactive band from the gel and soak the crushed gels in 20 mM Tris-HCl, pH 7.5, 200 mM NaCl, 1 mM EDTA. Shake the tube at 4° overnight.
6. Purify the DNA fragment by an Elutip-d column (Schleicher & Schuell) according to the manufacturer's instruction.
7. Add the labeled DNA fragment (50,000 cpm per reaction) to the phosphorylation reaction mixture of ResD and ResE and incubate at room temperature for 20 min. The reaction could include RNAP with σ^A, both at 50 nM, α (0.75 to 3 μM), or α-CTD (6.25 to 25 μM) protein.
8. Treat the reaction mixture with 10 ng of DNase I at room temperature for 20 s for free probes and 40 s for reactions containing proteins.
9. Use the same labeled primer for sequencing of the template DNA with a Thermo Sequenase cycle sequencing kit (USB).
10. Run the footprinting reactions with the sequencing reactions in a 6 to 8% polyacrylamide-urea gel, dry a gel, and analyze by PhosphorImager.

Acknowledgments

Research in our laboratory was supported by Grant MCB0110513 (to M.M.N.) from the National Science Foundation and Grant GM45898 (to P.Z.) from National Institutes of Health. We are deeply indebted to Shunji Nakano for construction of the α-CTD library and expression system of the subunits of RNAP. We also thank past and current members in our laboratory for support and enthusiasm.

References

Akaike, T., Yoshida, M., Miyamoto, Y., Sato, K., Kohno, M., Sasamoto, K., Miyazaki, K., Ueda, S., and Maeda, H. (1993). Antagonistic action of imidazolineoxyl N-oxides against endothelium-derived relaxing factor/NO through a radical reaction. *Biochemistry* **32**, 827–832.

Baruah, A., Lindsey, B., Zhu, Y., and Nakano, M. M. (2004). Mutational analysis of the signal-sensing domain of ResE histidine kinase from *Bacillus subtilis*. *J. Bacteriol.* **186**, 1694–1704.

Busby, S., and Ebright, R. H. (1999). Transcription activation by catabolite activator protein (CAP). *J. Mol. Biol.* **293**, 199–213.

Chen, H., Tang, H., and Ebright, R. H. (2003). Functional interaction between RNA polymerase alpha subunit C-terminal domain and sigma70 in UP-element- and activator-dependent transcription. *Mol. Cell* **11**, 1621–1633.

Cruz Ramos, H., Hoffmann, T., Marino, M., Nedjari, H., Presecan-Siedel, E., Dressen, O., Glaser, P., and Jahn, D. (2000). Fermentative metabolism of *Bacillus subtilis*: Physiology and regulation of gene expression. *J. Bacteriol.* **182**, 3072–3080.

Duport, C., Zigha, A., Rosenfeld, E., and Schmitt, P. (2006). Control of enterotoxin gene expression in *Bacillus cereus* F4430/73 involves the redox-sensitive ResDE signal transduction system. *J. Bacteriol.* **188**, 6640–6651.

Gaal, T., Ross, W., Blatter, E. E., Tang, H., Jia, X., Krishnan, V. V., Assa-Munt, N., Ebright, R. H., and Gourse, R. L. (1996). DNA-binding determinants of the alpha subunit of RNA polymerase: Novel DNA-binding domain architecture. *Genes Dev.* **10**, 16–26.

Geng, H., Nakano, S., and Nakano, M. M. (2004). Transcriptional activation by *Bacillus subtilis* ResD: Tandem binding to target elements and phosphorylation-dependent and -independent transcriptional activation. *J. Bacteriol.* **186**, 2028–2037.

Geng, H., Zhu, Y., Mullen, K., Zuber, C. S., and Nakano, M. M. (2007). Characterization of ResDE-dependent *fnr* transcription in *Bacillus subtilis*. *J. Bacteriol.* **189**, 1745–1755.

Goldstein, S., Russo, A., and Samuni, A. (2003). Reactions of PTIO and carboxy-PTIO with *NO, *NO2, and O2-*. *J. Biol. Chem.* **278**, 50949–50955.

Härting, E., Geng, H., Hartmann, A., Hubacek, A., Münch, R., Ye, R. W., Jahn, D., and Nakano, M. M. (2004). *Bacillus subtilis* ResD induces expression of potential regulatory genes *yclJK* upon oxygen limitation. *J. Bacteriol.* **186**, 6477–6484.

LaCelle, M., Kumano, M., Kurita, K., Yamane, K., Zuber, P., and Nakano, M. M. (1996). Oxygen-controlled regulation of flavohemoglobin gene in *Bacillus subtilis*. *J. Bacteriol.* **178**, 3803–3808.

Larsen, M. H., Kallipolitis, B. H., Christiansen, J. K., Olsen, J. E., and Ingmer, H. (2006). The response regulator ResD modulates virulence gene expression in response to carbohydrates in *Listeria monocytogenes*. *Mol. Microbiol.* **61**, 1622–1635.

Miller, J. H. (1972). "Experiments in Molecular Genetics." Cold Spring Harbor Laboratory, Cold Spring Harbor, NY.

Moore, C. M., Nakano, M. M., Wang, T., Ye, R. W., and Helmann, J. D. (2004). Response of *Bacillus subtilis* to nitric oxide and the nitrosating agent sodium nitroprusside. *J. Bacteriol.* **186**, 4655–4664.

Nakano, M. M. (2002). Induction of ResDE-dependent gene expression in *Bacillus subtilis* in response to nitric oxide and nitrosative stress. *J. Bacteriol.* **184**, 1783–1787.

Nakano, M. M., Dailly, Y. P., Zuber, P., and Clark, D. P. (1997). Characterization of anaerobic fermentative growth in *Bacillus subtilis*: Identification of fermentation end products and genes required for the growth. *J. Bacteriol.* **179**, 6749–6755.

Nakano, M. M., Geng, H., Nakano, S., and Kobayashi, K. (2006). The nitric oxide-responsive regulator NsrR controls ResDE-dependent gene expression. *J. Bacteriol.* **188,** 5878–5887.

Nakano, M. M., Hoffmann, T., Zhu, Y., and Jahn, D. (1998). Nitrogen and oxygen regulation of *Bacillus subtilis nasDEF* encoding NADH-dependent nitrite reductase by TnrA and ResDE. *J. Bacteriol.* **180,** 5344–5350.

Nakano, M. M., Marahiel, M. A., and Zuber, P. (1988). Identification of a genetic locus required for biosynthesis of the lipopeptide antibiotic surfactin in *Bacillus subtilis. J. Bacteriol.* **170,** 5662–5668.

Nakano, M. M., Zheng, G., and Zuber, P. (2000a). Dual control of *sbo-alb* operon expression by the Spo0 and ResDE systems of signal transduction under anaerobic conditions in *Bacillus subtilis. J. Bacteriol.* **182,** 3274–3277.

Nakano, M. M., and Zhu, Y. (2001). Involvement of the ResE phosphatase activity in down-regulation of ResD-controlled genes in *Bacillus subtilis* during aerobic growth. *J. Bacteriol.* **183,** 1938–1944.

Nakano, M. M., Zhu, Y., Haga, K., Yoshikawa, H., Sonenshein, A. L., and Zuber, P. (1999). A mutation in the 3-phosphoglycerate kinase gene allows anaerobic growth of *Bacillus subtilis* in the absence of ResE kinase. *J. Bacteriol.* **181,** 7087–7097.

Nakano, M. M., Zhu, Y., LaCelle, M., Zhang, X., and Hulett, F. M. (2000b). Interaction of ResD with regulatory regions of anaerobically induced genes in *Bacillus subtilis. Mol. Microbiol.* **37,** 1198–1207.

Nakano, M. M., Zhu, Y., Liu, J., Reyes, D. Y., Yoshikawa, H., and Zuber, P. (2000c). Mutations conferring amino acid residue substitutions in the carboxy-terminal domain of RNA polymerase α can suppress *clpX* and *clpP* with respect to developmentally regulated transcription in *Bacillus subtilis. Mol. Microbiol.* **37,** 869–884.

Newberry, K. J., Nakano, S., Zuber, P., and Brennan, R. G. (2005). Crystal structure of the *Bacillus subtilis* anti-alpha, global transcriptional regulator, Spx, in complex with the alpha C-terminal domain of RNA polymerase. *Proc. Natl. Acad. Sci. USA* **102,** 15839–15844.

Pragman, A. A., Yarwood, J. M., Tripp, T. J., and Schlievert, P. M. (2004). Characterization of virulence factor regulation by SrrAB, a two-component system in *Staphylococcus aureus. J. Bacteriol.* **186,** 2430–2438.

Qi, Y., and Hulett, F. M. (1998). PhoP-P and RNA polymerase σ^A holoenzyme are sufficient for transcription of Pho regulon promoters in *Bacillus subtilis*: Pho-P activator sites within the coding region stimulate transcription *in vitro. Mol. Microbiol.* **28,** 1187–1197.

Rigaud, J. L., Pitard, B., and Levy, D. (1995). Reconstitution of membrane proteins into liposomes: Application to energy-transducing membrane proteins. *Biochim. Biophys. Acta* **1231,** 223–246.

Ross, W., Schneider, D. A., Paul, B. J., Mertens, A., and Gourse, R. L. (2003). An intersubunit contact stimulating transcription initiation by *E coli* RNA polymerase: Interaction of the alpha C-terminal domain and sigma region 4. *Genes Dev.* **17,** 1293–1307.

Schau, M., Eldakak, A., and Hulett, F. M. (2004). Terminal oxidases are essential to bypass the requirement for ResD for full Pho induction in *Bacillus subtilis. J. Bacteriol.* **186,** 8424–8432.

Sun, G., Birkey, S. M., and Hulett, F. M. (1996a). Three two-component signal-transduction systems interact for Pho regulation in *Bacillus subtilis. Mol. Microbiol.* **19,** 941–948.

Sun, G., Sharkova, E., Chesnut, R., Birkey, S., Duggan, M. F., Sorokin, A., Pujic, P., Ehrlich, S. D., and Hulett, F. M. (1996b). Regulators of aerobic and anaerobic respiration in *Bacillus subtilis. J. Bacteriol.* **178,** 1374–1385.

Throup, J. P., Zappacosta, F., Lunsford, R. D., Annan, R. S., Carr, S. A., Lonsdale, J. T., Bryant, A. P., McDevitt, D., Rosenberg, M., and Burnham, M. K. (2001). The *srhSR* gene pair from *Staphylococcus aureus*: Genomic and proteomic approaches to the identification and characterization of gene function. *Biochemistry* **40,** 10392–10401.

Yarwood, J. M., McCormick, J. K., and Schlievert, P. M. (2001). Identification of a novel two-component regulatory system that acts in global regulation of virulence factors of *Staphylococcus aureus*. *J. Bacteriol.* **183,** 1113–1123.

Youngman, P., Poth, H., Green, B., York, K., Olmedo, G., and Smith, K. (1989). Methods for genetic manipulation, cloning, and functional analysis of sporulation genes in *Bacillus subtilis*. *In* "Regulation of Procaryotic Development" (I. Smith, R. A. Slepecky, and P. Setlow, eds.), pp. 65–87. American Society for Microbiology, Washington, DC.

Zhang, X., and Hulett, F. M. (2000). ResD signal transduction regulator of aerobic respiration in *Bacillus subtilis*; *ctaA* promoter regulation. *Mol. Microbiol.* **37,** 1208–1219.

Zhang, Y., Nakano, S., Choi, S. Y., and Zuber, P. (2006). Mutational analysis of the *Bacillus subtilis* RNA polymerase alpha C-terminal domain supports the interference model of Spx-dependent repression. *J. Bacteriol.* **188,** 4300–4311.

[24] Detection and Measurement of Two-Component Systems That Control Dimorphism and Virulence in Fungi

By JULIE C. NEMECEK, MARCEL WÜTHRICH, AND BRUCE S. KLEIN

Abstract

Systemic dimorphic fungi include six phylogenetically related ascomycetes. These organisms grow in a mold form in the soil on most continents around the world. After the mold spores, which are the infectious particles, are inhaled into the lung of a susceptible mammalian host, they undergo a morphological change into a pathogenic yeast form. The ability to convert to the yeast form is essential for this class of fungal agents to be pathogenic and produce disease. Temperature change is one key stimulus that triggers the phase transition from mold (25°) to yeast (37°). Genes that are expressed only in the pathogenic yeast form of these fungi have been identified to help explain how and why this phase transition is required for virulence. However, the regulators of yeast-phase specific genes, especially of phase transition from mold to yeast, have remained poorly understood. We used *Agrobacterium*-mediated gene transfer for insertional mutagenesis to create mutants that are defective in the phase transition and to identify genes that regulate this critical event. We discovered that a hybrid histidine kinase senses environmental signals such as temperature and regulates phase transition, dimorphism, and virulence in members of this fungal family. This chapter describes our approach to the identification and analysis of this global regulator.

Introduction

The systemic dimorphic fungi include six phylogenetically related ascomycetes: *Blastomyces dermatitidis, Coccidioides immitis, Histoplasma capsulatum, Paracoccidioides brasiliensis, Sporothrix schenkii*, and *Penecillium marnefiii*. These primary pathogens are capable of converting from a nonpathogenic morphotype in the environment to a pathogenic morphotype after infectious spores are inhaled into the lungs of human or other mammalian hosts. These fungi collectively cause over a million new infections a year in the United States alone and remain latent after prior infection in tens of millions people worldwide in whom they may reactivate if the host becomes immune deficient (Ajello, 1971; Chiller *et al.*, 2003; Galgiani, 1999; Klein *et al.*, 1986; Wheat *et al.*, 1990).

METHODS IN ENZYMOLOGY, VOL. 422
Copyright 2007, Elsevier Inc. All rights reserved.

0076-6879/07 $35.00
DOI: 10.1016/S0076-6879(06)22024-X

The morphologic conversion of dimorphic fungi from mold to yeast is required for virulence. In experimental studies of *H. capsulatum*, transition to the yeast form is required for the establishment of disease. Treatment of mycelia with the sulfhydryl inhibitor *p*-chloromercuriphenylsulfonic acid (PCMS) permanently and irreversibly prevents the transition to yeast at 37°. PCMS-treated *H. capsulatum* failed to cause illness in a mouse model of lethal experimental infection. In addition, no fungal colonies were recoverable from spleens of the infected mice (Medoff *et al.*, 1987). This evidence suggests that the conversion to yeast is necessary for virulence in *H. capsulatum*. Conversion to yeast also offers protection against killing by neutrophils, monocytes, and macrophages. Drutz and colleagues (1985) showed that *B. dermatitidis* yeast are too large to be ingested by polymorphonuclear neutrophils (PMNs), unlike the much smaller conidia. In addition, both PMNs and peripheral blood monocytes are more efficient at killing conidia than yeast (Drutz and Frey, 1985).

The phenotypic switch from an environmental mold morphotype to a pathogenic yeast morphotype results in a change not only in cell shape, but also in the cell wall composition, the presence of antigenic molecules, and the expression of virulence traits. In *B. dermatitidis*, the conversion from mold to yeast results in an increased content of α-(1,3)-glucan and a decreased β-(1,3)-glucan content (San-Blas and San-Blas, 1984). In the pathogenic yeast forms of several dimorphic fungi, including *B. dermatitidis*, the level of α-(1,3)-glucan in the cell wall correlates with the level of virulence (Hogan and Klein, 1994; Rappleye *et al.*, 2004). Additionally, as cells adapt to changes in temperature, multiple changes occur in the lipid composition of the plasma membrane, which leads to remodeling and reorganization of the membrane (Maresca and Kobayashi, 2000).

During conversion to the pathogenic form, dimorphic fungi express phase-specific antigenic molecules that are essential in virulence. In *B. dermatitidis*, only the yeast phase expresses the immunoreactive 120-kDa protein antigen BAD1 (formerly named WI-1) (Rooney *et al.*, 2001). BAD1 binds to chitin on the yeast cell wall, and about 4.7×10^6 molecules are estimated to be present on each individual yeast cell (Klein and Jones, 1990). This surface molecule functions as an adhesin and essential virulence factor that binds the fungus to macrophages and lung tissue. *BAD1* also alters the immune response of the host by downregulating production of the proinflammatory cytokine tumor necrosis factor (TNF)-α in phagocytes and upregulating production of the anti-inflammatory cytokine TGF-β, aiding in the progression of a pulmonary infection (Finkel-Jimenez *et al.*, 2001). Similarly, in *H. capsulatum*, only the pathogenic yeast phase expresses a released calcium-binding protein (CBP1) essential for survival in the host and pathogenicity (Batanghari *et al.*, 1998; Sebghati *et al.*, 2000).

H. capsulatum yeast also produce Yps3, a cell wall-localized protein that is not produced by the mycelial phase (Bohse and Woods, 2005). Yps3 expression has also been correlated with increased virulence in *H. capsulatum* (Keath *et al.*, 1989). The search for virulence determinants in other dimorphic fungi has led to the identification of phase-specific genes in *P. marneffei* and *C. immitis* that are important for both pathogenesis and survival in the host (Hung *et al.*, 2001, 2002; Pongpom *et al.*, 2005).

Two-component signaling systems are widespread in prokaryotes. Eukaryotes have been thought to rely mainly on serine, threonine, and tyrosine kinases for signal transduction, but histidine kinase two-component systems have been shown to play a role in environmental sensing and cell development in eukaryotes as well (Santos and Shiozaki, 2001), including in the opportunistic fungal pathogen *Candida albicans*, where they regulate filamentation and pathogenicity (Alex *et al.*, 1998; Yamada-Okabe *et al.*, 1999).

We uncovered the long-sought regulator that controls the switch from a nonpathogenic mold form to a pathogenic yeast form in dimorphic fungi. We found that a hybrid histidine kinase *(DRK1)* functions as the global regulator of dimorphism and virulence in both *B. dermatitidis* and *H. capsulatum* (Nemecek *et al.*, 2006) (Fig. 1). *DRK1* is required for phase transition from mold to yeast, expression of virulence genes, and pathogenicity *in vivo*. Disruption of *DRK1* locks *B. dermatitidis* in the mold form at temperatures that normally trigger phase transition to yeast. RNA silencing of DRK1 expression in *B. dermatitidis* results in decreased

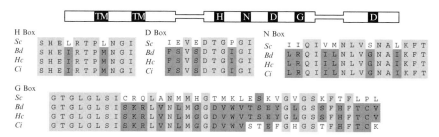

FIG. 1. Structure of the hybrid histidine kinase DRK1. DRK1 contains homology to histidine kinase domains by BLAST analysis and CD search. These necessary histidine kinase domains include the histidine-containing H box and aspartate-containing D box involved in phosporelay and the N and G boxes used in ATP-binding and catalytic function. Outside of these highly conserved domains, DRK1 has minimal homology to other hybrid histidine kinases. Additionally, DRK1 is predicted to have two transmembrane domains. The *B. dermatitidis* sequence is homologous to the hybrid histidine kinase SLN1 in *Saccharomyces cerevisiae* and to sequences in the genomes of *H. capsulatum* and *C. immitis*. Reprinted from Nemecek *et al.* (2006), with permission.

BAD1 expression, severe alterations in the cell wall, and reduction in transcription of α-(1,3)-glucan synthase and the yeast phase-specific gene *BYS1*. In *H. capsulatum*, *DRK1* also controls expression of the yeast phase-specific genes *CBP1* and *YPS3*. Hence, the hybrid histidine kinase *DRK1* functions as a sensor of environmental change in a major class of human fungal pathogens—dimorphic fungi—and dictates their adaptation to the change of environment inside mammalian hosts and their ability to cause disease. This chapter discusses various methods for analyzing these global effects of a two-component system on dimorphism and virulence in pathogenic fungi.

Experimental Approaches

RNA Interference Utilizing Agrobacterium-Mediated Gene Transfer

Generating targeted knockouts of individual genes is challenging in *B. dermatitidis* and related dimorphic fungi because of the low rate of homologous recombination owing to the preference in these organisms for illegitimate recombination (Brandhorst *et al.*, 1999). Although a *DRK1* knockout was successfully generated, it is locked in the mold phase and completely unable to sporulate and produce conidia, which are the infectious particles of the fungus. This property of being locked in a mold form makes some types of further analysis of the *DRK1* knockout technically challenging. The disruption of some histidine kinases in fungi may also result in a lethal defect (Ota and Varshavsky, 1993). As an alternative approach, RNA interference (RNAi) has been used successfully for gene silencing in *B. dermatitidis, H. capsulatum*, and other fungi. RNAi has been reviewed extensively elsewhere (Nakayashiki, 2005).

1. A vector containing an inverted repeat for the gene of interest will need to be generated (Fig. 2). RNA hairpin loops ranging from 500 to 4000 bp have been used successfully to silence genes in *B. dermatitidis* and *H. capsulatum* (Nemecek *et al.*, 2006; Rappleye *et al.*, 2004). This vector should also contain a selective marker—in *B. dermatitidis*, potential dominant markers include hygromycin and nourseothricin, and uracil is one available auxotrophic marker. To attempt to avoid problems with off-target silencing, it is best to design the RNA hairpins using two independent sequences from the target gene, for instance, one N- and one C-terminal fragment. The RNAi vector lacking a hairpin sequence can be used as a negative control.

2. The generated vector must then be transformed into *Agrobacterium tumefaciens*. We use *A. tumefaciens* strain LBA110 (Beijersbergen *et al.*, 1992). Mix electrocompetent *A. tumefaciens* cells with 2 μl of vector containing approximately 100 ng of DNA. Electroporate cells using a Bio-Rad gene pulser with the following settings: 2.5 kV, 25 μF, and 200 Ω

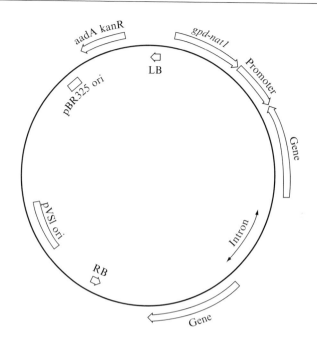

Fig. 2. Basic structure of a vector for RNAi in dimorphic fungi. The structure of a vector for RNA silencing. This example of a binary vector for *Agrobacterium*-mediated gene transfer includes a selectable marker (*gpd-nat1*: the nourseothricin resistance gene under the control of the glyceraldehyde-3-phosphate dehydrogenase promoter from *A. nidulans*), a promoter driving transcription of a hairpin of the gene of interest, and the left and right border (LB, RB) of the T-DNA. Additionally, the binary vector contains a kanamycin resistance cassette (aadA kanR), a pBR325 origin of replication, and a pVS1 origin of replication.

(Mattanovich *et al.*, 1989). Suspend cells in 1 ml of super optimal broth with catabolite repression (SOC) and plate a portion of the transformation mix (10–20 μl) on LB media containing 100 μg/ml of spectinomycin, 100 μg/ml of kanamycin, and 0.1% glucose. *Agrobacterium* transformants will emerge after 2 to 3 days of incubation at 28°. Pick and purify single colonies. Freshly restreak the *A. tumefaciens* strain on LB media containing 100 μg/ml of spectinomycin, 100 μg/ml of kanamycin, and 0.1% glucose before use in transformation.

3. Four days prior to transformation, spread a loop full of either *B. dermatitidis* or *H. capsulatum* cells on HMM agar plates in 100 μl of liquid HMM. Incubate plates at 37° for 4 days. Alternatively, cells may be plated on 7H10 agar slants and incubated as described previously. Additionally, strains can be inoculated into liquid HMM and grown at 37° while aerating at 250 rpm.

4. Two days prior to transformation, inoculate a loop full of the *A. tumefaciens* strain into 25 ml of *Agrobacterium* minimal medium (Hooykaas *et al.*, 1979). Incubate overnight at 28° while aerating at 250 rpm.

5. One day prior to transformation, determine the OD_{660} of the *A. tumefaciens* culture started in step 4. Dilute the culture to an OD_{660} of 0.05 in induction medium (IM) containing 200 μM acetosyringone (AS) (Bundock *et al.*, 2002). Incubate overnight as described in step 4.

6. On the day of transformation, measure and record the OD_{660} of the *A. tumefaciens* culture. In 5 ml of liquid HMM, harvest the cell material from the plates inoculated in step 3. Count the harvested cells and then centrifuge at 2500 rpm for 5 min. Remove the supernatant and resuspend the pellet in a volume of HMM so that 100 μl of culture contains between 1×10^6 and 1×10^7 cells.

7. Using sterile tweezers, place a sterile BioDyneA (Pall, East Hills, NY) membrane on an IM plate containing 200 μM AS.

8. To each plate, add 100 μl of the *A. tumefaciens* culture. Add 100 μl of the fungal cell suspension from step 4. Spread the mixture over the entire plate using a sterile glass rod.

9. Incubate at 28° for 3 days.

10. Using sterilized tweezers, transfer the membrane from the IM plate to a 3 *M* plate containing 200 μM cefotaxime and the appropriate antibiotic for selection of the RNAi vector. Incubate the plates at 37°. When using nourseothricin for selection, transformants often begin to appear after 8 to 12 days. When using hygromycin for selections, transformants usually begin to appear after 15 to 18 days.

11. Using a sterile toothpick, patch the transformants to a new 3 *M* plate containing 200 μM cefotaxime and the appropriate antibiotic for selection. Patches will need to be colony purified. Analyze transformants for the expected phenotypes resulting from RNA silencing of the target.

Chitin Assay Protocol

Chitin is a 1,4-β-linked polymer of *N*-acetylglucosamine found in the cell wall of fungi. The polymer provides structural integrity to the cell wall and, in some instances, offers an anchoring point for other cell wall components. In *B. dermatitidis*, the surface-displayed adhesin BAD1 binds to chitin in the cell wall (Brandhorst and Klein, 2000). Thus, chitin is necessary not only for the structure rigidity of the wall, but potentially plays a role in pathogenesis by promoting the surface localization and display of this virulence factor. *DRK1*-silenced strains of *B. dermatitidis* show varied defects in the cell wall, including decreased levels of chitin (Nemecek *et al.*, 2006).

Chitin Isolation from Cell Wall

1. Grow *B. dermatitidis* in liquid HMM at $37°$ to midlog phase. Spin down an equal number of cells for each sample and remove supernatant. For our analysis of wild-type *B. dermatitidis* strain ER3, 1×10^7 cells are sufficient (Baumgardner and Paretsky, 1999). Store dried pellets at $-20°$ until they can be assayed.

2. Wash lyophilized cells with $1\times$ phosphate-buffered saline (PBS), pH 7.4. Spin at full speed in a microcentrifuge for 5 min to pellet cells and remove supernatant.

3. Resuspend pellet in 4 ml of 3% sodium dodecyl sulfate (SDS).

4. For each sample, split the resuspended cell mixture into three 2-ml centrifuge tubes (samples are split into smaller tubes in order to fit into a heat block. This may not be necessary if you have a heat block that will accommodate larger tubes). Heat for 15 min at $100°$. Cool tubes to room temperature. Combine contents of tubes in a 15-ml conical tube. Centrifuge again to pellet cells and discard supernatant.

5. Generate a set of standard curve samples: chitin in water at concentrations ranging from 0.5 μg/ml to 5 mg/ml. (*Note*: chitin is not soluble in water, but this is not important at this step.) Handle these standard samples the same as the experimental samples for the remainder of this protocol.

6. Resuspend the cell pellet in 3 ml of KOH (120 g of KOH in 100 ml water).

7. Split each sample into two 2-ml tubes. Heat at $130°$ for 1 h.

8. Cool tubes and recombine the aliquots of each sample into a single 15-ml conical tube. Add 8 ml of ice-cold 75% ethanol to each sample. Shake until the KOH and ethanol form a single phase (it may require adding a few drops of water to form a single phase). Place tubes in an ice water bath for 15 min.

9. Add 300 μl of a Celite-545 slurry (1 g of Celite-545 in 10 ml of 75% ethanol, allowed to stand and form a slurry for 2 min before using) to a sample. Mix the sample and centrifuge the tubes at $1500g$ for 5 min at $2°$.

10. Wash the pellet once with ice-cold 40% ethanol. Wash the pellet twice with ice-cold water. The pellet can now be stored at $4°$ until assayed.

Chitin Assay

1. Bring each sample from the chitin isolation protocol up to 500 μl volume with distilled water.

2. To each, add 500 μl NaNO$_2$ (5%, w/v) and 500 μl of KHSO$_4$ (5%, w/v).

3. Gently mix three times over a total of 15 min, keeping samples at room temperature.

4. Centrifuge at 1500g for 2 min at 2°.

5. Take a 600-μl aliquot of the supernatant from the experimental sample tubes and bring to room temperature. Also, allow the standard curve samples to warm to room temperature.

6. To all tubes, add 200 μl of ammonium sulfate (12.5%, w/v) and mix vigorously once each minute for 5 min. (The addition of ammonium sulfate generates a lot of foam. A 2-ml tube or larger provides extra room in the tube to accommodate this additional volume.)

7. Add freshly made 200 μl 3-methyl-2-benzothiazoline hydrazone solution (50 mg in 10 ml of water). Mix well.

8. Heat tubes at 100° for 3 min. (Strongly positive samples turn light yellow.)

9. After cooling, add 200 μl of $FeC_{l3} \cdot 6H_2O$ (0.83%, w/v) and incubate samples for 25 min at room temperature. (You will usually see an immediate color change from blue to light green.)

10. Measure the optical density of samples at 650 nm. Samples may require dilution to bring within the range of the instrument. The samples can also be transferred to a 96-well plate and read using a plate reader. Use the chitin standard samples to generate a standard curve and then calculate the total amount of chitin present in the samples (Lehmann and White, 1975) (Fig. 3).

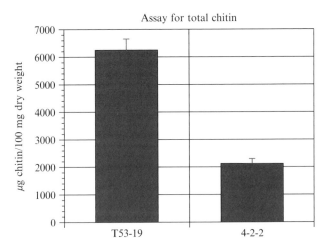

Fig. 3. Assay for chitin in the cell wall. Results of an assay for total chitin in the cell wall of a wild-type strain (T53-19) and a chitin-deficient mutant (4-2-2). Data represent the mean and standard deviation of results from three independent experiments.

Analysis of Major Components of Cell Wall

In addition to an altered level of chitin in the cell wall, *DRK1*-silenced strains of *B. dermatitidis* exhibit altered levels of the other major components of the cell wall. The primary components of the fungal cell wall, in addition to chitin, include α- and β-glucans, mannans, and proteins. Analysis of these components is done as follows.

Cell Wall Preparation (Aguilar-Uscanga and Franvβois, 2003; Boone et al., 1990; Dijkgraaf et al., 1996; Fonzi, 1999)

1. Begin with as much as 5 mg of lyophilized cell pellet, prepared as in step 1 of the chitin isolation from the cell wall protocol.
2. Suspend the cell pellet in 1 ml of 1× PBS, pH 7.4. Transfer the cell mixture to a 2-ml screw cap tube. Add approximately 300 μl of 0.45- to 0.50-mm glass beads.
3. Extract the supernatant in 0.5 ml of 0.7 M NaOH for 1 h at 75°. Pellet the cells and move the extract to a clean microfuge tube. This extract is the alkali-soluble glucan fraction.
4. Wash the remaining pellet once with 100 mM Tris, pH 7.5.
5. Wash the remaining pellet once more with 10 mM Tris, pH 7.5.
6. Suspend the pellet in 1 ml of 10 mM Tris (pH 7.5), 0.01% sodium azide, and 5 mg of Zymolyase 20 T (ICN Biomedical, Inc., Aurora, OH). Incubate for 16 h at 37° with gentle mixing.
7. Pellet the insoluble material (13,000 rpm in a table-top centrifuge for 15 min). This pellet is the zymolyase-insoluble glucan fraction.
8. Treat the supernatant with 1 unit of chitinase (Sigma-Aldrich, St. Louis, MO) at pH 6. Incubate samples at 25° for 2 h to overnight.
9. Divide the supernatant into two portions. Dialyze one portion of the eluate against distilled water using a pore size of 6000–8000 Da. The remaining fraction is alkali-insoluble β-(1-6)-glucan.
10. Analyze the remaining eluate for total glucan level to determine the amount of alkali-insoluble β-(1-3)-glucan.

Carbohydrate Analysis (Dubois et al., 1956)

1. To each 100-μl sample and 100-μl carbohydrate standard, add 2.5 μl of 80% phenol.
2. Add 250 μl of concentrated sulfuric acid. Allow to stand for 10 min at room temperature.
3. Shake the tubes and then place them in a water bath set between 25 and 30° for 20 min.
4. Determine the OD at 490 nm wavelength (Fig. 4).

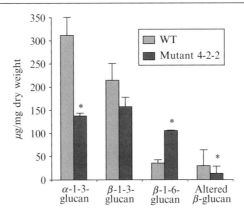

Fig. 4. Analysis of the major components of the cell wall. Results of an analysis of cell wall composition in *B. dermatitidis*. Data are shown for a wild-type strain of *B. dermatitidis* (WT) and a mutant strain with a defect in the cell wall (mutant 4-2-2). * $p < 0.05$, ANOVA test. Data are mean $\pm$ SD of three experiments. Reprinted from Nemecek *et al.* (2006), with permission. (See color insert.)

Calcofluor and Congo Red Sensitivity

The *DRK1*-silenced strains of *B. dermatitidis* show altered levels of the major components of the fungal cell wall, indicating that these strains have a severe defect in cell wall assembly and composition. Mutants with a defect in the cell wall often show altered sensitivity to the cell wall-binding compounds calcofluor and Congo red. Sensitivity to these chemicals can be analyzed in the yeast via analysis in liquid or solid media. Using the mold phase, sensitivity can be assayed using race tubes to measure mycelial growth.

Liquid Assays

1. Grow *B. dermatitidis* in liquid HMM at 37° to midlog phase.

2. Count cells using a hemocytometer. Passage the cells to an equal cell number per volume of HMM containing 20 μg/ml of Congo red or 20 μg/ml of calcofluor. As a negative control, maintain cells in HMM lacking either additive. The growth of *B. dermatitidis* strain ER3 has been analyzed in concentrations of calcofluor or Congo red ranging from 2 to 400 μg/ml. ER3 is resistant to effects of up to 200 μg/ml of calcofluor or Congo red. Record the initial concentrations of cells at the zero time point. Using strain ER3, a good starting point is 1–3 $\times$ 10^5 cell/ml.

3. Every 24 h, remove a small volume from each culture and count the cell number per milliliter using a hemocytometer (Fig. 5). With an initial culture of 3 $\times$ 10^5 cells/ml, strain ER3 will enter stationary phase

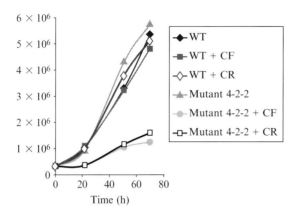

Fig. 5. Liquid assay for calcofluor and Congo red sensitivity. Results of an assay for sensitivity to the cell wall-binding agents calcofluor (CF) and Congo red (CR) in a wild-type strain of *B. dermatitidis* (WT) and a mutant strain with a defect in the cell wall (mutant 4-2-2). Data are the mean of three experiments. Reprinted from Nemecek *et al.* (2006), with permission. (See color insert.)

after 4 to 5 days at 37°. If cells are particularly aggregated, you can use a 20-gauge needle and syringe to break up the clumps of cells.

Solid Media Assays for Yeast Analysis

1. For each strain of interest, spot 10 μl of a 2×10^6 cell/ml solution onto HMM plates containing either 20 μg/ml of Congo red or 20 μg/ml of calcofluor. As a negative control, spot the same number of cells on HMM plates lacking either additive. On solid media, strain ER3 is resistant to up to 200 μg/ml of either Congo red or calcofluor.

2. Incubate plates at 37° for 2 to 3 days. Sensitive strains should show considerably reduced growth in the presence of Congo red or calcofluor. Plates may be incubated up to a week to show a more dramatic difference in growth rates. Sensitive strains will produce smaller colonies in the presence of Congo red or calcofluor.

Race Tubes for Mold Studies

1. Race tubes are horizontal test tubes containing a flat surface of agar. Mold is inoculated at the open end of the tube and mycelial growth is measured as it spreads across the length of the tube (Ryan *et al.*, 1943).

2. Harvest and count yeast phase cells using a hemocyometer. Wash cells once in 1× PBS and resuspend at a concentration of 2×10^6 cell/ml. Inoculate 10 μl of this suspension at the open end. Close race tubes loosely.

Allow the inoculum to dry before moving the tube. Mark the initial location of inoculum on the tube as a reference point.

3. Measure growth several times over 1 to 2 months, depending on the growth rate of the strain used. At each time point, mark the end point of the mycelia and measure the distance from the start point.

Analysis of Sporulation

The mold form of dimorphic fungi generates vegetative spores, which are the particles believed to establish a pulmonary infection. *DRK1*-silenced strains of *B. dermatitidis* show a decreased ability to sporulate. A defect in sporulation in a histidine kinase-defective strain may indicate that the identified histidine kinase is important in pathogenesis. Harvested spores from this protocol may be used to infect mice to analyze virulence (see later), as well as to quantify the ability of a strain to sporulate.

1. Begin with yeast phase cells grown on 7H10 slants at 37° for 3 to 4 days. Harvest the cells using 5 ml of $1 \times$ PBS.

2. Wash cells once using $1 \times$ PBS. Spin to pellet and remove the supernatant. Resuspend the cells in 0.5 to 1 ml of PBS.

3. Plate half of the cell suspension on each of two potato-flake agar plates. Spread the cell mixture using a sterile glass rod. Let the plates sit overnight in a laminar flow hood to dry at room temperature.

4. Incubate plates for 3 to 4 weeks at 22°.

5. To harvest spores, add 5 ml of PBS to each plate. Disrupt spores from mycelia using a sterile glass rod. Transfer the spore suspension to a tube. To compare spore yields, count the spores using a hemocytometer. To prepare spores for intratracheal infection in mice, spores must be passed one or two times through a 40-μm cell strainer to remove agar and mycelial fragments. After filtering, count the spores and suspend at the desired concentration (for infection of mice with *H. capsulatum*, we used 10^8 spores per 20 μl).

6. Alternatively, a scotch tape preparation can be used to visualize the spores on hyphae. First, fix the spores by adding 5 ml of 4% paraformaldehyde to each plate. Let sit for 10 min. Remove paraformaldehyde very gently so as not to disrupt the spores. Allow the plate to dry thoroughly. If desired, a drop of lactophenol cotton blue can first be placed on a clean glass slide. Press the adhesive side of a short strip of clear tape to the surface of the fixed plate. Spread the tape on the slide, over the lactophenol cotton blue if used. This preparation allows for clear visualization of spores still attached to mycelia.

Immunoblots for BAD1 Expression

BAD1 is a well-established virulence factor that is indispensable in the pathogenicity of *B. dermatitidis*. This protein is released extracellularly from yeasts and binds back to the cell, fixing itself to chitin fibrils and coating the yeast surface. *DRK1*-silenced strains of *B. dermatitidis* show greatly decreased expression of *BAD1* on the yeast cell surface. Alterations in the production of BAD1 protein can be analyzed rapidly and easily by a colony-overlay immunoblot.

1. For each strain of interest, spot 10 μl of a 2×10^6 cell/ml solution onto duplicate 3 M plates. Approximately 20 spots fit onto a standard plate without overlapping BAD1 immunoblot signals (Fig. 6).

2. Incubate both plates overnight at 37°.

3. Lay a sterile nitrocellulose membrane over one patch plate. Then incubate both patch plates for 48 h. (If the resulting BAD1 signal in the immunoblot is too high, this second incubation period can be reduced to 24 h.)

4. Lift the membrane from the plate using sterile tweezers and rinse off excess cell material with 1× Tris-buffered saline (TBS). Place the membrane colony-side up in a small box, sized to fit the membrane for immunoblotting.

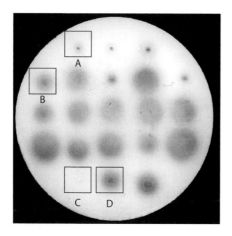

FIG. 6. Immunoblot for BAD1 expression. An example of a colony-overlay immunoblot for BAD1 expression in *B. dermatitidis*. This nitrocellulose filter was grown in contact with *B. dermatitidis* colonies for 48 h. (A) A transformed strain of *B. dermatitidis* with reduced expression of BAD1 protein. (B) A transformed strain of *B. dermatitidis* with wild-type levels of BAD1 expression. (C) Strain 55, the BAD1 knockout strain, has no detectable BAD1 protein. (D) A wild-type strain of *B. dermatitidis* expressing normal levels of BAD1 protein is shown as a positive control.

5. Block the membrane in 1× TBS–0.05% Tween 20 for 1 h at room temperature with shaking.

6. Remove the blocking solution. Blot the membrane with 1:500 anti-BAD1 monoclonal antibody DD5-CB4 supernatant in 1× TBS–0.05% Tween 20.

7. Remove the blotting solution. Rinse the membrane twice in 1× TBS–0.05% Tween 20.

8. Blot the membrane with 1:10,000 of 1 mg/ml goat antimouse IgG (H + L) (Promega, Madison, WI) in 1× TBS–0.05% Tween 20. Shake at room temperature for 1 h.

9. Remove the blotting solution. Rinse the membrane twice in 1× TBS–0.05% Tween 20.

10. Mix 5 ml of alkaline phosphate (AP) buffer with 33 μl of nitroblue tetrazolium chloride and 17 μl of 5-bromo-4-chloro-3'-indolyphosphate p-toluidine salt. This mixture must be used immediately. AP buffer is 100 mM Tris (pH 9.5), 100 mM NaCl, and 5 mM MgCl$_2$.

11. Wash membrane with AP buffer mixture and allow 5–30 min for color development. Stop the development of color by rinsing the membrane in distilled water. Air dry the membrane and photograph.

12. Photograph the duplicate spotted plate to serve as a control for colony growth.

Immunofluorescence Assay for Major Components of Cell Wall

An additional approach is available to identify and characterize altered expression of the major components in the cell wall. This involves immunofluorescence staining of the cell surface. Monoclonal antibodies exist for BAD1, α-(1,3)-glucan, and β-glucan. Chitin can be detected through the use of fluorescently labeled wheat germ agglutinin. These types of staining assays are quick, easy, and semiquantitative.

1. Harvest and count midlog phase cells. Wash 1×10^6 cells in 1 ml of 1× PBS–0.2% bovine serum albumin (BSA). If cells are particularly aggregated, use a 20-gauge needle and syringe to break up the clumps of cells. Centrifuge cells for 5 min at 2500 rpm at room temperature.

2. Incubate cells with 100 μl of a dilution of the appropriate primary antibody:
   a. BAD1: incubate with a 1:50 dilution of 1 mg/ml monoclonal antibody DD5-CB4 for 30 min at room temperature (Klein *et al.*, 1994).
   b. Chitin: incubate with 0.1 mg/ml dilution of FITC-labeled wheat germ agglutinin (Sigma-Aldrich).

c. α-(1,3)-Glucan: incubate with a 1:20 dilution of the monoclonal antibody MOPC 104e (Sigma-Aldrich) for 1.5 h at 37° (Hogan and Klein, 1994).

d. β-Glucan: incubate with the monoclonal antibody 744 (provided by Marta Feldmesser, Albert Einstein College of Medicine) at a concentration of 100 μg/ml for 1 h at room temperature (Hohl *et al.*, 2005).

3. Wash cells twice with 1 ml of 1× PBS–0.2% BSA. Spin as described in step 1.

4. Incubate cells with 100 μl of a dilution of the appropriate secondary antibody:

   a. BAD1: incubate with 5 μg/ml of Alexa Fluro 532-labeled IgG (H + L) (Molecular Probes, Eugene, OR) or with a 1:200 dilution of 1 mg/ml goat antimouse IgG (H+L) (Promega) for 30 min at room temperature.

   b. Chitin: no need for another antibody.

   c. α-(1,3)-Glucan: incubate with a 1:20 dilution of goat antimouse FITC IgG for 1.5 h at 37°.

   d. β-Glucan: incubate with a 1:200 dilution of goat antimouse FITC IgM for 1 h at room temperature.

5. Wash cells twice with 1 ml of 1× PBS–0.2% BSA. Spin as described in step 1.

6. Fix cells with 4% paraformaldehyde for 30 min at room temperature.

7. Wash cells with 1 ml of 1× PBS–0.2% BSA. Spin as described in step 1.

8. Resuspend cells in 100 μl of PBS for microscopy.

Detection of CBP via Ruthenium Red Staining

The *CBP1* gene in *H. capsulatum* encodes a released calcium-binding protein that is produced only by the yeast phase (Batanghari *et al.*, 1998). This protein may assist yeast in surviving in low calcium environments, such as the macrophage phagolysosomal compartment. CBP is produced during infection in mice and is essential for survival in the host and pathogenicity (Batanghari *et al.*, 1998; Sebghati *et al.*, 2000). The inorganic dye ruthenium red binds to calcium-binding proteins and can be used to detect secreted CBP on nitrocellulose filters. This assay can be performed using either supernatant from liquid cultures or via a colony overlay.

Supernatant Assay

1. Grow *H. capsulatum* to mid- to late log phase in liquid HMM at 37° with 5% CO_2.

2. Remove 10 μl of culture supernatant and spot onto a nitrocellulose filter. The location of each spot should be marked with a pencil. Nitrocellulose does not need to be sterile. Allow filter to air dry in a laminar flow hood.
3. Place nitrocellulose filter face up in a small container. Stain the filter in ruthenium red buffer (60 mM KCl, 5 mM MgCl$_2$, 10 mM Tris-HCl, pH 7.5) containing 25 mg/liter of ruthenium red. Staining may take anywhere from 10 to 30 min.
4. Allow stained filter to dry and then photograph to document.

Colony Overlay Assay

1. For each strain of interest, spot 10 μl of a 2 $\times$ 10^6 cell/ml solution onto HMM agar.
2. Incubate for 2 days at 37$°$ with 5% CO$_2$.
3. Overlay the plates with a sterile nitrocellulose filter using sterilized tweezers. Allow colonies to grow for another 3 days at 37$°$ with 5% CO$_2$.
4. Lift the filter from the plate using sterilized tweezers and transfer into a small container colony-side up. Rinse the filter with distilled water. Stain as described in steps 3 and 4 of the supernatant assay.

Experimental Infection with Spores

The mouse model offers features of infection that closely mimic disease in human hosts. Investigators have used varied routes to establish infection: respiratory, intraperitoneal, and intravenous. We favor the respiratory route, as this is the natural route of primary infection in the vast majority of patients. We prefer to instill the inoculum intratracheally to ensure that the cells enter the lower respiratory tract. The inoculum used for infection may be either spores or yeast. Most investigators favor the latter, as working with yeast does not require BL3 containment. However, spores are believed to be the infectious particle and infection with this form mimics the natural infection. For studies dealing with dimorphic fungi believed to be unable to undergo a phase transition, conidia should be used to establish infection and explore whether they can shift to yeast and establish infection and illness. For our studies then, we infect mice intratracheally with a lethal dose of *B. dermatitidis* or *H. capsulatum* spores.

Establishment of Infection

1. To determine the lethal dose for each fungal species and strain, conduct a pilot experiment and infect mice with a serial dilution of spores. Virulent strains of *B. dermatitidis* (ATCC strain 60636 and strain 14081,

which is a clinical isolate obtained from the Wisconsin State Laboratory of Hygiene) and *H. capsulatum* (strain KD, which is a clinical isolate obtained from the clinical microbiology laboratory of the University of Wisconsin Hospital and Clinics), require 10^3 to 10^4 and $>5 \times 10^7$ spores, respectively, to cause a uniformly lethal infection in naïve 5- to 6-week-old C57BL/6 males.

2. For intratracheal infection, anesthetize mice by intraperitoneal (ip) injection of Etomidate in a dose of 30 mg/kg weight of the mouse, mixed together with fentanyl in a dose of 0.02–0.06 mg/kg subcutaneously or ip (Green *et al.*, 1981). A suitable depth of anesthesia is determined by waiting until animals are still and unresponsive to toe pinching. Etomidate may cause myoclonus, which can be difficult to distinguish from twitching due to inadequate anesthesia. In such mice, a fraction of a second dose of the aforementioned anesthesia can be administered prior to surgery.

3. After mice are anesthetized, administer *B. dermatitidis* or *H. capsulatum* spores intratracheally under aseptic techniques. Clean skin over the neck with Betadine. Make a small incision through the skin over the trachea and separate the underlying tissue. Insert a device for delivery of the inoculum into the exposed trachea, which is visualized using a magnifying lamp. The device is a 30-gauge needle (Becton Dickinson, Rutherford, NJ) bent and attached to a tuberculin syringe (Becton Dickinson) containing *B. dermatitidis* (or other fungus) in PBS. After the needle is inserted into the trachea, dispense 20 μl of the inoculum. Ensure precise administration of inoculum volume using a Stepper device (Tridak, Brookfield, CT). After administration, withdraw the needle and close the skin with cyanoacrylate adhesive (Nexaband, Veterinary Products Laboratories). Complete surgery within 3–5 min.

4. For postoperative recovery, place animals under a warming lamp to prevent postoperative hypothermia. After a few minutes of warming, place mice in their cages to complete recovery, while prone and inclined at a 30° angle, head up. Check mice to ensure recovery. Because fentanyl is an opiate analgesic, no further analgesia is needed postoperatively.

Enumeration of Lung Colony-Forming Units (CFU), Survival, and Statistical Analysis

To assess virulence of test strains (e.g., wild-type vs RNAi-silenced strains or a gene knockout strain) two outcomes are measured: lung CFU and survival.

1. For *B. dermatitidis*, quantify lung infection by plating homogenized lung and enumerating yeast CFUs on brain–heart infusion (Difco Laboratories, Detroit, MI) agar. Place one harvested lung in a 10- to 15-ml tube

with 2 ml of sterile PBS and homogenize the tissue with a glass homogenizer. Depending on the degree of lung inflammation caused by the yeast burden, prepare 2–5 log dilutions of homogenized lung tissue in PBS. Try to have 50 to 500 CFU on a plate to allow easy enumeration. Plate out 200 μl of two dilutions likely to yield the targeted CFUs on a plate. As a frame of reference, assume that terminally ill animals harbor a multiple of 10^6 to 10^7 yeast CFU per lung. Yeast colonies take 4–7 days to emerge to a countable size when incubated at 37°. Lung CFU for a mouse is defined as total CFU per lung. The detection limit is 10 CFU per lung.

2. For *H. capsulatum*, enumerate the burden of lung infection by plating homogenized lung tissue on Mycosel (BD Biosciences) agar. To prepare 1 liter of Mycosel agar, add 36 g Mycosel agar, 8 g Bacto agar, 10 g dextrose, and 0.1 g cysteine. After autoclaved medium has cooled down to 56°, add 5 mg gentamycin or penicillin/streptomycin and 50 ml of defibrinated sheep blood (prewarmed to 37° before addition). Because *H. capsulatum* yeast will not grow on Mycosel agar at 37°, incubate the plates at 30° to allow mycelial growth of the organism. To prevent overgrowth and allow easy enumeration of mycelial colonies, only aim for 10 to 100 yeast per plate. For guidance as to which dilutions to plate, assume that moribund animals harbor 5×10^7 yeast CFU/organ. It takes 6–8 days for the colonies to emerge to a countable size. Lung CFU for a mouse is defined as total CFU per lung. The detection limit is 10 CFU per lung.

3. Alternatively, monitor the duration of survival. In experiments where *survival* is the end point, euthanize mice that show significant signs of illness to minimize their discomfort. Monitor mice for signs of illness (reduction of activity, huddling, diminished intake of food and water, ruffled coat) twice daily.

4. To determine the significance of differences in lung CFU and survival, analyze data statistically. Analyze differences in number of CFU using Wilcoxon rank tests for nonparametric data (Fisher and van Belle, 1993). Consider a two-sided p value of less than 0.05 statistically significant. Regard survival times of mice alive by the end of the study as censored. Analyze time data by the log rank statistic (Mantel–Haenzel test) (Mantel and Haenszel, 1959).

Histidine Kinase Assay

For the analysis of *DRK1* in *B. dermatitidis, DRK1* was cloned into the yeast expression vector pESC-TRP (Stratagene, La Jolla, CA) and placed under the control of a galactose-inducible promoter, with a FLAG tag at the C-terminal end. A control vector containing *SLN1* under the control of a galactose-inducible promoter, with a c-Myc tag at the C-terminal end, was generously provided by Jan Fassler (University of Iowa, Iowa City, IA). Both vectors were transformed into an *sln1Δ* strain of *S. cerevisiae* (JF 2007; provided by Jan Fassler), in which the otherwise lethal *sln1* defect is viable due to the presence of a plasmid containing a phosphatase gene *PTP2*. Ptp2p dephosphorylates the Hog1 protein that accumulates in the absence of the functional histidine kinase (Ota and Varshavsky, 1992). After lithium acetate transformation of JF2007 with an expression vector containing either *DRK1 or SLN1*, we selected against maintenance of the *PTP2* transgene via growth on 5-fluoroorotic acid. Transformants receiving either *SLN1* or *DRK1* survived the loss of PTP2, implying that *DRK1* functionally complements the *sln1* defect. Kinase protein immunoprecipitated from the transformed strains of *S. cerevisiae* were assayed for kinase activity through the use of a luminescence kinase assay to verify biochemically that DRK1 functions as a kinase.

Galactose Induction of S. cerevisiae

1. Inoculate yeast strain carrying the expression plasmid into glucose-free SC media containing 2% raffinose and the appropriate selective agents. Grow cultures to saturation at 30°, shaking at 200 rpm. This may take up to 48 h.
2. Dilute culture 1:50 in glucose-free SC media containing 4% raffinose. Grow cultures to an OD_{600} value of 0.5. This will take approximately 12 h.
3. Divide culture into two halves. To one-half of the culture, add 4% galactose for induction. The other half of the culture will serve as a negative control for induction. Incubate the cultures at 30°, shaking at 200 rpm, for 3 to 4 h.

Immunoprecipitation of Histidine Kinase from S. cerevisiae

1. Harvest cells by centrifugation: $5000g$ for 5 min at room temperature.
2. Resuspend cells in 500 μl of lysis buffer [10% glycerol, 50 mM Tris, pH 7.6, 150 mM NaCl, 1 mM EDTA, 0.1% β-mercaptoethanol, 50 μg/ml leupeptin, 1 μg/ml pepstatin, and 17.4 μg/ml phenylmethylsulfonyl fluoride [PMSF]). Add the leupeptin, pepstatin, and PMSF to the lysis buffer immediately before using. Transfer the cell mixture to a 2-ml screw cap for later bead beating.
3. Freeze the resuspended cell mixture at $-80°$ for at least 10 min. The cell mixture can be stored in the freezer for at least overnight, if needed.
4. Thaw the cell mixture. Add 1 mM PMSF and approximately 300 μl of 0.45- to 0.50-mm glass beads. Lyse the cells by bead beating three times— 1 min of bead beating followed by 1 min on ice for each repetition.

5. Clear the lysate by centrifuging at 13,200 rpm in a microcentrifuge for 5 min. Move lysate to a new centrifuge tube.

6. Add 5 μl of the appropriate antibody for immunoprecipitation for each 100 μl of lysate. Incubate at 4° for 2 h to overnight with orbital rotation.

7. Add 25 μl slurry of protein G-agarose beads to the mix, hydrated as per the manufacturer's instruction. Incubate at 4° for 2 h to overnight with orbital rotation.

8. Wash the beads three times with lysis buffer. Wash the beads four times with storage buffer (50 mM Tris, pH 7.6, 50 mM KCl, 5 mM MgCl$_2$, 0.1% β-mercaptoethanol, and 50% glycerol). Resuspend the beads in at least 50 μl of storage buffer. Store the bead mixture at $-20°$ until assayed.

9. Take a small portion of the beads (about 10 μl) and treat to remove the protein from the beads, as described by the bead manufacturer. Assay this protein mixture for total protein levels by the bicinchoninic acid (BCA) assay.

Kinase Assay

1. Assay the protein kinase for activity while still attached to the beads. A wide-bore pipette tip can be used to easily transfer the beads to a 96-well plate for assaying.
2. Follow the instructions for the Kinase-Glo luminescent kinase assay (Promega) to assay kinase activity from the bead slurry (Fig. 7).

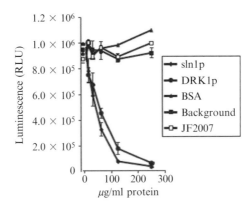

Fɪɢ. 7. Determination of kinase activity via a luminescent kinase assay. Decreasing relative light units (RLU) indicate increasing kinase activity. The protein was immunoprecipitated from *S. cerevisiae* JF2007 transformed with c-myc tagged *SLN1* expression vector (Sln1p), FLAG-tagged DRK1 expression vector (DRK1p), or untransformed JF2007 (JF2007) using anti-myc or anti-flag antibody. BSA and reaction buffer (background) are negative controls. Data are the mean ± SD of three experiments. Reprinted from Nemecek *et al.* (2006), with permission.

Acknowledgment

The work described in this chapter was supported by grants from the USPHS to B.S.K.

References

Aguilar-Uscanga, B., and Franvßois, J. M. (2003). A study of the yeast cell wall composition and structure in response to growth conditions and mode of cultivation. *Lett. Appl. Microbiol.* **37,** 274.

Ajello, L. (1971). "Distribution of *Histoplasma capsulatum* in the United States." Charles C. Thomas, Springfield, IL.

Alex, L. A., Korch, C., Selitrennikoff, C. P., and Simon, M. I. (1998). COS1, a two-component histidine kinase that is involved in hyphal development in the opportunistic pathogen *Candida albicans. Proc. Natl. Acad. Sci. USA* **95,** 7069–7073.

Batanghari, J. W., Deepe, G. S., Jr., Di Cera, E., and Goldman, W. E. (1998). Histoplasma acquisition of calcium and expression of CBP1 during intracellular parasitism. *Mol. Microbiol.* **27,** 531–539.

Baumgardner, D. J., and Paretsky, D. P. (1999). The *in vitro* isolation of *Blastomyces dermatitidis* from a woodpile in north central Wisconsin, USA. *Med. Mycol.* **37,** 163–168.

Beijersbergen, A., den Dulk-Ras, A., Schilperoort, A., and Hooykaas, P. J. (1992). Conjugative transfer by the virulence system of *Agrobacterium tumefaciens. Science* **256,** 1324–1327.

Bohse, M. L., and Woods, J. P. (2005). Surface localization of the Yps3p protein of *Histoplasma capsulatum. Eukaryot. Cell* **4,** 685–693.

Boone, C., Sommer, S. S., Hensel, A., and Bussey, H. (1990). Yeast KRE genes provide evidence for a pathway of cell wall beta-glucan assembly. *J. Cell Biol.* **110,** 1833–1843.

Brandhorst, T., and Klein, B. (2000). Cell wall biogenesis of *Blastomyces dermatitidis*: Evidence for a novel mechanism of cell surface localization of a virulence-associated adhesin via extracellular release and reassociation with cell wall chitin. *J. Biol. Chem.* **275,** 7925–7934.

Brandhorst, T. T., Wuthrich, M., Warner, T., and Klein, B. (1999). Targeted gene disruption reveals an adhesin indispensable for pathogenicity of *Blastomyces dermatitidis. J. Exp. Med.* **189,** 1207–1216.

Bundock, P., van Attikum, H., den Dulk-Ras, A., and Hooykaas, P. J. (2002). Insertional mutagenesis in yeasts using T-DNA from *Agrobacterium tumefaciens. Yeast* **19,** 529–536.

Chiller, T. M., Galgiani, J. N., and Stevens, D. A. (2003). Coccidioidomycosis. *Infect. Dis. Clin. North Am.* **17,** 41–57, viii.

Dijkgraaf, G. J., Brown, J. L., and Bussey, H. (1996). The KNH1 gene of *Saccharomyces cerevisiae* is a functional homolog of KRE9. *Yeast* **12,** 683–692.

Drutz, D. J., and Frey, C. L. (1985). Intracellular and extracellular defenses of human phagocytes against *Blastomyces dermatitidis* conidia and yeasts. *J. Lab. Clin. Med.* **105,** 737–750.

Dubois, M., Giles, K. A., Hamilton, J. K., Rebers, P. A., and Smith, F. (1956). Colorimetric method for determination of sugars and related substances. *Anal. Chem.* **28,** 350–356.

Finkel-Jimenez, B., Wuthrich, M., Brandhorst, T., and Klein, B. S. (2001). The WI-1 adhesin blocks phagocyte TNF-alpha production, imparting pathogenicity on *Blastomyces dermatitidis. J. Immunol.* **166,** 2665–2673.

Fisher, L. D., and van Belle, G. (1993). "Biostatistics: A Methodology for the Health Sciences," pp. 611–613. Wiley, New York.

Fonzi, W. A. (1999). PHR1 and PHR2 of *Candida albicans* encode putative glycosidases required for proper cross-linking of beta-1,3- and beta-1,6-glucans. *J. Bacteriol.* **181,** 7070–7079.

Galgiani, J. N. (1999). Coccidioidomycosis: A regional disease of national importance. Rethinking approaches for control. *Ann. Intern. Med.* **130,** 293–300.

Green, C. J., Knight, J., Precious, S., and Simpkin, S. (1981). Metomidate, etomidate and fentanyl as injectable anaesthetic agents in mice. *Lab. Anim.* **15,** 171–175.

Hogan, L. H., and Klein, B. S. (1994). Altered expression of surface alpha-1,3-glucan in genetically related strains of *Blastomyces dermatitidis* that differ in virulence. *Infect. Immun.* **62,** 3543–3546.

Hohl, T. M., Van Epps, H. L., Rivera, A., Morgan, L. A., Chen, P. L., Feldmesser, M., and Pamer, E. G. (2005). *Aspergillus fumigatus* triggers inflammatory responses by stage-specific beta-glucan display. *PLoS Pathog.* **1,** e30.

Hooykaas, P. J., Roobol, C., and Schilperoort, R. A. (1979). Regulation of the transfer of the Ti plasmids of *Agrobacterium tumefaciens*. *J. Gen. Microbiol.* **110,** 99–109.

Hung, C. Y., Yu, J. J., Lehmann, P. F., and Cole, G. T. (2001). Cloning and expression of the gene which encodes a tube precipitin antigen and wall-associated beta-glucosidase of *Coccidioides immitis*. *Infect. Immun.* **69,** 2211–2222.

Hung, C. Y., Yu, J. J., Seshan, K. R., Reichard, U., and Cole, G. T. (2002). A parasitic phase-specific adhesin of *Coccidioides immitis* contributes to the virulence of this respiratory fungal pathogen. *Infect. Immun.* **70,** 3443–3456.

Keath, E. J., Painter, A. A., Kobayashi, G. S., and Medoff, G. (1989). Variable expression of a yeast-phase-specific gene in *Histoplasma capsulatum* strains differing in thermotolerance and virulence. *Infect. Immun.* **57,** 1384–1390.

Klein, B. S., Chaturvedi, S., Hogan, L. H., Jones, J. M., and Newman, S. L. (1994). Altered expression of surface protein WI-1 in genetically related strains of *Blastomyces dermatitidis* that differ in virulence regulates recognition of yeasts by human macrophages. *Infect. Immun.* **62,** 3536–3542.

Klein, B. S., and Jones, J. M. (1990). Isolation, purification, and radiolabeling of a novel 120-kD surface protein on *Blastomyces dermatitidis* yeasts to detect antibody in infected patients. *J. Clin. Invest.* **85,** 152–161.

Klein, B. S., Vergeront, J. M., and Davis, J. P. (1986). Epidemiologic aspects of blastomycosis, the enigmatic systemic mycosis. *Semin. Respir. Infect.* **1,** 29–39.

Lehmann, P. F., and White, L. O. (1975). Chitin assay used to demonstrate renal localization and cortisone-enhanced growth of *Aspergillus fumigatus* mycelium in mice. *Infect. Immun.* **12,** 987–992.

Mantel, N., and Haenszel, W. (1959). Statistical aspects of the analysis of data from retrospective studies of disease. *J. Natl. Cancer Inst.* **22,** 719–748.

Maresca, B., and Kobayashi, G. S. (2000). Dimorphism in *Histoplasma capsulatum* and *Blastomyces dermatitidis*. *Contrib. Microbiol.* **5,** 201–216.

Mattanovich, D., Ruker, F., Machado, A. C., Laimer, M., Regner, F., Steinkellner, H., Himmler, G., and Katinger, H. (1989). Efficient transformation of *Agrobacterium* spp. by electroporation. *Nucleic Acids Res.* **17,** 6747.

Medoff, G., Kobayashi, G. S., Painter, A., and Travis, S. (1987). Morphogenesis and pathogenicity of *Histoplasma capsulatum*. *Infect. Immun.* **55,** 1355–1358.

Nakayashiki, H. (2005). RNA silencing in fungi: Mechanisms and applications. *FEBS Lett.* **579,** 5950–5957.

Nemecek, J. C., Wuthrich, M., and Klein, B. S. (2006). Global control of dimorphism and virulence in fungi. *Science* **312,** 583–588.

Ota, I. M., and Varshavsky, A. (1992). A gene encoding a putative tyrosine phosphatase suppresses lethality of an N-end rule-dependent mutant. *Proc. Natl. Acad. Sci. USA* **89,** 2355–2359.

Ota, I. M., and Varshavsky, A. (1993). A yeast protein similar to bacterial two-component regulators. *Science* **262,** 566–569.

Pongpom, P., Cooper, C. R., Jr., and Vanittanakom, N. (2005). Isolation and characterization of a catalase-peroxidase gene from the pathogenic fungus, *Penicillium marneffei. Med. Mycol.* **43,** 403–411.

Rappleye, C. A., Engle, J. T., and Goldman, W. E. (2004). RNA interference in *Histoplasma capsulatum* demonstrates a role for alpha-(1,3)-glucan in virulence. *Mol. Microbiol.* **53,** 153–165.

Rooney, P. J., Sullivan, T. D., and Klein, B. S. (2001). Selective expression of the virulence factor BAD1 upon morphogenesis to the pathogenic yeast form of *Blastomyces dermatitidis*: Evidence for transcriptional regulation by a conserved mechanism. *Mol. Microbiol.* **39,** 875–889.

Ryan, F. J., Beadle, G. W., and Tatum, E. L. (1943). The tube method of measuring the growth rate of *Neurospora. Am. J. Bot.* **30,** 784–799.

San-Blas, G., and San-Blas, F. (1984). Molecular aspects of fungal dimorphism. *Crit. Rev. Microbiol.* **11,** 101–127.

Santos, J. L., and Shiozaki, K. (2001). Fungal histidine kinases. *Sci. STKE* 2001, RE1.

Sebghati, T. S., Engle, J. T., and Goldman, W. E. (2000). Intracellular parasitism by *Histoplasma capsulatum*: Fungal virulence and calcium dependence. *Science* **290,** 1368–1372.

Wheat, L. J., Connolly-Stringfield, P. A., Baker, R. L., Curfman, M. F., Eads, M. E., Israel, K. S., Norris, S. A., Webb, D. H., and Zeckel, M. L. (1990). Disseminated histoplasmosis in the acquired immune deficiency syndrome: Clinical findings, diagnosis and treatment, and review of the literature. *Medicine (Baltimore)* **69,** 361–374.

Yamada-Okabe, T., Mio, T., Ono, N., Kashima, Y., Matsui, M., Arisawa, M., and Yamada-Okabe, H. (1999). Roles of three histidine kinase genes in hyphal development and virulence of the pathogenic fungus *Candida albicans. J. Bacteriol.* **181,** 7243–7247.

[25] Using Two-Component Systems and Other Bacterial Regulatory Factors for the Fabrication of Synthetic Genetic Devices

By Alexander J. Ninfa, Stephen Selinsky, Nicolas Perry, Stephen Atkins, Qi Xiu Song, Avi Mayo, David Arps, Peter Woolf, and Mariette R. Atkinson

Abstract

Synthetic biology is an emerging field in which the procedures and methods of engineering are extended living organisms, with the long-term goal of producing novel cell types that aid human society. For example, engineered cell types may sense a particular environment and express gene products that serve as an indicator of that environment or affect a change in that environment. While we are still some way from producing cells with significant practical applications, the immediate goals of synthetic biology are to develop a quantitative understanding of genetic circuitry and its interactions with the environment and to develop modular genetic circuitry derived from standard, interoperable parts that can be introduced into cells and result in some desired input/output function. Using an engineering approach, the input/output function of each modular element is character-ized independently, providing a toolkit of elements that can be linked in different ways to provide various circuit topologies. The principle of mod-ularity, yet largely unproven for biological systems, suggests that modules will function appropriately based on their design characteristics when combined into larger synthetic genetic devices. This modularity concept is similar to that used to develop large computer programs, where indepen-dent software modules can be independently developed and later combined into the final program. This chapter begins by pointing out the potential usefulness of two-component signal transduction systems for synthetic biol-ogy applications and describes our use of the *Escherichia coli* NRI/NRII (NtrC/NtrB) two-component system for the construction of a synthetic genetic oscillator and toggle switch for *E. coli*. Procedures for conducting measurements of oscillatory behavior and toggle switch behavior of these synthetic genetic devices are described. It then presents a brief overview of device fabrication strategy and tactics and presents a useful vector system for the construction of synthetic genetic modules and positioning these modules onto the bacterial chromosome in defined locations.

METHODS IN ENZYMOLOGY, VOL. 422
Copyright 2007, Elsevier Inc. All rights reserved.
0076-6879/07 $35.00
DOI: 10.1016/S0076-6879(06)22025-1

Using Two-Component Signal Transduction Systems in
  Synthetic Biology Approaches

Two-component signaling systems have been studied intensively since the mid-1980s and provide numerous examples of systems where the cellular physiology of the regulatory phenomena and the activities of the signal transduction components are reasonably well understood (Hoch and Silhavy, 1995). Certain aspects of these signal transduction systems make them particularly useful for synthetic biology purposes. Foremost among these is that many two-component systems are not essential for viability under most growth conditions and instead control fairly small numbers of genes that are only required under some stress condition that need not be applied. Thus, "closed" systems can be envisioned, where all of the natural stress response genes and regulated promoters are deleted and the two-component system functions solely in the synthetic genetic device.

The mechanisms of two-component signal transduction are also useful from a synthetic biology perspective. These systems contain a transcriptional activator whose activity is controlled by reversible phosphorylation. This feature allows the activation activity to be "tuned" by experimental manipulation of the transmitter protein activities, leading to receiver phosphorylation and dephosphorylation. This can be obtained, for example, by manipulating growth conditions, by using mutant forms of the transmitter protein, or by using mutant forms of the receiver protein (transcription factor) that have different levels of constitutive activity. A variety of transmitter and receiver protein mutants with fixed output activities are available for some two-component regulatory systems, and for the others it should be possible to introduce mutations based on available data from these characterized systems.

The sensory and signal transduction properties of certain transmitter proteins, in particular EnvZ, are distinct functions, allowing facile reengineering for the development of new sensors. It has long been known that the transmitter module of EnvZ, when fused to the transmembrane portions of cellular chemotaxis receptors, allows OmpR-dependent gene transcription in response to the ligands of the chemotaxis receptors (Jin and Inouye, 1993). The generality of this phenomenon has been extended by linking the EnvZ transmitter module to a light-sensing receptor (Levskava et al., 2005) and may be extended further using altered chemosensory apparatus to widen the range of environmental stimuli that can control expression from OmpR-dependent promoters. Presumably, all of the many transmitter proteins that are structurally related to EnvZ could be reengineered for altered sensory activity using the same approaches, allowing coupling of a variety of stimuli to various receiver proteins.

The NRI/NRII (NtrC/NtrB) two-component system controlling nitrogen assimilation in many bacteria has a number of particularly appealing features. First, the transcriptional activator, NRI~P, acts from an enhancer that is relatively position independent (Ninfa *et al.*, 1987). The main positioning limitation is that the enhancer must be a minimal distance of about 70 bp from the site of polymerase binding (promoter). This position independence of the enhancer is very useful in that it simplifies the construction of promoters that are combinatorially regulated by different transcription factors. Furthermore, the concentration of activator required to drive gene expression depends largely on the strength of the enhancer sequences, allowing experimental manipulation of promoter activation by using natural strong or weak enhancers, or by mutation of a natural enhancer (Feng *et al.*, 1995). Thus, promoters in a synthetic genetic device can be activated sequentially, just as the natural Ntr promoters are sequentially activated by increasing NRI~P during the cellular response to nitrogen limitation (Atkinson *et al.*, 2002a). A second very useful property of NRI~P is that the activating species is either a hexamer or a heptamer of NRI subunits, thus providing a very high kinetic order of transcriptional activation (Lee *et al.*, 2003). A third useful property is that certain NRI~P-dependent promoters are completely silent in the absence of activator. Specifically, the *glnK* promoter of *Escherichia coli* exhibits essentially no basal expression in the absence of NRI~P, whereas it is a very strong promoter in the presence of a high concentration of NRI~P (Atkinson *et al.*, 2002a). Certain other Ntr promoters, such as the *glnA* promoter, have a powerful and tight Ntr promoter (*glnAp2*) coupled with a weak non-Ntr promoter (*glnAp1*) that provides a basal level of expression in the absence of NRI~P and is repressed by NRI~P (Reitzer and Magasanik, 1985). Thus, one has the choice of very tight or somewhat leaky transcription, and there is no reason to suspect that this property cannot be "tuned" by mutation to provide a desired ratio of basal/activated expression. A wide variety of mutant forms of the transmitter protein are available that differ in their ability to bring about the phosphorylation and dephosphorylation of NRI, thus allowing "tuning" of the phosphorylation state in the context of a synthetic genetic device (Atkinson and Ninfa, 1992, 1993; Pioszak and Ninfa, 2003a,b). Finally, NRI-dependent promoters utilize RNA polymerase containing the minor sigma factor σ^{54} and are completely silent in the absence of this sigma factor, which is not essential for viability (Hunt and Magasanik, 1985). This opens the possibility for even more elaborate devices in the future that use independent signaling pathways to control the presence of NRI~P and σ^{54}, resulting in AND gate function from an Ntr promoter.

Using the NRI/NRII System to Build a Synthetic Genetic Clock

The basic circuit topology for the synthetic genetic clock is shown in Fig. 1. The clock consists of two modules: activator and repressor. The activator module (Fig. 1, left) consists of a promoter that drives the expression of the activator, and is itself activated by the activator. The activator also drives the expression of the repressor module (Fig. 1, right), which produces the repressor. The repressor protein blocks the expression of the activator module. Modeling of this circuit indicated that it had the potential to produce a variety of oscillatory outputs, ranging from sinusoidal oscillators to relaxation oscillators, as well as nonoscillatory steady states, depending on parameters (Atkinson et al., 2003; unpublished data). Interestingly, modeling also indicated that for each set of parameters where the intact device produced oscillations, the isolated activator module would function as a toggle switch that displayed strong hysteresis of repression (Atkinson et al., 2003). The basic design of this clock is related to an earlier hypothetical clock (Barkai and Leibler, 2000), except that in the clock of Barkai and Leibler (2000), the repressor protein antagonizes the activity of the activator, as opposed to repressing activator expression.

The implementation of the design is depicted in Fig. 2 (Atkinson et al., 2003). The activator module (Fig. 2, left) consists of the E. coli glnA promoter region, driving the expression of NRI. The natural glnA promoter contains a strong enhancer (Fig. 2, unfilled boxes) and three low-affinity activator-binding sites that function as a band limiter or governor (Fig. 2, light gray boxes [Atkinson et al., 2002b]). The region also contains two promoters (depicted as bent arrows in Fig. 2): the upstream glnAp1 promoter is repressed by activator binding to the enhancer and the downstream glnAp2 utilizes the polymerase containing σ^{54} and is activated by NRI~P. To connect this module to the repressor module, "perfect" lac operators

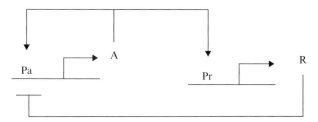

Fig. 1. Basic circuit topology of the synthetic genetic clock. The activator module (left) consists of a promoter, Pa, that drives the expression of activator, A. The repressor module (right) consists of a promoter, Pr, that drives the expression of repressor, R. Repressor block transcription from Pa.

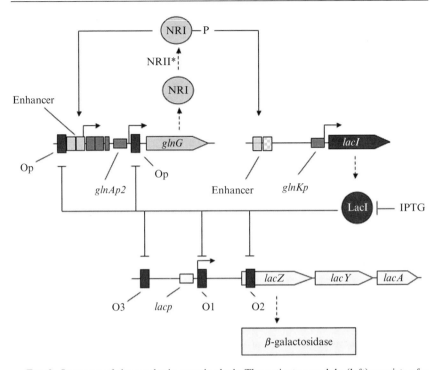

FIG. 2. Structure of the synthetic genetic clock. The activator module (left) consists of a modified version of the *glnA* promoter, driving the expression of *glnG* (*ntrC*). Light blue boxes in the *glnA* control region signify the *glnAp2* enhancer, dark blue boxes signify "perfect" *lac* operator sites, and gray boxes signify *glnAp2* governor sites. The relative positions of the *glnAp1* and *glnAp2* transcription start sites are shown by bent arrows. The product of *glnG* (*ntrC*), NRI, is converted to its active, phosphorylated form by NRII2302, which is provided in excess from a plasmid (not depicted). NRI~P increases its own expression by binding to the enhancer, whereupon it represses *glnAp1* and activates *glnAp2*. The repressor module (right) consists of the natural *glnK* promoter and translation initiation region fused to the *lacI* structural gene. The *glnK* promoter is associated with a weak enhancer (signified by a light blue box and a stippled light blue box). The product of the repressor module, LacI, blocks transcription from the activator module and from the native *lacZYA* operon (bottom). Repression of the *lacZYA* operon is due to repressor interaction with the three operators of this operon (dark blue boxes), as indicated. The product of *lacZ*, β-galactosidase, serves as a reporter for oscillatory behavior. (See color insert.)

(Fig. 2, dark gray boxes [Atkinson *et al.*, 2003]) were added at two positions: immediately downstream from the *glnAp2* transcriptional start site, imitating the position of the *lacO1* operator in the *lacZYA* operon, and immediately upstream from position −162. The intention of this design was to permit the Lac repressor to form a DNA loop and contact both operators simultaneously, as it does in repressing the *lacZYA* operon (Oehler *et al.*, 1990). However, subsequent results have suggested that a repression DNA loop does not form, although the upstream operator is still required for good oscillatory function (unpublished data). The repressor module consists of the *glnK* promoter region and mRNA leader sequence, fused to the structural gene for the Lac repressor, *lacI*. To facilitate the fusion, the initiation codon for *lacI* was converted to AUG. The *glnK* promoter has essentially no basal expression in the absence of activator and requires a high concentration of activator for expression owing to its weak enhancer. Use of the Lac repressor as the clock repressor permits easy synchronization of the cells with the lactose analogue isopropyl-β-D-thiogalactoside (IPTG). To bring about phosphorylation of NRI, the strain containing activator and repressor modules was transformed with the pBR322-derived p3Y15, which contains the mutant *glnL2302* allele for the transmitter protein NRII, driven from its natural promoter. The product of the *glnL2302* allele, NRII2302, is defective in bringing about the dephosphorylation of NRI~P, but is fully functional in autophosphorylation and phosphotransfer to NRI, ensuring that the NRI is highly phosphorylated regardless of environmental conditions (Pioszak and Ninfa, 2003a).

The expected function of the circuit is as follows. When repressed, neither activator nor repressor is synthesized, and their levels will decrease by dilution as cells grow and divide. Eventually, the concentration of repressor will decrease below the threshold for repression of the *glnA* control region, and the *glnApI* promoter will drive the expression of NRI. This NRI will be phosphorylated quickly and will repress the *glnApI* promoter and activate the *glnAp2* promoter, leading to a dramatic increase in the concentration of NRI~P. Eventually, the activator will reach a high enough concentration to activate the *glnK* promoter, leading to a burst of repressor synthesis. This will continue until the concentration of the repressor is sufficient to repress expression of the activator module, and eventually the system will be returned to its repressed state. To monitor oscillations, the repression of the *lacZYA* operon and the activation of Ntr genes, such as the natural *glnA* gene, can be monitored by measuring the enzymes β-galactosidase and glutamine synthetase as the clock-containing strain grows in continuous culture.

Fabrication of Synthetic Genetic Clock

A unique aspect of our synthetic genetic clock is that the activator and repressor modules are not contained on plasmids in the cell, but rather are integrated into defined chromosomal locations, referred to as "landing pads" (Atkinson *et al.*, 2003). We expect that this should provide a stable copy number to the modules, and indeed allows subtle manipulation of the copy number by using different locations on the chromosome, as the copy number of genes in rapidly growing cells displays a gradient from the origin of replication to the terminus of replication. The landing pads were designed to have nearby selectable markers that facilitate transfer of the modules between strains using standard P1vir-mediated generalized transduction. The basic features of landing pads include bracketing the synthetic genetic module with transcriptional terminators to prevent transcription from outside promoters. Two landing pads, located in the *rbs* and *glnK* regions of the chromosome, were used to integrate the activator and repressor modules into the chromosome. These landing pads were described previously and are available in the plasmids pRBS3 and pDK11, respectively (Table I). Some of the features of these landing pads are provided in Table I. Since the fabrication of our synthetic genetic clock, we have devised a convenient plasmid system for construction of new landing pads and for modular construction of synthetic genetic devices in these landing pads. This vector system is described in a later section. Our original fabrication methods for building the synthetic genetic clock modules consisted of standard molecular biology methods, including polymerase chain reaction (PCR), site-specific mutagenesis, and so on and were described previously (Atkinson *et al.*, 2003). A later section describes improved methods and materials that should ease the construction of synthetic genetic devices.

Assembly of the clock strain is as follows (strains listed in Table I). The starting strain 3.300 contains a *lacI* null mutation. The *glnA::Tn5* mutation is then introduced by P1vir-mediated generalized transduction, with selection for kanamycin resistance and screening for glutamine auxotrophy, forming strain 3.300A. A deletion of *glnL* and *glnG* is then introduced by transducing 3.300A with phage grown on strain SN24 (*glnLglnG*), with selection for glutamine prototrophy (*glnA+*) and screening for kanamycin sensitivity, producing strain 3.300LG. Strain 3.300LG is thus *lacI glnL glnG* but *glnA+* and *lacZYA+* and serves as the host for the synthetic genetic clock.

To recombine the clock genetic modules onto their chromosomal landing pads, a *recD* mutant is transformed with linearized plasmid DNA, where the site of cleavage is within the plasmid vector sequences. We have had success using strain TE2680 (Elliott, 1992) and have observed that the *recD:: TN10* mutation (conferring tetracycline resistance) can be transduced from strain TE2680 into a variety of strain backgrounds, permitting transformation with linear DNA. We typically use electroporation of the linearized DNA to obtain a large number of transformants and use up to 1 μg of DNA per transfection. The drug resistance marker of the landing pad is selected (Table I), and the transformants are screened for resistance to ampicillin,

TABLE I
USEFUL STRAINS, MODULES, AND PLASMIDS

Bacterial strains for clock and toggle switch experiments

Strain	Relevant genotype	Source or reference
3.300	*lacI*	CGSC[a]
3.300LG	*lacI glnL glnG*	Atkinson *et al.* (2003)
3.300LG-Act	3.300, but *rbs::glnApOG6...gen*[r]	Atkinson *et al.* (2003)
NC12	3.300-Act, but *glnK::glnKp-lacI2...cam*[r]	Atkinson *et al.* (2003)
NC12/p3Y15	NC12, but containing p3Y15 (*glnL2302*)	Atkinson *et al.* (2003)
TE2680	*recD::Tn10*	Elliott (1992)
3.300-Rep	3.300, but *glnK::glnKplacI2...cam*[r]	Atkinson *et al.* (2003)
M7044	*lacY*	CGSC[a]
TS1	M7044, but *rbs::glnApOG6...gen*[r]	

Landing pads

Plasmid	Cloning	Linearization	Selection	Location
pDK11	*EcoRV* (blunt)	*PstI*	cam	*glnK*
pRBS	*PstI, SalI, XbaI, SacI*	*NotI, SpeI*	gent	*rbsK*
pStep1t-AraBk	Fig. 7	*PstI*	kan	*araB*
pStep1t-MalPk	Fig. 7	*PstI*	kan	*malP*
pStep1t-FucKk	Fig. 7	*PstI*	kan	*fucK*
pStep1t-MalPc	Fig. 7	*PstI*	cam	*malP*
pStep1-MalQc	Fig. 7	*PstI*	cam	*malQ*
pStep1t-FucKc	Fig. 7	*PstI*	cam	*fucK*
pStep2-kan	Fig. 7	*PstI*	cam	*kan*
pStep3-cam[b]	Fig. 7			
pStep4-kan[b]	Fig. 7			

Modules and plasmids

Name	Function	Feature
rbs::glnApOG6	Activator module	*glnAp* with proximal and distal *lacOp* driving expression of *glnG*
glnK::glnKplacI2	Repressor module	*glnKp* driving expression of *lacI*
p3Y15	Provides NRII2302	*glnL2302* driven from its natural promoter, derived from pBR233 (Atkinson and Ninfa, 2003)

[a] coli genetic stock center: http://cgsc.biology.yale.edu/.
[b] Under development.

which is encoded by the plasmid vector sequences. Transfectants that have the landing pad drug resistance but lack resistance to ampicillin are those in which the landing pad has recombined onto the chromosome. Transducing phages are then grown on these recombinants, and the integrated landing pad is introduced into desired strains by selection for the associated drug resistance of the landing pad. Thus, to build the genetic clock, strain 3.300 LG is sequentially transduced with *rbs:glnApOG6...gent* (introducing the activator module) and *glnK::pglnKlacI2...cam* (the repressor module), with selection for gentamycin resistance and chloramphenicol resistance, respectively. This strain, referred to NC12, is then made competent by standard procedures and transformed with p3Y15 (encoding NRII2302), with selection for ampicillin resistance, forming strain NC12/p3Y15, which is then used for clock studies. We have observed that the final plasmid-containing strain can be stored as a freezer culture at $-80°$, and reproducible clock experiments can be performed by streaking a few ice crystals containing the frozen cells on LB medium containing ampicillin immediately prior to the experiment and picking a single colony isolate for the experiment. This strain and the intermediate strains and modules (Table I) can be obtained from the authors.

Functions of Individual Clock Modules

Activator module and repressor module functions can be measured independently in intact cells. The activator module, when present in cells containing wild-type NRII and wild-type *lacI* encoding the Lac repressor, is predicted to form an N-IMPLIES logic gate with respect to ammonia and IPTG (Fig. 3). In the presence of wild-type NRII, the nitrogen-rich state brought about by the presence of ammonia causes formation of the NRII–PII complex and rapid dephosphorylation of NRI∼P. Furthermore, in the absence of IPTG, the constitutively present Lac repressor blocks expression of the activator module. Expression of a reporter consisting of the *glnK* promoter to *lacZ*, or any other Ntr gene, thus requires the presence of IPTG and the absence of ammonia, providing the N-IMPLIES logic function. After construction of the bacterial strain, we observed that the cells indeed only expressed β-galactosidase in the absence of ammonia and in the presence of IPTG.

The activator module also displays toggle switch function in the absence of the repressor module, as predicted by modeling (Atkinson *et al.*, 2003; Fig. 4). For these experiments, where only repression of the activator module is examined, the activator module is integrated into a strain deleted for both *glnL* and *glnG*, and the transmitter protein function is complemented by the plasmid p3Y15, encoding NRII2302. The strain also contains a wild-type *lacI* and *lacZ*, but contains a mutation of *lacY*. The mutation of *lacY* is essential to allow control of the internal IPTG concentration by variation in the external IPTG concentration (Novick and Weiner, 1957). The assembly of the strain is as follows. Strain M7044 (*lacY*) is transduced

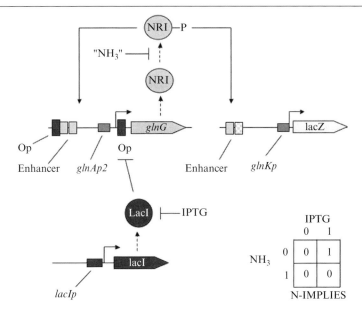

FIG. 3. The activator module is an N-IMPLIES gate with positive feedback. When in cells containing wild-type natural components of the Ntr system except for NRI, containing the activator module to provide wild-type NRI, containing the wild-type *lacI* gene, and containing a fusion of *lacZ* to the *glnK* promoter, the activator module provides an N-IMPLIES logic gate for the regulation of β-galactosidase by ammonia and IPTG. (See color insert.)

to *glnA::Tn5* by selecting for kanamycin resistance and checking for gluta-mine auxotrophy, forming strain M7044A. This is then transduced to glu-tamine prototrophy using phage grown on strain SN24 (*glnLglnG*), producing strain M7044LG. This strain is then transduced with the activator module by selection for gentamycin resistance of the integrated *rbs* landing pad, producing strain TS1. Finally, strain TS1 is made competent by stan-dard methods and transformed with plasmid p3Y15, with selection for ampicillin resistance.

Toggle switch experiments are performed by growing the strain over-night in the absence and presence of 0.1 mM IPTG, resulting in naive and induced cultures. The overnight cultures are then diluted one million-fold into media containing various concentrations of IPTG. Growth is continued for about 15 to 17 generations, until the cultures are in midlog phase and β-galactosidase and glutamine synthetase are measured. The glutamine synthetase measurement provides an indication of the level of activator, whereas the β-galactosidase measurement provides an indication of the level of repressor. It is expected that naive and induced cultures should

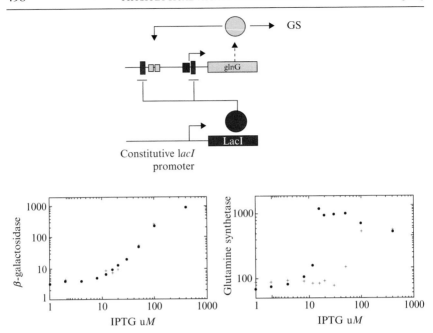

FIG. 4. The activator module can function as a toggle switch. The basic circuit topology for the toggle switch is shown at the top. The activator module produces activator that drives expression of GS and the activator module. LacI is produced from the natural chromosomal *lacI* gene and represses the activator module, as well as the natural *lacZYA* (not depicted). At the bottom, typical results for *lacZ* expression (left) and glutamine synthetase expression (right) are shown for induced (dot) and naive (+) cultures. The overlapping curves for *lacZ* expression data show that the steady state was reached; GS data show hysteresis of the activator module. An intuitive explanation for this hysteresis is as follows. When the level of NRI~P is high, as in the induced culture, it is difficult for the repressor to get control of the system, and thus repression is only achieved when the IPTG concentration is low. In the naive cell, the level of NRI~P is very low and thus it is considerably easier for the repressor to get control of the system. Thus, in naive cells, the system stays repressed even at fairly high concentrations of IPTG. (See color insert.)

show equivalent levels of β-galactosidase expression, but, if there is hysteresis in the activator module control, should show very different glutamine synthetase levels. As shown in Fig. 4, this behavior was observed. It should be noted that since the *glnL2302* mutation was present on p3Y15, a variety of growth media can be used for the experiment, including complex media such as LB or nutrient broth. To compensate for the loss of normal Ntr regulation, glutamine is included in all media at 0.2% (w/v). (Note that glutamine does not survive autoclaving and that filter-sterilized glutamine must be added to the medium after autoclaving.)

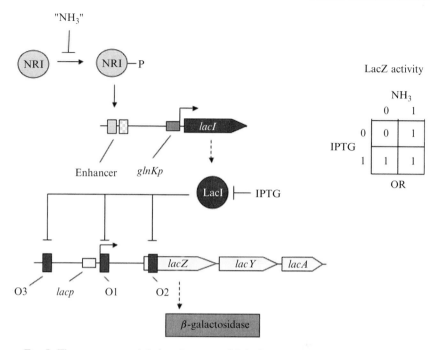

FIG. 5. The repressor module functions as an OR logic gate. When the repressor module is present in cells containing a mutation in the natural *lacI* gene, and with otherwise wild-type Ntr system and *lacZYA*, it provides OR gate function with regard to ammonia and IPTG for expression of *lacZYA*. (See color insert.)

The repressor module, when present in cells containing a deletion of the natural *lacI* and wild-type *lacZYA* and Ntr system, is predicted to display OR gate function with regard to ammonia and IPTG control of *lacZYA* expression (Fig. 5). This is because ammonia blocks the formation of the activator, while IPTG blocks the function of the Lac repressor produced from the repressor module. Consequently, either stimulus provides full expression of *lacZYA* (Fig. 5). The designed strain (3.300 containing the repressor module) was observed to indeed show OR gate function.

Procedures for Clock Experiments

We have used a variety of growth media for clock experiments; a very good medium providing rapid growth of cells and good oscillatory function is W-salts based glucose-glutamine-caseamino acids, with the following formula (per liter): 10.5 g K_2HPO_4, 4.5 g KH_2PO_4, 0.65 ml of 1 *M* $MgSO_4$, 0.04 g thiamine, 0.04 g tryptophan, 0.5 g glutamine, 10 g glucose, 5 g Bacto caseamino acids, 0.1 g ampicillin, and 1 ml of 34 mg/ml chloramphenicol. Because of its labile components, the medium is filter sterilized. A single colony of the clock strain is picked from an LB + ampicillin plate, inoculated into 2 ml of medium, and allowed to grow to saturation; 1.2 ml of the overnight culture is used to inoculate 120 ml of medium containing 0.5 mM IPTG. This culture is incubated until the turbidity reaches the desired turbidity for the clock experiment; typically a 10- to 12-h incubation provides an OD_{600} of ~0.6, which gives good results. The cells are harvested by centrifugation, resuspended in 120 ml of fresh medium lacking IPTG, repelleted, again resuspended in 120 ml of fresh medium lacking IPTG, and introduced into the continuous culture device. The volumes stated previously can be scaled as appropriate for a variety of working volumes required by different fermentors. The optical density of the culture can be corrected in the initial stages of the run by control of the fermentor nutrient pump. Best oscillatory behavior was observed when the continuous culture was run at OD_{600} of 0.5–0.6. Cells are grown in the continuous culture device at a constant optical density by controlling the nutrient flow. Samples are removed periodically (or obtained from the fermentor efflux) and assayed for β-galactosidase and glutamine synthetase.

We have used two methods for conducting the continuous culture experiments: one utilizing a standard laboratory fermentor that was not designed to function as a turbidostat and one using a custom-built turbidostat. Both methods are described briefly here. It is important to note that the clock strain does not grow at a constant rate in continuous culture; rather the presence of the synthetic genetic clock causes growth to slow down as the activator module and *lacZYA* expression are derepressed and to speed up as the activator module and *lacZYA* are repressed. Thus, to maintain a constant culture optical density, the nutrient flow must be adjusted continuously. When using a standard laboratory fermentor not designed to function as a turbidostat, this is accomplished by sampling the fermentor efflux periodically, measuring OD_{600}, and adjusting the nutrient flow rate manually so as to hold the culture at constant turbidity. Because clock experiments typically are run for 80 h or more, the experiments require lots of coffee and/or several people to take shifts minding the fermentor. Nevertheless, good oscillatory function can be easily observed in such experiments, despite the fairly crude control of the culture turbidity (Atkinson *et al.*, 2003; Fig. 6).

For experiments with a typical laboratory fermentor, the starting culture at approximately the desired working OD_{600} is introduced directly into the reactor, and the optical density is corrected to the desired optical density by manipulation of the nutrient pump. Filtered air is pumped into the reactor and vigorous stirring is used to provide good aeration. Medium

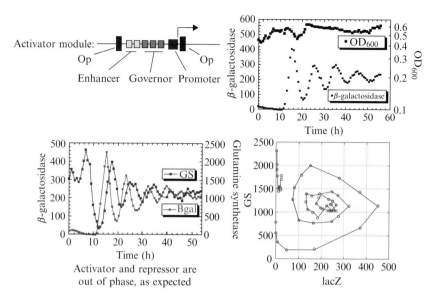

FIG. 6. Results of clock experiments using a standard laboratory continuous culture device. The activator module used in the experiments is depicted at top left. Experiments were conducted as described in the text, with manual control of the nutrient pump to maintain stable culture turbidity. At top right, results are shown from an experiment where OD_{600} and β-galactosidase were monitored. At bottom left, another experiment where glutamine synthetase was also measured is shown; for clarity OD_{600} data are not shown for this experiment. As predicted by the model, the expression of glutamine synthetase and β-galactosidase was out of phase. At bottom right, the phase diagram for β-galactosidase and glutamine synthetase expression data is shown. Note that if the system contained a perfect oscillator, the phase diagram would reveal a stable orbit instead of spiraling inward to a steady state. (See color insert.)

is pumped into the reactor from a 4-liter reservoir, to which fresh sterile medium is added aseptically as needed. Samples are collected from the efflux line directly into a clean disposable cuvette; the OD_{600} of the culture is measured, and aliquots are used for the β-galactosidase and glutamine synthetase assay.

To automate the clock experiments, we developed a turbidostat as depicted in Fig. 7. The reactor is a 500-ml Erlenmeyer flask containing a 120-ml culture kept at constant optical density by varying the media flow appropriately. A close-up view of the reactor is shown in Fig. 8; it has four connections to the outside: one line to pour media in, one line to suck culture out when the volume slightly exceeds 120 ml (waste line), one line to take samples, and one line to pump in filtered air. The air pump is a large fish tank aerator connected to a standard 0.22-μm filter to provide sterility. The reactor is contained within a standard small laboratory incubator, and

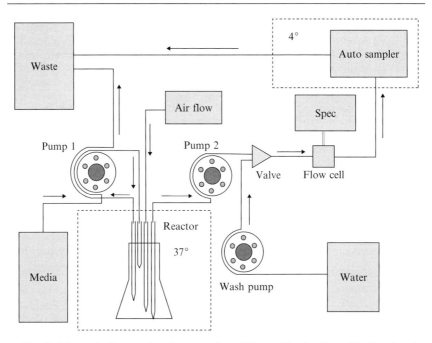

Fᴵɢ. 7. Schematic diagram for a home-made turbidostat. For details and further description, see the text. (See color insert.)

good mixing of the culture is provided by placing the flask on a standard stir plate and including a magnetic stir bar in the vessel (not depicted). In addition to stirring, the pumping in of air contributes to mixing within the reactor.

The medium is pumped in from a media reservoir at a variable speed. To maintain sterility and accommodate the large volumes of medium required, we use a home-made medium "tree" that consists of up to six 1-gal bottles connected in a daisy chain by sterilizable Tygon tubing. This apparatus is sterilized dry under a tin-foil tent and is placed into a bacteriological hood while still hot, and the filter sterilized medium is introduced after it cools sufficiently. The waste line sucks culture out and has twice the flow rate of incoming media; it sucks culture out only when the volume goes higher than 120 ml. The waste line needs to have a stronger flow rate than the media line in order to keep the volume constant. The depicted design was chosen, as opposed to having the same flow rate of media in and culture out, because fluctuations in the flow rate of the lines, even if both lines are connected to the same pump, make control of the volume more difficult.

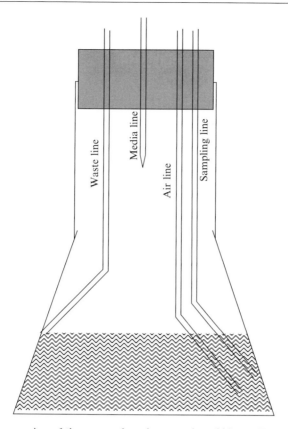

Fig. 8. Close-up view of the reactor for a home-made turbidostat. For explanation, see the text.

Any slight difference in flowing rates may lead to volume accumulation or reduction, and because each experiment lasts several days, this constitutes a problem. The waste flow rate is made double to the incoming medium flow rate using tubing of twice the diameter and the same pump (Fig. 7).

The sampling pump (pump 2 in Fig. 7) intermittently removes culture from the fermentor at a rate that is considerably less than the nutrient flow. Every 5 min a small aliquot of culture is pumped out through the sampling line to a flow cell (Fig. 7). The flow rate of media is changed as a function of this OD_{600} in order to keep the optical density constant. After each OD_{600} reading, the sampling pump is shut off, and the flow cell is washed with sterile water, using water that is pumped into the sampling line from a water reservoir by the wash pump (Fig. 7). This wash routine was found to be important, as continuous pumping of the culture through the flow cell and autosampler leads to formation of a biofilm on the surfaces, fouling the reading of absorbance and contaminating the samples and instruments. Every 30 min, or six OD_{600} readings, a 2-ml sample of the culture is directed to waste to clear the lines and a 1-ml sample is collected in a well of a deep well 96-well microtiter plate. The automatic sampler allows the sampling line to point to any of the 96 wells in the microtiter plate, or to waste.

The system depicted in Fig. 7 is fully computer controlled. The media, sampling, and wash are each pumped by a computer-controlled peristaltic pump (media and sampling lines are pumped by VS-series Alitea peristaltic pumps, water is pumped by a S-series Alitea peristaltic pump), while the spectrometer (Ocean Optics SD2000) and autosampler (AIM 3200) also have an interface to the computer. The flow cell to measure absorbance, as well as the pumps, autosampler, and spectrometer, is from FIAlab Instruments (Bellevue, WA).

The software used to control the system was implemented in Labview (National Instruments, Austin, TX). Labview is a graphical programming language that eases the interface of the computer with instruments and data acquisition devices. Some instruments, such as the autosampler and valves, have an interface to the computer via the serial port and an ascii language, whereas others, such as the spectrometer or a data acquisition card, have some other way of interfacing with the computer, but they offer a basic library of functions in Labview to operate them. Essentially any instrument can be controlled from Labview. In addition to its ease of interfacing with instruments, as a programming language, Labview offers the capabilities of any other programming language (such as C/C++ or Java). Using Labview, we designed the following algorithm to run the system (collect samples and keep the OD constant).

1. Set the target OD, the media flow rate (measured as the percentage of the top speed of the pump, initially 50%), and make an initial OD measurement by pumping culture to the flow cell.
2. Every 5 min:
   a. Pump culture from the flask to the flow cell and make an OD measurement.
   b. Wash the flow cells with sterile water.
   c. Adjust the media flow rate according to the formula:

$$new\_flow = flow - \frac{100V}{max\_flow \cdot \Delta t} Ln\left(\frac{tOD \cdot pOD}{OD^2}\right) \quad (1)$$

   where V is the volume of the of the culture, max_flow is the physical flow when the pump is at top speed (typically

5.7 ml/min), Δt is the time interval between readings (5 min), tOD is the target OD, pOD is the previous OD reading, and OD is the current OD reading.

3. Every half an hour (every six readings), pump 2 ml of culture to waste and collect 1 ml of culture in a deep well 96-well plate.

Conducting automated experiments requires that the samples be maintained at low temperature (4°) to prevent growth of the cultures and changes in the level of β-galactosidase and glutamine synthetase. We observed that samples held at 4° displayed stable levels of these two enzymes for several days. To provide uniformity, we routinely maintain all samples at 4° for at least 4 h before conducting the assays. The autosampler is contained within a standard small refrigerator, which we adapted to our purpose by drilling a small entry port into one side and a small "overflow" port into its base. The entire apparatus, including refrigerator, incubator, computer, pumps, and flow cell, fits easily onto one standard laboratory bench.

The automated system utilizes miniaturized assays and uses samples that have been stored at 4°. We observe that the miniaturized assays are somewhat noisier than the standard assays and that some of this may be due to inconsistent resuspension of the settled cells from the stored samples. Thus, we recommend special attention to resuspending the settled cells by both vortexing the collection plates extensively and pipetting the samples into and out of the well several times before taking the aliquots for measurement. This noisiness notwithstanding, oscillatory behavior can be easily observed using an automated system with samples stored at 4° and miniaturized assays (Fig. 9).

Procedures for Measurement of Glutamine Synthetase and β-Galactosidase for Clock and Toggle Switch Experiments

Glutamine Synthetase Microassay

1. Assemble the reaction mix as follows:

Reagent	Amount (ml)
0.45 M imidazole, pH 7.33	35
0.3 M NH$_2$OH·HCl	7
0.01 M MnCl$_2$	3.5
0.06 M KAs$_2$O$_4$, pH 7.2	35
6 mM ADP, pH 7.0	7
1.5 mg/ml CTAB	7
2.9% glutamine	12

Correct the pH to 7.27 with KOH.

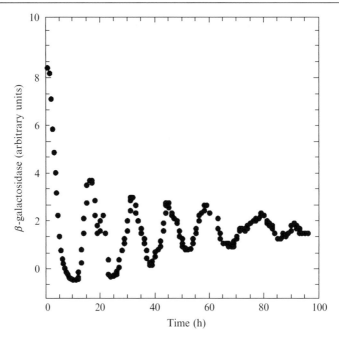

FIG. 9. Results of a clock experiment employing an automated system and miniaturized β-galactosidase assay.

2. Transfer 400 μl of each sample to be assayed to a well in a deep well 96-well plate. Centrifuge to pellet the cells and remove the supernatant carefully and discard. Resuspend the cells in 40 μl of CTAB cell wash buffer and transfer to a standard flat-bottom 96 well microtiter plate. The CTAB cell wash buffer is 5 mM imidazole, pH 7.15, 0.1 mg/ml CTAB, and 0.27 mM MnCl$_2$.

3. Pipette 100-μl aliquots of the reaction mix into the wells; this starts the reaction. Typically, add the mix aliquots at 20-s intervals using an eight-channel micropipettor; later the reactions are stopped in the same sequence at 20-s intervals. When all samples have been added, incubate the plate at 37° in a standard microbiological incubator. In a separate plate, set up a sample from an early time point in the clock experiment (where the glutamine synthetase activity is expected to be high) in triplicate in three separate wells of a microtiter plate. This plate is used to determine the appropriate length of time to incubate the reactions. At 30-min intervals, add stop solution to one of the three wells in the test plate, and allow the reactions to continue until a dark brown color is obtained upon stopping the test reaction. Typically, a 1-h incubation is sufficient for measurement of the activity. The stop solution is (per liter) 55 g FeCl$_3 \cdot$6H$_2$O, 20 g TCA, and 21 ml HCl.

4. To stop the reactions, add 100 μl of stop solution to each well in the same sequence and time interval that the reactions were started. Upon addition of the stop solution, the reaction mixtures that contain high glutamine synthetase activity will turn dark brown. Because the cell samples should all have similar OD_{600}, the oscillations in GS level should become immediately apparent as the reactions are stopped.

5. Measure the OD_{540} of the stopped reactions directly in a plate reader. In a separate plate, read the OD_{600} of the original cell samples directly. The glutamine synthetase activity, in arbitrary units, is simply the OD_{540} value divided by the OD_{600} value.

6. For toggle switch experiments, pellet cells from 1 ml of the cultures in microcentrifuge tubes and resuspend in 0.1 ml of CTAB cell wash buffer. Then add 20- and 50-μl aliquots to the reaction mix in microtiter plates and follow the procedure described earlier.

β-Galactosidase Microassay

1. Load the wells of a deep-well 96-well plate with 400 μl of Z buffer, 10 μl of 0.1% SDS, and 10 μl of chloroform (add last). The formula for Z buffer is (per liter) 16.1 g Na_2HPO_4 $7H_2O$, 5.5 g NaH_2PO_4 H_2O, 0.75 g KCl, and 0.246 g $MgSO_4$ $7H_2O$. Immediately prior to the assay, add 0.27 ml of β-mercaptoethanol/100 ml of Z buffer.

2. Add 100 μl of the culture sample progressively to each well and mix thoroughly by pipetting (the chloroform must be well mixed into the suspension for uniform cell permeabilization). We typically use an eight-sample multichannel micropipettor to minimize the difference in exposure of samples to the chloroform. Incubate the reaction mixtures for 10 min at room temperature.

3. To start the reactions, add 100 μl of ONPG (4 mg/ml in 0.1 mM phosphate buffer) to the wells in sequence and at fixed intervals and mix thoroughly by pipetting.

4. Allow the reaction to proceed until the appearance of yellow color in some of the wells. Note that oscillations in the level of β-galactosidase become readily apparent at this stage, and some of the wells should never turn bright yellow, as they contain very low levels of β-galactosidase. Typically, a 5- to 10-min incubation is required.

5. Stop reactions in the sequence they were started and at the same intervals by adding 200 μl of 1 M Na_2CO_3.

6. Allow the stopped reactions to sit for 10 min for full color development and settling of the cells, transfer 150-μl aliquots (from the top) from each well to a 96-well flat-bottom plate, and record the OD_{410} and OD_{560}. In a separate series of wells, measure the OD_{600} of the original culture samples.

7. The β-galactosidase activity is $[OD_{410} - (1.75 \times OD_{560})]/(OD_{600} \times time)$.

Improved Procedures for Fabrication of Synthetic Genetic Modules and Integration of These Modules into Chromosomal Landing Pads

During our synthetic biology studies, we have developed fabrication methods in concert with development of the clock, such that early versions of the clock do not incorporate the most useful aspects of fabrication methodologies developed later. This chapter presents our most recent vector system for the fabrication of synthetic genetic modules and incorporation of the fabricated modules into chromosomal landing pads. The vector system should allow placement of any module in any position and on either strand of the chromosome and permit the genetic modules to be constructed easily from diverse natural components.

To develop new landing pads, the vectors shown in Fig. 10 can be used. A nonessential and easily scored gene, such as a sugar utilization gene, is chosen as the target for chromosomal integration. Two fragments of the gene of ~90 bp each are cloned between the *Eco*RI and the *Sac*I sites and between the *Xho*I and the *Hin*DIII sites. The landing pad will be localized by recombination of these sequences with their chromosomal counterparts; we typically choose "right" and "left" target sequences from within the same gene. Next, we introduce an antibiotic resistance gene flanked by strong transcriptional terminators between the *Sac*I and the *Not*I sites, as depicted in Fig. 11 (step 1 plasmids). We have used kanamycin resistance and chloramphenicol resistance in step 1 plasmids; we refer to this marker as "AB1" Step 1 plasmids are functional landing pads that can be used to integrate modules into the target gene; those available and under construction are listed in Table I. One clones the genetic module to be placed onto

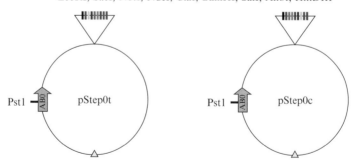

Fig. 10. pStep0 plasmids used to create new landing pads. The vector backbone consists of the *Eco*RI–*Sap*I fragment of pBR322 containing the origin of replication (triangle) and ampicillin resistance gene (AB0). Plasmids contain a linker with multiple unique sites in either of two sequences, denoted "t" and "c." (See color insert.)

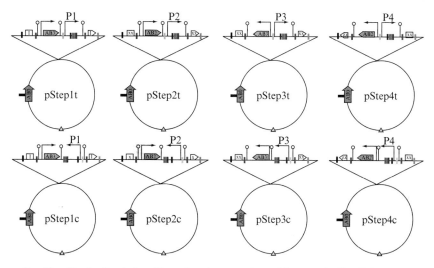

Fig. 11. pStep1–pStep4 plasmids used to place genetic modules onto the *E. coli* chromosome. For pStep1 plasmids, T signifies the target for recombination with the chromosome, P1 signifies a module promoter, and lollipop symbols signify transcriptional terminators. The selective marker for pStep1 recombination is AB1. For pStep2 plasmids, AB1 fragments constitute the target for recombination with the chromosome, P2 signifies a module promoter, and the selective marker for recombination is AB2. For pStep3 plasmids, AB2 fragments constitute the target for recombination with the chromosome, P3 signifies a module promoter, and the selective marker for recombination is AB1. For pStep4 plasmids, AB1 fragments constitute the target for recombination, P4 signifies a module promoter, and AB2 provides the selective marker for recombination. Note that the "t" and "c" series of plasmids differ in the arrangement of internal sites of the linker. For all plasmids, AB0 is the ampicillin resistance gene, containing the *Pst*I site that may be used for plasmid linearization.

the chromosome between the remaining five unique sites flanked by the recombination target fragments. One could introduce the module as a unit after assembly in another plasmid or add portions sequentially. Typically, we use the *Nde*I site for the beginning of the structural gene, as *Nde*I sites contain an ATG initiation codon. If the structural gene is cloned as an *Nde*I–*Sal*I fragment or as an *Nde*I–*Bam*HI fragment, a strong terminator can be added between the *Sal*I and the *Xho*I sites, isolating the gene from extraneous transcription from both directions. Next, the promoter and mRNA leader sequence are added as a *Not*I–*Nde*I fragment, completing the module. Typically, this fragment is assembled in separate plasmids or added as a synthesized DNA fragment by annealing long oligonucleotides that have been designed to have the appropriate overhangs. We have worked with synthesized fragments of 100 bp in length, which permits the construction of many combinatorial promoter/leader sequences. All of our vectors contain a unique *Pst*I site in the vector sequence, which allows easy linearization of the plasmid for electroporation into a *recD* mutant. In the event that there is a *Pst*I site in the module itself, several other unique sites are also present in the vector portion of the molecule that might be used.

Synthetic genetic devices can be built up by placing different modules into different chromosomal landing pads, as was done for our initial clock. The problem with this approach is that each module is linked to a distinct antibiotic resistance marker, and one quickly runs out of antibiotic resistance selections that offer very strong selection in the presence of each other. Tetracycline resistance provides a strong selective marker, but we exclude its use from our system as we use the *recD::Tn10* (tet resistant) mutant to recombine modules onto the chromosome. Similarly, our plasmid vectors encode ampicillin resistance, and thus we do not use this drug resistance marker elsewhere in the system. We have found that kanamycin resistance and chloramphenicol resistance provide very strong selection (Table I) and that gentamycin resistance, as found in our pRBS landing pad (Table I), does not. Thus, integrating three separate modules onto the chromosome is a problem. To overcome this, two approaches are taken. First, we are currently investigating additional antibiotic resistance markers, such as spectinomycin resistance, for use as "AB1" in step 1 plasmids. Second, we designed the vector system such that each landing pad can be used in a reiterative fashion by the sequential integration of modules, which end up next to each other on the bacterial chromosome. The plasmids used for these successive integrations, pStep2, pStep3, and pStep4, are shown in Fig. 11. As before, there are two versions of each of these to permit either orientation of the genetic module. Target sequences of the pStep2–pStep4 plasmids consist of internal segments of the drug resistance genes used in the prior step. Thus, if one uses a pStep1 plasmid with kanamycin as the selectable marker (AB1 in Fig. 11), the corresponding pStep2 plasmid will integrate into the AB1 gene and contain as a selectable marker chloramphenicol resistance (AB2). The pStep3 plasmid will contain AB1 as the selectable marker and integrate into the AB2 gene, and the pStep4 plasmid will contain the AB2 gene as a selectable marker and integrate into the AB1 gene. Thus, the drug resistance of the module is switched between AB1 and AB2 upon each successive integration, and the synthetic genetic constructs are each flanked by terminators and located next to one another on the chromosome. A simple PCR test of the final strain can then confirm the expected structure. The sequence of nesting is depicted in Fig. 12, as well as the predicted end product of an assembly where four promoter-gene fusions (G1–G4) are positioned in a single landing pad.

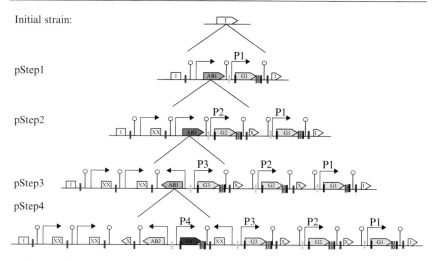

FIG. 12. The sequence of nesting that occurs upon successive use of pStep1–pStep4 plasmids for integration of modules into the chromosome. Symbols are as in Fig. 11; G1–G4 depict structural genes within the modules, expressed from module promoters P1–P4. Recursive use of pStep1–pStep4 plasmids to modify a chromosomal site results in side-by-side integration of modules on the chromosome, separated by transcriptional terminators.

Fabricating Genetic Modules

Most of the genetic modules that we fabricate consist of a regulated promoter that drives the expression of a regulatory gene. The minimal components therefore consist of a bacterial promoter, regulatory sites for control of the promoter, an mRNA leader sequence containing translational initiation sequences, the structural gene for the desired regulatory protein, and a transcriptional terminator sequence. For genetic isolation, it is advisable to also include transcriptional termination sequences upstream from the regulated promoter. Fabrication consists of choosing the sequences to be used and assembling the sequences. Modules pertinent to the clock and toggle switch experiments are presented in Table I. A much more extensive registry of parts for synthetic biology constructions, including parts developed by over 40 laboratories (including our own), is the registry of BioBricks (http://parts. mit.edu). This registry contains a large variety of promoter, operator, mRNA leader, and terminator elements, as well as a variety of structural genes encoding transcription factors and reporters. Standard cloning methods can be used to introduce any element from the BioBricks collection into our vectors by either performing a blunt end ligation or using PCR to add appropriate sites to the BioBrick. We are currently developing a modified series of landing pad vectors that are able to directly accept any Biobrick fragment or construct. Although cloning from BioBricks or other sources is an appropriate way to build modules, we have found that in many cases, simply cloning annealed long oligonucleotides of the desired sequence is the most direct way to build modules.

References

Atkinson, M. R., Blauwkamp, T. A., Bondarenko, V., Studitsky, V., and Ninfa, A. J. (2002a). Activation of the *glnA*, *glnK*, and *nac* promoters as *Escherichia coli* undergoes the transition from nitrogen excess growth to nitrogen starvation. *J. Bacteriol.* **184,** 5358–5363.

Atkinson, M. R., and Ninfa, A. J. (1992). Characterization of *Escherichia coli glnL* mutations affecting nitrogen regulation. *J. Bacteriol.* **174,** 4538–4548.

Atkinson, M. R., and Ninfa, A. J. (1993). Mutational analysis of the bacterial signal-transducing kinase/phosphatase nitrogen regulator II (NRII or NtrB). *J. Bacteriol.* **175,** 7016–7023.

Atkinson, M. R., Pattaramanon, N., and Ninfa, A. J. (2002b). Governor of the *glnAp2* promoter of *Escherichia coli. Mol. Microbiol.* **46,** 1247–1257.

Atkinson, M. R., Savageau, M. A., Meyers, J. T., and Ninfa, A. J. (2003). Development of genetic circuitry exhibiting toggle switch or oscillatory behavior in *Escherichia coli. Cell* **113,** 597–607.

Barkai, N., and Leibler, S. (2000). Circadian clocks limited by noise. *Nature* **403,** 267–268.

Elliott, T. (1992). A method for constructing single-copy *lac* fusions in *Salmonella typhimurium* and its application to the *hemA-prfA* operon. *J. Bacteriol.* **174,** 245–253.

Feng, J., Goss, T. J., Bender, R. A., and Ninfa, A. J. (1995). Activation of transcription initiation from the *nac* promoter of *Klebsiella aerogenes. J. Bacteriol.* **177,** 5523–5534.

Hoch, J. A., and Silhavy, T. J. (1995). "Two-Component Signal Transduction." American Society for Microbiology, Washington, DC.

Hunt, T. P., and Magasanik, B. (1985). Transcription of *glnA* by purified *Escherichia coli* components: Core RNA polymerase and the products of *glnF*, *glnG*, and *glnL*. *Proc. Natl. Acad. Sci. USA* **82,** 8453–8457.

Jin, T., and Inouye, M. (1993). Ligand binding to the receptor domain regulates the ratio of kinase to phosphatase activities of the signaling domain of the hybrid *Escherichia coli* transmembrane receptor, Taz1. *J. Mol. Biol.* **232,** 484–492.

Lee, S. Y., De La Torre, A., Yan, D., Kustu, S., Nixon, B. T., and Wemmer, D. E. (2003). Regulation of the transcriptional activator NtrC1: Structural studies of the regulatory and AAA+ ATPase domains. *Genes Dev.* **17,** 2552–2563.

Levskava, A., Chevalier, A. A., Tabor, J. J., Simpson, Z. B., Lavery, L. A., Levy, M., Davidson, E. A., Scouras, A., Ellington, A. D., Marcotte, E. M., and Voigt, C. A. (2005). Synthetic biology: Engineering *Escherichia coli* to see light. *Nature* **438,** 441–442.

Ninfa, A. J., Reitzer, L. J., and Magasanik, B. (1987). Initiation of transcription at the bacterial *glnAp2* promoter by purified *E. coli* components is facilitated by enhancers. *Cell* **50,** 1039–1046.

Novick, A., and Weiner, M. (1957). Enzyme induction as an all-or-none phenomenon. *Proc. Natl. Acad. Sci. USA* **43,** 553–566.

Oehler, S., Eismann, E. R., Kramer, H., and Muller-Hill, B. (1990). The three operators of the lac operon cooperate in repression. *EMBO J.* **9,** 973–979.

Pioszak, A. A., and Ninfa, A. J. (2003a). Genetic and biochemical analysis of phosphatase activity of *Escherichia coli* NRII (NtrB) and its regulation by the PII signal-transduction protein. *J. Bacteriol.* **185,** 1299–1315.

Pioszak, A. A., and Ninfa, A. J. (2003b). Mechanism of the PII-activated phosphatase activity of *Escherichia coli* NRII (NtrB): How the different domains of NRII collaborate to act as a phosphatase. *Biochemistry* **42,** 8885–8899.

Reitzer, L. J., and Magasanik, B. (1985). Expression of *glnA* in *Escherichia coli* is regulated at tandem promoters. *Proc. Natl. Acad. Sci. USA* **82,** 1979–1983.

Author Index

A

Abe, H., 234
Abeles, F. B., 270
Abergel, C., 41
Abhinandan, K. R., 50, 67
Abian, J., 308
Abulencia, C. B., 36
Acuna, G., 5, 15, 18
Adachi, J., 304
Adams, P. D., 110
Adeishvili, N., 307
Adler, J., 192, 196, 200, 212, 423,
    430, 433, 434
Agapow, P. M., 381
Agrawal, R., 381
Aguilar-Uscanga, B., 473
Aguirre, A., 380
Agula, H., 278
Ahmer, B. M. M., 247, 256, 257
Aiba, H., 33, 95, 363
Ajello, L., 465
Akaike, T., 452
Akimitsu, N., 235, 237
Akins, D. R., 422, 423
Alam, M., 11, 191, 194, 202, 214, 215
Albanesi, D., 397
Albertini, A. M., 116, 398, 404
Aldrich, R. W., 76, 82, 83, 87, 90
Alex, L. A., 3, 15, 33, 56, 306, 467
Alexander, R. P., 3, 14, 34, 41, 192
Alexandre, G., 21, 27
Allaire, M., 246
Allen, E. E., 33, 34, 37, 39, 43
Alley, M. R., 411
Alloni, G., 116, 398, 404
Alm, E., 116
Alm, R. A., 5, 15, 18, 20
Almeida, J. S., 373
Alon, U., 124, 133, 362, 440
Altabe, S., 396, 397
Altschuh, D., 76
Altschul, S. F., 6, 78, 145, 295

Amberg, D. C., 273
Ames, G. F.-L., 256
Ames, P., 439
Amin, D. N., 193, 194, 222, 223, 224
Amiot, N. C., 65
Amjadi, M., 36
Amrein, K. E., 387
Anand, G. S., 61
Anantharaman, V., 13, 63
Andersen, K. K., 396, 397
Anderson, I. J., 33, 34, 39, 40
Anderson, J. B., 51
Andersson, L., 147
Andrews, B. W., 123, 132, 136
Anfinrud, P. A., 313
Angly, F. E., 37
Annamalai, R., 245
Annan, R. S., 449
Aoki, K., 234
Aoki-Kinoshita, K. F., 95
Aono, R., 386
Apweiler, R., 51, 145
Arai, M., 60
Araki, M., 95
Aravind, L., 5, 11, 13, 36, 53, 63, 66
Arcas, B. A. Y., 440
Arents, J. C., 77
Argos, P., 60
Arighi, C. N., 51
Arisawa, M., 467
Arkin, A. P., 127
Arkin, K., 116
Arminski, L., 145
Armitage, J. P., 4, 5, 15, 17, 18,
    439, 440, 445
Arndt, U. W., 311
Arnold, E., 108
Arps, D., 488
Asahi, N., 235
Asai, K., 396, 411
Ashburner, M., 381
Ashida, M., 235
Aslund, F., 235

513

Subject Index

A

Adenylate cyclases, identification from sequence analysis, 66

Aer
  aerotaxis overview, 191
  assays of aerotaxis
    apparatus
      gas flow cell, 197–198
      gas proportioner, 198–200
      microscope, 198
    buffer, 196–197
    capillary assays, 203–208
    Escherichia coli growth, 196, 200
    microscopy, 201–203
    motile bacterial strains, 195–196
    smear preparation, 201
    soft agar plate assays, 208–214
    spatial-gradient assays, 194, 203–205, 208–214
    temporal-gradient assays, 194–195, 201–203
  periplasmic accessibility studies, 227
  preferred partial oxygen pressure determination, 205–208
  redox taxis assays
    spatial assays, 214–215
    temporal assays, 215–217
  signaling, 192
  structure
    disulfide cross-linking analysis
      bifunctional sulfhydryl-reactive linkers, 223–224
      copper phenanthroline cross-linking, 220–210
      intradimeric versus interdimeric bond determination, 221–222
      overview, 218–219
      site-directed mutagenesis for cysteine replacement, 219–220
    membrane topology accessibility studies

5-iodoacetamidofluorescein accessibility, 225–226
    membrane vesicle preparation, 224–225
    methoxy polyethylene glycol maleimide accessibility, 226–227
    overview, 224
  overview, 192–194, 218
Allelic exchange mutagenesis, *see* Chemotaxis, *Borrelia burgdorderi*
Antibiotic screening, *see* YycG–YycF system
Aranorosinol B, YycG autophosphorylation inhibition, 390
ArcA
  model building in receiver domain classification using protein interaction surfaces, 147–148, 150
  operon targets, 362
ArcB
  function, 289
  histidine protein phosphatase, *see* SixA

B

BAD1
  virulence role, 466, 470
  Western blot, 477–478
Biliverdin, *see* Photoactive yellow protein-phytochrome
BLAST
  chemotaxis component analysis in bacteria, 6
  gene prediction, 39
  sensory domain identification in histidine kinases, 58
  SixA homolog analysis, 295
  Spo0A output domain identification, 63–64
Bode plot, chemotaxis control analysis, 129–132
BtuB, TonB interaction assays, 263–264

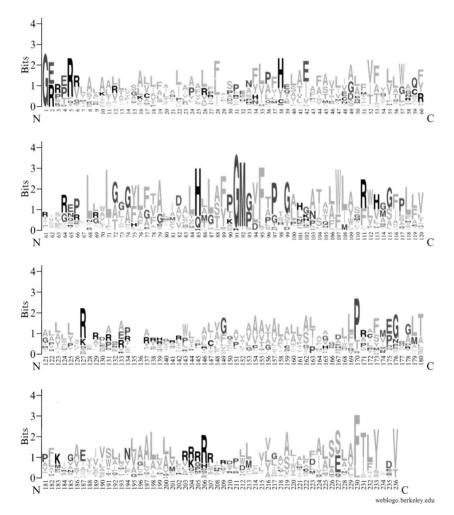

GALPERIN AND NIKOLSKAYA, CHAPTER 3, FIG. 1. Sequence logo of the MASE3 domain generated using the WebLogo (http://weblogo.berkeley.edu) tool from a multiple alignment of 35 different sequences of the MASE3 domain aligned to the *Methanosarcina acetivorans* histidine kinase MA3481. Residue numbering starts from Gly-4 of the MA3481 sequence.

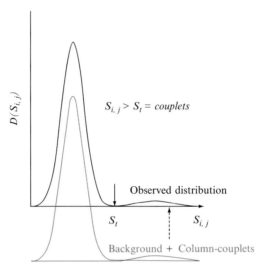

WHITE *ET AL.*, CHAPTER 4, FIG. 1. Schematic diagram of threshold determination. Couplets are defined to be column pairs that have corrected MI values, $S_{i,j}$, above the onset of an anomaly, S_t, in an assumed background distribution.

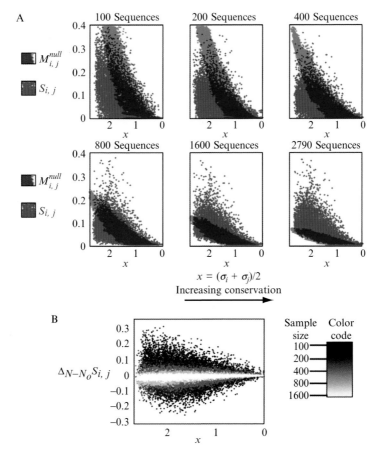

WHITE ET AL., CHAPTER 4, FIG. 2. Sample size analysis of corrected MI. (A) Overlaid plots of the corrected MI value $S_{i,j}(x)$ (blue) and the sample-size correction $M_{i,j}^{null}(x)$ (grey) for increasingly large subsets of the set of functionally coupled domains, where x is the average entropy between the columns i and j. For alignments consisting of 100 sequences, the sample-size correction is greater than the corrected values with larger contributions between columns with large entropy, that is, highly variable. The relative invariance of the corrected MI value over the range of sample sizes, even when the correction varies widely, suggests a useful correction. (B) Direct comparison of each value between the largest sample size available and the smaller sets $\Delta_{N-N_0}S_{i,j} \equiv S_{i,j}(N,x) - S_{i,j}(N_0,x)$, where N is the sample size and $N_0 = 2790$. If possible, a perfect correction would show no change between $S_{i,j}$ values for different sample sizes. The fact that the fluctuations around $\Delta_{N-N_0}S_{i,j} = 0$ are roughly symmetric for sample sizes greater than 200 suggests that the deviations from zero are not caused by a systematic problem with the correction. Rather, the fluctuations are likely to be caused by other sources, such as sampling of the different areas of the phylogenetic landscape being represented by the random selection process of the subsets.

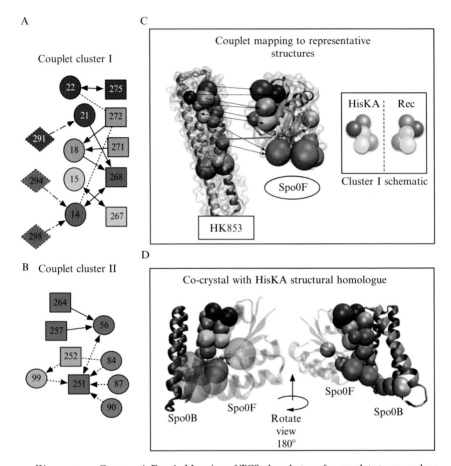

WHITE *ET AL.*, CHAPTER 4, FIG. 4. Mapping of TCS phosphotransfer couplets to exemplary structures. (A and B) The two distinct clusters formed in the couplet network. The assignment of arrows (or lack thereof) in the networks is achieved by the "best friend" transformation described in the text. Circles correspond to positions in the Rec domain and are numbered according to the Spo0F protein. Rectangles represent positions in the HisKA domain and are numbered according to the HK853 kinase (Marina *et al.*, 2005). Diamond shapes denote HisKA residues that cannot be reliably mapped to the Spo0B phosphotransferase structure through the HisKA HMM. Solid edges between nodes denote residue pairs that have a minimum distance d < 8 Å when mapped to the Spo0B/Spo0F complex (Zapf *et al.*, 2000), dotted edges d > 8 Å, and dash dotted edges are unknown. Coloring for the nodes is achieved by selecting best friends of the HisKA nodes; if they share a best friend they are colored identically. (C) Mapping of couplet network to HK853 and Spo0F with residue colorings (in bead format using VMD 1.8.5 [Humphrey *et al.*, 1996]) derived from node coupling coloring. Edges correspond to edges in the couplet diagram, and the thick red edge connects the His-Asp phosphotransfer residues. (Inset) Schematic of the mirror symmetry of couplets. (D) Front and back views of couplets mapped to the phosphotransfer complex of phosphotransferases Spo0B/Spo0F. In both C and D the His-Asp residues where the phosphotransfer takes place are shown in red "licorice" representation.

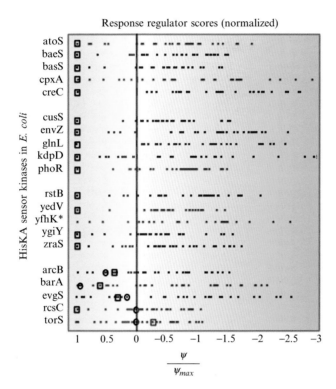

WHITE ET AL., CHAPTER 4, FIG. 6. Graphic representations of all Rec domain scores with each HisKA domain in *E. coli*. For each HisKA domain, scores are normalized by the maximum score for that domain. The blue (red) region represents all positive (negative) scores. Each dot is a Rec domain score. Dots with surrounding squares represent the known mate. Dots with circles surrounding are self-scores of the hybrid TCS proteins. The known mate of yfhK is the most negative score off the scale of this graph.

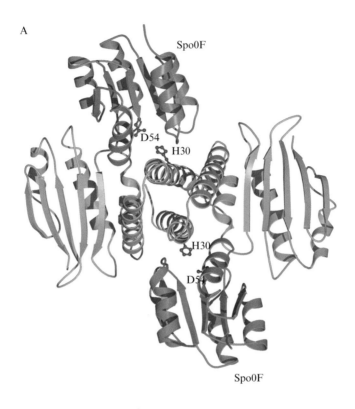

A

Spo0F

D54

H30

H30

D54

Spo0F

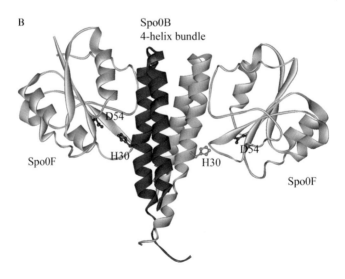

B

Spo0B
4-helix bundle

Spo0F

D54

H30

H30

D54

Spo0F

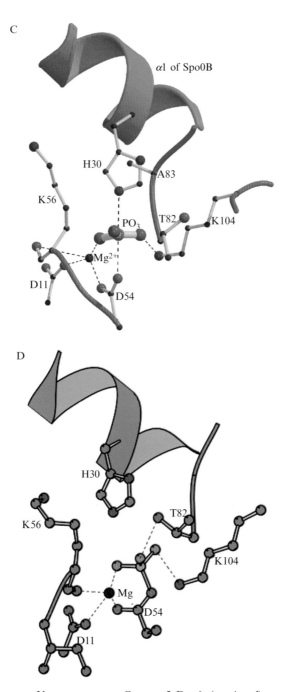

VARUGHESE *ET AL.*, CHAPTER 5, FIG. 6. (*continued*)

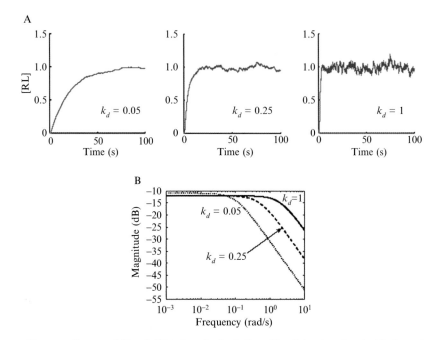

YI ET AL., CHAPTER 6, FIG. 5. Filtering of noise in ligand levels by changing the binding rate constants. (A) The dynamics of receptor–ligand complex, [RL], was simulated in response to a unit step ligand input that was corrupted by additive Gaussian noise with zero mean and 0.5 variance. The K_d was kept constant ($K_d = k_d/k_a = 1$) while the values of k_d (and k_a) were varied from 1 to 0.05. For the slower binding kinetics the amount of noise in [RL] levels was reduced because the noise in ligand levels was averaged over a longer time period. (B) Bode plot of sensitivity function of [RL] to the input [L]. Reducing k_d shifts the position of the roll-off to lower frequencies: $k_d = 1$ (solid line), $k_a = 0.25$ (dashed line), and $k_d = 0.05$ (dotted line). This change in the Bode plot explains the enhanced filtering caused by the slower binding dynamics.

VARUGHESE ET AL., CHAPTER 5, FIG. 6. Association of SpoOF with SpoOB and phosphory transfer. (A) A view of the SpoOB:SpoOF complex down the axis of the four-helix bundle. (B) A view of the SpoOF:SpoOB complex perpendicular to the four-helix bundle. For clarity, the C-terminal domains of SpoOB are omitted. Residues His30 of SpoOB and Asp54 of SpoOF are in close proximity for phosphoryl transfer. (C) A model for the transition state intermediate, created by placing a phosphoryl group between His30 and Asp54. The phosphorus atom forms partial covalent bonds with O^δ of Asp and N^ε of His and is in a penta-coordinated state. Negative charges on the phosphoryl oxygens are compensated through interactions with Mg^{2+} and Lys104 (reproduced with permission from Zapf *et al.*, 2000). (D) Active site interactions in the crystal structure of beryllofluoride SpoOF with SpoOB (reproduced with permission from Varughese *et al.*, 2006).

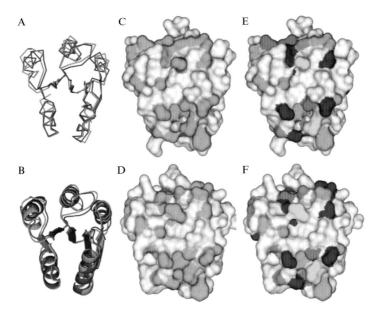

KOJETIN *ET AL.*, CHAPTER 7, FIG. 3. Modeled vs experimentally determined structure of PhoB from *E. coli.* Cα alignment between the modeled (blue) and the solved (red) structure of PhoB shown in (A) ribbon and (B) cartoon diagrams. Comparison of the hydrophobic surfaces of the model (C, E) and experimentally determined structure (D, F) of PhoB using a (C, D) single color for all hydrophobic residues or (E, F) a color-coded hydrophobic scale as described in the text.

PhoP

Strip analysis
  –strip 1: $(1 + 1)/2 = 1$
  –strip 2: $(5 + 4 + 5)/3 = 4.7$
  –strip 3: $(2 + 2 + 1)/3 = 1.7$

α1/α5 interface hydrophobic content
  –R: 2, O: 1, Y: 0, B: 2, G: 3

ResD

Strip analysis
  –strip 1: $(1 + 1)/2 = 1$
  –strip 2: $(5 + 4 + 5)/3 = 4.7$
  –strip 3: $(3)/1 = 3$

α1/α5 interface hydrophobic content
  –R: 2, O: 1, Y: 1, B: 0, G: 2

YycF

Strip analysis
  –strip 1: $(1 + 2)/2 = 1.5$
  –strip 2: $(5 + 4 + 5)/3 = 4.7$
  –strip 3: $(1 + 4)/2 = 2.5$

α1/α5 interface hydrophobic content
  –R: 2, O: 2, Y: 0, B: 1, G: 2

KOJETIN *ET AL.*, CHAPTER 7, FIG. 7. Example of color-coded hydrophobic surface classification. Strip plots of the receiver domains of PhoP, ResD, and YycF from *B. subtilis* are shown. Details about the breakdown during the strip analysis and α-helix1/α-helix5 interface hydrophobic content are provided for each protein.

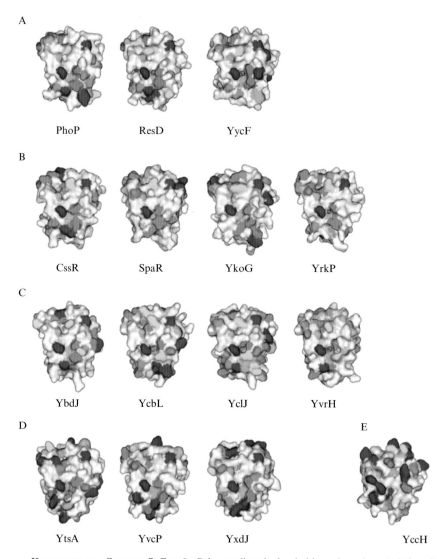

KOJETIN *ET AL.*, CHAPTER 7, FIG. 8. Color-gradient hydrophobic surface characteristics of *B. subtilis* comparative models. Models are categorized into the subclasses listed in Table VI (A–E). The perspective is such that the region comprising the α-helix 1 and α-helix 1/α-helix 5 interface is visible.

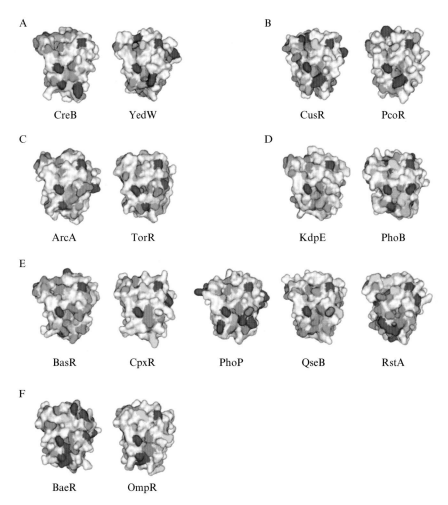

KOJETIN *ET AL.*, CHAPTER 7, FIG. 9. Color-gradient hydrophobic surface characteristics of *E. coli* comparative models. Models are categorized into the subclasses listed in Table VI (A–F). The perspective is such that the region comprising the α-helix 1 and α-helix 1/α-helix 5 interface is visible.

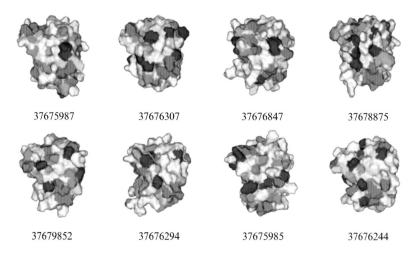

37675987   37676307   37676847   37678875

37679852   37676294   37675985   37676244

Kojetin *et al.*, Chapter 7, Fig. 10. Examples of *V. vulnificus* subclass 1 models. The perspective is such that the region comprising the α-helix 1 and α-helix 1/α-helix 5 interface is visible.

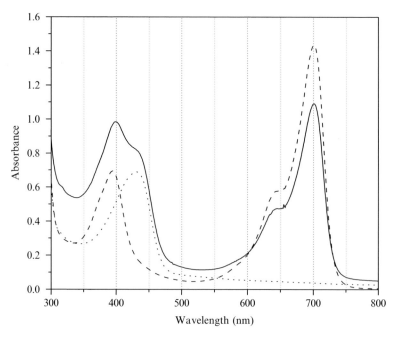

Chung *et al.*, Chapter 9, Fig. 1. Absorbance spectra of Ppr reconstituted with different chromophores in the dark. Ppr with both chromophores, including activated 4-hydroxycinnamic acid and biliverdin (solid line). Ppr with biliverdin (dashed line). Ppr with the activated 4-hydroxycinnamic acid (dotted line).

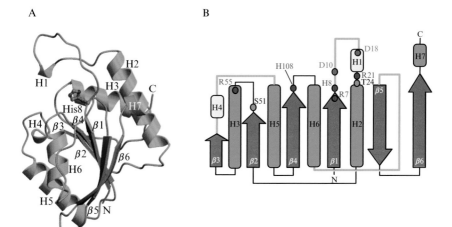

HAKOSHIMA AND ICHIHARA, CHAPTER 14, FIG. 2. Structure of the protein histidine phosphatase SixA. (A) Ribbon model of the SixA structure. The side chain of active residue His8 is shown as a stick model, and the α helices (green), 3_{10} helices (blue) and β strands (red) are labeled. (B) Folding topology of SixA. Color codes are the same as in A. Loops H4-H5, β1-H2, and H6-β5 that frequently contain insertions in other RHG phosphatases are colored orange (see text).

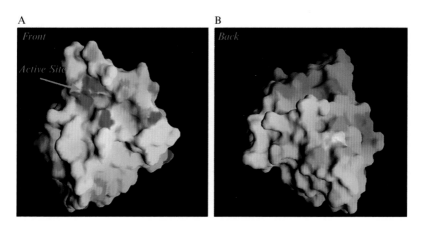

HAKOSHIMA AND ICHIHARA, CHAPTER 14, FIG. 3. Electrostatic molecular surface of SixA. (A) Front view of the electrostatic molecular surface of SixA. Regions of the surface that possess positive electrostatic potentials are colored blue, while those possessing negative potentials are colored red (blue = 10 k_BT, red = −10 k_BT, where k_B is Boltzmann's constant and T is the absolute temperature). The tungstate ion is shown as a stick model. Acidic and basic residues are labeled. (B) Back view of the electrostatic molecular surface of SixA.

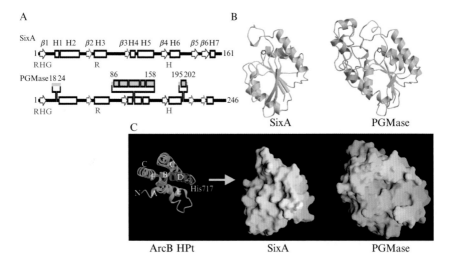

HAKOSHIMA AND ICHIHARA, CHAPTER 14, FIG. 4. Comparison of SixA and PGMase. (A) Comparison of the linear arrangement of structural elements and insertion. Helices and β strands are represented with arrows and rectangles, respectively. Inserted segments forming the helical domain are colored orange as in Fig. 2B. (B) Comparison of the RHG phosphatase folds found in SixA (PDB code 1A0B) and PGMase (PDB code 5PGM), the prototype of the RHG enzymes. Ribbon representation of the corresponding RHG phosphatase folds colored pale green. Each active histidine residue is shown as a ball-and-stick model. The extra helical domains forming the active sites of the RHG phosphatase domains are colored orange. (C) The convexity of the ArcB HPt domain (with ribbon representation) is matched with the grooved molecular surface of SixA (with surface representation), whereas the α-helical subdomain of yeast PGMase generates an active site that forms a deep pocket for binding to its small substrate 3-phosphoglycerate and prevents docking to the HPt domain. The active histidine residue (His717) of the HPt domain is shown as a ball-and-stick model. The extra helical domain is colored yellow. The active histidine residues of SixA and PGMase are colored blue-green.

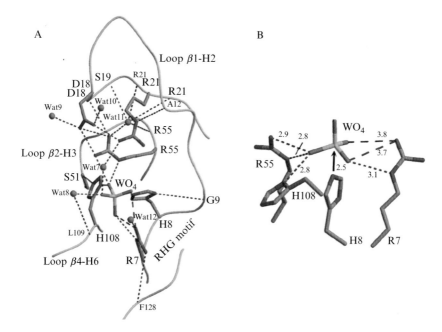

HAKOSHIMA AND ICHIHARA, CHAPTER 14, FIG. 5. Active site of SixA. (A) The active site of the tungstate-bound form of SixA. Side chains are shown as stick models with large labels. Main-chain tracings are shown with tubes colored light green. Hydrogen bonds are indicated by dotted lines. Water molecules are shown with small labels (Wats). Residues that participate in hydrogen-bonding interactions through their main chains are indicated with small labels at the main-chain tracings. (B) Close-up stereo view of residues interacting with the tungstate ion at the active site. Hydrogen-bonding and ion-pairing interactions are indicated by dotted lines and broken lines, respectively. Bonding distances are also shown. The contact between the tungsten atom and the N_ε nitrogen atom of His8 is indicated by an arrow.

\<Bacteria\>

alpha-proteobacteria

Rickettsia	-
Wolbachia	-
Pelagibacter	-
Anaplasma	-
Ehrlichia	-
Neorickettsia	-
Mesorhizobium loti	+ (2)
Mesorhizobium sp.	+
Sinorhizobium	+ (2)
Agrobacterium	+
Rhizobium	+ (2)
Brucella	+
Bradyrhizobium	+
Rhodopseudomonas	+
Nitrobacter	+
Bartonella	+
Caulobacter	+
Silicibacter	+
Rhodobacter	+
Jannaschia	+
Roseobacter	+
Zymomonas	-
Novosphingobium	+
Sphingopyxis	+
Erythrobacter	+
Gluconobacter	+
Rhodospirillum	+
Magnetospirillum	-

beta-proteobacteria

Neisseria	-
Chromobacterium	+
Ralstonia	+
Burkholderia	+
Bordetella	-
Rhodoferax	+
Polaromonas	+
Nitrosomonas	+
Nitrosospira	+
Azoarcus	+
Dechloromonas	+
Thiobacillus	+
Methylobacillus	+

gamma-proteobacteria

Escherichia	+
Salmonella	+
Yersinia	+
Erwinia	+
Photorhabdus	+
Buchnera	-
Wigglesworthia	-
Blochmannia	-
Sodalis	+
Haemophilus	+
Pasteurella	+
Mannheimia	+
Xylella	+
Xanthomonas	+
Vibrio	+
Photobacterium	+
Pseudomonas	+
Psychrobacter	+
Acinetobacter	+
Shewanella	+
Idiomarina	+
Colwellia	+
Pseudoalteromonas	+
Saccharophagus	+
Coxiella	+

Legionella	+
Methylococcus	-
Francisella	-
Nitrosococcus	-
Hahella	+ (2)
Chromohalobacter	-
Alcanivorax	+
Baumannia	-

delta/epsiron-proteobacteria

Geobacter	+
Pelobacter	-
Desulfovibrio	-
Lawsonia	-
Bdellovibrio	+
Desulfotalea	+
Anaeromyxobacter	+
Myxococcus	+
Syntrophus	+
Helicobacter	+
Wolinella	+
Thiomicrospira	+
Campylobacter	+

Actinobacteria

Mycobacterium	+
Corynebacterium	+
Nocardia	+
Rhodococcus	+
Streptomyces	+
Tropheryma	-
Leifsonia	-
Propionibacterium	+
Thermobifida	+
Frankia	+
Bifidobacterium	+
Symbiobacterium	-
Rubrobacter	-
marine actinobacterim clade	-

Cyanobacteria

Synechocystis	+
Synechococcus	+
Thermosynechococcus	+
Gloeobacter	+
Anabaena	+
Nostoc	+
Prochlorococcus	-
Trichodesmium	+

Bacteroidetes/Chlorobium

cytophaga	+
Porphyromonas	-
Salinibacter	-
Chlorobium	+
Pelodictyon	+

Spirochaetales

Borrelia	-
Treponema	-
Leptospira	+

Deinococcus-Thermus group

Deinococcus	-
Thermus	+

Aquificae

Aquifex	+

Fibrobacters/Acidobacteria

Acidobacteria	+

Fusobacteria

Fusobacterium	-

Planctomices

Rhodopirellula	-

Chlamydiae

Chlamydia	-
Chlamydophila	-
Parachlamydia	-

Thermotogae

Thermotoga	-

Chloroflexi

Dehalococcoides	-

Firmicutes

Bacillus	-
Oceanobacillus	-
Geobacillus	-
Staphylococcus	-
Listeria	-
Lactococcus	-
Streptococcus	-
Lactobacillus	-
Enterococcus	-
Clostridium	-
Carboxydothermus	-
Desulfitobacterium	+
Thermoanaerobacter	-
Moorella	-
Mycoplasma	-
Ureaplasma	-
Phytoplasma	-
Mesoplasma	-

\<Archaea\>

Euryarchaeota

Methanococcus	-
Methanosarcina	-
Methanococcoides	-
Methanospirillum	-
Methanobacterium	-
Methanosphaera	-
Methanopyrus	-
Archaeoglobus	+
Halobacterium	-
Haloarcula	-
Haloquadratum	-
Natronomonas	-
Thermoplasma	-
Picrophilus	-
Pyrococcus	-
Thermococcus	-

Crenarchaeota

Aeropyrum	+
Sulfolobus	+
Pyrobaculum	+

Nanoarchaeota

Nanoarchaeum	-

HAKOSHIMA AND ICHIHARA, CHAPTER 14, FIG. 6. A list of bacteria and archaea in which SixA homologs were found. Each genus name is assigned + or − to indicate the presence or absence, respectively, of a SixA homolog. Species that contain two copies of SixA homologs are marked (2) and include α-proteobacteria *Mesorhizobium loti MAFF 303099*, *Rhizobium etli CFN 42*, and *Sinorhizobium meliloti 1021* and γ-proteobacteria *Hahella chejuensis KCTC 2396*. Classification of bacteria and archaea is according to NCBI taxonomy (http://www.ncbi.nlm.nih.gov/).

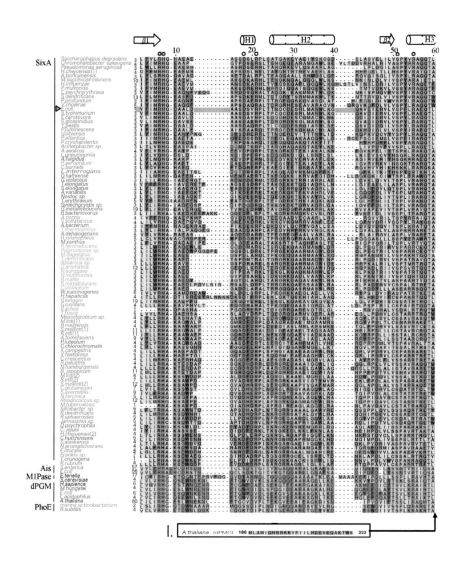

HAKOSHIMA AND ICHIHARA, CHAPTER 14, FIG. 7. (*continued*)

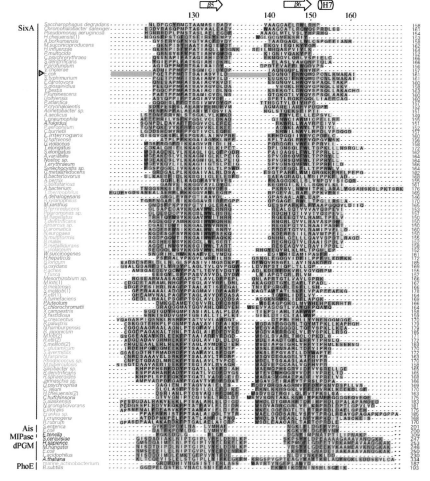

HAKOSHIMA AND ICHIHARA, CHAPTER 14, FIG. 7. Multiple alignment of SixA homologs with some well-characterized RHG phosphatases. Sequence data with significant similarity to *E. coli* SixA were collected following database searches using BLAST with E-value <0.001 and PSI-BLAST employing five iterations with E-value <0.005 at the NCBI site (http://www.ncbi.nlm. nih.gov/blast/psiblast.cgi). Following alignment of the collected sequences by MAFFT (Katoh *et al.*, 2002, 2005), alignment was slightly modified by visual inspection. Bacterial proteins are colored magenta (α-proteobacteria), orange (β-proteobacteria), red (γ-proteobacteria), blue (δ/ε-proteobacteria), cyan (actinobacteria), green (cyanobacteria), ruby (bacteroidetes/chlorobium), and other colors (others). Archaeal and eukaryotic proteins are in gray and black, respectively. Numbers at the beginning and end of each sequence indicate amino acid position of the protein. Aromatic (W, F, Y) residues are colored red, aliphatic (V, L, I, M) pink, acidic and amide (D, E, N, Q) blue, basic (K, R, H) green, cysteine (C) yellow, and others (S, T, P, G, A) gray.

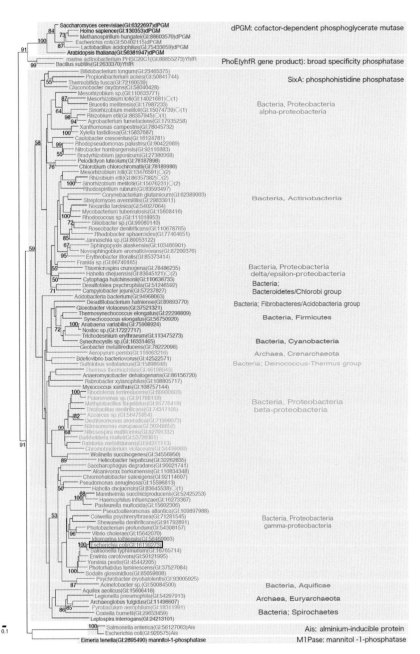

HAKOSHIMA AND ICHIHARA, CHAPTER 14, FIG. 8. (*continued*)

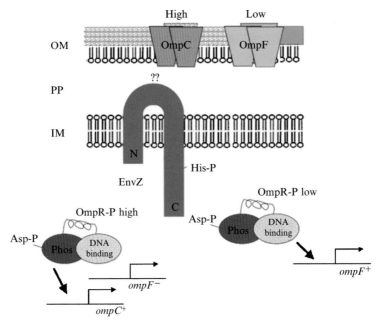

THOMAS KING AND KENNEY, CHAPTER 17, FIG. 1. General scheme for osmoregulation in *E. coli*. The sensor kinase, EnvZ, is an inner membrane protein. The response regulator OmpR consists of two domains, an N-terminal phosphorylation domain and a C-terminal DNA-binding domain, separated by a flexible linker. The phosphorylation site is indicated by Asp~P. At low osmolarity, OmpF is the major porin in the outer membrane ("OmpR~P low"). At high osmolarity, the concentration of OmpR~P increases ("OmpR~P high"), activating *ompC* transcription and repressing *ompF* transcription. OmpC is the major porin in the outer membrane at high osmolarity. OM, outer membrane; PP, periplasm; IM, inner membrane; high, high osmolarity; low, low osmolarity.

HAKOSHIMA AND ICHIHARA, CHAPTER 14, FIG. 8. An unrooted NJ tree of SixA and some well-characterized RHG phosphatases. Each sequence is indicated by the source name and the GI number in NCBI. For clarity, each genus contains one representative species. From the multiple alignments, 114 unambiguously aligned sites were used for the calculation of evolutionary distances. The evolutionary distance between every pair of aligned sequences was calculated as the maximum likelihood (ML) estimate (Felsenstein, 1996) using the JTT model (Jones *et al.*, 1992) for the amino acid substitutions. Based on these distances, an NJ tree was constructed for all of the sequences included in the alignment. The statistical significance of the NJ tree topology was evaluated by bootstrap analysis (Felsenstein, 1985) with 1000 iterative tree constructions. Bootstrap probability of a cluster is only shown at the root node of the cluster when the value is equal to or greater than 50%. The bar under the tree corresponds to 0.1 amino acid substitutions/site. Phylogenetic analysis and tree drawing were carried out using the XCED program package(http://www.biophys.kyoto-u.ac.jp/~katoh/programs/align/xced/).

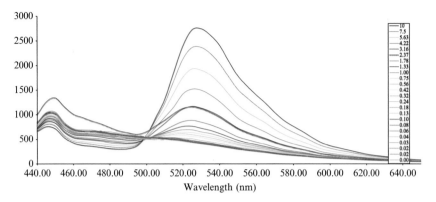

THOMAS KING AND KENNEY, CHAPTER 17, FIG. 3. Fluorescence resonance energy transfer (FRET) with EnvZ-GFP to fluorescein-conjugated OmpR. EnvZ-GFP was overexpressed and spheroplasts were prepared and lysed in cold H_2O according to Osborn *et al.* (1972). Fluorescent-labeled OmpR (fluorescein or rhodamine) ranged from 0 to 10 μM in the presence of 250 nM EnvZ-GFP. A control experiment was performed with 0 to 1000 nM unconjugated rhodamine (or fluorescein) in the presence of 250 nM EnvZ-GFP to measure the amount of nonspecific interaction of the donor (GFP) and acceptor (rhodamine or fluorescein) fluorophores. The OmpR concentration is shown in the boxed inset.

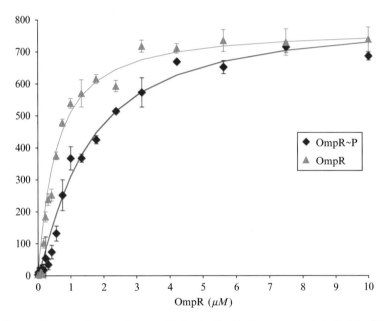

THOMAS KING AND KENNEY, CHAPTER 17, FIG. 4. The binding curve was a result of the OmpR-fluorescein and EnvZ-GFP FRET signal at 530 nm subtracted from the control titration of EnvZ-GFP and the fluorescein signal of the unconjugated probe and gave a similar value to a binding curve resulting from the decrease in GFP emission at 450 nm subtracted from the control titration of EnvZ-GFP and the rhodamine signal of the unconjugated probe. Data obtained were fit to the equation $F = F_{max}[OmpR]H/K_d + [OmpR]H$ using the sum of least squares. The K_d for OmpR and for OmpR~P is 519 nM and 1.6 μM, respectively. The Hill coefficient was 1.17.

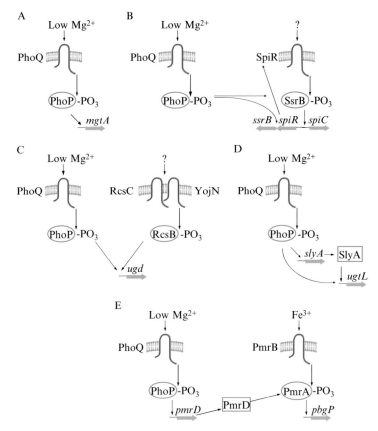

ZWIR *ET AL.*, CHAPTER 18, FIG. 1. The PhoP/PhoQ system controls the expression of a large number of genes in a direct or indirect fashion. (A) The PhoP protein recognizes a direct hexanucleotide repeat separated by five nucleotides, which has been termed the PhoP box, activating the *mgtA* promoter of *Salmonella*. (B) The PhoP/PhoQ uses a transcriptional cascade mediated by the SsrB/SpiR two-component system to regulate the *spiC* promoter. (C) The PhoP/PhoQ system works cooperatively with the RcsB/RcsC system to activate the ugd promoter. (D) The PhoP/PhoQ system utilizes a feed-forward loop mediated by the SlyA protein to activate the *ugtL* promoter in *Salmonella*. (E) The PhoP/PhoQ system controls the *pbgP* promoter at the posttranslational level, where the PhoP-dependent PmrD protein activates the regulatory protein PmrA.

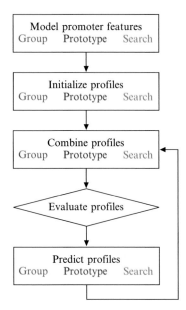

ZWIR *ET AL.*, CHAPTER 18, FIG. 2. The GPS method. GPS is a machine learning technique that *models* promoter features as well as relations between them, uses them to describe promoters, *combines* such characterized promoters into groups termed profiles, *evaluates* the resulting profiles to select the most significant ones, and performs genome-wide *predictions* based on such profiles. To accomplish this task, GPS carries out three basic operations: *grouping* observations from the data set; *prototyping* such groups into their most representative elements (centroid); and *searching* in the set of optimal solutions (i.e., Pareto optimal frontier) to retrieve the most relevant profiles, which are used to describe and identify new objects by similarity with the prototypes.

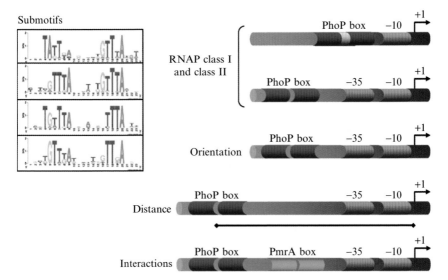

Submotifs

RNAP class I
and class II

Orientation

Distance

Interactions

PhoP box   −10   +1

PhoP box   −35   −10   +1

PhoP box   −35   −10   +1

PhoP box   −35   −10   +1

PhoP box   PmrA box   −35   −10   +1

Zwir *et al.*, Chapter 18, Fig. 3. Schematics of PhoP-regulated promoters harboring different features analyzed by GPS. GPS performs an integrated analysis of promoter regulatory features, initially focusing on six types of features for describing a training set of promoters: *submotifs*, which model the studied transcription factor-binding motifs; *RNA pol sites*, which characterize the RNA polymerase motif, the class of $\sigma 70$ promoter that differentiates *class I* from *class II* promoters, and the distance distributions (*close, medium*, and *remote*) between RNA polymerase and transcription factor-binding sites in activated and repressed promoters; *activated/repressed*, where we learn activation and repression distributions by compiling distances between binding sites for RNA polymerase and a transcription factor; *interactions*, where we evaluate motifs for several transcription factor-binding sites and model the distance distributions between motifs colocated in the same promoter regions; and *expression*, which considers gene expression levels.

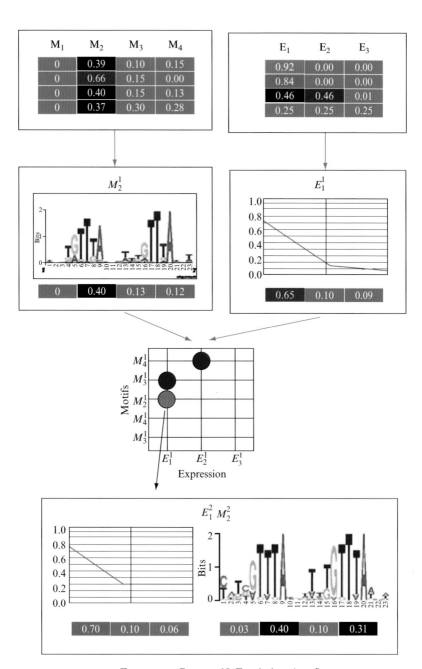

ZWIR *ET AL.*, CHAPTER 18, FIG. 4. (*continued*)

ZWIR *ET AL.*, CHAPTER 18, FIG. 4. Using GPS to build promoter profiles. GPS generation of the profile $E_1^2 M_2^2$ is shown here. It partially corresponds to the highlighted substructure of the lattice shown in Fig. 5. GPS starts by using information from databases and microarray data to construct a family of models for each feature (e.g., expression levels E_1 to E_3, PhoP box submotif M_1 to M_4, as well as other features [not shown]). The promoters are described using the modeled features, the degree of matching between features and promoters being encoded as a vector of independent values, where 1 (red color) corresponds to maximum matching and 0 (green color) corresponds to the absence of the feature. For each feature, the promoters are then grouped into subsets that share similar patterns using fuzzy clustering. Each subset shown in the initial panel is prototyped by locating the centroid that best represents the group to generate the initial, level 1 profiles (e.g., $E_1^1 M_2^1$ and I_3^1). The centroids are encoded as a vector and also visualized by graphical plots for the "expression" and the "interactions" features and by a sequence logo (Crooks *et al.*, 2004) for the "submotifs" feature. These level 1 profiles are combined to generate level 2 profiles (e.g., $E_1^2 M_2^2$ and $M_2^2 I_3^2$) by the intersection of the ancestor profiles and then prototyped. (Blue circles represent profiles containing other subsets of promoters. The absence of a circle signifies that no promoters are classified into these profiles.) Further navigation through the feature-space lattice generates the level 3 profiles (for example, $E_1^3 M_2^3 I_3^3$) after incorporating the "interactions" feature (Fig. 5). Note that the vectors of the daughter profiles are built anew from the constituent promoters and are slightly different than those of their ancestors because of the refinement that takes place during the profile learning process.

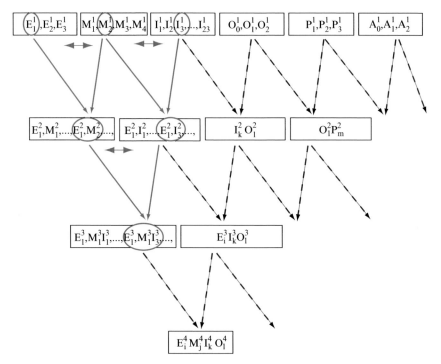

ZWIR *ET AL.*, CHAPTER 18, FIG. 5. GPS navigates through the feature-space lattice, generating and evaluating profiles. For analysis of promoters regulated by the PhoP protein, we identified up to five models for each type of feature, which are used to describe the promoters. Then, GPS generates profiles, which are groups of promoters sharing common sets of features. (Subscripts denote the different profiles for each feature, whereas superscripts denote the level in the lattice of the profile.) For example, is a particular "expression" profile that differs from and . These level 1 profiles of each feature are combined to identify level 2 profiles; similarly, level 2 profiles are combined to create level 3 profiles. In addition, because of the fuzzy formulation of the clustering, any promoter that was initially assigned to a specific profile can participate in profile of level t where is involved (i.e., indicated as a double-headed arrow). Thus, observations can migrate from parental to offspring clusters (i.e., hierarchical clustering) and among sibling clusters (i.e., optimization clustering). Here we show a small part of the complete lattice, where the part that is highlighted in red is also described in Fig. 4.

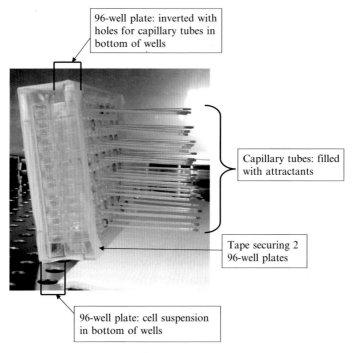

Motaleb et al., Chapter 21, Fig. 1. Ninety six-well chemotaxis chambers. One type of chemotaxis chamber is shown in which the cell suspension is placed in the wells of one 96-well plate; holes are created in the corresponding wells of another plate, and the plates are taped together with the bottom of both sets of wells facing outward. Capillary tubes containing attractants are placed through the holes into the cell suspension. The chamber is then placed in an incubator, positioned so that capillary tubes are approximately horizontal.

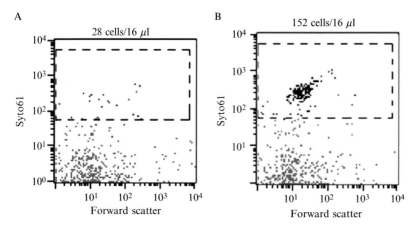

Motaleb et al., Chapter 21, Fig. 2. Flow cytometry analysis. Following incubation in a chemotaxis chamber with *B. burgdorferi*, capillary tubes containing no attractant (A) or 0.1 *M* *N*-acetylglucosamine (B) were removed and the contents prepared for flow cytometry as described. The *X* axis is forward scattering, a measure of particle complexity, and the *Y* axis is Syto61 fluorescence intensity. The dashed line indicates regions "gated" or regions in which events (*B. burgdorferi* cells) were counted during a 16-μl sampling period. Approximately five times more cells entered the capillary tube containing *N*-acetylglucosamine than the tube containing no attractant (negative control).

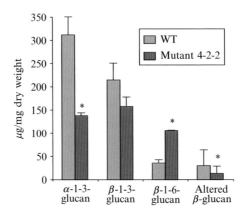

NEMECEK *ET AL.*, CHAPTER 24, FIG. 4. Analysis of the major components of the cell wall. Results of an analysis of cell wall composition in *B. dermatitidis*. Data are shown for a wild-type strain of *B. dermatitidis* (WT) and a mutant strain with a defect in the cell wall (mutant 4-2-2). *$p < 0.05$, ANOVA test. Data are mean ± SD of three experiments. Reprinted from Hung *et al.* (2002), with permission.

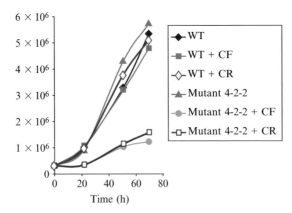

NEMECEK *ET AL.*, CHAPTER 24, FIG. 5. Liquid assay for calcofluor and Congo red sensitivity. Results of an assay for sensitivity to the cell wall-binding agents calcofluor (CF) and Congo red (CR) in a wild-type strain of *B. dermatitidis* (WT) and a mutant strain with a defect in the cell wall (mutant 4-2-2). Data are the mean of three experiments. Reprinted from Hung *et al.* (2002), with permission.

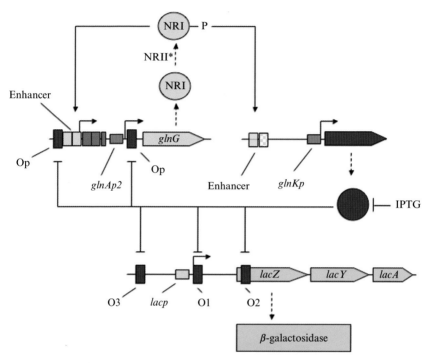

NINFA ET AL., CHAPTER 25, FIG. 2. Structure of the synthetic genetic clock. The activator module (left) consists of a modified version of the *glnA* promoter, driving the expression of *glnG* (*ntrC*). Light blue boxes in the *glnA* control region signify the *glnAp2* enhancer, dark blue boxes signify "perfect" *lac* operator sites, and gray boxes signify *glnAp2* governor sites. The relative positions of the *glnAp1* and *glnAp2* transcription start sites are shown by bent arrows. The product of *glnG* (*ntrC*), NRI, is converted to its active, phosphorylated form by NRII2302, which is provided in excess from a plasmid (not depicted). NRI~P increases its own expression by binding to the enhancer, whereupon it represses *glnAp1* and activates *glnAp2*. The repressor module (right) consists of the natural *glnK* promoter and translation initiation region fused to the *lacI* structural gene. The *glnK* promoter is associated with a weak enhancer (signified by a light blue box and a stippled light blue box). The product of the repressor module, LacI, blocks transcription from the activator module and from the native *lacZYA* operon (bottom). Repression of the *lacZYA* operon is due to repressor interaction with the three operators of this operon (dark blue boxes), as indicated. The product of *lacZ*, β-galactosidase, serves as a reporter for oscillatory behavior.

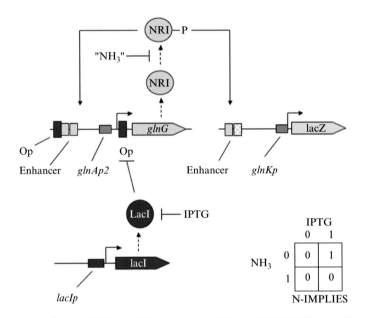

NINFA *ET AL.*, CHAPTER 25, FIG. 3. The activator module is an N-IMPLIES gate with positive feedback. When in cells containing wild-type natural components of the Ntr system except for NRI, containing the activator module to provide wild-type NRI, containing the wild-type *lacI* gene, and containing a fusion of *lacZ* to the *glnK* promoter, the activator module provides an N-IMPLIES logic gate for the regulation of β-galactosidase by ammonia and IPTG.

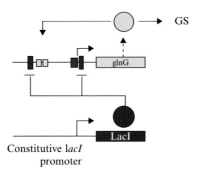

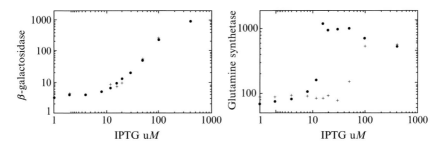

NINFA *ET AL.*, CHAPTER 25, FIG. 4. The activator module can function as a toggle switch. The basic circuit topology for the toggle switch is shown at the top. The activator module produces activator that drives expression of GS and the activator module. LacI is produced from the natural chromosomal *lacI* gene and represses the activator module, as well as the natural *lacZYA* (not depicted). At the bottom, typical results for *lacZ* expression (left) and glutamine synthetase expression (right) are shown for induced (dot) and naive (+) cultures. The overlapping curves for *lacZ* expression data show that the steady state was reached; GS data show hysteresis of the activator module. An intuitive explanation for this hysteresis is as follows. When the level of NRI~P is high, as in the induced culture, it is difficult for the repressor to get control of the system, and thus repression is only achieved when the IPTG concentration is low. In the naive cell, the level of NRI~P is very low and thus it is considerably easier for the repressor to get control of the system. Thus, in naive cells, the system stays repressed even at fairly high concentrations of IPTG.

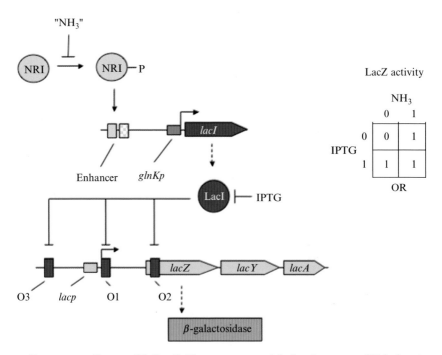

NINFA ET AL., CHAPTER 25, FIG. 5. The repressor module functions as an OR logic gate. When the repressor module is present in cells containing a mutation in the natural *lacI* gene, and with otherwise wild-type Ntr system and *lacZYA*, it provides OR gate function with regard to ammonia and IPTG for expression of *lacZYA*.

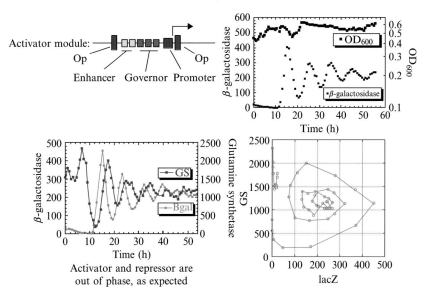

Ninfa *et al.*, Chapter 25, Fig. 6. Results of clock experiments using a standard laboratory continuous culture device. The activator module used in the experiments is depicted at top left. Experiments were conducted as described in the text, with manual control of the nutrient pump to maintain stable culture turbidity. At top right, results are shown from an experiment where OD_{600} and β-galactosidase were monitored. At bottom left, another experiment where glutamine synthetase was also measured is shown; for clarity OD_{600} data are not shown for this experiment. As predicted by the model, the expression of glutamine synthetase and β-galactosidase was out of phase. At bottom right, the phase diagram for β-galactosidase and glutamine synthetase expression data is shown. Note that if the system contained a perfect oscillator, the phase diagram would reveal a stable orbit instead of spiraling inward to a steady state.

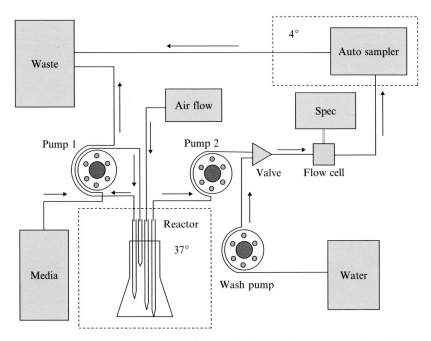

NINFA *ET AL.*, CHAPTER 25, FIG. 7. Schematic diagram for a home-made turbidostat. For details and further description, see the text.

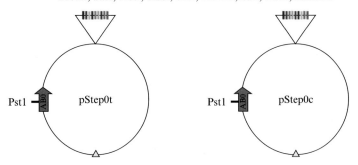

NINFA *ET AL.*, CHAPTER 25, FIG. 10. pStep0 plasmids used to create new landing pads. The vector backbone consists of the *Eco*RI–*Sap*I fragment of pBR322 containing the origin of replication (triangle) and ampicillin resistance gene (AB0). Plasmids contain a linker with multiple unique sites in either of two sequences, denoted "t" and "c."